**METHODS OF
BIOCHEMICAL ANALYSIS**

Volume 38

HPLC IN ENZYMATIC ANALYSIS

SECOND EDITION

Edward F. Rossomando
University of Connecticut Health Center

with the collaboration of
  Zdenek Deyl, Academy of Sciences of the Czech Republic
  Jan Kehr, Karolinska Institute
  David Lambeth, University of North Dakota
  Ivan Mikšík, Academy of Sciences of the Czech Republic
  Franco Tagliaro, University of Verona
  Kathi J. Ulfelder, Beckman Instruments

WILEY-INTERSCIENCE

A JOHN WILEY & SONS, INC., PUBLICATION

New York • Chichester • Weinheim • Brisbane • Toronto • Singapore

This book is printed on acid-free paper. ∞

Copyright © 1998 by John Wiley & Sons, Inc. All rights reserved.

Published simultaneously in Canada.

No part of this publication may be reproduced, stored in a retrieval system or transmitted in any form or by any means, electronic, mechanical, photocopying, recording, scanning or otherwise, except as permitted under Sections 107 or 108 of the 1976 United States Copyright Act, without either the prior written permission of the Publisher, or authorization through payment of the appropriate per-copy fee to the Copyright Clearance Center, 222 Rosewood Drive, Danvers, MA 01923, (508) 750-8400, fax (508) 750-4744. Requests to the Publisher for permission should be addressed to the Permissions Department, John Wiley & Sons, Inc., 605 Third Avenue, New York, NY 10158-0012, (212) 850-6011, fax (212) 850-6008, E-Mail: PERMREQ @ WILEY.COM.

Library of Congress Cataloging-in-Publication Data:

High performance liquid chromatography in enzymatic analysis :
  applications to the assay of enzymatic activity / Edward F.
  Rossomando ; with the collaboration of Zdenek Deyl . . . [et al.] --
  2nd ed.
     p. cm.
  Includes index.
  ISBN 0-471-10340-3 (cloth : alk. paper)
  1. Enzymatic analysis. 2. High performance liquid chromatography.
I. Rossomando, Edward F. II. Deyl, Zdeněk.
QP601.R783 1998
572'.7--dc21   97-21334
                                                                      CIP

Printed in the United States of America.

10 9 8 7 6 5 4 3 2 1

*To my wife,
Nina,
and our children,
Natasha and Michelle,
with affection;
and
to my collaborators and colleagues,
with appreciation.*

CONTENTS

Preface	xix
Preface to the First Edition	xxi
Collaborators	xxiii

1. Application of HPLC to the Assay of Enzymatic Activities   1

Overview		*1*
1.1	Introduction	1
1.2	Anatomy of an Enzyme Assay	2
1.3	Classification of Enzymatic Assay Methods	3
	1.3.1 Continuous Methods	3
	1.3.2 Coupled Method	4
	1.3.3 Discontinuous Method	5
	1.3.4 HPLC as a Discontinuous Method	6
1.4	Criteria for the Selection of an Assay Method	10
	1.4.1 Separation and Detection of Components	10
	1.4.2 The Reaction Mixture	10
	1.4.3 The Enzyme Sample	11
1.5	Summary and Conclusions	11
General References		12

2. Concepts and Principles of High Performance Liquid Chromatography   13

Overview		*13*
2.1	Introduction	13
2.2	The Introduction of HPLC	14
2.3	Basic Components and Operation	15
2.4	Coupling the Components: On the Perils of Ferrules	16
2.5	The Chromatogram	18
2.6	Interpretation of the Chromatogram	19
2.7	Selection of the Stationary Phase: Some Help from an Understanding of the Process of Separation	23
	2.7.1 Gel Filtration Chromatography	24
	2.7.2 Reverse-Phased Chromatography	25
	2.7.3 Ion-Exchange Chromatography	30

2.8	Composition and Preparation of the Mobile Phase	35
2.9	Column Maintenance	36
2.10	Monitoring Column Performance	37
2.11	Summary and Conclusions	37
General References		39

3. Concepts and Principles of High Performance Capillary Electrophoresis   41
with Franco Tagliaro, Zdenek Deyl, Ivan Mikšík, and Kathi J. Ulfelder

Overview   *41*

- 3.1 Introduction   41
- 3.2 HPCE: Definition, History, and Literature   41
- 3.3 Basic Components and Operations   42
- 3.4 The Process of Electrophoretic Separation   43
    - 3.4.1 Electrophoretic Separation   43
    - 3.4.2 Electroosmosis   44
- 3.5 Instrumentation in Detail   46
    - 3.5.1 Injection   46
        - 3.5.1.1 Hydrodynamic Injection by Pressure/Vacuum Application   47
        - 3.5.1.2 Electrokinetic Injection   47
    - 3.5.2 The Capillary   48
    - 3.5.3 Power Supply   49
    - 3.5.4 Detection   50
        - 3.5.4.1 Absorbance Detection   50
        - 3.5.4.2 Fluorimetric, Electrochemical Detection, and other Detection Modes   51
        - 3.5.4.3 Mass Spectrometric Detection   51
        - 3.5.4.4 Indirect Detection   52
- 3.6 Separation Efficiency and Resolution   52
    - 3.6.1 Theoretical Plate Number and Resolution   52
    - 3.6.2 Practical Hints   54
- 3.7 Methods   55
    - 3.7.1 Capillary Zone Electrophoresis (CZE)   55
    - 3.7.2 Micellar Electrokinetic Chromatography (MEKC)   55
    - 3.7.3 Capillary Isotachophoresis (CITP)   58
    - 3.7.4 Capillary Gel Electrophoresis (CGE)   59
    - 3.7.5 Capillary Isoelectric Focusing (CIEF)   60
    - 3.7.6 Chiral Separations   60
- 3.8 Summary   61
- References   62

4. Strategy for Design of an HPLC System for Assay of Enzyme Activity — 64

Overview — 64
- 4.1 Setting Up the Assay — 64
  - 4.1.1 Analysis of the Primary Reaction — 64
  - 4.1.2 Analysis of Secondary Reactions — 65
  - 4.1.3 Selection of the Stationary Phase and Method of Elution — 65
  - 4.1.4 Modification of Reaction Conditions for the HPLC Assay Method — 68
  - 4.1.5 Understanding and Dealing with Secondary Reactions — 68
  - 4.1.6 Components of the Reaction Mixtures Can Cause Problems: Effects of Metals on Separation — 71
  - 4.1.7 Terminating the Reaction — 73
  - 4.1.8 Setting Up the Reaction Conditions — 76
  - 4.1.9 Detector Sensitivity — 77
  - 4.1.10 Summary and Conclusions — 79
- 4.2 The Use of HPLC to Establish Optimal Conditions for the Enzymatic Reaction — 81
  - 4.2.1 Initial Decisions: Composition of the Reaction Mixture — 81
  - 4.2.2 Obtaining Initial Rate Data — 82
  - 4.2.3 Quantitative Analysis of the Reaction — 83
  - 4.2.4 Initial Rate Determination at Low Substrate Concentrations — 85
  - 4.2.5 The "Sensitivity Shift" Procedure — 86
  - 4.2.6 Substrate Analogs: Their Use in Limiting Secondary Reactions — 86
  - 4.2.7 Summary and Conclusions — 87

References — 90
General References — 90

5. Strategy for the Preparation of Enzymatic Activities from Tissues, Body Fluids, and Single Cells — 92

Overview — 92
- 5.1 Introduction — 93
  - 5.1.1 The First Goal: Selection of the Biological Starting Point — 93
  - 5.1.2 The Second Goal: Determining the Extent of the Purification, or End Point — 95

5.2	Preparation and Assay of Enzymatic Activities in Samples of Tissues, Organs, and Biological Fluids			97
	5.2.1	Separation of Cellular from Extracellular Compartments		97
		5.2.1.1	Samples Obtained Directly from an Organism	97
		5.2.1.2	Samples Obtained from Tissue or Organ Culture	98
		5.2.1.3	Samples Obtained from Biological Fluids	99
		5.2.1.4	Samples Obtained from Cell Cultures	100
	5.2.2	Assay of Activities in the Extracellular Compartment		100
	5.2.3	Assay of Activities in the Cellular Compartment		100
5.3	Preparation and Assay of Activities in Intact Cells			103
5.4	Preparation and Assay of Activities in Subcellular Samples			103
5.5	Initial Purification and Assay of Activities in Cell-Free Lysates			105
5.6	HPLC for Purification of Enzymes: A Brief Background			106
5.7	Strategy for Use of HPLC in the Purification of Activities			107
5.8	Problems Related to the Assay of Activities Following Their Purification by HPLC			112
5.9	Summary and Conclusions			113
General References				114

6. Microdialysis: An In Vivo Method for the Analysis of Body Fluids — 115

Overview			*115*
6.1	Introduction		116
	6.1.1	Principle of In Vivo Microdialysis	116
	6.1.2	Extracellular Space	116
	6.1.3	Microdialysis Probe	117
	6.1.4	Dialysis Recovery	118
6.2	Technical Aspects of Microdialysis		119
	6.2.1	Microdialysis Instrumentation	119
	6.2.2	HPLC Analysis	121
	6.2.3	Performing a Microdialysis Experiment on a Rat	122
	6.2.4	Performing a Microdialysis Experiment on a Human	122
6.3	Applications of Microdialysis/HPLC in Enzymatic Analysis		123
	6.3.1	Body Fluids Sampled by Microdialysis	123
		6.3.1.1 Blood	123

		6.3.1.2	Cerebrospinal Fluid (CSF)	124
		6.3.1.3	Vitreous Humor	124
		6.3.1.4	Synovial Fluid	124
		6.3.1.5	Perilymph	124
		6.3.1.6	Bile	124
	6.3.2	Typical Analytes: Small Molecules		125
	6.3.3	Estimating Enzymatic Activities		126
		6.3.3.1	Studying Enzymatic Activities in Almost Intact Environments/Cellular Compartments	126
		6.3.3.2	Measuring the Entire Time Course in a Single Experiment	126
		6.3.3.3	Investigating the Effects of Cofactors and/or Drugs, in Small, Localized Tissue Structures	128
		6.3.3.4	Testing Drugs That Do Not Penetrate the Blood–Brain Barrier	128
		6.3.3.5	Estimating Enzymatic Activities Under Various Physiological/Pathological Stimuli	128
		6.3.3.6	Microdialysis Sampling of Enzymes	129
6.4	Conclusions			129
References				130

7. Fundamentals of the Polymerase Chain Reaction and Separation of the Reaction Products 137
with Kathi J. Ulfelder

Overview			*137*
7.1	Introduction: Polymerase Chain Reaction		137
7.2	Principles of Nucleic Acid Separation		139
	7.2.1	Separation Mechanism	139
	7.2.2	Classical Methods of DNA Analysis	141
7.3	CE Methods: Principles and Strategies for Nucleic Acids		141
	7.3.1	CE Principles Related to Nucleic Acids	141
	7.3.2	Buffer Systems	141
	7.3.3	Intercalators	142
	7.3.4	Typical Instrument Parameters	144
	7.3.5	Detection	144
	7.3.6	Data Analysis	146
	7.3.7	Sample Preparation and Injection Considerations	146
	7.3.8	Artifacts	150

7.4	PCR Applications		151
	7.4.1	Forensic Analysis	151
	7.4.2	Identification by Hybridization	151
	7.4.3	Quantitative Analysis	153
	7.4.4	Quantitative RNA-PCR	156
7.5	Conclusion: Future Applications		160
References			160

8. Applications of HPLC/HPCE in Forensics — 164
with Franco Tagliaro, Zdenek Deyl, and Ivan Mikšík

Overview			*164*
8.1	Introduction		164
8.2	Forensic Toxicology		165
8.3	HPCE Analysis of Illicit Drug Substances		165
8.4	Analysis of Gunshot Residues and Constituents of Explosives		172
8.5	Analysis of Pen Inks		175
8.6	Analysis of Proteins in Forensics		176
	8.6.1	Separation of Proteins by HPCE	176
	8.6.2	Separation of Protein Mixtures by Two-Dimensional Techniques	178
8.7	Analysis of Nonenzymatically Modified Proteins		179
	8.7.1	Nonenzymatic Modifications to Keratins by Ethanol	179
	8.7.2	Nonenzymatic Modifications to Collagen by Glucose	183
8.8	Enzymatic Activity Assay by Capillary Electrophoresis		185
8.9	Protein–drug Binding Assays		192
8.10	Nucleic Acids and Their Constituents		196
8.11	Conclusion		200
References			200

9. Survey of Enzymatic Activities Assayed by the HPLC Method — 207
with David Lambeth

Overview			*207*
9.1	Introduction		208
9.2	Catecholamine Metabolism		208
	9.2.1	Tyrosine Hydroxylase	208
	9.2.2	5-Hydroxytryptophan Decarboxylase	211
	9.2.3	Dopa Decarboxylase (L-Aromatic Amino Acid Decarboxylase)	212
	9.2.4	Dopamine β-Hydroxylase	215

	9.2.5	Catechol O-Methyltransferase	219
	9.2.6	Phenylethanolamine N-Methyltransferase	221
	9.2.7	Monoamine Oxidases A and B	222
	9.2.8	Arylsulfatase	224
	9.2.9	Monoamine Oxidase and Phenol Sulfotransferase	225
	9.2.10	N-Acetyltransferase	226
	9.2.11	Acetyl–CoA/Arylamine N-Acetyltransferase	229
9.3	Proteinase		229
	9.3.1	Vertebrate Collagenase	229
	9.3.2	Dipeptidyl Carboxypeptidase (Angiotensin I Converting Enzyme, EC 3.4.15)	231
	9.3.3	Luteinizing Hormone–Releasing Hormone Peptidase	235
	9.3.4	Papain Esterase	235
	9.3.5	Plasma Carboxypeptidase N (Kininase I, Bradykinin-Destroying Enzyme, EC 3.4.12.7)	237
	9.3.6	Dipeptidase	238
	9.3.7	Aminopeptidase	239
	9.3.8	Enkephalinases A and B	239
	9.3.9	Rhinovirus 3c Protease	241
	9.3.10	Stromelysin	243
	9.3.11	Dipeptidyl Peptidase IV/Amino Peptidase-P	244
	9.3.12	Carboxypeptidase N	244
	9.3.13	Renin	246
9.4	Amino Acid and Peptide Metabolism		247
	9.4.1	Ornithine Aminotransferase	247
	9.4.2	Glutamine Synthetase, Glutamate Synthetase, and Glutamate Dehydrogenase	249
	9.4.3	Asparagine Synthetase	251
	9.4.4	Tryptophanase	253
	9.4.5	Dihydroxyacid Dehydratase	255
	9.4.6	Glutaminyl Cyclase	256
	9.4.7	Leucine 2,3-Aminomutase	257
	9.4.8	Diaminopimelate Epimerase and Decarboxylase	259
	9.4.9	Lysine–Ketoglutarate Reductase	259
	9.4.10	γ-L-Glutamylcyclotransferase	261
	9.4.11	γ-Glutamylcysteine Synthetase and Glutathione Synthetase	261
	9.4.12	Glutamic Acid Decarboxylase	262
	9.4.13	Histamine N-Methyltransferase	262
	9.4.14	Amino Acid Decarboxylase	263
	9.4.15	Aromatic L-Amino Acid Decarboxylase	264

	9.4.16	D-Amino Oxidase	264
	9.4.17	Threonine/Serine Dehydratase	265
	9.4.18	Tryptophan Dioxygenase	265
	9.4.19	Tryptophan 2,3-Dioxygenase	267
	9.4.20	Kynureninase	267
	9.4.21	Kynurenine 3-Monoxygenase	268
	9.4.22	N^5-Methyltetrahydrofolate-homocysteine Methyltransferase	269
	9.4.23	L-Alanine: Glyoxylate Aminotransferase	270
	9.4.24	Tyrosinase	270
	9.4.25	δ-(L-α-Aminoadipyl)-L-Cysteinyl-D-Valine Synthetase	271
9.5	Polyamine Metabolism		272
	9.5.1	Ornithine Decarboxylase	272
	9.5.2	Spermidine Synthetase	273
	9.5.3	Polyamine Oxidase	275
	9.5.4	Diamine Oxidase	275
9.6	Heme Metabolism		276
	9.6.1	δ-Aminolevulinic Acid Synthetase	276
	9.6.2	5-Aminolevulinate Dehydrase	278
	9.6.3	Uroporphyrinogen Decarboxylase	278
	9.6.4	Heme Oxygenase	279
	9.6.5	Ferrocheletase	280
	9.6.6	Protoporphyrinogen Oxidase	281
9.7	Carbohydrate Metabolism		283
	9.7.1	β-Galactosidase	283
	9.7.2	Lactose-Lysine β-Galactosidase	284
	9.7.3	Arylsulfatase B (N-Acetylgalactosamine 4-Sulfatase)	285
	9.7.4	Galactosyltransferase	287
	9.7.5	Uridine Diphosphate Glucuronosyltransferase	287
	9.7.6	α-Amylase (1,4-α-D-Glucaglucanohydrolase EC 3.2.2.1)	290
	9.7.7	Lysosomal Activities	291
	9.7.8	Sialidase	293
	9.7.9	Cytidine Monophosphate–Sialic Acid Synthetase	294
	9.7.10	Succinyl–CoA Synthetase	295
	9.7.11	α-Ketoglutarate Dehydrogenase	299
	9.7.12	Sucrose Phosphate Synthetase	300
	9.7.13	6-Phosphogluconate Dehydratase	300
9.8	Steroid Metabolism		301
	9.8.1	Δ^5-3 β-Hydroxysteroid Dehydrogenase	301
	9.8.2	11-β-Hydroxylase and 18-Hydroxylase	302

	9.8.3	25-Hydroxyvitamin D_3-1α-Hydroxylase	304
	9.8.4	Cholesterol 7α-Hydroxylase	304
	9.8.5	3β-Hydroxy-Δ^5-C_{27}-steroid Oxidoreductase	306
	9.8.6	Cytochrome P450$_{SCC}$	306
	9.8.7	Steroid 17α-Hydroxylase/C_{17-20} Lyase (Cytochrome P-450$_{21SCC}$)	307
9.9	Purine Metabolism		309
	9.9.1	Nicotinate Phosphoribosyltransferase	309
	9.9.2	5'-Nucleotidase	310
	9.9.3	Alkaline and Acid Phosphatase	312
	9.9.4	Adenosine Deaminase	317
	9.9.5	AMP Deaminase	317
	9.9.6	Cyclic Nucleotide Phosphodiesterase	320
	9.9.7	ATP Pyrophosphohydrolase	320
	9.9.8	Hypoxanthine Guanine Phosphoribosyltransferase	322
	9.9.9	Nucleoside Phosphorylase	323
	9.9.10	Creatine Kinase	325
	9.9.11	Adenosine Kinase	326
	9.9.12	Adenylate Cyclase	327
	9.9.13	cAMP Phosphodiesterase	330
	9.9.14	Adenylate Kinase	333
	9.9.15	Adenylosuccinate Synthetase	334
	9.9.16	Dinucleoside Polyphosphate Pyrophosphohydrolase	336
	9.9.17	NAD Glycohydrolase	337
	9.9.18	Assay of Enzymes Involved in Cytokinin Metabolism	338
	9.9.19	Xanthine Oxidase	339
	9.9.20	Phosphoribosylpyrophosphate Synthetase	340
	9.9.21	Guanase	342
	9.9.22	Urate Oxidase	344
	9.9.23	Glutamine: 5-Phosphoribosyl-1-pyrophosphate Amidotransferase	344
	9.9.24	Thiopurine Methyltransferase	345
	9.9.25	NAD Pyrophosphorylase	345
	9.9.26	Nucleoside Diphosphate Kinase	346
	9.9.27	ATPase	348
9.10	Oxygenations		348
	9.10.1	Acetanilide 4-Hydroxylase	348
	9.10.2	Ceruloplasmin	349
	9.10.3	Aryl Hydrocarbon Hydroxylase (EC 1.14.14.2)	351
	9.10.4	Hepatic Microsomal Testosterone Hydroxylase	352

9.11	Pterin Metabolism		353
	9.11.1	Folic Acid Cleaving Enzyme	353
	9.11.2	Dihydrofolate Reductase	353
	9.11.3	Guanosine Triphosphate Cyclohydrolase I	357
9.12	Lipid Metabolism		360
	9.12.1	Retinal Oxidase	360
	9.12.2	Serum Cholinesterase	361
	9.12.3	Carnitine Palmitoyltransferase I	362
	9.12.4	Fatty Acid ω-Hydroxylase	363
	9.12.5	Acyl–CoA: Alcohol Transacylase	363
	9.12.6	Lipase	364
9.13	Modification of Proteins and Peptides		365
	9.13.1	Tyrosine Protein Kinase	365
	9.13.2	Adenosine Diphosphate–Ribosylarginine Hydrolase	366
	9.13.3	Peptidylglycine α-Amidating Monoxygenase	367
	9.13.4	Myosin Light Chain Kinase	369
	9.13.5	Transglutaminase	369
	9.13.6	Phosphotyrosyl Protein Phosphatase	370
	9.13.7	Phosphotyrosine Phosphatases	371
	9.13.8	Protein Phosphatase 2B (Calcineurin)	371
9.14	Vitamin Metabolism		372
	9.14.1	Thiamine Triphosphatase	372
	9.14.2	Lipoamidase	372
	9.14.3	Pyridoxal Kinase, Pyridoxamine Oxidase, and Pyridoxal-5'-phosphate Phosphatase	373
	9.14.4	Pyridoxine Kinase	373
	9.14.5	Biotinidase	374
9.15	Xenobiotic Metabolism		374
	9.15.1	ATP-Sulfurylase	374
	9.15.2	Sulfotransferase	375
	9.15.3	Glutathione S-Transferase	376
	9.15.4	Adenosine 3'-Phosphate 5'-Sulfophosphate Sulfotransferase	380
	9.15.5	Phenolsulfotransferase	380
	9.15.6	Aryl Sulfotransferase	382
	9.15.7	Cysteine Conjugate β-Lyase	383
	9.15.8	UDP–Glucuronyl Transferase	384
	9.15.9	UDP–Glucosyltransferase	385
	9.15.10	Ethoxycoumarin O-Deethylase	385
	9.15.11	Cytochrome P450$_{2E1}$	386
	9.15.12	Flavin-Containing Monoxygenase	387

9.16	Pyrimidine Metabolism	388
	9.16.1 Dihydropyrimidine Dehydrogenase	388
	9.16.2 Dihydroortic Acid Dehydrogenase	389
	9.16.3 Cytidine Deaminase	389
	9.16.4 β-Ureidopropionase	390
	9.16.5 Dihydroorotase	391
	9.16.6 Thymidylate Synthetase	391
9.17	Metabolism of Complex Saccharides and Glycoproteins	392
	9.17.1 α-L-Fucosidase	392
	9.17.2 α-N-Acetylgalactosaminyltransferase	392
	9.17.3 GM_1 Ganglioside β-Galactosidase	393
	9.17.4 Aspartylglycosylaminase	394
	9.17.5 β-Galactosidase and Glycosyltransferase	395
	9.17.6 Glucose-1-phosphate Thymidylyltransferase	396
	9.17.7 CMP-N-Acetylneuraminic Acid: Glycoprotein Sialyltransferase	396
	9.17.8 Thyroxine: UDP–glucuronosyltransferase	397
	9.17.9 trans-p-Coumaroyl Esterase	397
9.18	Miscellaneous	399
	9.18.1 Carboxylases	399
	9.18.2 Carbonyl Reductase	400
	9.18.3 6-Pyruvoyl Tetrahydropterin Synthetase	400
	9.18.4 Pteroylpolyglutamate Hydrolase	401
	9.18.5 Nitrogenase	402
	9.18.6 Strictosidine Synthetase	403
	9.18.7 Anhydrotetracycline Oxygenase and Tetracycline Dehydrogenase	403
9.19	Nucleic Acid Modification and Expression	405
	9.19.1 DNA Topoisomerase	405
	9.19.2 Chloramphenicol Acetyltransferase	405
9.20	Summary and Conclusions	407
References		409
General References		416

10. Multienzyme Systems — 418

Overview — 418
- 10.1 Introduction — 418
- 10.2 Assay of Two Activities Forming Different Products from the Same Substrate — 419
- 10.3 Assay of Two Activities Forming the Same Product from the Same Substrate — 420
- 10.4 Formation of Two Separate Products from Two Separate Substrates by the Same Activity — 426

10.5	Assay of a Multienzyme Complex by the Reconstitution Method		428
	10.5.1	The Salvage Pathway: The Formation and Fate of IMP	428
	10.5.2	The Degradation of IMP to Inosine	429
	10.5.3	The Conversion of IMP to AMP	430
	10.5.4	The Return of AMP to IMP	432
10.6	Assay of a Multienzyme Complex Using the HPLC Method		432
10.7	Summary and Conclusions		435
References			435
General References			435

Subject Index 437

PREFACE

Among the various products of genes, enzymes are unique because they act as catalysts for the processes that define living systems. Given their central role, it is no wonder that advances in technologies for measuring enzymatic activities are of interest to a broad spectrum of scientists. High performance liquid chromatography (HPLC) is one such advance. In 1987 the first edition of *High Performance Liquid Chromatography in Enzymatic Analysis* was published with the following goals: to explain how enzymatic activities could be assayed by HPLC, and to serve as a reference source by cataloguing activities for which HPLC had been used. Given these goals, and the decade that has elapsed since the first edition, it is reasonable to ask whether the book has been successful.

One measure of success is anecdotal: At more than one scientific meeting, graduate students and colleagues have told me how pleased they were with the book. By following the directions provided, they have been able to set up HPLC assays, and they have found the book most useful as a reference source. A more quantitative measure would be the number of investigators who rely on HPLC to monitor enzymatic activities. If we use publications as a measure, we find that the 1987 edition reported HPLC assays of 62 activities. In contrast, the 1998 edition cites 169 activities involving assay by HPLC. Thus, 11 years brought an almost threefold increase in the number of activities assayed by HPLC.

And finally, more types of activity are being studied. The first edition listed activities in only 11 categories. This volume discusses activities in 18 categories, including enzymes for metabolism of lipids, vitamins, xenobiotics, pyrimidines, complex saccharides, and glycoproteins, and activities that modify proteins and peptides, as well the modification of nucleic acids and their expression.

Using these criteria as yardsticks, the first edition has been a success. It is now obsolete, however: Because of the increases in both number and types of activities assayed, it is no longer an accurate catalog of enzymatic activities investigated by means of the HPLC method. For this work to continue to serve as a reference source, it would need updating. While it was the obsolescence of the first edition that in part prompted the development of a second edition, there were other considerations as well. These included the introduction of high performance capillary electrophoresis (HPCE) as a method for separation, the development of microdialysis as a method for collection of samples in situ, the application of HPLC/CE to the field of forensics, and finally the wide application of the polymerase chain reaction (PCR).

To cover these new areas in the second edition in a complete, scholarly and professional manner, I enlisted collaborators. Franco Tagliaro, Zdenek Deyl, and Ivan Mikšík not only contributed to the chapter on CE, but developed the chapter on forensics. Kathi Ulfelder also contributed to the chapter on CE and developed the chapter on PCR. Jan Kehr contributed the chapter on microdialysis. David O. Lambeth updated and extended the scope of the chapter surveying enzyme activities assayed with HPLC. I hope that this new material, by augmenting the first edition, will make this second edition of value to researchers and especially to students.

The goals of the second edition become expanded versions of those articulated in the first: to demonstrate how enzymatic activities can be assayed by HPLC/CE and microdialysis; to show how HPLC/CE can be used with PCR; and finally to provide a reference source to determine whether an HPLC assay has been developed for your activity.

No work of this scope and magnitude can be completed alone. I thank my collaborators and the investigators who have agreed to have their work cited in this volume. I also express my appreciation to Katherine O'Conner for organizing and typing sections of this edition. Finally, I thank the editors at John Wiley for working with me in the production of this second edition.

EDWARD F. ROSSOMANDO

Farmington, Connecticut
January 1998

PREFACE TO THE FIRST EDITION

The importance of the introduction of high performance liquid chromatography (hplc) to studies in the life sciences is now widely recognized. Since its introduction, this method has been rapidly accepted by biochemists and more recently by biologists and clinicians. Such rapid acceptance should not be surprising, since advances in separation and analysis have usually been readily assimilated.

It is the ability of hplc to accomplish separations completely and rapidly that led to its original application to problems in the life sciences, particularly those related to purification. An analysis of the literature revealed that this technique was used primarily for the purification of small molecules, macromolecules such as peptides and proteins, and more recently antibodies. This application to purification has all but dominated the use of the method, and there has been a plethora of books, symposia, and conferences on the use of hplc for these purposes. However, it was only a matter of time before others began to look beyond and to explore the possibilities that result from the capacity to make separations quickly and efficiently.

What emerged from these early studies was the idea that hplc might be used as a method for the analysis of enzymatic activities rather than its traditional use as a tool for separation. This change in emphasis is particularly attractive to those who wish to make use of the activity of an enzyme as an indicator of cellular function, a determinant of a given stage of differentiation (or dedifferentiation), or even as a measure of gene function. In the past, because contaminating activities led to conflicting results, tedious purification of the enzyme was often necessary to clarify the results of ambiguous activity determinations brought about in part as a consequence of methods that measure only one component of the reaction mixture. Such ambiguous results will occur much less often with the hplc method, since its ability to separate quickly a group of related compounds allows for the assay of one activity in the presence of several others. Thus, the advantage of analyzing an activity after only a minimal amount of purification is inherent to the hplc technique.

This book describes the hplc method and explains and illustrates its use. Each chapter deals with a different aspect of the method, beginning with an overview and ending with a detailed summary. Throughout, an attempt has been made to focus on questions related to the assay of the activity of an enzyme rather than its purification. More detailed discussions on the theory of hplc and on its use for purification, particularly for the purification of proteins, will be found in the references at the end of each chapter.

No task of this magnitude can be completed without the guidance, inspiration, and help of many others. I wish to thank Jessica Hodge Jahngen, who introduced me to the possibilities and potential of hplc. She and E. G. Jahngen have provided much of the work in this volume from my laboratory. More recently, data and assistance have been provided by Jane Hadjimichael, whose persistence and insistence in the final stages helped it all come together. A special note of appreciation and thanks to Vickey Shockley for organizing and typing the manuscript together with Pamala Vachon, and to Sherry Perrie for the original artwork.

My appreciation is also extended to my colleagues Edward J. Kollar and James A. Yaeger for editing the early drafts of the manuscript. More recently, editing assistance has been provided by Cynthia Beeman, Mina Mina, and David Richards. Finally, I wish to thank Phyllis Brown for encouraging me to complete this task in the way I thought best and Stanley Kudzin for his interest in the subject and for providing the opportunity to write the book. Also, this work could not have been completed without the contributions of numerous investigators who consented to have their work cited here. My thanks to them as well.

EDWARD F. ROSSOMANDO

Farmington, Connecticut
January 1986

COLLABORATORS

Zdenek Deyl, PhD, Institute of Physiology, Academy of Sciences of the Czech Republic, Videnska 1083, CZ-14220 Prague 4, Czech Republic

Jan Kehr, PhD, Department of Neuroscience, Karolinska Institute, S-171 77 Stockholm, Sweden

David Lambeth, PhD, Department of Biochemistry and Molecular Biology, University of North Dakota School of Medicine and Health Sciences, Grand Forks, North Dakota 58202-9037

Ivan Mikšík, PhD, Institute of Physiology, Academy of Sciences of the Czech Republic, Videnska 1083, CZ-14220 Prague 4, Czech Republic

Franco Tagliaro, MD, PhD, Institute of Forensic Medicine, University of Verona, 37134 Verona, Italy

Kathi J. Ulfelder, A.B., Advanced Technology Center, Beckman Instruments, Inc., 2500 Harbor Boulevard, Fullerton, California 92834

CHAPTER 1

Application of HPLC to the Assay of Enzymatic Activities

OVERVIEW

This chapter describes the anatomy of an enzyme assay, focusing on the significance of separation and detection in the assay procedure. A classification of the methods used in the assay of enzymatic activities is developed, using the separation step as the criterion for the grouping. Having placed the high performance liquid chromatography (HPLC) method within this classification, we then examine the question of when to use it and discuss some strategies developed for its use. The chapter also identifies and comments on the parts of the enzyme assay that will be affected by the selection of HPLC as the method of analysis.

1.1 INTRODUCTION

Increasingly, investigators in the life sciences have expressed interest in the application of HPLC to the assay of enzymatic activities. This method not only provides a method to enhance the separation of reaction components, it also allows extensive and complete analysis of the components in the reaction mixture during the reaction. In addition, it can employ sensitive detectors, and it can be used for purification.

A number of questions must first be addressed, however, concerning the biochemical reaction catalyzed by the enzyme, the assay conditions normally used for this enzyme, and the enzyme itself. This chapter explores and answers these questions.

Section 1.2 presents the anatomy of the enzymatic assay, and from a dissection of its components, it is possible to obtain an appreciation of how HPLC can be used. Section 1.3 develops a classification of enzyme methods that allows the advantages and limitations of the HPLC method to be presented fairly. Section 1.4 is devoted to criteria for the selection of HPLC as an assay

2 APPLICATION OF HPLC TO THE ASSAY OF ENZYMATIC ACTIVITIES

system. Wherever possible, these points are illustrated with examples taken from work carried out in the author's laboratory.

1.2 ANATOMY OF AN ENZYME ASSAY

The assay of an enzymatic activity is composed of several discrete steps or events (Fig. 1.1). The first is *preparation* of both the reaction mixture and the

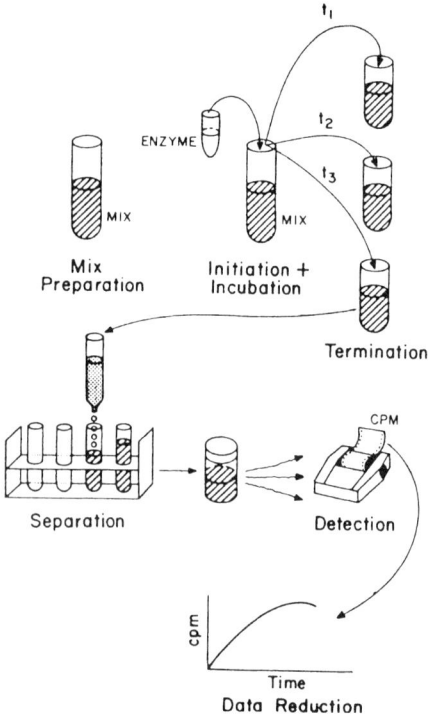

Figure 1.1 Schematic of a representative enzymatic assay. The reaction mixture is prepared (Mix Preparation) and the reaction can be started (Initiation) by the addition of the enzyme. During the reaction (Incubation), samples are removed at intervals labeled t_1, t_2, and t_3, and the reaction is stopped (Termination) by inactivating the enzyme. The incubation mixture is fractionated (the illustration shows a traditional chromatographic column), and the product is isolated from the substrate (Separation). In this assay, a radiochemical was used as the substrate and therefore the amount of product that formed is determined by its collection, the addition of scintillation fluid, and the measurement of radioactivity by scintillation counting (cpm: Detection). The progress of the reaction is given by the amount of radioactive product recovered (Data Reduction).

enzyme. The reaction mixture usually contains such components as the buffer used to establish the correct pH, the substrate, and any cofactors (e.g., metals) that may be required for catalysis. Preparation of the reaction mixtures involves mixing these ingredients in a reaction vessel such as a test tube or, for some assay methods, a cuvette. In some cases the reaction mixture is brought to the required temperature prior to initiation of the reaction. The enzyme must also be prepared. This complex topic is discussed in detail in Chapter 5.

In most cases, the second step in the assays comprises *initiation* and *incubation*. A reaction can be initiated by the addition of the enzyme preparation to the substrate in the reaction mixture, or vice versa. This step is considered the start of the reaction, and all subsequent time points are related to this time.

Many reactions require *termination*, which is the step that brings about the cessation of catalysis and thus stops the reaction. Termination may be achieved in several ways, usually via inactivation of the enzyme.

Termination is often followed by *separation* of the components in the reaction mixture. Most often separation involves isolating the substrate from the reaction product.

Detection, the fifth step, refers to that process by which the amount of product formed by the enzyme during a specific incubation interval is determined.

The last step in an assay involves *reduction* of the data. This step includes all procedures in which the data are analyzed and graphed to determine initial rates as well as kinetic constants.

Not all steps are involved in all assay methods, and in some methods one or more of the steps may be complex. The introduction of HPLC as an enzymatic assay method has improved the separation and detection steps primarily, although its use may also affect the preparation and termination steps.

1.3 CLASSIFICATION OF ENZYMATIC ASSAY METHODS

The methods in use for the assay of enzymatic activities may be divided into three groups. These will be referred to as the continuous, coupled, and discontinuous methods (see Table 1.1).

1.3.1 Continuous Methods

Continuous methods do not require a separation step prior to detection. For assays using this method, the substrate and product must differ in some property such that either one may be measured directly in the incubation solution. For example, the activity of an enzyme catalyzes the conversion of 4-nitrophenyl phosphate (4NP), a colorless compound, to 4-nitrophenol, which is yellow and has an absorption maximum at 510 nm. Since the substrate does not absorb in this region of the spectrum, the reaction can be carried out

TABLE 1.1 Classification System for Enzymatic Assay Methods

Assay Methods	Characteristics	Example
Continuous	Separation of substrate(s) from product(s) not required	4NP → 4N + P_i colorless yellow
Coupled	Separation not required for detection	PEP + ADP → pyruvate + ATP pyruvate + NADH → lactate + NAD
Discontinuous	System for separation of substrate(s) from product(s) required for detection	ATP + AA $\xrightarrow{enz}$ Enz – AA – AMP + PP_i Enz – AA – AMP + tRNA → tRNA – AA – AMP

directly in a cuvette (Fig. 1.2), and the amount of product formed may be determined continuously by measuring the change in optical density with time at this wavelength.

1.3.2 Coupled Method

In the second category of assays, the coupled assay method, activity is measured indirectly. In this method two reactions are involved. The first is the reaction of interest, such as A → B, second, the reaction that converts B to C, might be referred to an *indicator reaction,* not only because it uses the product of the first reaction (i.e., B) as a substrate, but also because the

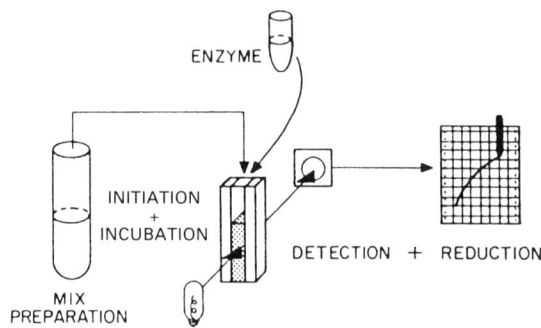

Figure 1.2 The assay of an enzymatic activity by the continuous assay method. In the illustration, the reaction mixture is transferred to a cuvette, which is shown in place in the light path of the spectrometer. The addition of the enzyme directly to the cuvette initiates the reaction. Product formation results in a change in absorbance, which is monitored continuously by the detector. This change signals a deflection on a recorder. Note that product formation requires neither termination of the reaction nor separation of the substrate from the product.

formation of C may be assayed by a continuous method—that is, without a separation step. In this way, the two reactions are coupled, the product of the first reaction, B, acting as the substrate for the second reaction.

For example, pyruvate kinase may be assayed by such a method. This enzyme catalyzes the reaction

$$\text{Phosphoenolypyruvate (PEP)} + \text{ADP} \rightarrow \text{pyruvate} + \text{ATP}$$

This, of course, is the reaction of interest that cannot be assayed directly by the continuous method. However, when a second enzyme, a dehydrogenase, such as lactate dehydrogenase, is added as the indicator together with pyruvate and NADH to the reaction mixture, a second reaction occurs and NAD forms in the cuvette as follows:

$$\text{Pyruvate} + \text{NADN} \rightarrow \text{lactate} + \text{NAD}$$

The formation of NAD may be followed in a continuous manner by the decrease in absorbance at 340 nm, and therefore the progress of the kinase reaction of interest may be followed through this coupling of the formation of pyruvate to the formation of NAD.

1.3.3 Discontinuous Method

The discontinuous method measures activity by separating the product from the substrate. Assays characteristic of this group usually require two steps, since separation often does not include detection. Thus, first, the substrate and the product are separated, and usually the amount of product formed is measured. Assays that use radiochemical substrates are included in this group, since radiochemical detectors are unable to differentiate between the radiolabel of the substrate and that of the product. Examples of enzymes whose assay methods fall into this category are legion, and these approaches characterized by a separation step.

As an illustration, consider the assay to measure the activity of the tRNA synthetases. These enzymes catalyze the covalent attachment to tRNA of an amino acid, usually radioactive as follows:

$$\text{ATP} + {}^*\text{AA} + \text{Enz} \rightarrow \text{Enz-AMP-}{}^*\text{AA} + \text{PP}_i \quad (1)$$

$$\text{Enz-AMP-}{}^*\text{AA} + \text{RNA} \rightarrow \text{RNA-}{}^*\text{AA} + \text{AMP} \quad (2)$$

(By convention, radioactivity is indicated by an asterisk preceding one labeled substance, here the amino acid AA.)

The activity is usually followed by measuring the amount of RNA-*AA, the product of reaction (2) formed during the incubation. Since the radiochemical detector cannot differentiate the free radioactive amino acid used as the substrate from that bound covalently to the RNA, the free and the bound amino acids must be separated prior to the detection or quantitation step.

This separation step requires first the addition to the sample of an acid such as trichloroacetic acid (TCA), which also serves to terminate the enzymatic reaction. However, since TCA also precipitates the RNA and any radioactive amino acid covalently linked to it, the reaction product RNA-*AA will be precipitated as well. And since the precipitate can be separated from the soluble components by a sample filtration step, the separation of the bound from the free amino acid can be accomplished. As illustrated in Figure 1.3, the reaction product, which is trapped on the filter as a precipitate, can be detected by transferring the filter to a scintillation counter for quantitation and, of course, measuring the amount of product formed. Since assays of this design usually focus on one component at a time, no information is obtained about the amount of ATP, AMP, PP_i, or free amino acid during the course of the reaction.

1.3.4 HPLC as a Discontinuous Method

Within the framework of the scheme just described, the HPLC method would be classified as discontinuous, since a separation step is part of the procedure. However, because termination can be accomplished by injecting the sample directly onto the column, the HPLC detection is usually "on-line," that is, carried out continuously with separation. Thus, the separation and detection steps merge into a single operation, which for all practical considerations means that it is a "continuous" method.

In addition, unlike many other discontinuous assays that focus on only one of the components of the reaction, the HPLC assay offers the potential to monitor several. For example, consider adenosine kinase, the enzyme that uses two substrates and forms two products according to the reaction Ado + ATP → AMP + ADP. Since HPLC can readily separate all four compounds (see Fig. 1.4), and all four compounds can be detected at 254 nm, it is apparent that with the HPLC method, the level of each component can be monitored during the course of the reaction, providing a complete analysis of each "time point."

Having a complete analysis of the contents of the reaction vessel during the incubation can be helpful in another way: It provides information on what is not present as well; and since most other assay methods are designed to detect only one component, it is often difficult to account for a result that occurs unexpectedly during a study. For example, consider the results obtained during the purification of the enzyme E-1, which catalyzes the conversion of substrate A to product B. Consider also that the method used to follow activity measures only the amount of B in the incubation mixture. As illustrated in panel I of Figure 1.5, when the activity E-1 is assayed in the crude sample, the formation of substantial product (B) is observed (graph line 1).

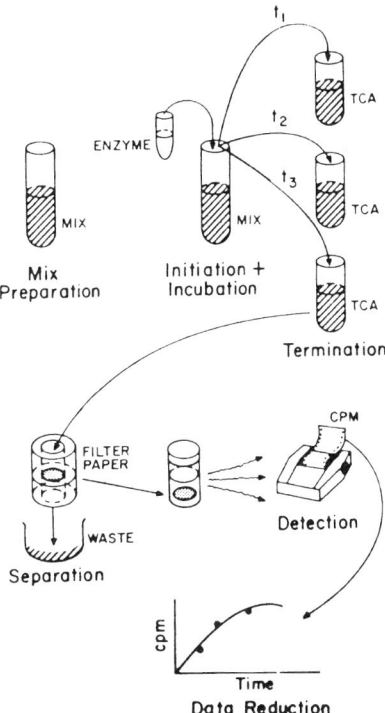

Figure 1.3 The assay of an enzymatic activity by the discontinuous assay method. In assays of this type, the reaction mixture is prepared, and usually the reaction is started by the addition of the enzyme. Samples are removed at intervals t_1, t_2, t_3, and the reaction is terminated by transferring the sample to a solution that inactivates the enzyme. Here a radioactive substrate is converted to radioactive product that is precipitable in trichloroacetic acid, (TCA), while the substrate remains soluble in the acid. Thus two components can be separated by filtration. The product is shown being collected on the filter while the unreacted substrate flows through into the filtrate. The amount of radioactivity trapped on the filter is determined by scintillation counting. When these data are graphed as a function of sample time (i.e., t_1, t_2, t_3, they provide the kinetics of product formation.

Imagine now that the enzyme sample is purified further and the purified enzyme is assayed for the same activity E-1 by the same method. However, in this case, following the addition of the substrate (A), the formation of product (B) during the course of the reaction is greatly reduced (graph line 2). While this result might indicate a true loss of E-1 activity, it might also be a result of an increase in the activity of E-2, a second enzyme that catalyzes the degradation of B to C (Fig. 1.5, panel II). In the absence of data on the level of the substrate (A) during the course of these reactions, the second possibility cannot be excluded.

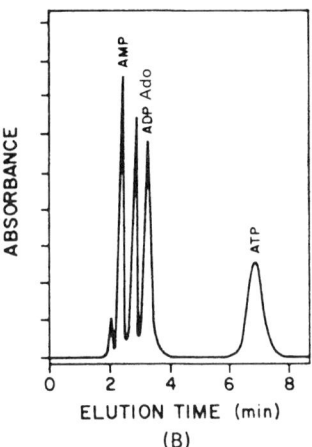

Figure 1.4 Separation of substrates and products of an adenosine kinase reaction on ion-paired reversed-phase HPLC. The separation was carried out on a prepacked C_{18} (μBondapak) column with a mobile phase of 65 mM potassium phosphate (pH 3.7) containing 1 mM tetrabutylammonium phosphate and 5% methanol. The column was eluted isocratically, and the detection was at 254 nm. Four relative elution positions (elution times) are shown.

To test this second possibility will require an assay for E-2, which in turn will call for a method for the measurement of the amount of C. Therefore, a reaction mixture optimized for E-2 will have to be prepared, and determinations of C formation in both the crude and purified samples will have to be carried out. If the data obtained in these determinations appeared as shown in panel II of Figure 1.5, these results would show that in the crude sample E-2 activity was indeed lower than that observed in the purified sample, thus indicating that the purification of E-2 could account for the loss of activity of E-1 during the purification. This example clearly shows that an assay method that measures only the levels of a single compound such as the product may provide very limited results.

In contrast, with a well-designed HPLC assay for activity E-1, capable of separating A, B, and C, the levels of each may be obtained from an analysis of a single sample from the reaction mixture of both the crude and purified samples (panel III of Fig. 1.5). In fact, the data obtained provide information not only on what compounds are present but also on what are not. The availability of such negative information can provide the "data" leading to the exclusion of alternative explanations that had been proposed to explain unexpected results.

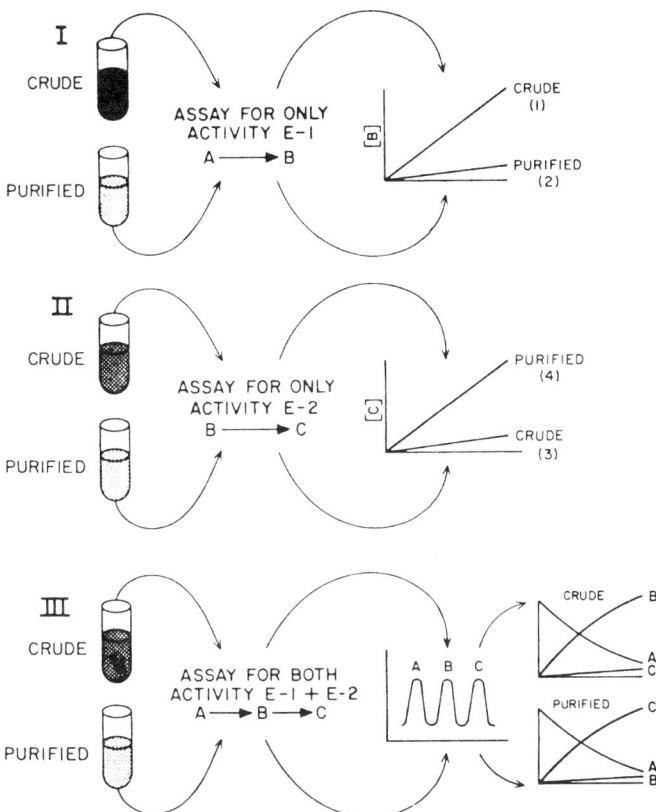

Figure 1.5 Comparison of the advantage of the HPLC assay method to traditional methods of assay for following the activity of an enzyme during purification. (*I*) The assay of the hypothetical enzyme, called E-1, which catalyzes the conversion of the substrate A to the product B. In this traditional method, an assay would be developed to follow the production of the product B. Note that in the crude extract E-1 produced B at the rate shown by line (1). However, after purification, E-1 produced B at a much slower rate, as indicated by line (2). To understand the reduction in rate, a second hypothetical enzyme was proposed that would convert B, the product of E-1, to a new product C. However, another traditional assay was used, and only the formation of C was measured. (*II*) E-2 activity was measured in both fractions, and while activity was found in both the crude and purified fractions, the rate of C formation in the purified fraction shown by line (4) was significantly greater than in the crude as shown by line (3). The activity of the enzymes E-1 and E-2 is shown being measured by the HPLC method. This assay was developed to separate A, B, and C simultaneously, and therefore it was possible to measure the activity of both E-1 and E-2 simultaneously. The results of assays carried out on both the crude and the purified preparations are shown. The plots at the right display the difference in levels of A, B, and C during the course of the incubation with both the crude and purified preparations.

1.4 CRITERIA FOR THE SELECTION OF AN ASSAY METHOD

The HPLC assay may not always be the procedure of choice, and several points should be considered before deciding. These points are summarized in Table 1.2.

1.4.1 Separation and Detection of Components

To utilize HPLC for an enzymatic assay requires first a system for separating the components. This involves the selection of a solid phase (the column packing), a mobile phase (the column eluent), and a method of elution from the solid phase by the mobile phase. Two procedures are generally used for elution: the isocratic and the gradient methods. In the isocratic method the mobile phase composition remains constant throughout elution, while in the gradient method the mobile phase varies in some parameter and in some fixed manner (e.g., a linear increase in salt concentration during elution). In addition, the HPLC method requires a monitor for the detection of the product. While a variety of detectors are available for monitoring various properties of molecules, a number of compounds cannot be detected on-line at present.

1.4.2 The Reaction Mixture

It is important to determine early on whether the reaction conditions previously developed for the assay of a given activity can be adapted for use with HPLC assay. For example, is the reaction mixture of sufficient volume to permit the withdrawal of multiple samples? For assays carried out in volumes of a few microliters, it is virtually impossible to withdraw samples of sufficient volume for analysis on the HPLC system. Thus, unless dilutions can be made after sampling, HPLC analysis must be ruled out in such cases.

Other factors should be considered as well. These include whether the reaction mixture contains any components that might make using HPLC

TABLE 1.2 Questions to Be Considered Prior to the Selection of HPLC for the Assay of an Enzymatic Activity

1. *Separation and detection*
   Must product be separated from substrate for analysis?
   Are detectors available?
2. *The reaction mixture*
   Are there limits to the total volume of the incubation mixture?
   Are cofactors such as metals a problem?
   How will the reaction be terminated?
3. *The enzyme*
   Is the enzyme pure, or will contaminating activities be present and affect product levels?

difficult. Such components include metals, which often hinder the interpretation of chromatograms. While the problem of metals can be solved easily by the addition of chelators, for other problems the solution may be more complex. For reactions that must be terminated prior to injection, for example, the termination process itself often alters the incubation mixture. Termination by acids (e.g., trichloroacetic, perchloric) reduces the pH of the sample. Since differences between the pH of the sample and the mobile phase can produce discrepancies in chromatographic profiles, any reduction in pH brought about by the termination may cause problems in interpretation. Also, termination increases the possibility of producing a precipitate, which will have to be removed before the sample can be injected into the column, to prevent clogging of the system. While not a difficult task, the removal of precipitates does introduce an additional step into the assay procedure.

1.4.3 The Enzyme Sample

A final point to be considered in the use of HPLC as an assay procedure is the enzyme itself. Will the activity be a pure enzyme? Will it be part of a rather crude cell-free extract? Or will it be present in a fermentation broth? In the latter two cases, the presence of contaminating activities must be considered. While as mentioned above, these activities, by affecting the recovery of the product or even by affecting substrate levels during the course of the reaction, could easily cause problems with other assay procedures, they are not a problem for the HPLC assay method. Thus, HPLC should be considered first when activity in some crude extracts is to be assayed.

1.5 SUMMARY AND CONCLUSIONS

The assay of the activity of an enzyme can be subdivided into several steps: formation of a reaction mixture, preparation of an enzyme sample, combination of the two to initiate the reaction, incubation of the reaction, termination of catalysis, separation of components, their detection, and finally, reduction or processing of the data.

Not all assays require a separation step, and this fact may be used to develop a classification scheme for assay methods. Assays that require no separation have been grouped under the heading "continuous assay methods," while "discontinuous methods" incorporate those that do.

The need to use a discontinuous assay method does not automatically mean that the HPLC method is the procedure of choice. For HPLC to be suitable, it must be possible to separate the components, and some method for detection and quantitation must be available. Next, neither the ingredients in the reaction mixture nor those used to terminate the reaction should produce problems for the separation and detection. Finally, the enzyme itself should be considered.

Excess protein can contaminate columns, and extraneous enzymes can cause problems in quantitation.

GENERAL REFERENCES

Reviews of liquid chromatography

Ettre LS (1983) The evolution of modern liquid chromatography. *LC Mag* **1:**108.
Freeman DH (1982) Liquid chromatography in 1982. *Science* **218:**235.

Classification of enzyme assay methods

Dixon M, Webb EC, Thorne CJR, Tipton KF (1979), *Enzyme techniques.* In *Enzymes,* 3rd ed. Academic Press, New York, Ch. 2

CHAPTER 2

Concepts and Principles of High Performance Liquid Chromatography

OVERVIEW

This chapter introduces some of the basic concepts and principles of liquid chromatography, providing background on the development of high performance liquid chromatography (HPLC) and briefly describing the basic system components.

The chromatogram is introduced as the record of the separation, and we identify the information it does and does not contain. Examples of different chromatographic profiles are presented and interpreted.

A strategy for the selection of the stationary phase is developed based on a discussion of the mechanism underlying the separation involved in gel filtration, reverse-phased, and ion-exchange chromatography.

The mobile phase will be considered, including its composition, preparation, and use.

The problem of column maintenance, particularly when the column is used for enzyme assays, is discussed, cleaning solutions are recommended, and a method for monitoring column performance described.

2.1 INTRODUCTION

Chromatography, the separation of classes or groups of molecules, in principle requires two phases. In liquid chromatography one of the phases is liquid and the other a stationary phase, often bonded to a solid. In days gone by, a stationary phase widely used in biochemistry laboratories was potato starch, and graduate students often found themselves up to their elbows in white "gooey stuff" making large batches of starch. After mixing, the starch would be poured into rectangular forms, where it hardened for later use as the solid phase for chromatographic analyses.

The sample was applied to one end of the block (or column if the stationary phase was poured into a vertical cylinder) and eluted from the other end (the bottom of the column) by allowing a mobile phase or solvent to flow through the stationary phase. Since the molecules of the sample are carried along by the mobile phase, the time required for a group of molecules to emerge from the stationary phase, other things being equal, was a property of the packing material. Those that emerged in the shortest time were considered not to have been affected by or to have interacted with the stationary phase, whereas those that emerged later did interact. Emergent time—or, as it is more often called, elution time—could be affected by two parameters: the distance (i.e., the path) that the molecules traverse as they pass through the packing, and the rate (velocity) at which they travel through the packing. These two parameters may be expressed in the relationship

$$\text{Emergent time} = \frac{\text{distance traveled}}{\text{rate of travel}}$$

2.2 THE INTRODUCTION OF HPLC

It has been known since the early days of liquid chromatography that the size of the particle used for the stationary phase affects the separation, or resolution, in a rather direct way: the smaller the particle, the better the separation.

However, with columns that used gravity to pull the mobile phase around the particles, a lower limit to particle size was reached, since the smaller the particles, the tighter they packed, eventually cutting off the flow of solvent. Thus, a pressurized solvent delivery system able to pump the mobile phase through the packing became necessary. Of course, as such pumps were developed, the old packings were found to collapse with the increased pressure, and new packing materials were required that could withstand these pressures. Together, pumps and new packing materials provided resolution and separation not achieved with earlier methods. As a fringe benefit, the new technique considerably shortened the time required to carry out separations. Separations that took hours or even days are now accomplished in minutes.

Another fringe benefit is the increased sensitivity provided by the detectors. Thus, it is no longer necessary to resort to the old ploy I was shown rather jokingly during my graduate school days. I was examining the results of an experiment in which radiolabeled compounds had been used, and I remarked to a fellow student on the incredibly low number of counts obtained during the experiments, expressing the wish for more counts. This prompted my fellow student to turn to the counter and change the counting-time dial from 1 minute to 10 minutes. Needless to say, a recount of the same samples produced more counts. In HPLC it is not necessary to resort to such tactics to obtain increased sensitivity. Rather, the geometry and low volume of the flow cells used in the detectors work in our favor.

2.3 BASIC COMPONENTS AND OPERATION

An HPLC system, shown schematically in Figure 2.1, consists of a *solvent reservoir,* which contains the eluent or mobile phse; a *pump,* often called a solvent delivery system; an *injector* through which the sample is introduced into the system without a drop in pressure or change in flow rate; the *analytical column,* which is usually stainless steel and contains the solid packing or stationary phase; and a suitable *detector* to monitor the eluent.

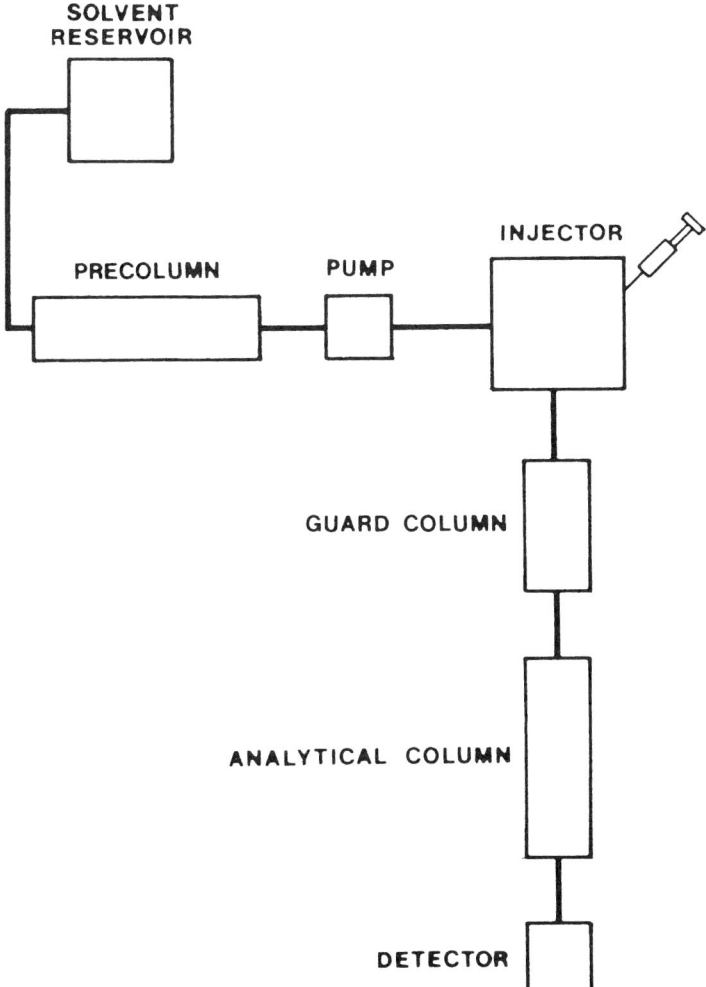

Figure 2.1 The components of an HPLC system: Solvent flow is from top to bottom. This diagram is representative only; the boxes are not drawn to the scale of actual system components.

Also shown in Figure 2.1 are two other components: a precolumn and a guard column. The *precolumn,* located in the system between the solvent reservoir and the pump, acts to filter out any impurities in the mobile phase before they reach the pump heads and the analytical column. In the case of analytical columns made of silica, impurities in the solvents can result in leaching of the silica. If solvents are made daily and are filtered (see below), a precolumn may not be needed.

The *guard column,* by virtue of its location between the injector and the analytical column, functions to remove any insoluble material and other debris that might have been injected and would otherwise clog the analytical column. For example, when used with enzyme assays, the guard column will remove precipitated proteins or other insoluble material carried over from the incubation mixtures. Since guard columns themselves can get clogged, if the pump is to maintain a constant flow rate it must generate greater pressures to drive the solvent through the clogged filter. This increase in pressure, referred to as "back pressure," can be eliminated by cleaning and repacking the guard column.

To operate an HPLC system, the sample is introduced through the injector into the system and is then pushed through the analytical column by the constant pumping of solvent (or mobile phase) from the reservoir through the system (Fig. 2.1). The mobile phase can be delivered in two ways: *isocratically,* that is, at constant composition, or in the form of a *gradient,* when the composition is varied. Chapter 4 explains how to decide which form to use and also describes injectors in more detail. Additional information about the operation of pumps can be obtained by consulting the references.

Following its emergence from the other end of the column, the eluent flows through the detector. Detectors operate on various principles. For example, some monitor the ultraviolet, visible, or fluorometric properties of molecules: others monitor radioactivity; and still others monitor differences in oxidation–reduction potential and refractive index. These detectors are listed in Table 2.1 together with some examples of the specific reactions with which they have been used.

2.4 COUPLING THE COMPONENTS: ON THE PERILS OF FERRULES

While one of the most confusing steps for the new user of HPLC is deciding what equipment to order, an even more difficult and frustrating step occurs after the equipment has arrived and connections must be made between the solvent delivery system (the pump), the various columns, and the detector. For the new user quickly discovers that connections are not made with the more familiar, easy-to-use flexible plastic, but with stainless steel tubing, which cannot be cut with scissors or easily coupled with plastic connectors. Thus special tools must be used for cutting, and connections must be made with

TABLE 2.1 Detectors and Their Applications

Detector	Reaction analyzed[a]
1. UV spectrometer	ATP → ADP + P_i
	IMP → Ino + P_i
2. Fluorometer	FoTP → cFoMP
	cFoMP → FoMP
3. Radiochemical	Hyp + PRPP → IMP + PP_i
	AMP + ATP → 2ADP
4. Electrochemical	L-Dopa → dopamine
	L-5′-Hydroxytryptophan → serotonin
5. Refractive index	Maltoheptose → oligosaccharides

[a] FoTP; formycin 5-triphosphate.

nuts and bolts and fittings called ferrules. Unfortunately, ferrules are not all the same, and once in place they are not easily removed.

The ferrule shown schematically in Figure 2.2 is swaged, that is, attached, when the end of one piece of stainless steel tubing is coupled to another. The coupling itself involves a bolt (B) and a nut (N), which are assembled as follows. The bolt is placed on the tubing, followed by the ferrule, and the nut is threaded to the male bolt, trapping the ferrule between and thus swaging the ferrule to the tubing. The appearance of the ferrule on the tubing before compression is shown in the upper panel of Fig. 2.2, and its appearance within the fitting after compression in shown in the lower panel. To avoid damage to the ferrules, these fittings should not be overtightened. Keep in mind, as

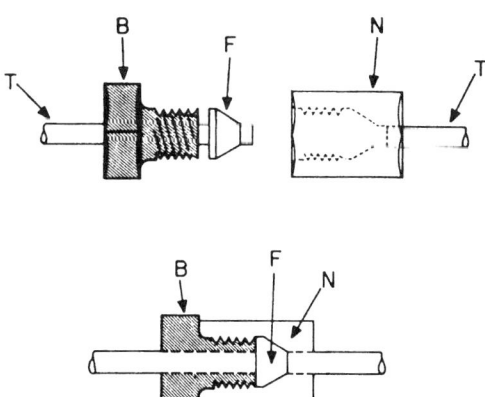

Figure 2.2 Cross-sectional diagram of the components used to couple two pieces of stainless steel tubing. *Top:* Units before swaging. T, tubing to be joined; B, "male" bolt; F, female; N, "female" nut. *Bottom:* After swaging. The pressure of the bolt on the nut has forced the ferrule to seal the joint between the two ends of the tubing.

2.5 THE CHROMATOGRAM

Information about the separation is displayed on a chromatogram, which is obtained by converting the detector output to an electrical signal and following this signal on a recorder as a function of the time after the loading of the sample. Figure 2.3 shows a representative HPLC chromatogram of a sample containing two species of compounds, A and B. In this example, while both enter the column at the same time (with the injection of the sample), compound B traverses the column at a faster rate than compound A. As shown in Figure 2.3, compound B will emerge and be detected first, followed by A. The time of injection of the sample, marked on the chromatogram by the arrow, is taken as zero time, and the time after injection is determined from the speed of the recorder. Of course, the rate of fluid flow is held constant and is controlled by the pump setting. Under these conditions, the chromatogram will show the elution of A and B as a function of time after loading the sample (injection time). Many investigators change the variable elution time to the more useful parameter *elution volume* by multiplying the flow rate, expressed in milliliters per minute, by the reciprocal of the chart speed, expressed in minutes per centimeter, to give the new unit of milliliters per centimeter. This maneuver allows the length unit (cm) on the chart to be converted to volume. The elution volume is especially useful if it becomes necessary to change the flow rate from run to run or if chromatograms obtained in different laboratories under different flow rates are to be compared.

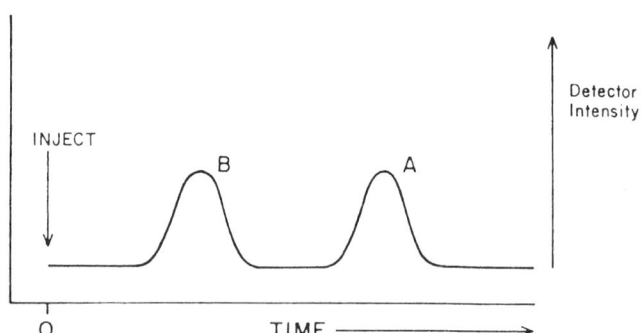

Figure 2.3 A representative HPLC chromatogram showing the separation of compounds A and B. The time of injection is taken as zero time, and the elution position is shown as a function of time after injection. The amount of each compound in the original sample is given by the peak height or area, as represented on the tracing by the letters B and A.

Figure 2.4, which shows chromatograms of the same sample obtained at different flow rates, illustrates the usefulness of the volume unit. Whereas in Figure 2.4A the sample was eluted at a flow rate of 2 mL/min, in Figure 2.4B the flow rate was 1 mL/min. A superficial analysis of these data would suggest that the two peaks in A and the two peaks in B represented four different compounds. However, if the same data are expressed as a function of elution volume, as shown in Figure 2.5, the two sets of peaks are easily seen to have similar retention volumes. Thus, based on this criterion, they are the same.

2.6 INTERPRETATION OF THE CHROMATOGRAM

In addition to showing that species such as A and B have been separated, chromatograms provide other information. For example, the shape of a curve provides information about the efficiency of the separation. With a system operating at high efficiency, peaks will be narrow and spikelike (Fig. 2.6A), while broad-based peaks suggest low efficiency (Fig. 2.6B). These results may be due in part to such factors as a luck of uniformity in either the size or homogeneity of the particles used in the stationary phase. Alternatively, broad peaks may indicate heterogeneity in the sample, as is often observed when the pH of the mobile phase is too near the pK of the molecules being separated.

The appearance of the peaks on chromatograms can also provide information about the quality of the resolution. Thus, if the compounds are well separated, the second peak will emerge only after the detector has completely

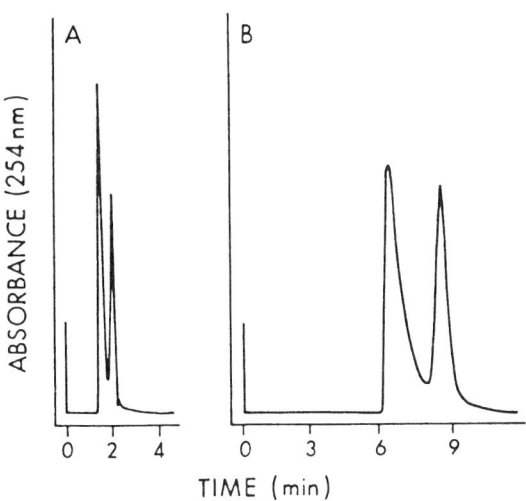

Figure 2.4 A comparison of the chromatography of the same two compounds carried out at flow rates of 1 mL/min (A) and 2 mL/min (B).

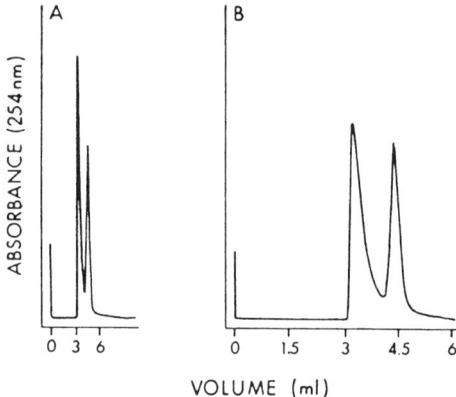

Figure 2.5 Data from Figure 2.4 replotted as a function of elution volume. The volume was determined by the multiplication of the flow rate (mL/min) by the reciprocal of the chart speed (cm/min). Expressed in this manner, each unit of distance is converted to a unit of volume.

returned to the baseline (Fig. 2.7A). Failure to achieve baseline separation (Fig. 2.7B) indicates poor resolution and suggests that something must be done to allow the second component to be retained longer—either slow its rate or increase the distance it must travel.

The resolution of any two components, therefore, is a ratio relating the distance between the apex of the peaks and the distance between their bases. With baseline separation, the bases of the peaks do not overlap. In the absence of baseline separation, however, the apex of each peak may be separate while the bases overlap. A mathematical expression can be written to describe this

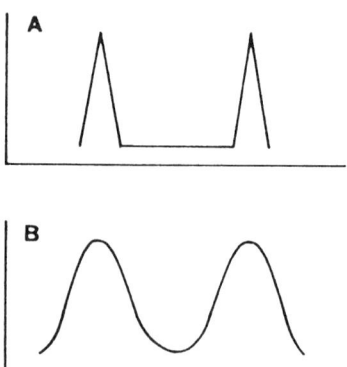

Figure 2.6 HPLC profiles of two components separated on two columns operating (A) at high efficiency and (B) at low efficiency.

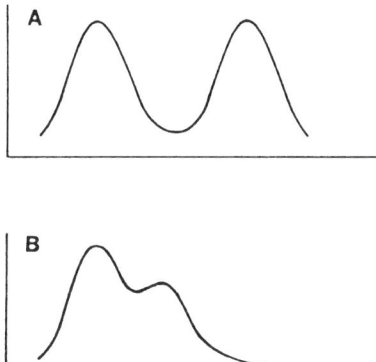

Figure 2.7 HPLC profiles of two compounds separated on column showing (A) good separation (resolution) and (B) poor resolution.

relationship be dividing the distance between the peaks, shown by the symbol delta (Δ) in Figure 2.8, by half the sum of the width of the bases, giving a numerical value for resolution (Fig. 2.8).

The symmetry of each peak can provide information about the sample. Tailing (Fig. 2.9A) suggests some heterogeneity in the sample—either real or introduced by the chromatographic conditions. Flat-topped peaks (Fig. 2.9B) suggest that the capacity of the column has been exceeded.

Of course, the magnitude of the signal from the detector can be used as a measure of the relative amount of each sample. While arbitrary units of area

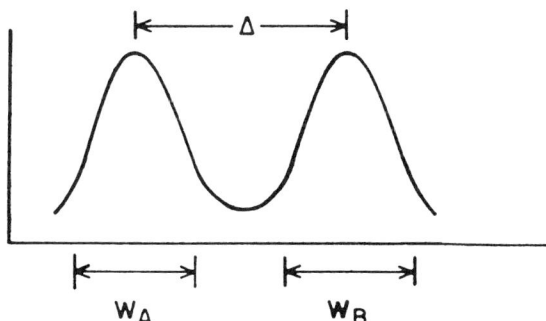

Figure 2.8 Representative HPLC chromatogram to illustrate a method for calculation of resolution R. The separation of two components labeled A and B) is shown. The width of each peak is shown by arrows and the symbol W, while the distance between peaks is shown by the symbol Δ. Resolution may be defined as

$$R = \frac{\Delta}{(1/2)(W_A + W_B)}$$

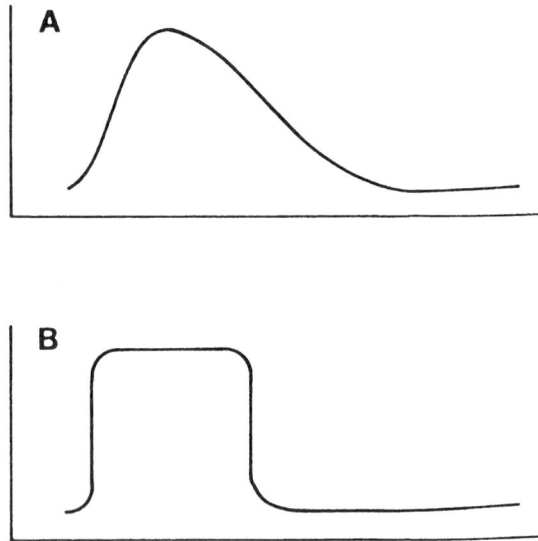

Figure 2.9 HPLC chromatograms of a single component showing (*A*) an example of "tailing" and (*B*) the profile obtained when a column is overloaded.

can be used, the conversion of these to traditional concentration units can be carried out easily after a calibration curve has been constructed.

Finally, the retention time (or volume) provided by a chromatogram can be used to identify an unknown compound. For example, comparison of the retention time of the unknown to the retention time of a series of standards (i.e., known compounds) is often sufficient to identify the unknown. However, a word of caution. Since any two compounds may coelute merely by coincidence, it is often necessary to apply criteria other than retention time before feeling certain about the identity of an unknown. Sometimes recourse to other methods, such as spectral analysis, is required to obtain a more definitive identification.

Enzymes themselves are often of use in identification of an unknown. Figure 2.10 shows a compound that had been tentatively identified as inosine 5'-phosphate (IMP) on the basis of its retention time. This conclusion was subjected to further testing using the enzyme 5'-nucleotidase, with the expectation that if the compound was IMP, the enzyme would catalyze the removal of the phosphate and the formation of inosine. The chromatogram obtained following the addition of the enzyme and incubation for about 20 minutes is shown in Figure 2.10. The chromatogram now shows in addition to the IMP, which is reduced in amount, a new peak with the correct elution time expected for inosine, the amount of which increases with incubation time (Fig. 2.10). These data add credibility to the claim that the starting material was IMP.

2.7 SELECTION OF THE STATIONARY PHASE

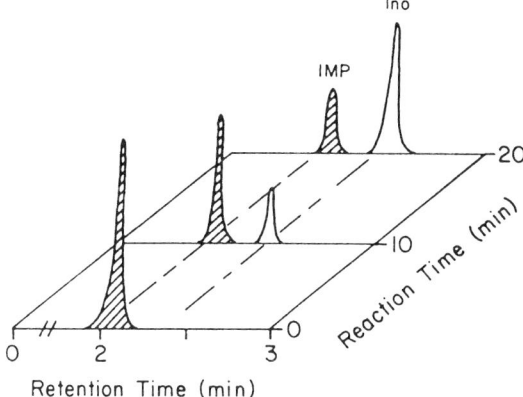

Figure 2.10 The use of enzymes to identify an unknown compound. The compound tentatively identified as IMP, based on its retention time of 2 minutes (chromatogram obtained at zero reaction time), was incubated with a commercially available preparation of 5′-nucleotidase. Samples of the incubation mixture were removed and analyzed by HPLC. The chromatograms, obtained at 10 and 20 minutes of reaction time, showed a reduction in the area of the IMP peak and an increase in the area of the inosine (Ino) peak, confirming that the original peak was IMP.

2.7 SELECTION OF THE STATIONARY PHASE: SOME HELP FROM AN UNDERSTANDING OF THE PROCESS OF SEPARATION

While the selection of a stationary phase to be used in the analytical column may appear complex, the decision can be greatly simplified by considering the three basic methods of separation currently in use. These are gel filtration or size-exclusion separation, reversed-phase or hydrophobic separation, and ion-exchange separation. In general, each type of separation uses a different kind of packing material, and since each type of separation exploits a different property of the molecules, the choice of packing really comes down to which property of the molecules would be most useful in achieving the separation.

For example, to use size-exclusion chromatography, the compounds to be separated must differ in size, shape, or both, while to use solubility or charge, the compounds must differ in polarity or net charge, respectively (Table 2.2).

TABLE 2.2 Selection of the Stationary Phase

Property	Separation mode
Size, shape	Gel filtration
Solubility, polarity	Reversed-phase
Charge, polarity	Ion exchange

2.7.1 Gel Filtration Chromatography

To understand gel filtration chromatography, imagine an analytical column packed with beads as shown in Figure 2.11A. If a single bead were examined

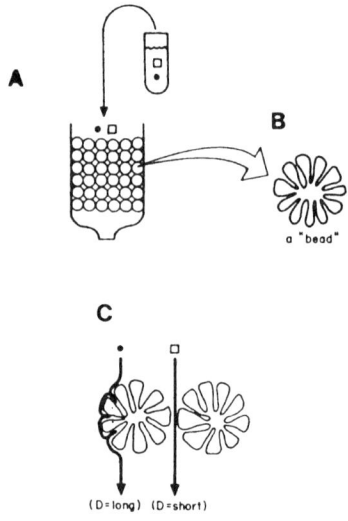

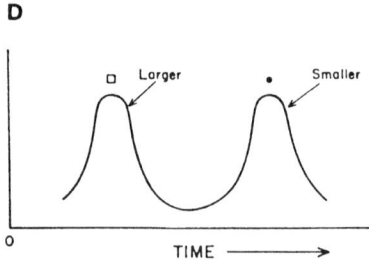

Figure 2.11 Gel filtration chromatography applied to a sample containing two compounds that differ in size larger (□) and smaller (●). (A) The two compounds are loaded onto a column packed with beads, which, when viewed at the ultrastructural level, would appear as depicted in (B). The larger compound passes between the beads, while the smaller enters the crevices of each bead (C). The chromatographic profile (D) illustrates that the larger molecule will emerge from the column in less time than the smaller.

by scanning electron microscopy, its image might be represented by Figure 2.11*B*. Each bead would be seen to have an irregular surface through and around which the mobile phase can enter and exit, in effect making the interior of the bead accessible to the mobile phase.

What about the sample? Imagine a sample to be composed of two types of compound, the molecules of which can be classed as "larger" and "smaller," as in Figure 2.11. Now imagine that because of these differences, the smaller molecules can follow the mobile phase as it meanders through and around the irregularities of the beads (FIg. 2.11*C*), while the larger cannot. Thus, the larger molecules will be excluded from taking the longer path, and as a result of this exclusion the path, or distance *D,* followed by the larger molecules through the column will be short, and the larger or excluded molecules will exit the column first. The volume of solvent requires for these molecules to emerge is spoken of as the *included volume.* The smaller molecule will follow a longer path and will emerge later. Because, in the ideal case, none of the molecules will interact with beads, the difference in times of emergence reflects the additional distance traversed by the smaller compound. Of course, altering the size of the irregularities of the beads will alter the size of the compounds excluded and therefore change the operating range of the analytical column.

2.7.2 Reversed-Phase Chromatography

We have seen that chromatography requires two phases: one solid and localized to the analytical column, the other mobile—the eluent or buffer that flows around and through the packing. The packing used by early workers was made of a material that was basically polar, while the mobile phases were nonpolar organic solvents. This arrangement of a polar stationary phase and a nonpolar mobile phase is, by virtue of tradition, referred to as *normal-phase* liquid chromatography. Fortunately, the type of liquid chromatography performed when the situation is reversed [i.e., the analytical columns are packed with a stationary phase that is nonpolar and eluted with polar (aqueous) buffers] was not referred to as "abnormal." Instead, since the phases have been reversed, this type of chromatography came to be known as *reversed phase.* More recently the term "hydrophobic" has been suggested.

In describing the underlying mechanism of operation of reversed-phase chromatography, it is convenient to focus again on the compounds in the sample and their movement through the analytical column. However, unlike gel filtration, where the order of elution of the compounds is determined by their path or the distance, traveled, in reversed-phase chromatography all the compounds in the sample travel the same path. In this case it is the rate at which they move through the column packing that determines the order of elution. Thus, a molecule that moves at a slower rate is said to be retained, and its time of elution is referred to as its *retention time.*

To understand the operation of the reverse phase, it is useful to consider the illustration in Figure 2.12, where oil droplets (or beads) are suspended in a column filled with an aqueous buffer. The water represents the mobile phase,

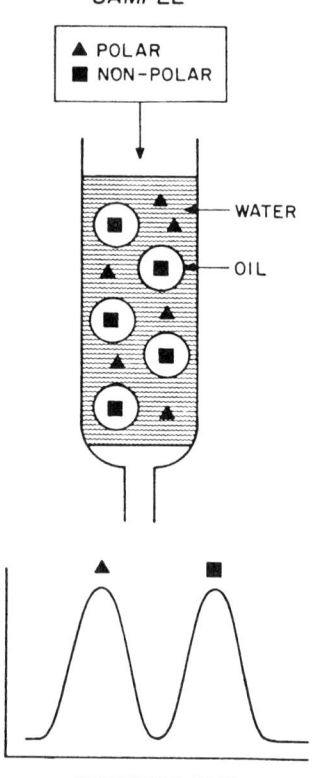

Figure 2.12 Reversed-phase HPLC of a sample composed of two compounds, one polar, the other nonpolar. The column packing (stationary phase) is symbolized by spheres and labeled "Oil" and the mobile phase as wavy lines labeled "Water." The polar molecules are shown remaining in the mobile phase (water), while the nonpolar molecules "enter" the stationary (oil) phase. Finally, the chromatographic profile illustrates that in this case the polar molecule will not be retained and will emerge with a shorter retention time than the nonpolar molecule.

while the oil represents the stationary phase in the analytical column. We load the sample, in this case molecules of compounds A and B, onto the surface of the column, and the question becomes whether the compounds will remain in the water (mobile phase) as they flow through the column or whether they will enter the beads of oil. Of course, the rate of progress through the column of a compound that remains in the water is effectively the flow rate of the mobile phase. There will be no interaction, and this compound will not be retained. It will emerge soon after loading and will have a short retention time.

In contrast, by entering the oil, a compound leaves the mobile phase and interacts with the beads: Its rate of passage through the system is in effect slowed. The compound will be retained, and it will have a longer retention time than a compound that does not interact.

A great deal of time and effort has been spent in trying to predict the retention time of compounds in the reversed-phase system. While some rules have emerged and some generalizations have been made, to date the best approach remains a few trial runs.

The most useful parameters to consider when developing a feel for the operation of this type of chromatography are polarity and the related parameter solubility. Values of both parameters have been published for many of the compounds used in biological systems.

In what follows, I introduce the notion of polarity and, after differentiating it from the net charge of a molecule, use it to explain the retention time for some classes of compounds.

Polarity should not be confused with any net charge a molecule might have. For example, in some cases, highly polar molecules contain no net charge. Polarity is a result of an electrical asymmetry that is due primarily to distribution of electrons. A case in point is the water molecule (Fig. 2.13). The polarity of a molecule of water is measurable when the two positive hydrogen atoms are localized on one side of the oxygen, resulting in negative and positive sides to the molecule, and the value is often expressed as a function of its dielectric constant: the greater the dielectric constant, the more polar the

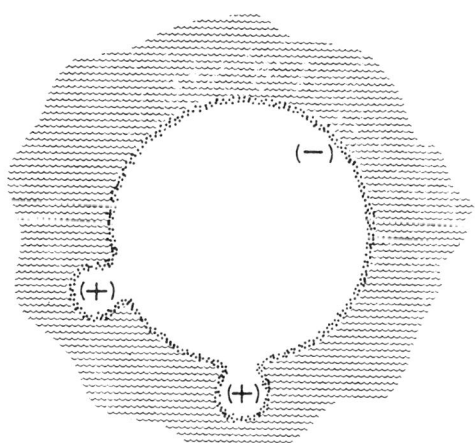

HOH MOLECULE

Figure 2.13 A representation of a water molecule to illustrate polarity. The positions of each of the two positively charged protons is shown as (+). The position of the negative charge of the oxygen atom is shown as (−). The asymmetric distribution of the positive and negative charges produces the polarity.

molecule. For example, the dielectric constant of water is about 81, while less polar molecules like alcohols have lower values. Thus, if retention time can be predicted from polarity or dielectric constant, such tables might prove useful.

Polar molecules are generally more soluble in water than nonpolar molecules, and therefore solubility values can also be useful in predicting retention times. For example, some amino acids (e.g., glycine, alanine, and others containing nonpolar side chains) are not very soluble in water; thus, on reversed-phase columns washed with only aqueous buffers, such compounds would be expected to interact with the nonpolar packing and be retained. Elution would be promoted by increasing the organic composition of the elution buffer.

Similarly, a comparison of the solubilities of some nucleobases in carbon tetrachloride and water will show that adenine is more soluble than guanine in organic solvents. Therefore, adenine will more likely enter the oil phase of the analytical column and be retained longer than guanine. Additional data obtained by measuring the distribution (or solubility) of a compound in either octanol or water show that adenosine is about 10 times more soluble than inosine in octanol. Again based on these findings, we might expect adenosine to have a longer retention time than inosine, and in fact it does.

Similarly, consider the compounds adenosine and ATP. As is well known to most biologists, ATP has greater solubility than adenosine in aqueous buffers. This knowledge, therefore, can be of value in predicting the behavior of such compounds in a reversed-phase system. Figure 2.14 shows the elution sequence of ATP and adenosine on a reversed-phased column eluted with an aqueous buffer: ATP elutes significantly before adenosine. In fact, such a short retention time suggests that ATP has great difficulty interacting with the nonpolar stationary phase.

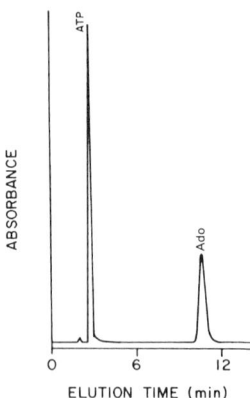

Figure 2.14 The separation of ATP and adenosine by reversed-phase HPLC. The prepacked column was C_{18} (μBondapak), and the mobile phase was a 10 mM potassium phosphate buffer (pH 5.5) containing 20% methanol. The column was eluted isocratically and monitored at 254 nm. The flow rate was 2 mL/min.

2.7 SELECTION OF THE STATIONARY PHASE

Adenosine, however, being less soluble (or more nonpolar) than ATP, will enter the stationary phase, and this is reflected in a longer retention time (Fig. 2.14). However, if a more nonpolar mobile phase were used to elute the column, the adenosine would remain longer in the mobile phase, hence its retention time would shorten. Thus, if a mobile phase was being used in which adenosine had as short a retention time as ATP, the separation of the two would be encouraged by reducing the amount of organic solvent in the mobile phase and causing the retention time of the adenosine to increase relative to that of ATP.

An interesting and useful variant of reversed-phase HPLC is called *ion-paired reversed-phase HPLC*. In such a system the analytical columns are packed with the same material, but a compound such as tetrabutylammonium is added to the mobile phase. The separation of ATP and adenosine on such a system is shown in Figure 2.15. A comparison of this profile to that shown for the same compounds in Figure 2.14 immediately highlights the change in the elution sequence. Whereas without ion pairing, the order is ATP followed by adenosine, with ion pairing the order is adenosine followed by ATP.

An explanation for the difference in retention times can be developed if one imagines the tetrabutylammonium compound, which is positively charged, paired with the negatively charged ATP molecule. While this pairing will, in fact, reduce the net charge, the reduction in net charge will also reduce the polarity of the ATP molecule. Since the short retention time initially was a result of the polarity, any reduction in polarity would be expected to increase retention time. Thus, coming full circle, the effect of the tetrabutylammonium salt on retention times might be explained by its effect (reduction) on polarity.

In a series of experiments designed to explore further the role of polarity in affecting retention time in reversed-phase chromatography, we developed chemical procedures for the condensation of molecules of known polarity,

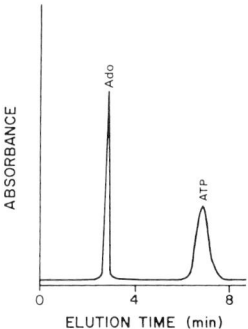

Figure 2.15 Separation of ATP and ADO on ion-paired, reversed-phase HPLC. The column was C_{18} (μBondapak), and the mobile phase was 65 mM potassium phosphate (pH 3.7) with 5% methanol and 1 mM n-tetrabutylammonium phosphate. The column was eluted isocratically, and the eluent was monitored at 254 nm.

expecting, for example, the joining two polar molecules to produce a relatively nonpolar molecule. In our first experiment we coupled the very polar nucleoside monophosphate AMP to lysine, an amino acid with a very polar side chain. The behavior of the two starting compounds in reversed-phase HPLC is shown in Figure 2.16A. Both have relatively short retention times, consistent with their polar character. However, when the retention time of the conjugate was determined, it was found to be longer than that of either of the starting compounds (Fig. 2.16A). Thus, the combination of two polar compounds can produce a compound more nonpolar than either of the parent compounds.

Similar experiments were undertaken joining AMP to a dipeptide hippuryllysine. This particular dipeptide was used because a comparison of the retention times of lysine and hippuryllysine revealed that the addition of the hippuric acid to the lysine reduced the polarity of the latter. However, a determination of the retention times of the dipeptide and AMP on a reversed-phase column (Fig. 2.16B) reveals both to be polar. Nevertheless, their conjugate has a longer retention time than either of the starting materials. Note, however, that the decrease in polarity of this conjugate is very much less than what was observed following the summation of the AMP and lysine (Fig. 2.16A).

Finally, the AMP was coupled to the tetrapeptide tuftsin, which has the amino acid sequence Thr-Lys-Pro-Arg. Based on its extremely long retention time on a reversed-phase column, the tuftsin can be considered a nonpolar molecule, a conclusion supported by its rather low solubility in aqueous systems and the requirement for 40% methanol to elute it from the column. When AMP is condensed onto the tuftsin, usually a single AMP per molecule of tuftsin, the polarity of the tuftsin is significantly decreased, as indicated by the decrease in the retention time of the conjugate. As shown in Figure 2.16C, the conjugate has a retention time much closer to that of AMP than to that of tuftsin. This finding suggests that the addition of the polar AMP to the nonpolar tuftsin decreases the overall electrical asymmetry but does not eliminate it completely. Thus, while the combination of two polar molecules can produce a nonpolar molecule, the combination of a polar with the nonpolar molecule can produce a molecule more polar than its nonpolar parent.

It should be noted that reversed-phase HPLC has been used to deduce polarity as, for example, in a study of cAMP and its analogs, as well as to predict partition coefficients and lipid solubility.

2.7.3 Ion-Exchange Chromatography

In ion-exchange chromatography, as in reversed-phase HPLC, the rate at which a molecule moves through the analytical column and its interaction with the packing determine the order of elution of given compounds. In this case, both the number and the magnitude of the charge influence interactions. The principles and method of operation in ion-exchange HPLC are similar to those of the more conventional ion-exchange systems.

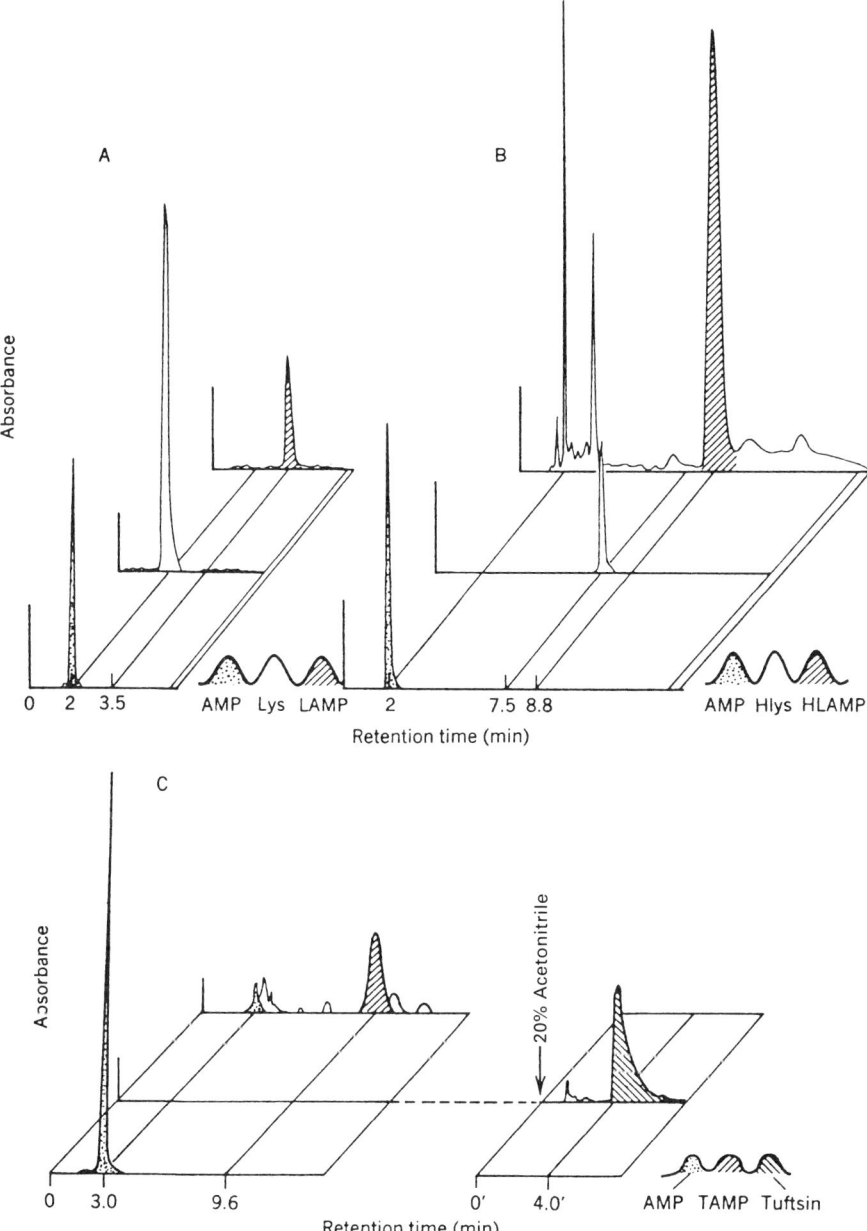

Figure 2.16 Effects of polarity on retention time. HPLC chromatography carried out on reversed-phase C_{18} (μBondapak) column. (*A*) Chromatographic profiles of AMP, lysine, and the lysyl–AMP conjugate obtained using a mobile phase of 65 mM potassium phosphate (pH 3.6) and 2% acetonitrile. (*B*) Profiles of AMP, hippuryllysine, and the hippuryllysyl–AMP conjugate eluted with a mobile phase of 10 mM potassium acetate (pH 7.2) containing 1% acetonitrile. (*C*) Chromatograms obtained with AMP, tuftsin, and tuftsin–AMP. Compounds were eluted with a mobile phase of 65 mM potassium phosphate (pH 3.6) and 2% acetonitrile for AMP and tuftsin–AMP. Tuftsin was eluted by 20% acetonitrile. Detection was at 230 nm for lysine and 254 nm for all others.

In general, the support (stationary) phase carries either a positive or a negative charge. During equilibration of the column with the eluent, a counterion is introduced. The molecules to be separated must also be charged, and when the sample is loaded, they bind to the fixed charges of the column packing and displace the counterion. Elution of the bound molecules is brought about by a second counterion, which is usually introduced as salt onto the packing by adding it to the elution buffer. The ability of the counterions (salts) to displace bound molecules relies on the difference in their affinities for the fixed charges of the stationary phase.

The interaction between the fixed charges of the stationary phase and the compounds adenosine, AMP, ADP, and ATP with zero, one, two, and three charges, respectively, is shown schematically in Figure 2.17. In anionic-

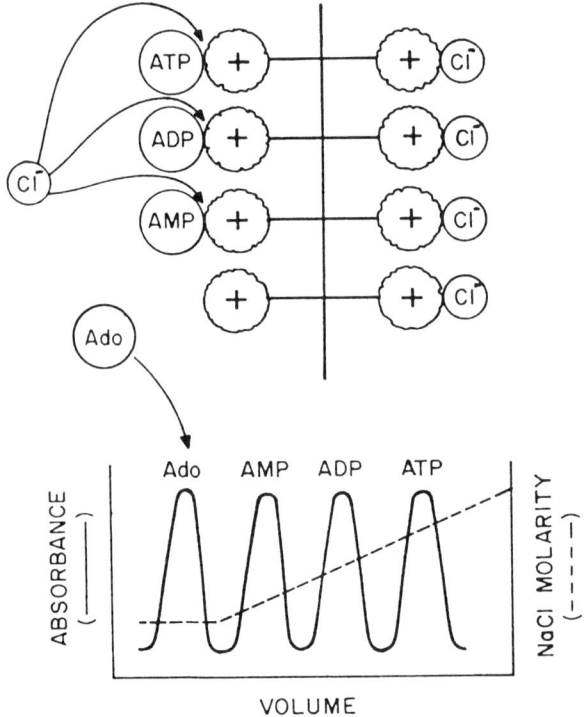

Figure 2.17 Representation of operation of ion-exchange chromatography. *Top:* Column functional groups, or fixed charges, are represented by serrated-edge circles carrying a positive charge, shown "fixed" to a lattice. The compounds to be fractionated are ATP, ADP, AMP, and Ado. The first three are shown bound to the fixed charges, while the Ado is shown unbound. The introduction of the counterion, a chloride, is shown displacing the bound molecules from the fixed charges. *Bottom:* The order of elution as a function of the NaCl molarity, which is represented by the dashed diagonal line. The relative elution position (volume) of each of the four compounds is shown.

2.7 SELECTION OF THE STATIONARY PHASE

exchange chromatography, adenosine, with no charge, is not retained; it will be eluted in the absence of the addition of any salt. Thus, as represented in Figure 2.17, adenosine (Ado) will have a short retention time.

Increasing concentrations of chloride will be required to displace and elute in series AMP, ADP, and even higher concentrations of ATP (Fig. 2.17). Thus the elution order will be Ado, AMP, ADP, and ATP, with Ado having the shortest retention volume of the four (see Fig. 2.17). Such an elution order is consistent with the explanation that increasing the number of charges of a molecule increases its interaction with the stationary phase packing, thereby reducing its flow rate through the analytical column.

A consideration of the effectiveness of the different salt cations (or anions) in displacing or exchanging bound molecules requires a discussion of the magnitude of the charge. Thus, molecules with one charge are not all equal when it comes to interacting with the fixed charges of the column packing. As a first approximation, the strength of the interaction may be considered in terms of the number of water molecules between or surrounding the salt ion; this is its hydrated radius.

For example, as represented in Fig. 2.18, cesium has only a few water molecules surrounding it compared to the lithium ion, which has many more (Fig. 2.18). One might think of these water molecules as a shield, with their elimination required for any interaction to take place between the ion and the packing. Clearly, it takes less energy to eliminate one molecule than several, and therefore it is not surprising that cesium is more effective than lithium at displacing molecules such as ATP bound to ion exchangers (see Fig. 2.18). This effectiveness is seen operationally in terms of the concentration of the counterion required to elute the bound sample. It is also not surprising that affinities can be affected by modifications that alter the water content of the system—for example, by increasing salt concentrations, temperature, or the organic solvent content of the mobile phase. However, on an ion-exchange HPLC column, the AMP is eluted at a lower salt concentration than cAMP, as illustrated in Figure 2.19. This difference in affinity might be explained by the model already described in which relative affinity is a function of the

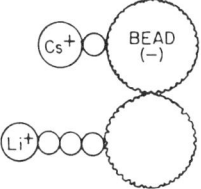

Figure 2.18 Effect of radius of hydration on distance between counterion and fixed charge of ion-exchange stationary phase. Cesium, with smaller radius of hydration, is shown with one water molecule (small circle) between it and fixed charge of the bead. Lithium is shown with three water molecules.

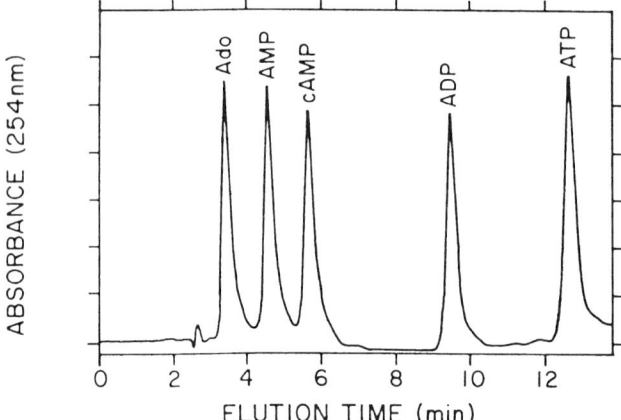

Figure 2.19 Separation of several nucleosides on ion-exchange HPLC was carried out on an ion-exchange column (AX-100) eluted isocratically with a mobile phase of 0.1 M sodium phosphate buffer (pH 7.3) containing 0.8 M sodium acetate. The column was monitored at 254 nm. A standard solution containing approximately 2 nmol each of adenosine, AMP, cAMP, ADP, and ATP was loaded onto the column.

distance between the mobile ion and the fixed ion. In this model, AMP, with less affinity, would have a greater distance between it and the fixed charge on the bead than cAMP would have. In both cases, as illustrated schematically in Figure 2.20A, the space would be occupied by water molecules. The net effect of this difference would be a lower concentration of the chloride required to displace the AMP, which would be eluted before cAMP, as illustrated in Figure 2.20B.

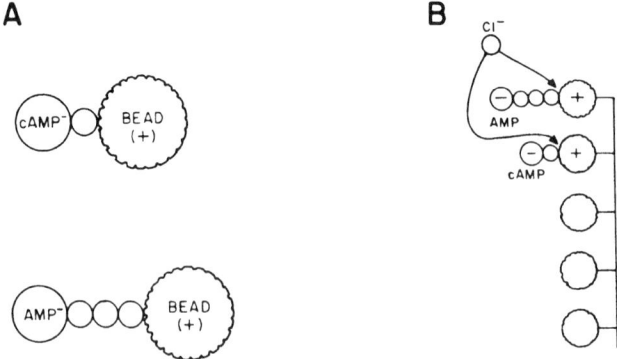

Figure 2.20 Representation of the interaction between the fixed charges of the ion-exchange beads and the cAMP and AMP molecules.

2.8 COMPOSITION AND PREPARATION OF THE MOBILE PHASE

A mobile phase containing salt and organic modifiers is commonly used to elute samples from a reversed-phase column. The salt is added to suppress ionic effects that could alter separation. However, with samples from enzymatic reactions, which are often at pH values different from that of the mobile phase, a buffer should be added to the mobile phase, as well. The buffering capacity of the mobile phase should be in excess of that in the incubation mixture. This excess will ensure that when the sample is injected, its pH will equilibrate to that of the mobile phase. These buffers should be made with distilled, deionized water that has been degassed to remove any trapped air. Degassing prevents bubble formation, which otherwise occurs in the pump head, particularly at high pressures. Bubbles are one of the major causes of variations in pump pressure, which, in turn, can produce artifacts in the chromatographic profile. Degassing, accomplished by vacuum aspiration, should be carried out with constant stirring until only a few bubbles are formed. Usually about 20 min/L is adequate.

Phosphate in concentration ranges of 10 to 100 mM can be used not only for buffering but also for ion suppression. However, some thought should be given to later uses contemplated for the compounds purified by the HPLC. For example, if after purification the component will be examined by phosphorus NMR, a mobile phase containing phosphate obviously should not be used. In addition, the use of phosphate in ion-exchange mobile phases can lead to high background absorbance in the UV range as a consequence of impurities in the phosphate buffers. Although methods have been described to reduce this background (see General References), where possible the phosphate should be eliminated. The pH can always be adjusted with KOH. With such bases, however, the choice of cation (e.g., K or Na), should be made with the composition of the sample buffer in mind. For example, if the buffer contains sodium dodecyl sulfate (SDS), potassium, which will precipitate SDS, should be avoided. Also, since halides such as chloride can cause corrosion of stainless steel tubing, they should be avoided. Finally, if the purified components are to be concentrated by an evaporation procedure, a buffer with volatile components should be chosen.

To reduce retention time of nucleotides on reversed-phase columns, organic modifiers such as methanol or acetonitrile may be added to the buffer. Acetonitrile is usually more effective than methanol in the sense that less is required to elute a given nucleotide with a specific retention time.

Once the buffer (mobile phase) of the appropriate composition has been made and the pH adjusted, it should be filtered through a 0.45 μm filter to remove particles that may clog the column head. Following filtration, the buffer may be used for about 2 to 3 days if stored at room temperature although the pH should be checked and precautions taken to keep the organic modifer from evaporating during storage.

At the conclusion of each work day it is advisable to wash the salt buffer thoroughly from the both the pump heads and the column. A 0.02% sodium azide solution prepared with degassed water should be used to wash the system free of salts. About 15 minutes of washing time is adequate with a flow rate of 2 mL/min. The sodium azide is used to control bacterial growth. A reversed-phase column should then be washed with and stored in a methanol–water solution (80:20).

Washing removes material from the guard column and the top of the analytical column, and when this material passes through the detector, it often causes changes in optical density. Therefore, the recorder should be left on during the washing to ensure complete removal of this debris. The recorder will return to a baseline reading when the washing is complete.

Note that water and methanol differ in viscosity. Thus the change in viscosity of the solutions accompanying a switch from one solvent to the other will produce an increase in back pressure that may be significant in magnitude; such changes are not worrisome, however. Also, a sodium azide solution moving through a detector set at 254 nm will usually produce an increase in optical density. Again these are normal changes and should not cause concern.

2.9 COLUMN MAINTENANCE

When reversed-phase columns are used for the analysis of enzymatic reactions, many of the components of the reaction may become bound to the packing material. As a result, the debris may alter the retention time, chromatographic profiles, or both of subsequently injected molecules. Types of column malfunction include peak splitting or the appearance of a shoulder; loss of baseline resolution, broadening of peaks, particularly at their base, or both; and an increase in back pressure. To some extent, all these symptoms may be traced to material that adhered to the column and was not removed during the methanol wash.

If divalent metals are suspected as the cause of the peak splitting, washing the column with 100 mL of 10 mM EDTA in 10 mM phosphate at pH 5.5 may help eliminate the problem by removing the metal.

Many components that bind can be eluted by changing the pH of the mobile phase. Thus, a wash consisting of 200 mL total volume, at 2 mL/min, of 100 mM phosphate solution ranging in pH from 2 to 8 is frequently useful. Finally, if an increase in back pressure is suspected to be a result of contamination from bound protein, washing the column with at least 100 mL of 6 M urea in 20 mM phosphate (pH 7.8) may eliminate the problem. Again, the return of back pressure to normal values can be taken as a sign of success of any one of these steps.

We have found urea to be a poor wash solution on columns packed with reversed-phase packings. Dimethyl sulfoxide (DMSO) has been useful in some cases.

In addition, a gradient progressing from 100% methanol through a series of less polar, more organic solvents such as carbon tetrachloride will serve to remove other reversibly bound contaminants. A reverse gradient should be used to reequilibrate the column to standard conditions.

Note that following any of the maintenance procedures listed above, reequilibrated the column must be to the original mobile phase, probably involving more time than would be needed for the normal change from water to the methanol solution routinely used to prepare the column for overnight storage.

As an illustration, consider the problem of the contamination of a reversed-phase column with a very sticky dextran sulfate material that had been added as an activator for an enzyme reaction. The compounds AMP, ADP, and ATP were being separated using a mobile phase containing phosphate buffer, acetonitrile, and tetrabutylammonium ion. The separation usually obtained is shown in Figure 2.21A. However, in the presence of dextran sulfate the separation was less than adequate (Fig. 2.21B). The column was regenerated by first washing with 6 M urea. Analysis of the sample produced is shown in Figure 2.21C. Some improvement in the separation is evident. Next the column was washed with toluene. The result, (Fig. 2.21D) shows complete restoration of the separation capabilities of the packing. Additional details may be obtained by consulting the General References at the end of this chapter.

2.10 MONITORING COLUMN PERFORMANCE

In general, new columns should be calibrated in the laboratory in which they will be used, and a standard mobile phase and a standard series of compounds should be available for this purpose. The resolution obtained under the standard conditions of the laboratory at the start of the useful life of the column should be recorded, together with the date of the analysis. All should be part of the record for that column. At the first sign of problems with this column, its performance should be checked against these records, using the same mobile phase and standards.

2.11 SUMMARY AND CONCLUSIONS

Chromatography involves the separation of classes or groups of molecules. Two phases are usually required: one is stationary or solid, and the other mobile—eluent or buffer.

If the solid phase is in the form of particles or beads, it is usually packed into a tube or column and the buffer or mobile phase is pulled through the packing by gravity or forced through with a pump.

The time required for a given compound to emerge from the column is a function of the packing material and the interactions between the compound and the packing. Transition time is affected by the distance the compound travels or the rate at which it travels a fixed distance.

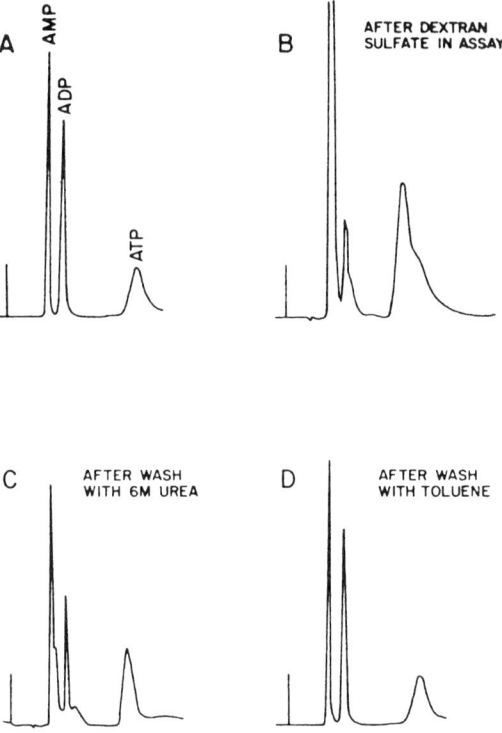

Figure 2.21 The effect of washing procedures on removal of debris (dextran sulfate) as measured by separation of nucleosides. Separations were carried out on a reversed-phase (C_{18}) column with a mobile phase of 65 mM potassium phosphate (pH 3.6), 2% acetonitrile, and 1 mM tetra-n-butylammonium phosphate. The flow rate was 2 mL/min; the column was eluted isocratically and monitored at 254 nm. (A) Separation routinely achieved with AMP, ADP, and ATP. (B), Separation observed after clogging the column with 10 mM dextran sulfate. (C) The separation observed after washing the column with 6 M urea. (D) The separation obtained after washing the column with toluene.

Other things being equal, resolution is enhanced by using smaller particles of the packing material. However, smaller particles result in tighter packing, which in turn requires higher pressures to push through the solvents. The combination of small particles with better pumps led to improved performance and the method called high performance liquid chromatography.

The basic equipment required for an HPLC system includes a solvent reservoir, a pump, an injector, an analytical column, a detector, and a recorder. The analysis of the sample is displayed as a chromatogram, with detector deflection presented usually as a function of time after loading the sample. By virtue of the shape of the curves, the distance between them, and their area, it is possible to determine whether the volume of the sample is too large, as well as the number of different compounds present and the amount of each compound occurring in the sample. From an understanding of the process of separation, it is possible to select the appropriate stationary phase. Separation by gel filtration requires compounds of different sizes and shapes, while reversed-phase HPLC will separate molecules that have different polarities. In contrast, ion-exchange HPLC separates molecules with different charges.

The mobile phase contains salts and organic modifiers. A buffer is also required with enzymatic reactions to ensure a constant pH during the separation step.

Column maintenance will often require washing the column with chelators, denaturants, organics, or salt solutions of high concentration, all designed to remove column-bound debris that is not removed by routine washing.

GENERAL REFERENCES

General references for HPLC

Brown PR (1973) *High Pressure Liquid Chromatography: Biochemical and Biomedical Applications.* Academic Press, New York.

Hearn MTW, Ed. (1985) *Ion-Pair Chromatography: Theory and Biological and Pharmaceutical Applications.* Dekker, New York.

Henschen A, Hupe K, Lottspeich F, Voelter W (1985) *High Performance Liquid Chromatography in Biochemistry,* VCH Publishers, Deerfield Beach, FL.

Krstulovic AM, Brown PR (1982) *Reversed-Phase High-Performance Liquid Chromatography: Theory, Practice and Biomedical Application.* Wiley-Interscience, New York.

Regnier FE (1983) *Science* **222**:245.

Snyder LR, Kirkland JJ (1979) *Introduction to Modern Liquid Chromatography.* Wiley, New York.

Polarity and solubility

Cohn EJ, Edsall JT (1939) *Proteins, Amino Acids and Peptides.* Hafner, New York.

Cullis PM, Wolfenden R (1981) *Biochemistry* **20**:3024.

Greenstein JP, Winitz M (1961). *Chemistry of the Amino Acids,* Vol. 1. Wiley, New York.

Hansch C, Leo AJ (1979) *Substituent Constants for Correlations Analysis in Chemistry and Biology.* Wiley, New York.

Kolassa N, Pfleger K, Rummel W (1970) *Eur J Pharm* **9**:265.

Nahum A, and Horvath C (1980) *J Chromatog.* **192**:315.

Plaut, GWE, Kuby SA, Lardy HA (1950) *J Biol Chem* **184**:243.

Reversed-phase chromatography

Hacky JE, Young AM (1984) *J Liquid Chromatog,* **7**:675.
Hammers WE, Meurs GJ, DeLigny CL (1982) *J Chromatog,* **247**:1.
Hancock WS, Ed. (1984) *Handbook of HPLC for Separation of Amino Acids, Peptides and Proteins, Vols 1 and II.* CRC Press, Boca Raton, FL.
Krstulovic AM, Brown PR (1982) *Reversed-Phase High-Performance Liquid Chromatography: Theory, Practice and Biomedical Applications,* Wiley-Interscience, New York.
Perrone P, Brown PR (1985) Ion-pair chromatography of nucleic acid derivatives. In *Ion-Pair Chromatography,* MTW Hearn, Ed. Dekker, New York.
Rossomando EF, Hadjimichael J (1986) *Int J Biochem* **18**:481.

Care and maintenance of columns

Runser DJ (1981) *Maintaining and Trouble Shooting HPLC Systems.* Wiley, New York.

Detectors

Henderson RJ, Jr, Griffin CA (1984) *J Chromotogr* **298**:231.

Ion-exchange chromatography

Jahngen JH, Rossomando EF (1983) *Anal Biochem* **130**:406.
Regnier F (1984) High-performance ion-exchange chromatography, In *Methods in Enzymology,* Vol. 104, WB Jakoby, Ed., p. 170. Academic Press, Orlando FL.

Preparation of mobile phase

Karkas JD, Germershauser J, Liou R (1981) *J Chromatogr* **214**:267.
Plunkett W, Hug V, Keating MJ, Chubb S (1980) *Cancer Res* **40**:588.

Pumps: operation and troubleshooting

Dolan JW, Berry VV (1983) *LC Mag* **1**:470
Dolan JW, Berry, VV (1984) *LC Mag* **2**:210.

CHAPTER 3

Concepts and Principles of High Performance Capillary Electrophoresis
with Franco Tagliaro, Zdeneck Deyl, Ivan Mikšík, and Kathi J. Ulfelder

OVERVIEW

This chapter introduces the basic concepts and principles of capillary electrophoresis (CE), presenting some background on electrophoresis and capillary electrophesis and describing the components of the system. The two main types of CE, capillary zone and micellar electrokinetic electrophoresis, are described, and a selection strategy, based on the two types of separation, electrophoretic migration and electroosmosis, is presented.

3.1 INTRODUCTION

HPLC uses a hydrodynamically driven flow of fluid to transport solutes. In contrast, in electrophoresis, where there is no transport of fluid, the solute molecules are transported by an electrical force or potential.

3.2 HPCE: DEFINITION, HISTORY, AND LITERATURE

High performance capillary electrophoresis is one form of free-solution electrophoresis. CE is useful to researchers and analysts working in areas in which traditional electrophoresis is customarily applied (e.g., biopolymer analysis) and also in disciplines not usually associated with electrophoretic analysis, such as inorganic ion analysis. The potential application areas of CE are vast, because this technique can separate a variety of ligates, from inorganic ions up to intact cells, using the same instrumental hardware designed for separations based on different physical–chemical mechanisms.

Since the late 1960s, capillaries have been used for free-solution electrophoresis, but instrumental limitations, such as capillary materials, the large internal diameters (i.d.) of capillaries, and detector sensitivity, hampered any direct application of this early technology. In the early 1980s Jorgenson and Lukacs (1981a, 1981b, 1983) advanced the technique by using thin (< 100 μm i.d.) fused silica capillaries and formally described the basic relationships between experimental parameters and separation performance. Since then, there has been an exponential growth in basic studies, instrumental improvements, and applications regarding CE. Only at the end of the past decade, however, has commercial instrumentation given CE the opportunity of spreading into the field of applied analytical science.

The main peculiarities of modern CE are great analytical versatility, high separation efficiency, high mass sensitivity, extremely low demands on sample volumes, short analysis times (usually < 30 min), minimum consumption of solvents (running buffer) and other consumables (e.g., capillaries), possibility of interfacing with different detection systems (including mass spectrometry), and simplicity of the basic instrumentation.

Several volumes (Grossman and Colburn, 1992; Mosher et al, 1992; Vindevogel and Sandra, 1992; Camilleri, 1993; Foret et al., 1993; Guzman, 1993; Khun and Hoffstetter-Kuhn, 1993; Jandik and Bonn, 1993; Weinberger, 1993; Li, 1994), fundamental reviews (Kuhr, 1990; Kuhr and Monnig, 1992; Monnig and Kennedy, 1994), and symposium proceedings have appeared.

3.3 BASIC COMPONENTS AND OPERATIONS

A representative CE system is illustrated schematically in Figure 3.1. The system consists of a narrow bore capillary tube, two buffer reservoirs, a high voltage power supply, a detector (usually on-column), and a data collection and analysis system (chart recorder or computer with software designed for peak integration and identification). In CE, the separation is carried out in the fused silica capillary, usually 20 to 100 mm i.d. and 20 to 100 cm long. Because of the large ratio of surface area to volume in a capillary, Joule heat is effectively dissipated during a run, allowing the application of voltages of up to 30 kV. The outer surface of the capillary, coated with polyimide to maintain flexibility, may be encased in a temperature-controlled chamber. The inner surface of the capillary may also be coated, depending on the application. The capillary is filled with an aqueous solution, and a minute amount of sample (< 20 NL) is introduced into the capillary by hydrodynamic or electrokinetic means. With hydrodynamic injections, the sample is pushed (pressure) or pulled (vacuum) onto the end of the capillary; an electrokinetic injection draws charged analytes into the capillary using an electric field. This latter method will cause a sampling bias, since analytes with a higher electrophoretic mobility will enter the capillary more readily than those of lower mobility.

Under aqueous conditions, the inner wall of the capillary becomes negatively charged as a result of the presence of acidic silanol groups. In an electric field, this negative charge results in a bulk flow of fluid in the capillary toward

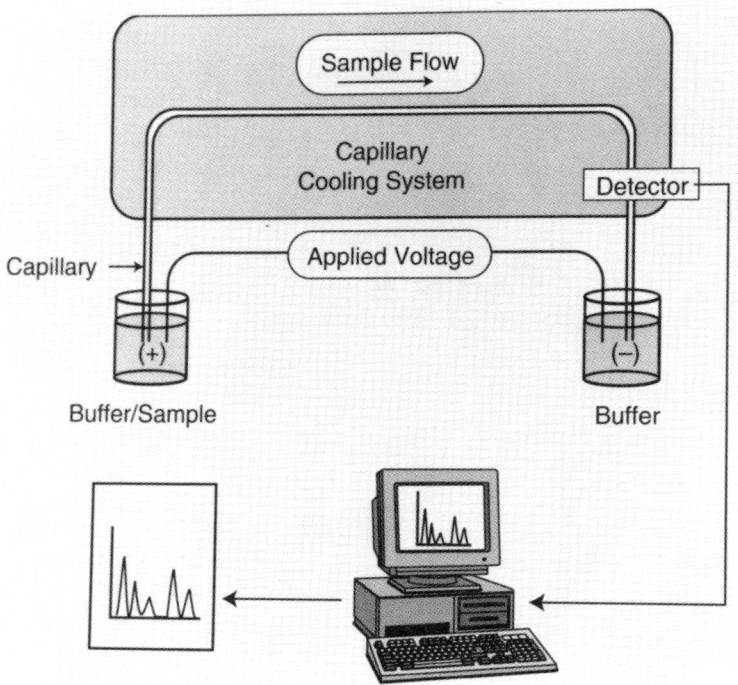

Figure 3.1 Diagram of a CE instrument with anode at injection side (normal polarity). (Reprinted with permission from Schwartz and Guttman, *Beckman Instruments Primer Series,* Vol. VII, 1995. Copyright: Beckman Instruments, Inc.)

the cathodic electrode. The magnitude of this force, known as *electroosmotic flow* (EOF), depends on many parameters: field strength, buffer ionic strength, and pH, to name a few. Coating of the capillary interior will reduce, reverse, or even eliminate EOF.

Detection is achieved by direct on-column monitoring through a window made in the capillary by removing 1 to 2 mm of the polyimide outer coating. Typical commercial detectors include UV absorbance, UV–visible scanning diode array, mass spectrometry interface, and fluorescence. Detectors of the latter type have been made highly sensitive through the use of a laser as a focused excitation source. Figure 3.2 shows the detail of the optical design used in a commercial laser-induced fluorescence (LIF) detector.

3.4 THE PROCESS OF ELECTROPHORETIC SEPARATION

3.4.1 Electrophoretic Separation

The velocity of migration of a charged analyte in an electric field depends on its electrophoretic mobility and on the magnitude of the applied electric field.

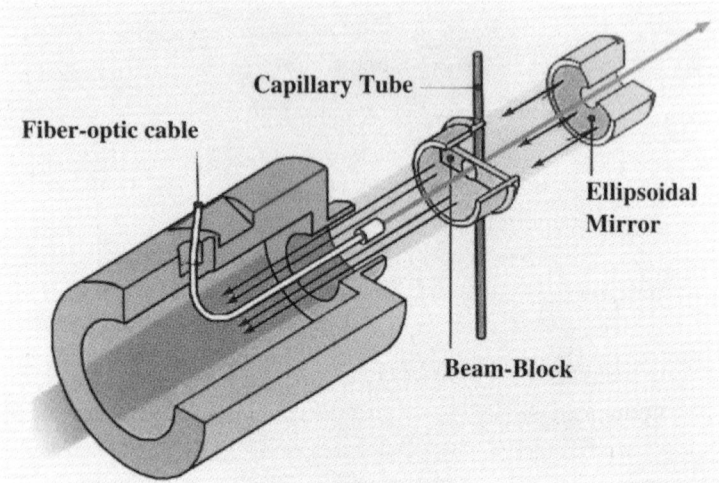

Figure 3.2 Schematic view of a commercial LIF detector for CE (from Beckman Instruments, Fullerton, CA). A fiber-optic cable transmits laser light from the laser to the detector and illuminates a section of the capillary. Fluorescence is collected by the ellipsoidal mirror and focused back onto the photomultiplier tube. To reduce unwanted laser light, a centered hole in the mirror allows most of the beam to pass. A beam block is used to attenuate scattered laser light. (Reprinted with permission from Schwartz and Guttman, *Beckman Instruments Primer Series,* Vol. VII, 1995. Copyright: Beckman Instruments, Inc.)

Electrophoretic mobility in turn increases with magnitude of the analyte's charge and decreases with its radius. That is, if you have two analytes of the same size, the one with the greater charge will move faster. But if you have two analytes of different size and the same charge, the smaller one will move faster.

3.4.2 Electroosmosis

Besides electrophoretic migration, analytes in CE move by a process called electroosmosis (or electroendoosmosis). This phenomenon, occurring also in slab gel electrophoresis, produces electroosmotic flow, the electrically driven flow of the liquid within the capillary. However, while in slab gel electrophoresis the gel matrix reduces EOF to an annoyance, in CE this liquid flow can have a significant effect on the separation process.

Electroosmosis originates from the charges present on the inner surface of the capillary. In the case of fused silica capillaries, the origin of these charges is the dissociation of the silanol (SiO^-) groups of the glass. This dissociation occurs at pH values higher than 2. These negative charges on the wall attract positively charged cations (e.g., Na, K in the buffer) to neutralize them, creating an ionic double layer and, consequently, a potential difference, the zeta potential. This situation obtains before the electrophoresis process begins.

When electrophoresis begins, the voltage difference set up in the capillary causes the migration of the mobile positive ions (cations) toward the cathode. This ionic movement in turn osmotically drags fluid, the water in the capillary, in the same direction. It is this movement of fluid that generates the EOF. The velocity of this flow is increased with the dielectric constant of the fluid and the magnitude of the zeta potential, and decreased by the solution's viscosity.

With bare (uncoated) silica capillaries, osmotic flow is from the anodic to the cathodic end. However, if the capillary is coated with a positive surface, the osmotic flow would be reversed. Although extremely variable in dependence of experimental conditions, EOF is generally in the order of fractions of milliliters per minute; it can be empirically measured by injecting a neutral marker (e.g., acetone).

A peculiarity of EOF is that originating at the walls of the capillary, its flow profile is almost flat, without the parabolic shape typical of pressure-generated laminar flow, as in HPLC (Fig. 3.3). It is of interest to note that in comparison to a parabolic profile, a flat profile limits the broadening of the zones during migration in the capillaries.

The usual arrangement of a capillary electropherograph necessitates that injection take place at the anodic end, with detection occurring close to the cathodic end of the capillary. Considering also that EOF is generally oriented toward the cathode and that it is greater than the electrophoretic velocity of most analytes, it follows that cationic, neutral, and anionic analytes will reach the detector in that order.

The actual migration velocity of cations and anions will result from the algebraic summation of the electrophoretic mobilities of the individual ionic

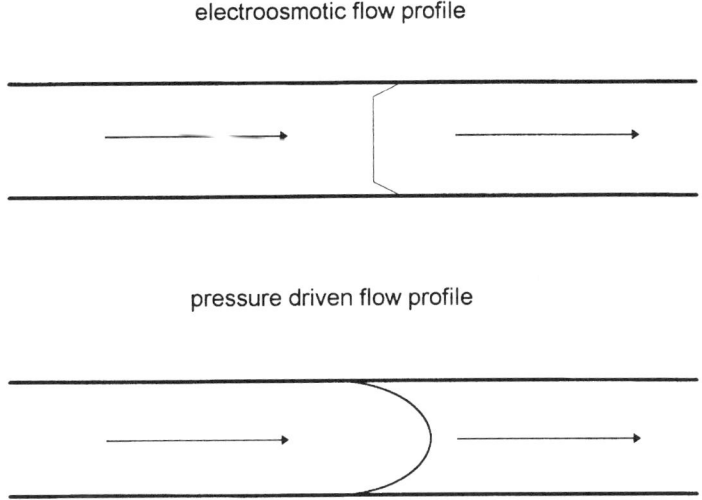

Figure 3.3 Schematic representation of the flow profiles of EOF and pressure-generated laminar flow.

species and that of EOF. All neutral species migrate at the velocity of the EOF, in a single zone (peak), and consequently, cannot be resolved from one another.

In the presence of the EOF, the migration velocity of the analytes will increase with the mobilities of EOF, the analyte, and the voltage and will decrease with capillary length.

EOF can be effectively controlled by changing several experimental conditions. For example, an increase in buffer pH produces an increase in EOF, as long as it increases the degree of dissociation of the wall silanols. Also the ionic strength of the buffer influences the zeta potential and, consequently, the EOF as follows: the less the ionic strength, the less the neutralization of wall charges and therefore the greater will be the zeta potential and the EOF. The addition of organic solvents to the buffer affects both the zeta potential and buffer viscosity. Also, increasing the applied voltage increases the Joule heating of the buffer, with consequent reduction in viscosity and increase of EOF. Buffer additives (e.g., surfactants, methylcellulose, polyacrylamide, quaternary amines) can change quantitatively and/or qualitatively the wall charges, thus reducing or, if cationic, even reversing EOF. Similarly, capillary wall coatings (e.g., polyacrylamide, proteins, amino acids, PEG, PVA, C_8, C_{18}) that covalently bond to the surface (often by silylation), and, in some cases, those that are only physically adsorbed (e.g., cellulose) can effect the EOF.

3.5 INSTRUMENTATION IN DETAIL

3.5.1 Injection

When analytes are introduced into a chromatographic system, a boundary, or zone, is created in the flow of the mobile phase. Separation of the analytes in the chromatographic system requires that these analytes move at different velocities; in a closed system, on the other hand, the magnitude of the separation (i.e., the resolution) depends on the length of the respective zones: It is the sample volume that contributes to the length of the zones. Sample volume is particularly important in CE, where the total volume of the chromatographic system is minimal. For example, for a capillary 50 cm long and having an internal diameter of 50 μm, the total volume will be 1.0 μL; in this capillary, the injection of a sample volume as little as 20 nL will produce a zone 10 mm long—that is, 2% of the total length, a percentage widely considered to be the upper limit of injection volume for a system that must maintain a high separation efficiency.

Two injection techniques currently in use are hydrodynamic injection and electrokinetic injection. Hydrodynamic injection is pressure driven, and therefore all components in the sample are injected simultaneously. In contrast, with electrokinetic injection, the entry of the components of the sample into the column depends on ion mobility, charge, and concentration.

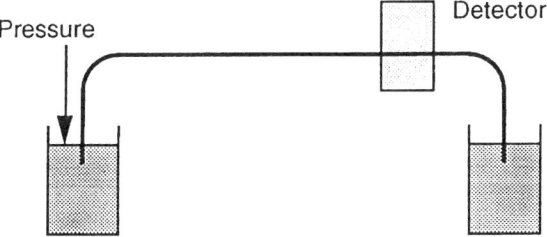

Figure 3.4 Schematic representation of injection by positive pressure.

3.5.1.1 Hydrodynamic Injection by Pressure/Vacuum Application In almost all the automated instruments, injection is carried out by application of pressure for a fixed time onto the sample vial into which the injection end of the capillary is dipped, or by application of vacuum at the opposite end (Figs. 3.4, 3.5).

Following Poiseuille's law, the amount of sample loaded will increase with an increase in pressure; the sample concentration and the injection time and will decrease with increases in solution viscosity and the length of the capillary.

3.5.1.2 Electrokinetic Injection In electrokinetic injection, the sample is introduced in the capillary by applying a voltage (in general, lower than that used for the separation), while the injection end is dipped in the sample (Fig. 3.6). Under these conditions, the analytes contained in the sample are injected by electromigration as well as by electroosmotic flow. The amount of sample loaded increases with the electrophoretic mobility of analyte, the electroosmotic flow mobility, the inner radius, the voltage, the sample concentration, and the injection time. The amount loaded will decrease with the capillary length.

Because the amount of analytes injected depends on both the ion mobility and the electroosmotic flow, variables that are difficult to control, electrokinetic injection is adopted only when hydrodynamic injection is not applicable even though it is theoretically superior in terms of selectivity. A potentially

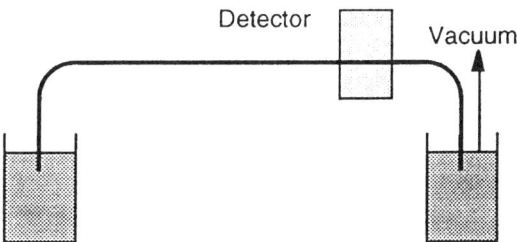

Figure 3.5 Schematic representation of injection by vacuum application.

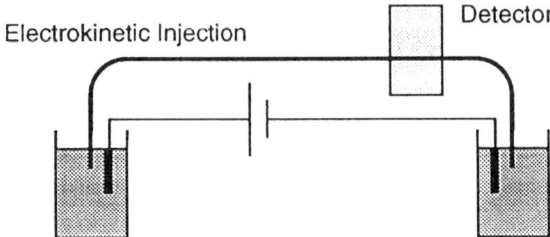

Figure 3.6 Schematic representation of electrokinetic injection.

useful application of electrokinetic injection is microsampling, in which injection can be carried out directly from discrete tissue regions or even from single cells (see Li, 1994).

3.5.2 The Capillary

Separation occurs in the capillary. Ideally, the capillary should be chemically and physically resistant, precisely produced with narrow internal diameters (typical dimensions for the internal diameter of these capillaries ranges from 20 to 100 μm), not prone to adsorb solutes, endowed with a high thermal conductivity, transparent to UV radiation, and inexpensive. Fused silica capillaries meet almost all the foregoing requirements (except that, in some instances, they tend to adsorb large proteins); for these reasons, they are the first choice in CE.

The internal surface of the capillaries can be uncoated, to allow the inner silica surface to be in contact with buffers and analytes, or coated (to shield the silica surface from interactions with analytes prior to filling). The external surfaces of the capillaries are coated with a layer of polyimide for flexibility. Where the capillary needs to be optically transparent to allow detection, the external coating is removed. In this window the capillary is very fragile but is transparent, even in the low UV region. When detection is carried out directly in the capillary, its "effective length" is from the injection end to the detection window, whereas the "total length" corresponds to the distance from the two ends of the capillary (Fig. 3.7). Capillaries may vary in length from 20 to 100 cm.

In summary, the silanol groups of the silica wall, ionized at pH values higher than 2 as SiO^-, will interact electrostatically with the ionized solutes. Buffer pH, ionic strength, and composition affect these equilibria and, consequently, the electrophoretic separation. Thus, the right choices of pH, ionic strength, and composition of the running buffer, as well as of the washing and conditioning steps are crucial.

Washing the capillary (in general with NaOH, 0.1–1.0 M) dissolves a thin film of silica from the inner wall, exposing a clean surface. A more sophisticated

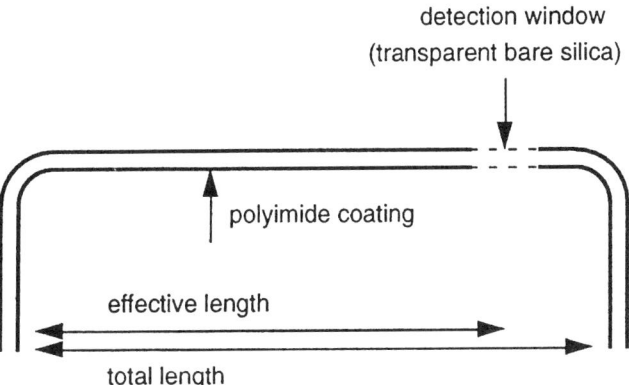

Figure 3.7 Schematic representation of the location of the detection window in a capillary for CE.

control of the capillary wall can be obtained by coating the silica with "modifiers," which can be physically adsorbed, chemically bound, or simply added to the running buffer (dynamic coating). The most common coatings are polyacrylamide, PEG, cellulose, PVA, amino acids, proteins (α-lactoglobulin), amines, surfactants, arylpentafluoro compounds, poly(vinylpyrrolidinone), and polyethyleneimine. Also, LC stationary phases (C_2, C_8, C_{18}) and GC-like capillary columns (PEG, phenylmethyl silicone) have been used.

3.5.3 Power Supply

A high voltage power supply for CE should deliver voltages up to 30 kV and currents up to 200–250 μA. Because of the direction of the EOF, which in silica capillaries is usually toward the cathode, the common polarity is with the anode at the injection end of the capillary and the cathode close to the detection window. CE separations are generally carried out at constant potential, but constant current may be preferable, especially when the control of temperature of the capillary is not efficient. In fact, increases in temperature due to uncontrolled Joule heating can determine changes in the buffer viscosity of analytes, and, consequently, cause changes in their in migration velocity. Under constant current conditions, the system tends to compensate for temperature changes, because to maintain a constant current in the system, increases of temperature, with consequent reduction of the electric resistance in the capillary, are automatically balanced by a corresponding decrease in the voltage. Gradients or steps in the voltage, though not very popular, may be useful in some cases to allow the analysis in a reasonable time of mixtures containing components that differ greatly in electrophoretic mobility.

Attention is also to be paid to the process of water decomposition taking place at both the electrodes: protons and oxygen are liberated at the anode, while at the cathode hydrogen and hydroxide ions are produced. It follows that care should be given to the buffering capacity of the background buffer, to avoid pH changes during an analysis or from analysis to analysis. Besides, venting of the electrode jars may be necessary, to avoid pressurization of the buffer jars by the released gases.

3.5.4 Detection

In general, detectors in CE have to cope with problems of three types: small mass (picogram levels) of analytes injected (due to the limitations in the volumes), which can be loaded into the capillary, limited peak volumes, and inadequately separated peaks.

3.5.4.1 Absorbance Detection UV(–visible) radiation absorption is universally present in CE. To avoid any possible postcolumn band broadening (which would minimize resolution and sensitivity), detection is accomplished through a window in the capillary. One disadvantage of this method is that the path length will be minimal, and; thus, described by Beer's law relating sensitivity to path length, a path length of 25 to 100 mm, such as would be found in CE, will provide less sensitivity than cases in which the optical path length can be about 100 times greater. Also, the cylindrical geometry of the capillary provides poor optical characteristics to the cell design. Two consequences of these limitations are that in this type of column UV detection, only moderate concentration sensitivity (e.g., 10^{-4}–10^{-6} M) and a narrow linear range ($\leq$ about 0.5 AU) will be obtained. In contrast because of nanoliter volumes injected, the mass sensitivity of CE can be high and, as a result, picogram amounts can be directly measured with UV detectors. This difference between sensitivity in terms of mass and sensitivity in terms of concentration is often a source of misunderstanding and possible disappointment for beginners in this field.

A way of increasing the optical path length, and sensitivity, is to use axial illumination (instead of the orthogonal illumination) of the capillary, with the light beam passing through a Z-shaped capillary. With a proper choice of the length of the middle branch of the capillary, which is axially illuminated (about 3 mm), the sensitivity can be increased substantially (i.e., 10 times or even more), with only a minimal loss of resolution (see Li, 1994). An alternative to increasing the optical path length without bending the capillary is the "bubble cell" design, which is now available commercially (Heiger, 1992). This design provides only a moderate increase in sensitivity (about 3–4 times), but, reportedly, there is no sacrifice in resolution.

Analogously to HPLC, photodiode array or multiwavelength fast scanning detectors can be used to increase the quantity and quality of information. These detectors allow the analyst to evaluate the on-line UV(–visible) spectra of the separated zones, and, by comparison with recorded reference spectra, to investigate peak purity and peak indentity.

3.5.4.2 Fluorimetric, Electrochemical Detection, and Other Detection Modes
Other detection techniques have successfully been implemented in CE, with advantages in sensitivity and/or selectivity. Laser excitation allows the focusing of high radiation energy into the capillary, and consequently laser-induced fluorescence achieves much better sensitivity: up to 10^{-12} M. Since, however, not all analytes are naturally fluorescent, fluorescence detection often requires derivatization (e.g., for amino acid analysis). The precolumn derivatization procedures used for HPLC are also used in CE. In contrast, the use of postcolumn derivatization in CE is limited because it is difficult to add and mix derivatizing reagents directly in the capillary after the separation without causing unacceptable band spreading.

Electrochemical detection has also been applied in CE. However, a problem not yet satisfactorily resolved is the need to keep the high voltage (with resulting high currents) used for separation from apart from voltages used in the detection compartment, where much smaller currents are generated and measured. This separation is generally accomplished by inserting a porous conductive joint before the detection end of the capillary. The joint is connected with the cathode, closing the high voltage circuit before the end of the capillary, whereas a carbon fiber, representing the working electrode of an amperometric detector, is inserted into the capillary end, a short distance downstream from this joint. Thus, the high electric current (up to 100–150 mA) applied to accomplish the electrophoretic separation does not interfere with the faradaic current generated at the electrode by the oxidation–reduction process. Unfortunately, the conceivable complexity of this scheme hampers the utility of electrochemical detectors, which otherwise would be sensitive (about 10^{-8} M).

Also, conductimetric detection has been used, by applying two electrodes into the capillary. In this detection mode, the separation of the high voltage part of the capillary (under dc) from the detection zone (under ac) can be accomplished also by electronic filtering. A capillary electropherograph, with a conductimetric detector, is commercially available.

A variety of other detection techniques have been applied in CE, including laser-based thermooptical detection, refractive index detection, radioisotope detection, and, notably, mass spectrometric (MS) detection.

3.5.4.3 Mass Spectrometric Detection
Electrospray has proven to be the method of choice for CE/MS (Smith et al., 1991). With electrospray, solutions containing the analytes are nebulized and subjected to high electric field. Following solvent evaporation and successive Coulombic explosions, pseudomolecular ions of the dissolved analytes are formed. These ions are focused into the mass spectrometer, which separates them according their mass-to-charge ratios. In this method, the end of the CE capillary is connected to the electrospray probe carrying 3 to 4 kV. Some limitations, regarding the choice of Ce buffers remain, however. For example, borate and phosphate salts are not recommended above 20 mM. Among the other MS interfacing

methods, ion spray–atmospheric pressure ionization and continuous flow–fast atom bombardment have been coupled to CE, also in a tandem MS (MS/MS) arrangement (Johansson et al., 1991).

3.5.4.4 Indirect Detection Indirect detection is suitable for the determination, without derivatization, of ionic compounds (e.g., inorganic ions, organic acids) that do not absorb UV radiation and are not fluorescent or electrochemically active. In this detection mode, the running buffer is added with an ionic compound, which is easily detected at low concentrations by the detector (including UV, fluorimetric, or electrochemical), thus determining a high background signal. To maintain the electroneutrality, during the CE separation, this additive is displaced from the zones occupied by the ionic analytes, resulting in "negative" peaks in the background signal. Thus, analytes are detected as "holes" of UV absorbance (or fluorescence or electrochemical activity) of the background electrolyte, without any need of derivatization. It is obvious that the choice of detection mode is not dependent on the characteristics of the analyte, but only the buffer additive used.

To optimize sensitivity in indirect detection, the concentration of the mobile phase additive should be kept as low as possible and the ratio of background signal to background noise (dynamic reserve) should be as large as possible. Another important factor is the transfer ratio, the number of molecules of the background buffer additive that are displaced by a molecule of analyte. For the best sensitivity, this ratio should be large.

Indirect detection is "universal," although limited to ionic compounds; however, its nonspecificity can also become a limitation, because of ionic components of a mixture can potentially interfere with the analytes of interest. Thus, lacking detection selectivity, the whole selectivity of the method relies on separation. Other drawbacks of this detection mode are inferior sensitivity (almost constantly lower than with the corresponding "direct" mode) and a rather narrow range of linearity.

3.6 SEPARATION EFFICIENCY AND RESOLUTION

3.6.1 Theoretical Plate Number and Resolution

Two parameters are used to describe performance in chromatography. These are theoretical plate number (N) and resolution (R). The magnitude of N, an expression of efficiency, is a measure of the ability of the system to restrict analyte diffusion during the separation process. The larger the N, the better, and the more efficient, the system. N is calculated directly from information provided by position of a peak on the chromatogram by first dividing the migration time of this peak by the peak width (measured at the half-peak height), then multiplying this ratio by itself (squaring the value) and finally multiplying this number by 5.54. Mathematically, N is expressed as $N = 5.54$

$(t_m/W_h)^2$. In contrast, R, is a measure of the ability of the system to separate two compounds from each other. Again, the greater the R, the better the system. R is also calculated directly from the chromatogram: that is, from the position of the two compounds being separated. In this case the migration time of compound 1 is subtracted from the migration time of compound 2. Next the peak width (at half-height) of compound 1 is subtracted from the peak width of compound 2. The migration time difference is divided by the peak width difference, and this result multiplied by 2 to give R.

In CE, these two separation parameters, theoretical plate number and resolution, are functions of both the electrophoretic mobility of the analytes and EOF mobility. N is increased with increases in electrophoretic mobility and applied potential, but it decreases with an increase in the diffusion coefficient. R in turn increases with electrophoretic mobility and applied voltage but decreases with diffusion coefficient. In general, both efficiency and resolution are higher at higher voltages and in the presence of substances having small diffusion coefficients.

As with traditional chromatography, the control of factors affecting peak broadening will be crucial for obtaining efficiency and resolution in CE. However, with CE, in addition to the traditional factors that affect peak broadening, some additional factors contribute to the broadening. These include diffusion, Joule heating, adsorption of analytes onto the capillary wall, siphoning between the two electrode jars, injection volume, higher conductivity of the injected plug than that of the running buffer, and "extracolumn" void volumes.

When Joule heating is uncontrolled in the capillaries, the temperature will increase more in the internal (axial) region of the capillary than in the periphery, where thermal dissipation is favored. Because viscosity is inversely related to temperature, one consequence of this temperature differential will be a differential in buffer viscosity, lower in the center and higher at the wall, producing a shift in the shape of the "front" from the flat profile expected from EOF to a more parabolic profile. This shift in front shape will broaden the zone of separation. Therefore, it is important to control Joule heating. Joule heating can be controlled in a number of ways, including the selection of narrow bore capillaries and low conductivity buffers, and by ensuring the efficiency of capillary thermostating (by stream of air or liquid).

Another major cause of loss of efficiency is adsorption of analytes to the capillary wall. To decrease interactions with the silica wall surface to a negligible level, the following measures can be used: increasing buffer ionic strength (i.e., adding ions competing with the analytes for interaction with wall charges); working at low pH (< 3) to suppress wall silanol ionization, or at high pH (> 9) to suppress the protonation of amino groups of analytes; and using buffer additives or coating the capillary surface to block ionizable free silanols.

Sample volume overloading can also reduce efficiency—for example, when the sample plug length is wider than the zone length determined by diffusion. This liability of high injectable volumes is often ignored, and efficiency is intentionally sacrificed to gain sensitivity.

Mismatched ionic conductivities between sample and running buffer can cause peak defocusing and peak distortion. These phenomena depend on local changes in the electric field and, consequently, in the migration velocity. However, a proper choice of the sample conductivity (lower than that of the buffer) allows in-capillary solute concentration with peak focusing.

3.6.2 Practical Hints

In capillary zone electrophoresis (CZE), the most important experimental factors controlling the separation are applied voltage, buffer pH and composition, additives, and capillary wall modifications.

In theory, the higher the voltage the better the efficiency and resolution. In practice, however, the consequences of increasing the voltage was complicated. For example, a higher voltage will increase Joule heating with consequent zone broadening. A higher voltage will also decrease the residence time of analytes in the capillary, reducing the time for separation. Thus, the choice of voltage must be decided by optimization of the system.

To be suitable for CZE, the background buffer should display:

Good buffering capacity at the chosen pH.

Low conductivity (to allow the use of high potentials without generating unacceptably high currents

Mobility of buffer ions matched with that of analytes (to avoid peak distortion)

Negligible interferences with the detection technique (low background absorbance at the chosen wavelength)

In general, high buffer concentrations have been shown to reduce the adsorption of analytes to the capillary walls. For this reason and because of a better buffering capacity, background buffers relatively high in molarity are preferred in CZE.

For most purposes, phosphate, borate, citrate, and phosphate–borate buffers can be used. Zwitterionic buffers (Tris, CHAPS, etc.) are sometimes necessary to enable work at high concentrations without excessive currents and consequently high temperatures in the capillary.

Buffer additives can affect selectivity and resolution, and they are often used to optimize the separation. The most important additives in CE are (1) organic solvents (methanol, acetonitrile, isopropanol, tetrahydrofuran, etc.) used at concentrations up to 50% to increase the solubility of the sample components in the running buffer; (2) anionic (e.g., SDS), cationic (e.g., CTAB), or neutral (e.g., Brij) surfactants to increase the solubility of poorly soluble substances and, by changing or masking surface charges, affect EOF; (3) organic amines (TEA, TEOHA) that affect wall charges, with consequences on EOF and on solute adsorption on the capillary wall and also act as counterions; (4) metal ions that compete with the wall charges, thus reducing adsorption; (5) urea, used for protein and double-stranded DNA denaturation and to interfere with hydrogen bonds in water increasing solubility of substances; (6) linear polymers (PEG, polyacrylamide, methylcellulose), to increase viscosity, mask wall charges and, at adequate concentrations, add a "gel sieving–like" selectivity to the system; and (7) complexing agents that can affect the separation of the compounds they interact with (e.g., borate for catecholamines and carbohydrates). Also, if these complexing agents establish stereoselective interactions—for example, with bile salts or cyclodextrins—they can produce a chiral selectivity during the CE separation (see Chiral Separations, Section 3.7.6).

3.7 METHODS

3.7.1 Capillary Zone Electrophoresis (CZE)

CZE is high voltage, free-solution electrophoresis carried out in a capillary. The capillary is filled with the running electrolyte (a buffer solution), and the ionic analytes are separated on the basis of the differences in their electrophoretic mobilities. The favorable ratio of surface area to volume allows the dissipation of the Joule heat from the capillary and the application of high electric fields with rapid and efficient separations. Also, the anticonvective characteristic of the capillary limits the process of zone diffusion, maintaining the efficiency of separation without the need of further anticonvective media such as gels.

3.7.2 Micellar Electrokinetic Chromatography (MEKC)

While in CZE charged molecules will undergo electrophoretic separation, neutral substances will migrate toward the detector at the same velocity as the EOF, without undergoing separation. To allow separation of nonionic compounds, micellar electrokinetic chromatography (MEKC) was introduced by Terabe (1984, 1985, 1989). In MEKC a micellar "pseudostationary" phase is added to the buffer, interacting with solutes according to partitioning mechanisms already described for HPLC. In the MEKC system, electroosmosis produces the flow of buffer known as EOF, which acts as the "mobile phase" in standard chromatography. EOF, with its "pluglike" flow profile, is ideal for chromatography because in contrast to the parabolic profile of hydrodynamically driven flow, EOF flow minimizes the broadening of zones during the separation process, maintaining a high efficiency for separation. In addition, EOF does not exhibit the pressure pulsation that occurs in HPLC with the use of reciprocating pumps.

The micellar "pseudostationary" phase is produced by adding a surfactant to the buffer at concentrations that exceed its critical micelle concentration (cmc). In most cases, the anionic surfactant sodium dodecyl sulfate (SDS) is used to create the micelles. Because the SDS micelles are anionic, they will be electrostatically driven toward the anode. However, because of the EOF, there remains a slow migration of buffer in the direction of the detector (toward the cathode). In addition, because the nonionic solutes interact with the micelles according to their partition coefficents, they will interact with the micelles to different degrees. It is this difference that alters the otherwise even "mobility" of nonionic solutes. Thus, a selectivity criterion is introduced that is based on partitioning of solutes in the lipophilic core of the micelles, which operates with nonionic substances in a "quasi-chromatographic" mode, with the micelles acting as the stationary phase and the EOF of the buffer (often containing additives) as the mobile phase. In general, because hydrophobic compounds will interact more strongly with the lipophilic core of the micelles, they will migrate more slowly and the more polar molecules will migrate faster; however, all nonionic substances will elute within a time window determined by the mobility of the EOF and that of the micelles (Fig. 3.8).

Retention, expressed by the capacity factor k' of nonionic analytes, is a function of the partition coefficient and the volume of pseudostationary phase (micelles), the volume of the mobile phase, the retention time of the analyte, the "dead time" (corresponding to the migration velocity of the EOF), and the retention time of the micelles. If the micelles were immobilized in the capillary (i.e., if the "pseudostationary" phase were "stationary"), the capacity factor would be similar to the standard equation of retention in chromatography.

Therefore, in MEKC, the only difference from chromatography, for nonionic solutes, is that the "pseudostationary phase" (the micelles) is not actually stationary, but slowly migrates toward the detector, eluting at a characteristic time. That time is determined experimentally by injecting a water-insoluble dye (e.g., Sudan III or Sudan IV), which is completely included in the micelles, and measuring its elution time.

For ionizable solutes, separation mechanisms rely not only on the partitioning of the nonionized forms in the micelles, but also on the electrophoretic mobility of the ionized form of the compounds in the aqueous phase. Electrostatic interactions between the ionic analytes and the charged surface of the micelles are also to be taken into consideration.

Common surfactants that have been used in MEKC, are listed in Table 3.1 with the respective critical micelle concentrations; the most popular are SDS, bile salts, and hydrophobic chain quaternary ammonium salts. Selectivity can also be modulated by the addition to the aqueous buffer of organic solvents (methanol, isopropanol, acetonitrile, tetrahydrofuran, up to a concentration of 50%). These agents will reduce the hydrophobic interactions between analytes and micelles in a way similar to reversed-phase chromatography. Organic modifiers also reduce the cohesion of the "hydrophobic core" of the micelles, increasing the mass transfer kinetics and, consequently, efficiency. Nonionic

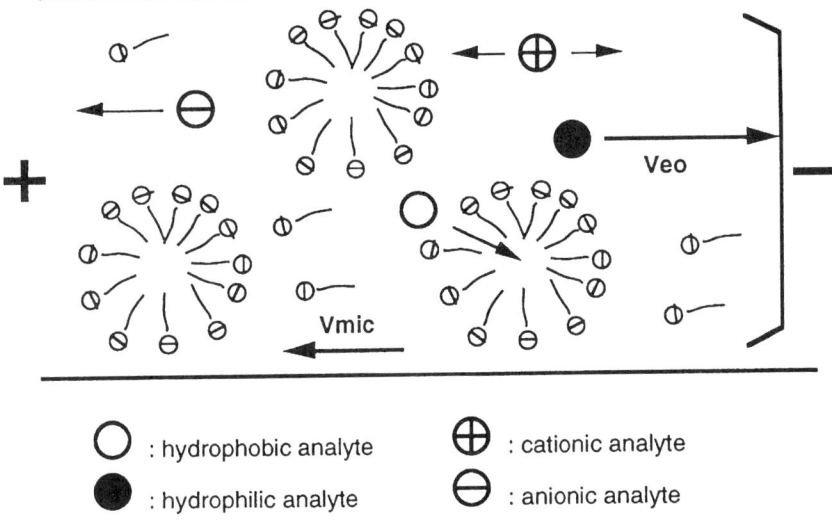

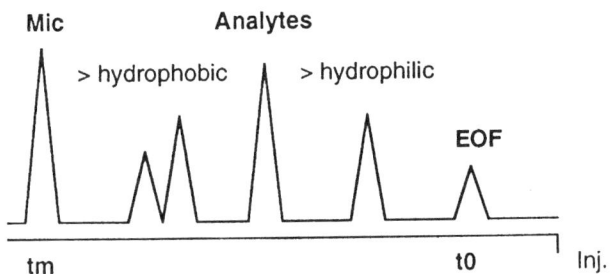

Figure 3.8 Schematic representation of MEKC (Veo, velocity of EOF; Vmic, velocity of micelles).

surfactants, added to the buffer, decrease the hydrophobic interactions of the analytes with the micelles and reduce the retention times.

When ion-pairing agents (surfactants, tetraalkylammonium salts, organic amines) are introduced into the system, their effects resemble those observed with ion-pairing chromatography. Urea (2–6 M) is known to increase the solubility in water of hydrophobic molecules; in MEKC separation by very lipophilic compounds was reported improved by highly concentrated urea.

Also used in MEKC are cyclodextrins (CDs), oligosaccharides with a hydrophilic external surface and a hydrophobic cavity, which attract other com-

TABLE 3.1 Micelle-Forming Additives in MEKC

Additive	Critical Micelle Concentration (mM)
ANIONIC	
Sodium decyl sulfate	33.0
Sodium dodecyl sulfate (SDS)	8.2
Sodium cholate	13.0
Sodium dehydrocholate	10.0
Sodium deoxycholate	5.0
Sodium taurocholate	15.0
CATIONIC	
CTAB	14.0
DTAB	1.3
ZWITTERIONIC	
CHAPS	8.0
NONIONIC	
Triton X-100	0.3

pounds by hydrophobic interaction. CDs are neutral, highly hydrophilic particles that migrate at the same velocity as the EOF. When added to micellar solutions, CDs decrease the interactions of hydrophobic compounds with the micelles, reducing their capacity factors. Because of their chiral structure and the stereospecific nature of their interactions, CDs are effective in enantiomer separation.

3.7.3 Capillary Isotachophoresis (CITP)

CITP, which resembles classical isotachophoresis, is an important separation mode used in CE. The ionic compounds migrate in discrete zones at the same velocity between two ionic solutions, one with the highest mobility (leading electrolyte), the other with lowest mobility (terminating electrolyte) among all the analytes. The different ionic analytes migrate in discrete zones after the leading electrolyte and before the terminating electrolyte, according to the individual mobilities. In this discontinuous buffer system, the electric field changes to ensure the maintenance of a uniform migration velocity of solutes, hence preventing mixing of the zones. Zones with higher mobility ions have lower electric fields than zones containing ions with lower mobilities. Thus, if an ion entered an adjacent zone (ahead) containing ions with a higher mobility, it would find a lower electric field than in its own zone, a condition that would slow down its migration velocity until the ion had come back to the previously occupied zone. Here, the "correct electric field" would restore its usual migration velocity. The opposite would happen if an ion entered a zone containing ions with a lower mobility than its own. In each zone, a constant ratio exists between concentration and mobility of ions. These principles are being used in CE not only for separation purposes, but also for sample preconcentration before CZE, obtaining concentration factors in excess of 100 times. For a comprehensive presentation of isotachophoresis, see Everaerts et al. (1976).

3.7.4 Capillary Gel Electrophoresis (CGE)

Gel electrophoresis is the main separation technique used in biomedicine, biochemistry, and biotechnology. As SDS–polyarcylamide gel electrophoresis (SDS–PAGE), it is the standard method for the separation and characterization of proteins, DNA sequencing (polyacrylamide gels), and DNA fragment mapping (agarose gels). However, gel electrophoresis (in slab gels) has always remained a manual, time-consuming technique, and the results depend on the expertise of the operator.

The separation mechanisms that operate in CGE highly resemble traditional agarose and polyacrylamide gel electrophoresis. The characteristics of a polyacrylamide gel can be described as follows:

$$\%T = \frac{\text{mg acrylamide} + \text{mg cross-linker}}{\text{mL buffer}} \times 0.01$$

$$\%C = \frac{\text{mg cross-linker}}{\text{mg acrylamide} + \text{mg cross-linker}} \times 100$$

Because in CGE the solutes have to travel for a considerable distance before reaching the detector, gels usually have a lower $\%T$ than in slab gels: polyacrylamide with 2 to 6%T and 3 to 6%C has widely been used.

However, problems were encountered with CE capillaries filled with cross-linked networks resembling slab gels. The first was that the EOF generated by the high electric fields used in CE was strong enough to extrude the gel from the capillary. So, it was necessary to bind covalently the polyacrylamide to the wall. Another problem was the formation of gas bubbles inside the capillary and at the ends, where it can be exposed to air. These phenomena caused unstable currents with consequent separation problems. Careful degassing of solutions, was mandatory, and it was essential to check for the formation of bubbles during the polymerization process. Nevertheless, if adequately produced and handled, gel capillaries can yield excellent separations, with typical 1% relative standard deviations (RSD) in migration times and exceptionally high efficiencies, up to millions of theoretical plates.

An alternative to cross-linked polymer gels is non-cross-linked polyacrylamide. The non-cross-linked polymers separate by means of a mechanism similar to traditional gels, but can be replaced several times in the capillary, overcoming the short lifetime (due to clogging at the injection end) of the traditional gel-filled capillaries. On this basis, agarose (0.05–1.2%) and linear polymers, such as polyacrylamide (0.1–20%) and hydroxyalkylcellulose (6–15%) have become very popular.

3.7.5 Capillary Isoelectric Focusing (CIEF)

CIEF is the replication in the capillary of slab gel isoelectric focusing, a widely used separation mode, particularly suitable for protein separation. In isoelectric focusing, substances are separated by applying an electric field in a buffer system forming a pH gradient: Analytes focus where the buffer local pH equals their isoelectric points.

Ampholines, zwitterionic compounds with an isoelectric point (pI) varying in a chosen pH range, are added to the buffer filling the capillary. The anodic end of the capillary is placed into an acidic solution, while the cathodic end is dipped in a basic solution. After the voltage has been applied, ampholines migrate in the capillary until they reach their isoelectric point, where they stop. Proteins, which are usually added together with ampholines, focus where they find a pH, determined by the ampholines, corresponding to their isoelectric point.

After focusing, the zones are moved toward the detector hydrodynamically (e.g., by lifting the height of one end of the capillary) or by adding sodium chloride to one reservoir, which, through a pH imbalance gradient, causes the migration of the separated zones. Recently, it has been demonstrated that EOF, which was abolished in the focusing step of CIEF, can be maintained to allow mobilization while the focusing process takes place.

3.7.6 Chiral Separations

CE is playing a major role in the separation of chiral compounds, a field that is gaining increasing attention in pharmaceutical sciences as well as in forensic toxicology (Lurie, 1994; Novotny et al., 1994; Ward, 1994). The chirally active selectors used in CE include optically active complexes such as Cu(II)-L-histidine, Cu(II)-aspartame, cyclodextrins, modified CDs, bile salts, crown ethers, and proteins (bovine serum albumin, α_1-acid glycoprotein, etc.).

Different mechanisms have been devised to achieve chiral resolution by electromigration, but often we have to deal with "mixed-mode" separations rather than pure processes. In any case, chiral resolution results from stereo-specific interactions of a chiral selector, with the enantiomers of the compound giving rise to a difference in migration velocity between the two entities. Chirally selective ligands, such as Cu(II)-L-histidine and Cu(II)-aspartame, have been used for derivatized amino acid mixtures.

Special mention should be made of CDs. These oligosaccharides have an external hydrophilic surface and a hydrophobic cavity, in which they can include other compounds by hydrophobic interaction. This inclusion mechanism is sterically selective: Analytes must fit the size of the cavity, which changes with the number of glucose units in the molecule's structure (6. 7, and 8, respectively for *a-, b-* and *g*-CDs). Native CDs are neutral and highly hydrophilic: therefore they migrate at the velocity of the electroosmotic flow. Of course, the migration velocity of the complexed form of an enantiomer will differ from that of the free molecule, because of the different size with the same charge. Recently introduced charged CEs can display a countermigration in relation to the electroosmotic flow, thus exerting a retardation effect on the compounds they interact with. Neutral CDs, when added to micellar solutions, decrease the interactions of hydrophobic compounds with the micelles, also with a retarding effect. Neutral CDs can also be incorporated into a polyacrylamide gel, which retards CD complexes on the basis of their different mass, thus exerting a chiral selectivity on the CD complexed analytes. CD copolymer gels have also been used for enantiomer separation.

Packed capillary columns with chirally selective stationary phases (e.g., a_1-acid glycoprotein), as well as wall-immobilized, CD-based stationary phases, have been successfully used in CE chromatographic separations. Also, macrocylic "crown" ethers, forming sterically selective complexes with the guest molecule, have been used for the resolution of optically active amines.

MEKC with chirally selective micelles has proved very powerful for enantiomer separation. Chiral surfactants (e.g., bile acid salts) forming chirally active micelles have widely been used. Mixed micellar solutions containing SDS and chiral surfactants or derivatized CDs are becoming more popular. Mixed-mode, chiral, and nonchiral, interactions increase the resolving power of these systems.

Among the many variables that can be adjusted to optimize CE chiral separation are chiral selector type and concentration, organic modifer concentration, ionic strength of the buffer, and temperature. A further possibility for achieving chiral separations in capillary electrophoresis, analogous to gas chromatography and high performance liquid chromatography, is the derivatization of the chiral analyte with chiral reagents followed by separation of the resulting diastereomers.

3.8 SUMMARY

This chapter discussed the concepts and principles of high performance capillary electrophoresis. A brief history of CE was followed by an analysis of the components of an CE system. Two types of separation were presented, electrophoretic migration and electroosmosis. Several types of CE were discussed in detail, including capillary zone and micellar electrokinetic electrophoresis.

REFERENCES

Camilleri P, Ed. (1993) *Capillary Electrophoresis: Theory and Practice. New-Directions in Organic and Biological Chemistry.* CRC Press, Boca Raton, FL.

Everaerts FM, Beckers JL, Verheggen TP (1976) Isotachophoresis: Theory, instrumentation and applications. *Journal of Chromatography Library,* Vol. 6, Elsevier, Amsterdam.

Foret F, Krivankova L, Bocek P (1993) *Capillary Zone Electrophoresis.* VCH, Weinheim.

Grossman PD, and Colburn JC, Eds. (1992) *Capillary Electrophoresis—Theory and Practice.* Academic Press, San Diego, CA.

Guzman NA, Ed. (1993) *Capillary Electrophoresis Technology.* (*Chromatographic Science Series.*) Dekker, New York.

Heiger DN (1992) *High Performance Capillary Electrophoresis,* pp. 100–101. Hewlett-Packard, Waldbronn, Germany.

Jandik P, Bonn G (1993) *Capillary Electrophoresis of Small Molecules and Ions.* VCH Publishers, New York.

Johansson IM, Pavelka R, Henion JD (1991) Determination of small drug molecules by capillary electrophoresis–atmospheric pressure ionization mass spectrometry. *J Chromatogr* **559**:515–528.

Jorgenson JW, Lukacs KD (1981a) Zone electrophoresis in open-tubular glass capillaries. *Anal Chem* **53**:1298–1301.

Jorgenson JW, Lukacs KD (1981b) High resolution separations based on electrophoresis and electroosmosis. *J Chromatogr* **218**:209–216.

Jorgenson JW, Lukacs KD (1983) Capillary zone electrophoresis. *Science* **222**:266–272.

Khun R, Hoffstetter-Kuhn S (1993) *Capillary Electrophoresis: Principles and Practice.* Springer-Verlag, New York.

Kuhr WG (1990): Capillary electrophoresis. *Anal Chem* **62**:403R–414R.

Kuhr WG, Monnig CA (1992) Capillary electrophoresis. *Anal Chem* **64**:389R–407R.

Li SFY (1994) *Capillary Electrophoresis, Principles, Practice and Applications,* 2nd ed. Elsevier, Amsterdam.

Lurie IS (1994) Analysis of seized drugs by capillary electrophoresis. In *Analysis of Addictive and Misused Drugs,* JA Adamovics, Ed., pp. 151–219. Dekker, New York.

Monnig CA, Kennedy RT (1994) Capillary electrophoresis. *Anal Chem* **66**:280R–314R.

Mosher RA, Saville DA, Thormann W, Eds. (1992) *The Dynamics of Electrophoresis.* VCH, Weinheim.

Novotny M, Soini H, Stefansson M (1994) Chiral separation through capillary electromigration methods. *Anal Chem* **66**:646A–655A.

Smith RD, Udseth HR, Baringa CJ, Edmonds CG (1991) Instrumentation for high-performance capillary electrophoresis–mass spectrometry. *J Chromatogr* **559**:197–208.

Terabe S, Otsuka K, Ichikawa K, Tsuchiya A, Ando T (1984) Electrokinetic separations with micellar solutions and open-tubular capillaries. *Anal Chem* **56:**111–113.

Terabe S, Otsuka K, Ando T (1985) Electrokinetic chromatography with micellar solution and open-tubular capillaries. *Anal Chem* **57:**834–841.

Terabe S (1989) Electrokinetic chromatography: An interface between electrophoresis and chromatography. *Trends Anal Chem* **8:**129–134.

Vindevogel J, Sandra P (1992) *Introduction to Micellar Electrokinetic Chromatography.* Hüthig Verlag, Heidelberg.

Ward TJ (1994) Chiral media for capillary electrophoresis. *Anal Chem* **66,** 633A–640A.

Weinberger R (1993) *Practical Capillary Electrophoresis.* Academic Press, San Diego, CA.

CHAPTER 4
Strategy for Design of an HPLC System for Assay of Enzyme Activity

OVERVIEW

This chapter presents a strategy for the design of an HPLC assay system. Section 4.1, Setting up the Assay, focuses on the enzymatic reaction and the steps leading to the development of the assay. These steps are previewed in Table 4.1. Section 4.2, The Use of HPLC to Establish Optimal Conditions for the Enzymatic Reaction, discusses the procedure for monitoring the activity of an enzyme with the HPLC method and the use of HPLC assays to determine the parameters required for obtaining optimal activity.

4.1 SETTING UP THE ASSAY

4.1.1 Analysis of the Primary Reaction

As indicated in Table 4.1, the design of an HPLC assay system for an enzymatic activity begins with a complete analysis of the primary reaction—the reaction catalyzed by the enzyme under study. To begin this analysis, indicate all substrates, products, and cofactors of the reaction. If metals are required for catalysis, include them. In the case of the metals, however, it is useful to note whether they are an integral part of the substrate (e.g., when the complex MgATP is the substrate) or whether they are required for some other function (e.g., activation of the enzyme). It is also useful to indicate the pH of the reaction as well as the type and concentration of the buffer to be used. The goal of this analysis is to list all the components present in the reaction mixture before the start of the reaction.

To illustrate this approach, consider the assay of a pyrophosphohydrolase, an enzyme that catalyzes the reaction

$$\text{MgATP} \rightarrow \text{AMP} + \text{PP}_i \qquad (1)$$

TABLE 4.1   Steps in Design of HPLC Assay for Enzymatic Reaction

1. Analyze the primary reaction.
2. Analyze all secondary reactions.
3. Select the mode of HPLC (size-exclusion, ion-exchange, reversed-phase) that will allow for separation of substrates from products.
4. Make initial selection of mobile phase (pH, buffer, salt concentration) and method of delivery (isocratic or gradient elution).
5. Select appropriate detector. Determine whether it will be necessary to collect fractions.

MgATP is the substrate, and AMP and pyrophosphate (PP_i) are the products. Since this activity is usually assayed at a pH of 7.5 using a Tris-HCl buffer system, the reaction tube will contain ATP, Mg, and Tris-HCl as illustrated in Figure 4.1.

4.1.2 Analysis of Secondary Reactions

The HPLC method can be used to advantage in the assay of activities in crude extracts or in preparations only partially purified. Such samples usually contain activities other than the one under study that will engage the substrate of the primary reaction or its product, or both. These other activities are referred to as secondary reactions. What are secondary reactions? They are reactions catalyzed by enzymatic activities other than the activity under study. A secondary reaction may use as substrate either the original substrate or the product of the primary reaction.

For example, AMP, the product of the primary reaction, reaction (1), may undergo secondary reactions to form adenosine and phosphate or IMP and ammonia. Other secondary reactions (e.g., the degradation of ATP to ADP) could involve ATP. These secondary reactions are summarized in Figure 4.1 in the step marked Incubation. While secondary reactions can be eliminated or their significance minimized, they should not be overlooked in the analysis and design of the assay system.

4.1.3 Selection of the Stationary Phase and Method of Elution

When the list of reactants, cofactors, and reaction conditions has been compiled, the stationary phase can be selected.

To select the appropriate stationary phase, it is necessary to examine the reactants to determine how they differ. The selection of the stationary phase should exploit this difference. For example, do the reactants differ in size, charge, or solubility? Examination of the modes of operation of the stationary phase materials presented in Chapter 2 reveals gel filtration to be ideal for

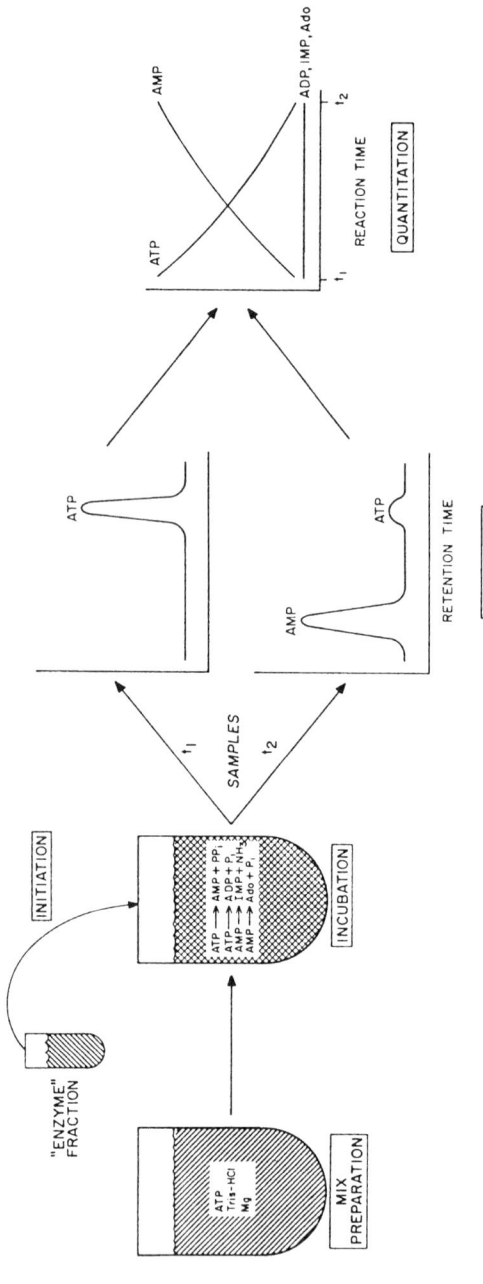

Figure 4.1 Overview of strategy for design of an HPLC method to determine enzymatic activity. The reaction tube contains a mix preparation to measure the activity of an ATP pyrophosphohydrolase, which catalyzes the formation of AMP and PP_i from ATP. The mix contains the substrate, ATP; the buffer, Tris-HCl; and magnesium, a metal cofactor. The addition of a sample from the "enzyme" fraction initiates the primary reaction and also several secondary reactions. Samples of the incubation mixture are withdrawn at intervals (t_1 and t_2), and the reaction is terminated by injection of the samples onto the HPLC column. A representative analysis of each sample is shown. The amount of each component can be calculated from the area of its peak and is graphed as a function of reaction time.

separation by size, while ion-exchange and reversed-phase HPLC are suitable for separation by charge and solubility, respectively. With the reactions illustrated in Figure 4.1, the reactants ATP, ADP, and AMP, and adenosine (Ado) must be separated. Since these substances differ in charge, they can be separated by ion exchange. But they differ in solubility also, and therefore a reversed-phase HPLC assay can be chosen instead. The choice is based on the consideration of another parameter, the elution procedure. As ion-exchange stationary phase may call for the use of gradient elution, an elution procedure that varies in salt concentration. With reversed-phase columns, an elution buffer of constant composition can be chosen. Since the latter is the easier of the two methods to use, the reversed-phase HPLC was chosen. With a reversed-phase HPLC system, however, the compounds involved in the primary and secondary reactions above emerge in the order ATP, ADP, AMP, and, as shown in Figure 4.2A, it is difficult to separate ATP from ADP. Since one of the goals of the separation procedure is the resolution of these two phosphates, it was decided to use ion-paired, reversed-phase HPLC (see Chapter 2). With this change, the ADP is eluted before ATP, as illustrated in Figure 4.2B. Therefore, when selecting the stationary phase, it is wise to consider the method of elution as well.

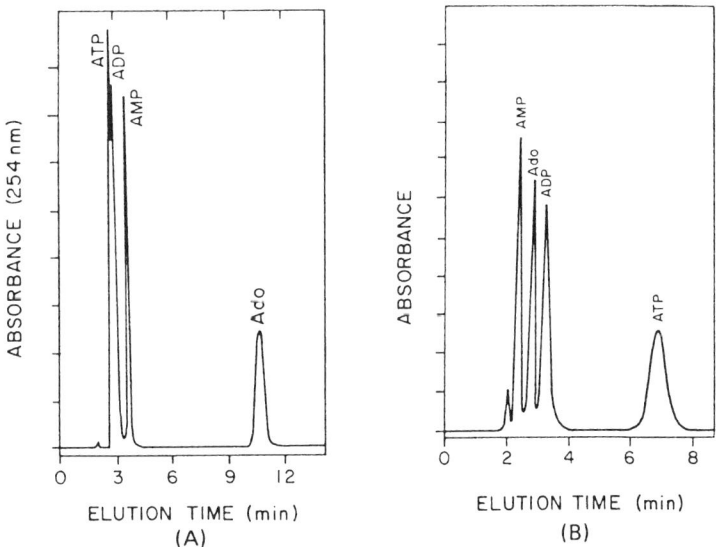

Figure 4.2 Analysis of Ado, AMP, ADP, and ATP by HPLC. Separations were carried out on reversed-phase C_{18} columns, eluted isocratically. The detection was at 254 nm. (A) The mobile phase was composed of 10 mM potassium phosphate (pH 5.5) and 20% methanol as the mobile phase. (B) The mobile phase contained 65 mM potassium phosphate (pH 3.7), 5% methanol, and 1 mM n-tetrabutyl ammonium phosphate.

4.1.4 Modification of Reaction Conditions for the HPLC Assay Method

The reaction conditions appropriate for other methods may have to be modified when the HPLC assay procedure is used. It may be necessary to change the concentration of the various components of the system being examined, such as metals, hydrogen ions, or enzyme, particularly if the samples for analysis are taken directly from the incubation mixture and injected using equipment arranged as in Figure 4.3. Note that when a sample is removed directly from the incubation mixture and injected onto the HPLC column for analysis, it brings with it everything present in the reaction mixture, including excess protein and metals, two components that can clog the column or alter the performance of the column packing. Most, if not all, of these problems can be solved by terminating the reaction prior to injection. The direct analysis of samples may require reducing the metal concentration, working at lower enzyme concentrations, or even compromising with pH values to balance reaction conditions with HPLC assay conditions.

4.1.5 Understanding and Dealing with Secondary Reactions

The importance of understanding secondary reactions cannot be overemphasized. This knowledge is invaluable to the interpretation of chromatographic

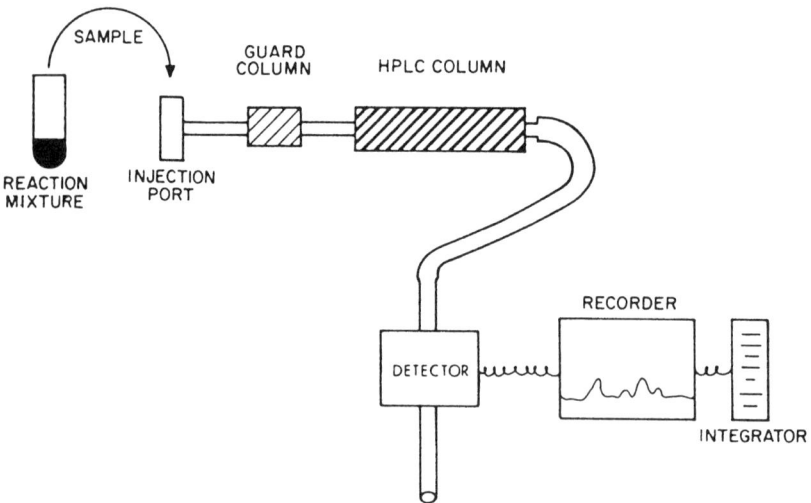

Figure 4.3 Arrangement of HPLC equipment for termination of reaction by direct injection of sample. A sample is removed from the reaction mixture and transferred directly to the injection port for introduction onto the column. The HPLC column is protected by a guard column, which removes debris. The eluent flows through the detector, from which a signal is displayed on a recorder. The area of each peak is electronically integrated.

profiles of enzymatic reactions. "Beware of secondary reactions" is a rule that should always be kept in the forefront. This rule should be remembered even when purified enzymes obtained from commercial sources are being subjected to HPLC analysis. Suppose that we want to study the activity of a commercially available preparation of alkaline phosphatase with AMP as the substrate, and we have used a reversed-phase column to separate the substrate from one of the expected products, adenosine. The addition of enzyme initiated the reaction, and samples were taken at intervals and analyzed by HPLC. Figure 4.4 shows several of the chromatograms obtained.

These chromatograms were studied with the expectation that the enzyme was pure. Therefore, we thought the chromatogram would show only the

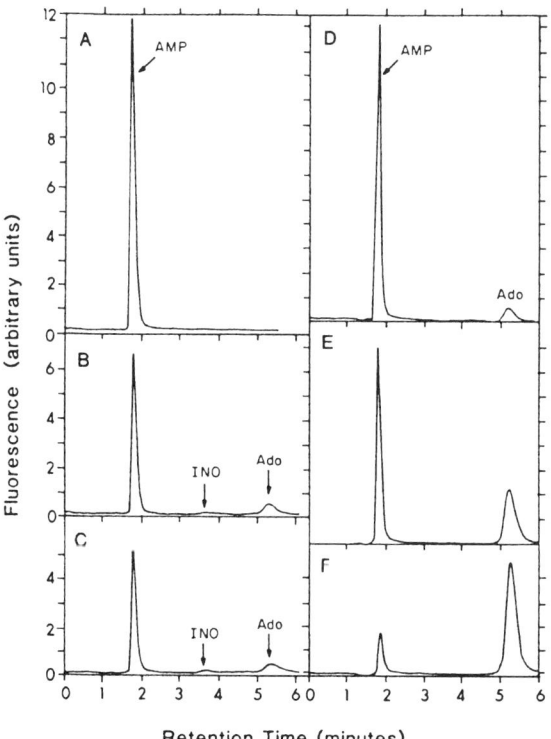

Figure 4.4 The HPLC analysis of a reaction mixture containing AMP and alkaline phosphatase. Separations were carried out on a reversed-phase column with a mobile phase of potassium phosphate (pH 5.5) and 10% methanol. The column was eluted isocratically, and the detection was at 254 nm. Two sets of tracings were obtained, according to the following schedules. For the original reaction mixture: (A) immediately after the addition of enzyme, (B) after 10 minutes, and (C) after 15 minutes. For the reaction mixture to which had been added EHNA (5 μM), an inhibitor of adenosine deaminase, the suspected contaminant: (D) after 2 minutes, (E) after 10 minutes, and (F) after 40 minutes. (From Rossomando et al., 1981.)

substrate AMP, and, as illustrated in Figure 4.4A, this was the case. Chromatograms obtained late in the incubation were expected to show the reaction product adenosine (Ado). However, later chromatograms showed three peaks (Fig. 4.4B,C): two were easily identified—one was AMP, and the other adenosine; the third peak was ignored, since its area appeared to be insignificant. Yet when we measured the amount of adenosine recovered (Fig. 4.4C), it did not equal the amount of AMP lost.

The possibility that more of the reaction product Ado had remained on the column was ruled out by using different mobile phases to elute all bound material. The chromatograms accounted for all the products. Since we had not expected any side reactions, we quantitated the yield of reaction products on the basis of area, assuming the presence of only adenosine-containing compounds. The formation of inosine, with a 50% reduction in extinction coefficient, could account for the apparent lack of recovery. Therefore, we considered the presence of secondary reactions. Either AMP had been converted to IMP, or adenosine was converted to inosine (Ino). By comparing the retention time of the third peak to authentic standards, we ruled out IMP as a product, and thus the identity of peak 3 was established as inosine. This led us to conclude that the commercial preparation of alkaline phosphatase was contaminated with a second activity, adenosine deaminase.

To follow up the observation, a second reaction mixture was prepared that was similar in composition to the first but also contained an inhibitor of adenosine deaminase. After the start of the reaction, samples were removed and analyzed, and the chromatograms obtained (Fig. 4.4D–F) illustrated the loss of AMP and quantitative recovery of the adenosine.

How can secondary reactions be handled? The procedures presented in Table 4.2 include purifying the activity of the primary reaction to homogeneity. This may not always be possible, however, since activities must be assayed in crude extracts. Therefore some other solution must be found. The use of analogs is one such solution. For example, if an analog of the substrate is used, then an analog of the product will be formed. If the latter is not a suitable substrate for the secondary enzyme, no secondary reactions will occur.

TABLE 4.2 Secondary Reactions

WHAT ARE THEY?

Secondary reactions are the result of enzymatic activities present in the sample that lead to destruction of the substrate and/or the formation of additional products, which can alter the appearance of the chromatogram.

HOW CAN THEY BE AVOIDED?

1. By purifying the enzyme to homogeneity.
2. By using analogs whose products are not substrates for secondary reactions.
3. By adjusting reaction conditions to minimize activity of secondary reactions.

Alternatively, one can try to adjust the reaction condition to ensure that the enzymes catalyzing the secondary reactions will not be active. For example, if the primary reaction does not require metals but the secondary reaction does, adding a chelator will inhibit the latter.

4.1.6 Components of the Reaction Mixtures Can Cause Problems: Effects of Metals on Separation

Many enzymes require metals for activity, and, unfortunately for HPLC use, the presence of the metal can occasionally have significant effects on a separation. For an explanation of this problem, return to the reaction illustrated in Figure 4.1, the degradation of ATP to form AMP and PP_i. The first HPLC method developed to assay this activity was carried out on a reversed-phase system with a mobile phase chosen for the exclusive separation of ATP from AMP (see Fig. 4.5). Since ADP was not involved, no thought was given to

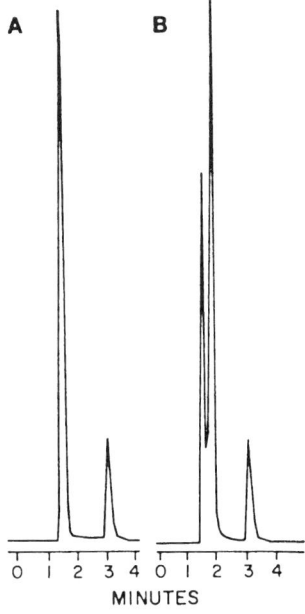

Figure 4.5 HPLC analysis of enzymatic assay with ATP in free and metal-bound forms. Separations were carried out on a reversed-phase C_{18} column with a potassium phosphate mobile phase containing 10% methanol. The flow rate was 2 mL/min. Assay mixture of 100 μL contained 8 mM Tris-HCl (pH 7.5), 2 mM ATP, and 2 mM $MgCl_2$ and enzyme preparation containing ATP pyrophosphohydrolase (10 μg of protein). Chromatograms of 20 μL samples illustrate incubation of (A) zero time and (B) 2 hours. (From Jahngen and Rossomando, 1983).

its separation. Later, having decided to study the metal requirements of this reaction, we performed a series of experiments for this purpose.

A reaction mixture was prepared that contained a metal at concentrations in excess of that of ATP. The reaction was started by the addition of the enzyme, and samples were taken and analyzed by the HPLC method. Surprisingly, the chromatograms for the experiments that included metals were different from the ones obtained earlier. In the original experiments, only two peaks were present, those representing ATP and AMP. However, in the experiment that included the metal calcium, the chromatograms showed the elution of an additional peak jsut after the ATP emerged (Fig. 4.5B). Further studies showed that the new peak had a retention time identical to that of ADP, and therefore we assumed that this second peak was ADP. From these findings, we speculated that the metal had stimulated the activity of an enzyme that catalyzed the formation of ADP.

Additional studies were performed. We analyzed samples of the reaction mixture before we added the enzyme and very soon afterward, before any enzymatic reaction could have taken place. Interestingly, these chromatograms (Fig. 4.6) also showed two peaks, with peak II identical in retention time to the presumptive ADP peak. In the absence of any metal, a single ATP peak (peak I) was observed, suggesting that the metal had altered the chromatographic properties of ATP.

Additional studies confirmed this possibility. For example, when the effect of metals on the chromatographic behavior of ATP was studied in more detail, the profiles illustrated in Figure 4.7 were obtained. Figure 4.7A shows the profile at the lowest metal concentration used in the experiment. Two peaks of ATP, labeled I and II, are clearly resolved. Increasing the metal concentration (Fig. 4.7B–F) led to an increase in the area of peak II and a corresponding decrease in peak I. This result suggests that the second peak was formed by the metal binding to the ATP.

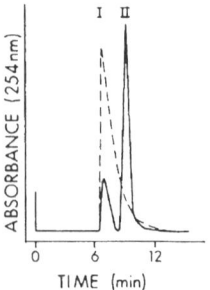

Figure 4.6 HPLC separation of ATP/Ca^{2+} mixture. Separations were carried out on a C_{18} (μBondapak) microcolumn (4.6 mm × 25 cm) with a mobile phase of 10 mM KH_2PO_4 (pH 5.5) with 4% methanol; flow rate was 0.5 mL/min. The sample volume was 20 μL, and detection was at 254 nm. Dashed line, 40 nmol ATP; solid line, 40 nmol ATP plus 160 nmol $CaCl_2$. (From Jahngen and Rossomando, 1983.)

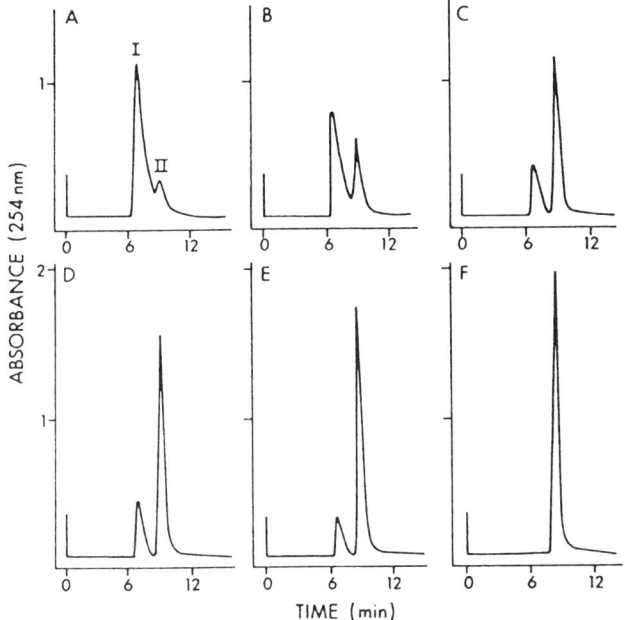

Figure 4.7 Effect of increasing amounts of Ca^{2+} in a mixture of this cation and ATP. Analysis by HPLC as described in Figure 4.6. Chromatograms are of 20 µL samples containing 40 nmol ATP and $CaCl_2$ at (A) 60 nmol, (B) 100 nmol, (C) 120 nmol, (D) 160 nmol, (E) 200 nmol, and (F) 400 nmol. (From Jahngen and Rossomando, 1983.)

This conclusion was tested further by the addition of a radiolabeled metal, in this case calcium-45, together with the ATP in the reaction mixture. A sample was injected, and the two peaks were collected in separate fractions. Following an analysis of the fractions, the ^{45}Ca elution profile was plotted on the chromatogram as shown in Figure 4.8. The calcium coelutes exclusively with the ATP in peak II (Fig. 4.8), confirming the conclusion that the second peak was indeed a metal–ATP.

4.1.7 Terminating the Reaction

In designing an assay for an enzyme, it is often necessary to introduce a termination step into the protocol (see Chapter 1). This is often done when protein is present in the incubation mixture at a concentration that would clog the column. There are a variety of ways to accomplish this termination process. Ideally, it would not be necessary to add reagents that might otherwise clog the column or alter its performance. For example, consider the changes that occur in the incubation mixture when the reaction is terminated by acid. The addition of trichloroacetic acid (TCA) will reduce the pH of the incubation

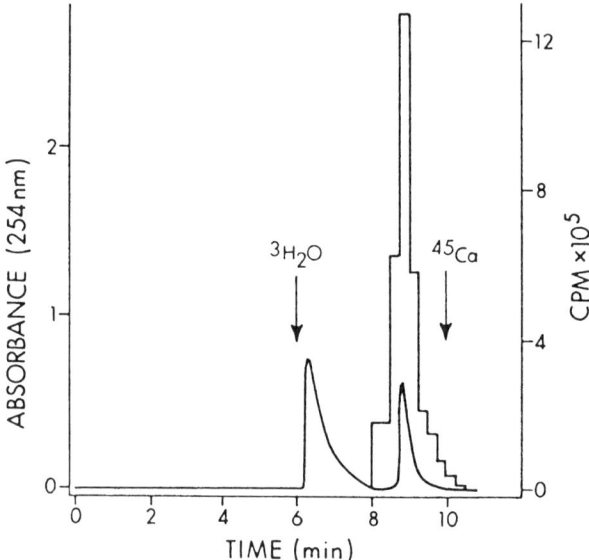

Figure 4.8 Coelution of ^{45}Ca with ATP/Ca^{2+} mixture. Injection volume and analytical conditions were as described in Figure 4.6. (Radioactive tracer approximately 3 million cpm ^{45}Ca) was added to 40 nmol ATP and 80 nmol $CaCl_2$. Fractions of 0.25 nL were collected, and the amount of radioactivity was determined by liquid scintillation counting. Radioactivity for ^{45}Ca with ATP/Ca^{2+} is shown in bars. Retention times of ^3H_2O and ^{45}Ca injected alone are shown by arrows. (From Jahngen and Rossomando, 1983.)

solution to a value unsafe for the stationary phase packing. In addition, the TCA has an absorption profile in the UV region, and therefore its presence will interfere with the use of UV spectrophotometers as detectors. Both problems can be solved by removing the TCA prior to analysis. The freon–alamine extraction method introduced by Khym, Bynum, and Volkin (1977) has been found to be useful for this purpose. While the removal of TCA is not a difficult step, it is clear that the inclusion of this acid in the assay complicates the original procedure.

A method that offers an alternative to the use of acids and bases is the addition of chelators such as EDTA. This technique is suitable only for reactions in which the enzymatic activities have an absolute requirement for a metal whose removal will terminate the activities.

Another alternative method we have found useful for terminating reactions is to heat the incubation mixture to a temperature that results in inactivation of the enzyme. Usually temperatures in excess of 100°C are required. One of the techniques often used is to immerse the reaction tube in a bath of boiling water. This method is not employed in my laboratory, however, because the incubation mixture cannot be brought to 100°C quickly enough to effect instantaneous termination.

We have tried using commercially available heating blocks. These heaters were also found unsuitable, for the rather trivial reason that test tubes being used for the incubation did not fit the holes in the block.

Finally, a simple device—a sand bath—was found to be effective in terminating reactions instantly. We filled a stainless steel rectangular pan (about 8 in. × 10 in.) with about 2 in. of sand and placed it on a hot plate, as illustrated in Figure 4.9A. The temperature of the sand bath is easily brought to 155°C,

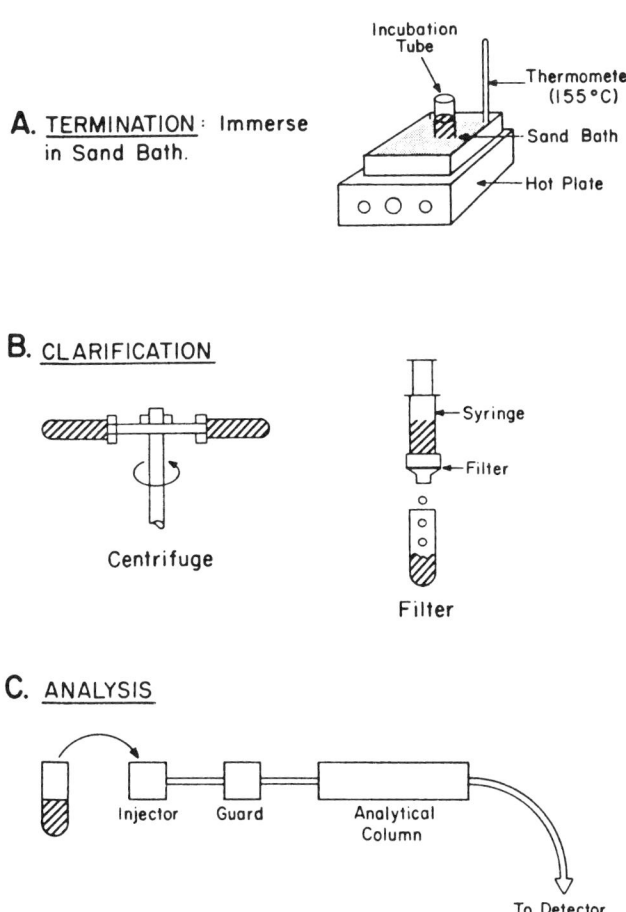

Figure 4.9 The preparation of a sample from a reaction mixture that contains an excess amount of protein prior to injection and analysis by HPLC. (A) Termination carried out by immersion of reaction sample in a sand bath maintained at 155°C. (B) Removal of the precipitated protein by either centrifugation or filtration. (C) Injection and analysis of the clarified solution.

and this temperature can be maintained throughout the working day without fear of evaporation. There is never a problem of fitting the tubes—one simply thrusts any size tube directly into the sand. When a tube containing as much as 500 μL of incubation mixture was inserted into the sand, the temperature inside the solution reached 100°C "instantly," thus terminating the reaction.

Termination of most enzymatic reactions with heat results in precipitation of any proteins present in the reaction mixture. Because this precipitation is irreversible, and because the amount of protein may be considerable when crude extracts are being assayed, it is often necessary to remove the precipitate prior to sampling.

The precipitate is removed either by filtration or by centrifugation (Fig. 4.9B). Because of the small volumes usually used in the reaction mixture (viz. 100–500 μL), centrifugation has always been difficult. Filters with dead volumes of about 50 μL now make the removal of precipitate less tedious. Following the removal of the precipitate, a sample may be removed from the filtrate and injected into the HPLC for analysis, as illustrated in Figure 4.9C.

Internal standards, compounds added at any stage of the analytical procedure, can be useful in calibrating and/or calculating the effect of that procedure on the recovery of the substrate or product of the reaction. The compounds chosen as internal standards should elute close to the substrate or product and should have similar detection characteristics.

4.1.8 Setting Up the Reaction Conditions

Some additional modifications of the reaction conditions used with other assay methods may be required to proceed with an HPLC assay. One change is related to the total volume of the reaction mixture. Because the HPLC assay is basically a discontinuous technique, obtaining kinetic data requires multiple samples, each one representing a single time point. Traditionally, reactions requiring multiple sampling have been arranged in one of two ways. In one arrangement (Fig. 4.10A), separate reaction mixtures are set up, each one representing a single time point. In this case, the total volume required for a single reaction mixture is the volume required for a single injection. The number of incubation tubes would be determined by the number of time points required by the experiment. In the second arrangement (Fig. 4.10B), a single incubation mixture is prepared, and samples are removed from it at suitable intervals for analysis. In this arrangement, the volume required for the reaction mixture would be determined as the product of the volume needed for each injection multiplied by the total number of injections.

Since with both arrangements the volume of a single injection is the important variable, it would appear that once this value has been determined, the overall reaction volume can be established. However, another variable, the type of injector to be used, must also be considered.

Injectors are of two basic designs. Those of the type illustrated in Figure 4.11A require that a sample of known volume be removed from the reaction

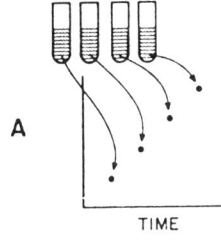
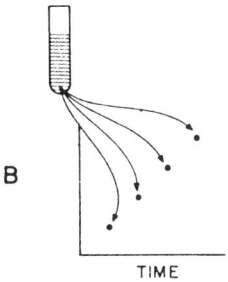

Figure 4.10 Representation of two procedures used to obtain multiple samples for analysis. (*A*) Several identical reaction mixtures are prepared, and the enzyme will be added to each to start the reaction. Each tube is sampled only once during the incubation. (*B*) Only one reaction mixture is prepared, and the enzyme is added to start the reaction. The incubation mixture is sampled repeatedly during the course of the reaction. Note that the volume of the reaction mixture in arrangement (*B*) is usually greater than in arrangement (*A*).

and injected. Thus, with this type of injector, the volume to be injected is determined by the volume drawn into the syringe.

In the second type, the injector unit contains a device called a "loop," which is a coiled tube of precise volume. The loop is loaded with sample, and once filled, its contents can be injected onto the column. It is always necessary to overfill such a loop to ensure complete loading. To overfill, the loop is loaded with sample until it overflows; the excess sample spills out one of the injection valve vents. Although injectors differ in the amount of overfilling required, experience leads us to recommend overfilling by as much as 30%, since it is critical to load the injector loop completely. Partially filled loops will result in abnormal chromatographic profiles. There is waste with injectors of the type shown in Figure 4.11*B*, and if material is in short supply or expensive, the type of injector (Fig. 4.11*A*) might be preferred.

4.1.9 Detector Sensitivity

Another potential problem concerns the selection of the range of substrate concentrations to be used throughout the study. Considering the sensitivity of most HPLC detectors and the apparent K_m values of most enzyme activities, the selection of the upper limit of concentration is usually not a problem. A problem will develop, however, when rate determinations are made at low substrate concentrations, since at these concentrations the amount of product formed during the course of the reaction will be small and may be below the monitor's level of detection.

Therefore, prior to executing any experimental protocols dealing with low substrate concentrations, it is prudent to determine what product concentra-

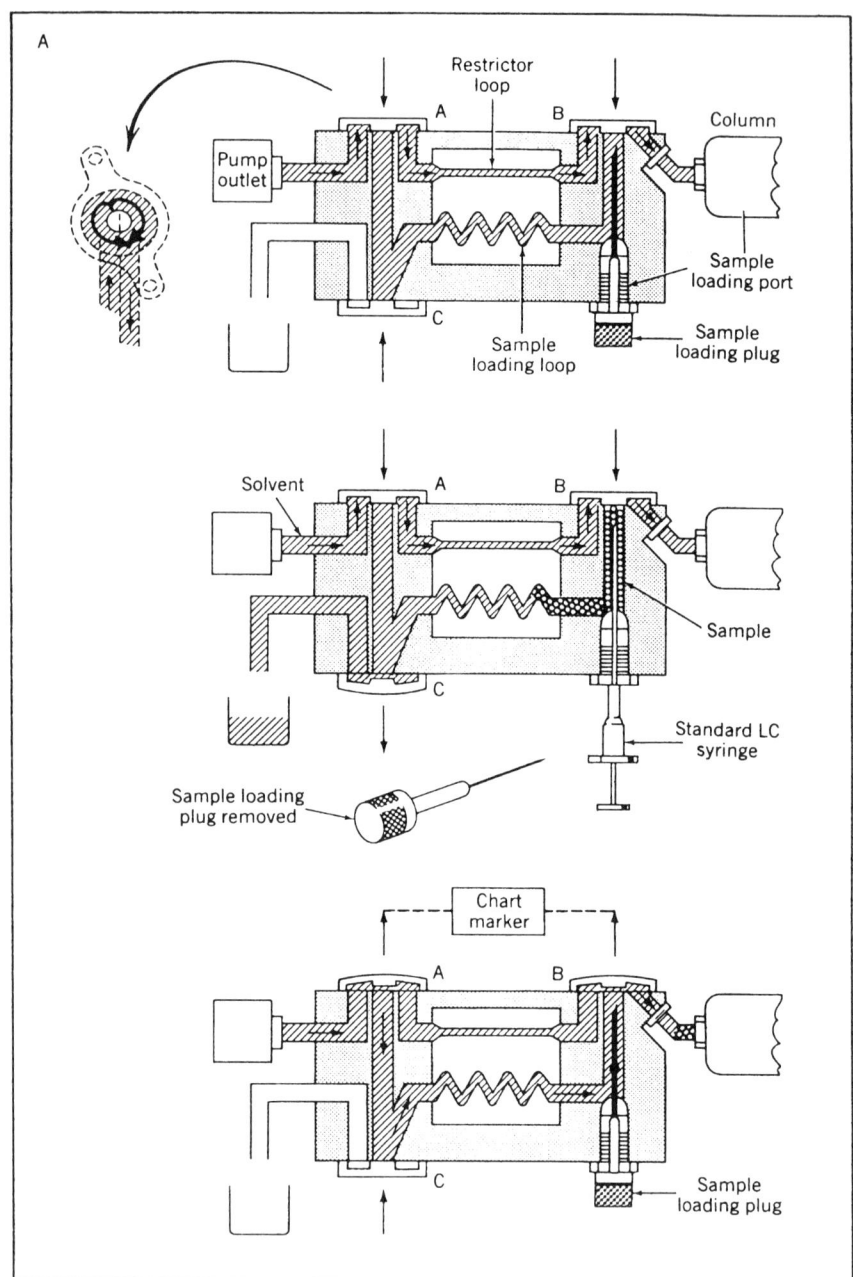

Figure 4.11 High-pressure injection valves. (*A*) The universal injector (Waters, Inc.). A known volume of sample is injected, displacing a corresponding volume from the "sample loading loop" shown in top diagram. After sample loading plug is replaced, the loading loop is placed "in line," and the eluent from pump moves sample onto column.

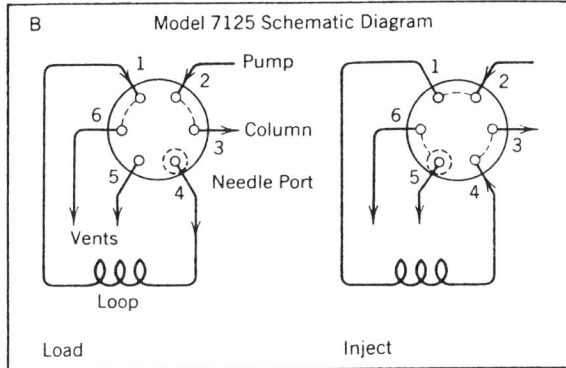

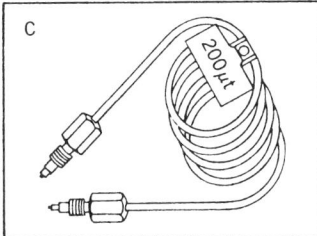

Figure 4.11 *(Continued)* (B) The Rheodyne injector (Rheodyne, Inc.) in which the volume injected is determined by the sample loop. Sample loops can be obtained commercially in a number of fixed volumes or made to any size. (C) A 200-μL loop.

tions can be detected by ascertaining the lower limits of the detector being used.

4.1.10 Summary and Conclusions

The strategy used to design an HPLC assay for enzyme activity begins with an analysis of the primary reaction, the reaction catalyzed by the activity of interest. (Figure 4.12 shows the steps in the assay design.) This step should be followed by an analysis of secondary reactions: the reactions catalyzed by activities that might be in the sample and are not of primary interest, although their presence might result in loss of substrate or products, thereby obscuring the results of the primary reaction.

The choice of the stationary phase can best be made after the analysis of both the primary and secondary reactions has been completed and the compounds to be separated have been listed. The selection of the stationary phase is guided by the number of compounds in the reactions, and the kinds present. With an understanding of the differences between these compounds, be it their size, solubility, or charge, the stationary and mobile phases can be selected.

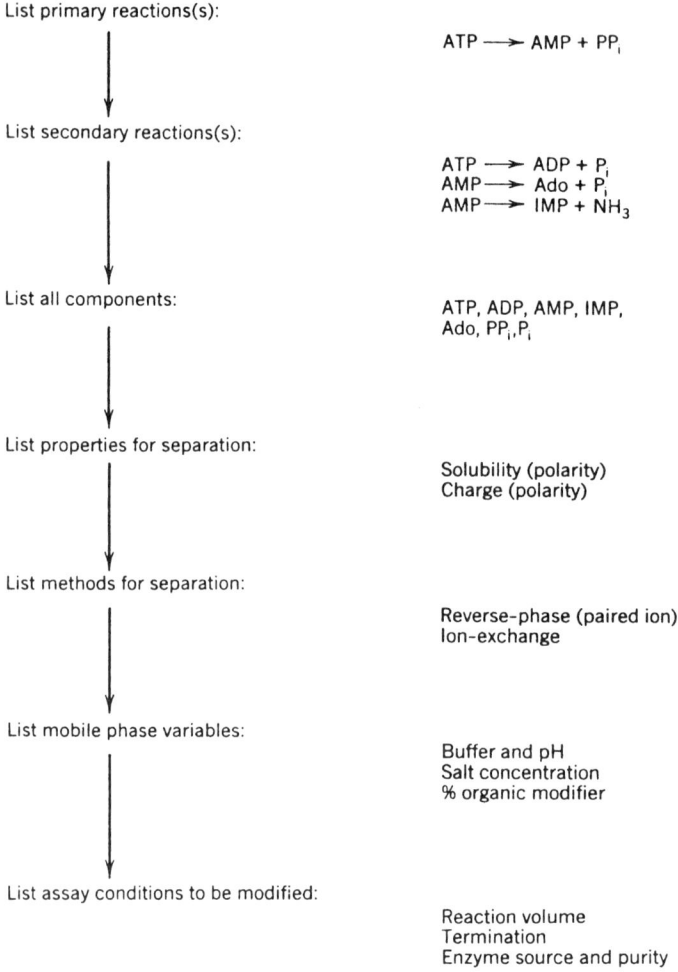

Figure 4.12 Steps in the design of an HPLC assay.

The use of the HPLC method to assay the activity of an enzyme may require some modifications in the composition of the reaction mixture. For example, the presence of metals in the reaction mixture can cause problems, since a metal complex may form and produce new peaks on the chromatogram. Complications associated with the requirement for termination of the reaction and for dealing with the small amounts of product that appear during early stages of the reaction may require changes in the reaction conditions as well.

Termination of a reaction is best accomplished by using heat to inactivate the enzymes. Numerous procedures have been employed to heat the incubation mixture; one of the most convenient is a sand bath. Heating will result

TABLE 4.3 Conditions That Allow the Use of HPLC for the Assay of an Enzymatic Activity

1. A system (stationary and mobile phase) can be found for the separation of substrate(s) from product(s).
2. A detection system for substrate(s) or product(s) is available.
3. Suitable standards exist for calibration and identification of reaction components.
4. Sufficient product is formed to permit the analyst to follow the course of the reaction.

in precipitated protein that must be removed. This can be accomplished by centrifugation and/or filtration prior to injection of the sample onto the analytical column.

In setting up an assay, some thought should be given to the incubation volume, since the type of injector may necessitate larger volumes than might be available.

Finally, the sensitivity of the detector should be examined. A calibration curve should be constructed using the product of the primary reaction to determine the sensitivity limit.

It is reemphasized that HPLC cannot always be used to assay an activity. A number of criteria must be met, and these are summarized in Table 4.3.

4.2 THE USE OF HPLC TO ESTABLISH OPTIMAL CONDITIONS FOR THE ENZYMATIC REACTION

4.2.1 Initial Decisions: Composition of the Reaction Mixture

The HPLC assay method is particularly useful when it is necessary to obtain initial rate data for a study of an enzymatic activity. Optimal assay conditions for the HPLC must be established first. Usually, the optimization process involves the determination of several variables, such as the optimal substrate concentration, pH, temperature, and enzyme concentration. It is assumed that the reader is familiar with the problems associated with assay conditions such as pH, buffer, and temperature. This chapter discusses only factors that might present problems for the HPLC assay method. For additional information, see the works cited in the General References.

When studying an enzymatic activity for which an apparent K_m value can be obtained from the literature, the determination of the optimum substrate concentration becomes somewhat easier. Thus, a concentration two to three times the K_m value (assuming the absence of "substrate inhibition") is adequate for use in early experiments. The literature can provide descriptions of other parameters, such as the pH range, the requirement for activators, and the optimal temperature for the incubation. Armed with this information, and with the incubation conditions determined, the analyst can prepare the reaction mixture.

Remaining to be determined prior to the initiation of the reaction by addition of the enzyme are the amount of enzyme to be added, the time course of the reaction, the time between samplings of the reaction mixture, and the volume of these samples. All four questions can be answered by trial and error using the following scheme. First, an arbitrary enzyme concentration is selected. While any concentration can be used, it should be remembered that in work involving a crude extract, excess protein can clog the column. Therefore, choose the lowest concentration possible.

Having chosen an enzyme concentration, add enzyme meeting this specification to the reaction mixture to start the reaction. Sampling of the reaction mixture, by the withdrawal of a predetermined quantity for analysis, can begin at any time.

The chromatogram obtained from this single sample is examined for a new peak—the product. Two outcomes are possible: Either product is present or it is not. If a product peak is detected, and its area is very small compared to the area of the substrate peak, a second sample can be withdrawn from the incubation mixture and injected for analysis.

Again the areas of the substrate and product peaks should be compared. If the area of the product peak is more than 50% of that of the substrate, the reaction has progressed too far, and it is necessary to start again by preparing a new reaction mixture. To obtain more time points, slow the reaction rate by using less enzyme.

Alternatively, in the absence of the formation of any product, continue incubation and withdraw samples every hour for analysis. The incubation can be continued for several hours with the expectation that detectable product may yet emerge. In the continued absence of any detectable product, prepare a new reaction mixture that contains more enzyme than the first. If this does not result in the formation of detectable product, the possibility should be considered that the fraction being assayed contains no activity.

4.2.2 Obtaining Initial Rate Data

The significance of obtaining rate data for the study of enzymes has been discussed elsewhere, and the reader is referred to the General References for additional information. Although usually relevant to the in-depth study of the mechanism of an enzyme reaction, such concerns are beyond the scope of the present discussion. Of concern in this text are the problems associated with obtaining initial rate data with the HPLC assay method.

The experiments described in Section 4.2.1 will yield values for two parameters: the amount of the enzyme required to form sufficient detectable product and the incubation time required to form this amount of product. Additional experiments now are required to refine the values of both parameters. Keep in mind that to generate the straight line needed to obtain the initial rate, you must have at least three time points. To obtain these points, proceed as follows. First, note that if the rate of product formation is too rapid—that is, if the reaction rate becomes nonlinear before three or four samples can be analyzed—the rate should be slowed by decreasing the amount of enzyme. Alternatively, if the rate of the reaction is too slow, so that it takes all day to form product, the enzyme concentration should be increased so that three or four samples may be analyzed in about 2 hours. If the assay procedure selected does not include a step for terminating a reaction, the overriding parameter that governs the sampling interval is the time needed to complete the HPLC analysis of each sample. It will not be possible to inject a second sample until the first has been completely eluted and the column prepared for the second sample. If the elution and preparation step takes 15 minutes, then this value establishes the minimum sampling time. Thus, the time required for the HPLC analysis of a single sample will determine the concentration of the enzyme used in the reaction.

Once a suitable concentration of enzyme has been established so that three or four samples can be analyzed, the quantitative data can be obtained. A reaction is started, the reaction mixture is sampled at intervals, chromatograms are obtained, and the amount of product formed is determined directly from the chromatogram by means of either peak height or electronic integration of the peaks. These values should be plotted as a function of reaction time.

Next, a second and third series of reaction mixtures should be prepared, with enzyme added at concentrations of half and twice the value used in the first. These reactions are started and sampled, chromatograms are obtained, and the data are plotted as a function of reaction time. At this early stage in the optimization of the assay, it is advisable to continue sampling one of the incubations until the rate of product formation becomes nonlinear or the amount of substrate present is exhausted. This prolonged incubation provides information about the extent of the primary reaction and also allows any secondary reactions to take place and form enough products to be detectable.

4.2.3 Quantitative Analysis of the Reaction

The chromatograms used to obtain the initial rate data described in Section 4.2.2 may be examined to provide information on the fate of the substrate during the course of the incubation. For example, assuming that no other reactions involving the substrate have taken place, the amount of product that was formed should equal the amount of substrate that was lost. It is important to determine this point first.

A careful visual inspection of the chromatogram should be sufficient to indicate the presence (or absence) of any peaks other than the substrate and the product. If only peaks of the latter type are present, the absence of secondary reactions involving either substrate or product is suggested. Second, a visual estimation of the area of both the substrate and product peaks should indicate whether the two are equal (assuming an equivalence of their extinction coefficients). Estimations will be more difficult if the widths of the substrate peak and the product peak differ.

It is also useful at this stage to make the data more quantitative by converting the amount of product formation from the "machine units," arbitrary integration units, or percentage obtained, directly to units of amount. Such a conversion requires access to a calibration curve that relates the machine units to more specific units of amount. An example of such a calibration curve for adenosine is shown in Figure 4.13. Having made the conversion, the initial rate of product formation determined earlier can now be plotted as a function of enzyme concentration as part of the optimization process.

These procedures will provide a graphical representation of the rate of product formation. Such data can be analyzed visually or subjected to statistical analysis (see General References for details). In addition, data collected at several substrate concentrations can be used to obtain kinetic constants. Again, these data, collected by the HPLC method, can be manipulated by standardized methods (see General References).

Obtaining initial rate data is, of course, a first step in the kinetic analysis of an enzyme-catalyzed reaction, and the reader is referred to the General References for several reviews and monographs describing the methods for this analysis.

For the two-substrate–two-product reactions, the BiBi reactions in Cleland's nomenclature, the HPLC assay method is particularly useful. Its application, however, will require the development of methods for the separation of the two substrates and the two products. Initial rate data generated by the

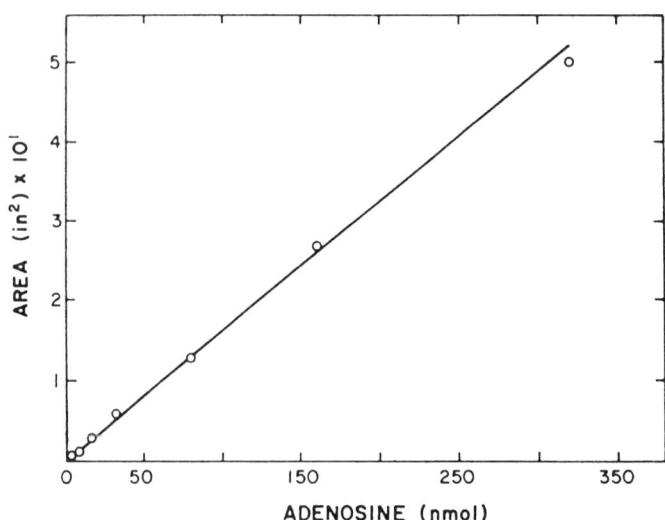

Figure 4.13 Representative calibration curve illustrating the relationship between peak area and the concentration of adenosine.

application of the HPLC assay method can be used to produce double-reciprocal plots [1/(initial velocity) vs. 1/(substrate concentration)], which can in turn be used to differentiate between sequential and ping-pong reaction mechanisms. Thus, as described by Alberty (1953), for a sequential mechanism the several lines on such a plot will converge or intersect at a common point, whereas for the ping-pong mechanism the lines in the series are nonconvergent or parallel.

The application of the HPLC assay method to studies on reaction mechanisms has been limited, and the reader is referred to the work of Sloan (1984). Sloan and his colleagues studied the formation of IMP or GMP (and pyrophosphate) from the substrates phosphoribosylpyrophosphate (PRPP) and either hypoxanthine or guanine. These reactions, catalyzed by hypoxanthine/guanine phosphoribosyltransferase (GHPRTase), were studied by HPLC after a method was developed to separate all the reactants and products simultaneously.

The strategy used was as follows. First the initial rates of IMP and GMP formation were studied separately, and from an inspection of the double-reciprocal plots obtained from these data, it was determined that product formation proceeds by a sequential kinetic mechanism. Next the formation of IMP and GMP was studied for several concentrations of both bases. Finally, the rate of formation of the common product, pyrophosphate, was examined over a series of fixed ratios of hypoxanthine to guanine.

On the basis of a graphical analysis of these data, it was concluded that the reaction proceeds by an ordered BiBi mechanism. According to this mechanism, first the substrate PRPP binds to the enzyme and then either base binds to the same site on the complex.

While such a study might have been carried out with conventional methods, the use of HPLC facilitated the work considerably by allowing all reactants and products to be measured in one analysis. Clearly, the HPLC assay method should be considered when the kinetics of a multisubstrate enzyme reaction are to be studied.

4.2.4 Initial Rate Determination at Low Substrate Concentrations

As in the case of any assay procedure, determination of the rate of product formation becomes difficult at the lower limits of substrate concentration. However, changes that can be made both in the assay system and in the chromatographic equipment can alleviate this problem.

The first change is, of course, to increase the sensitivity of the detector. Most HPLC detectors contain range switches that make this a simple matter. When range switching is carried out, it is useful to determine whether calibration curves constructed at one range setting are still valid at another.

Next, the amount of product being detected may be increased by increasing the volume of the reaction mixture that is injected for analysis. With fixed-loop injectors this requires changing the loop, a rather simple procedure. If the loop is changed, the additional volume must be entered when the total volume of the incubation mixture is calculated.

There is an upper limit to the volume of an injection, usually around 200 μL. The injection of larger volumes results in spreading of the peak, a phenomenon that decreases resolution. However, it is possible to make loops of any volume and thus control this upper volume limit. If separation is adequate and resolution is not a problem, the loop volume may be increased.

At times, obtaining enough reaction product requires the concentration of an entire reaction mixture. Following concentration, the residue is resuspended in a small volume of buffer and analyzed.

Finally, it is always possible to increase sensitivity by using as substrates analogs, such as radiochemicals, whereupon the amount of radioactive product that has formed is determined. For experiments of this type, the eluent may be carried through a radiochemical detector; in the absence of such an instrument, fractions may be collected and their radioactive content determined.

4.2.5 The "Sensitivity Shift" Procedure

In most enzymatic reactions it is not uncommon for a sample to contain a concentration of substrate 100- to 1000-fold greater than that of the reaction product. This situation often arises with enzymes that require high concentrations of substrate (low K_m) for reactivity and have low rates of product formation. It also occurs soon after the start of the reaction, when only small amounts of product have been formed.

The analysis of a sample from an incubation mixture that contains significant differences in substrate and product concentrations presents some problems, since the detector must be set so that the substrate peak is on scale, but also it must be set to detect small amounts of product. Usually different sensitivity settings are required to put both components on scale. For these cases we have adopted a procedure that might be called the "sensitivity shift." In this procedure, the injection is made with the sensitivity set at a value that allows for the detection of one of the compounds. If this is the product, the detector is set at its maximum sensitivity. As soon as the product has emerged, the sensitivity of the detector is changed (either manually or electronically by computer) to a value that allows the substrate peak to appear completely on scale.

Assuming that the settings on the detector are proportional, there should be no difficulty in relating the areas of both peaks to the calibration curve. These problems can be avoided with electronic integrators, which measure total absorbance and are not affected by sensitivity settings.

4.2.6 Substrate Analogs: Their Use in Limiting Secondary Reactions

Enzymes that catalyze unwanted secondary reactions are to be avoided in HPLC analysis. When such secondary reactions make it difficult to quantitate the primary reaction, one solution is to use an analog as the substrate for the latter. The analog should be chosen such that while it is a substrate for the primary reaction, the product formed is not a substrate for the secondary reaction.

Analogs can be used in another way. Consider the development of an assay procedure for adenosine kinase, the enzyme that catalyzes the primary reaction Ado + ATP → AMP + ADP. Problems will arise during the assay of this activity in crude extracts, since other enzymes may be present that can form AMP directly from ATP.

Radiochemical analogs such as radiolabeled adenosine are ideal for solving this problem, because if the formation of radiolabeled AMP is monitored, it is possible to distinguish the AMP formed from adenosine from that formed from ATP, which, of course, would not be labeled.

Alternatively, the same reaction can be assayed if adenosine is replaced by formycin A (FoA) (Fig. 4.14), a fluorescent analog. With this substrate, one product of the adenosine kinase reaction would be FoMP, the fluorescent analog of AMP, while AMP formed directly from ATP would not be fluorescent. Therefore, by monitoring both the fluorescence and the ultraviolet absorbance, using equipment arranged as shown in Figure 4.15, the analyst could follow both the kinase reaction and any secondary reactions.

4.2.7 Summary and Conclusions (Fig. 4.16)

The concentration of the substrate and other components of the reaction mixture must be established before the reaction is initiated. Decisions regarding enzyme concentration and sampling intervals must be made empirically with some information about the amount of expected product formation.

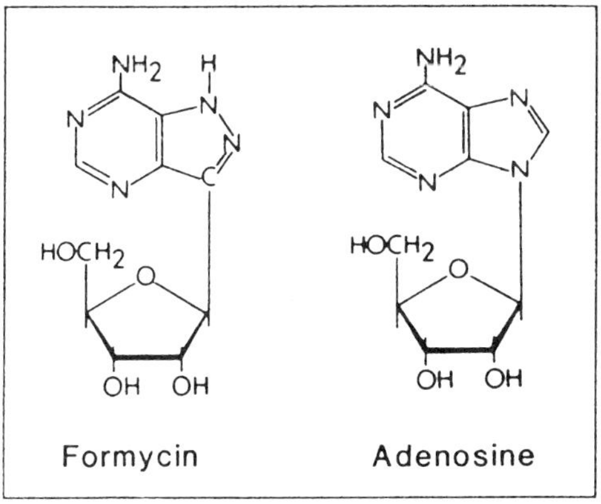

Figure 4.14 Comparison of structures of adenosine and its fluorescent analog formycin A.

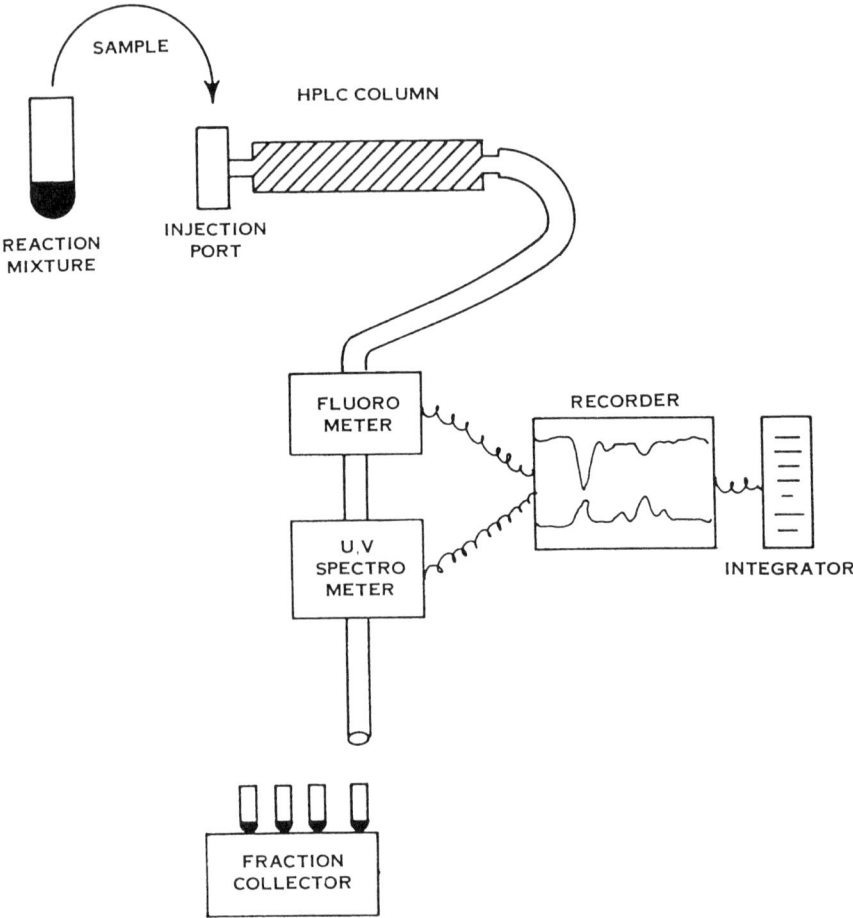

Figure 4.15 Arrangement of equipment showing relative position of multiple detectors and fraction collector.

These data will provide limits for the enzyme concentration and incubation period. Also, if a reaction is not to be terminated, the time required to elute a single sample must be established. This value will provide an estimate of the minimum interval between samples.

To obtain initial rate data, it is necessary to have a minimum of three samples of a reaction carried out under optimized conditions. The formation of product at several enzyme and substrate concentrations will yield data for the calculation of kinetic constants.

The chromatogram obtained from each analysis provides information about product formation and loss of substrate. In the absence of secondary reactions, the two values should be additive.

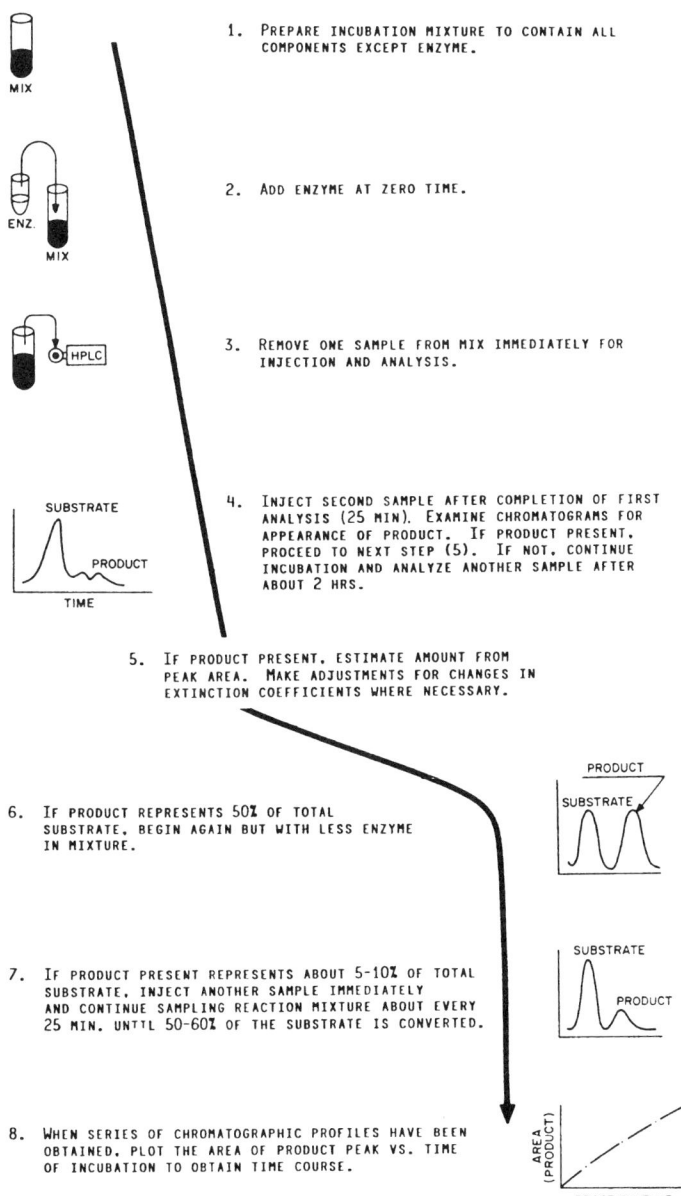

Figure 4.16 Strategy for optimizing an enzyme assay.

The use of the HPLC method provides some unique solutions to the problem of determination of product formation at low substrate concentrations. For example, the sensitivity of the detector can be enhanced even during the course of an analysis. Next, the volume of the incubation mixture assayed may be increased to provide more product, and finally, it is possible to use analogs such as radiolabeled or fluorescent compounds, for which there is greater detector sensitivity.

REFERENCES

Introduction to kinetic mechanisms

Alberty RA (1953) *J Am Chem Soc* **75**:1928.

Terminating reactions

Khym JX, Bynum JW, Volkin E (1977) *Anal Biochem* **77**:446.

Determination of kinetic mechanisms by the HPLC assay method

Sloan DL (1984) Kinetic analysis of enzymatic reactions using high performance liquid chromatography. In *Advances in Chromatography,* Vol. 23, JC Giddings, E Grushka, J Cazes, PR Brown, eds.) Dekker, New York.

GENERAL REFERENCES

General

Brown, PR (1970) *J Chromatogr* **52**:257.
Scoble, HA, and Brown PR (1983) Reversed-phase chromatography of nucleic acid fragments. In *High Performance Liquid Chromatography, Advances and Perspectives,* Vol. 3, C Horvath, Ed. Academic Press, New York.

Enzymes and reaction conditions

Bergmeyer HN, Gawehn K, Moss, DW (1974) *Methods of Enzymatic Analysis,* Vol. I. VCH Publishers, Deerfield Beach, FL.
Dixon M, Webb EC, Thorne CJR, Tipton KF (1979) *Enzymes,* Academic Press, New York, 1979, Ch. 2.

Secondary reactions

Palmer T (1985) *Understanding Enzymes,* 2nd ed. Wiley, New York.
Rossomando EF, Jahngen JH, Eccleston J (1981) *Anal Biochem* **116**:20.
Rossomando EF, Cordis GA, Markham GD (1983) *Arch Biochem Biophys* **220**:71.

Metals in enzyme reactions

Jahngen JH, Rossomando, EF (1983) *Anal Biochem* **130**:406.

Introduction to kinetic mechanisms

Cleland WW (1963) *Biochim Biophys Acta* **67**:104.
Cleland WW (1970) Steady state kinetics. In *The Enzymes,* 3rd ed., Vol. 2, PD Boyer, Ed. Academic Press, New York.
Fersht A (1985) *Enzyme Structure and Mechanism,* 2nd ed. Freeman, New York.
Fromm HJ (1979) Summary of kinetic reaction mechanisms. In *Methods in Enzymology,* Vol. 63A, DL Purich, Ed. Academic Press, Orlando, FL.

Statistical analysis of initial rate data

Cleland WW (1979) Statistical analysis of enzyme kinetic data. In *Methods in Enzymology,* Vol. 63A, DL Purich, Ed. Academic Press, Orlando, FL.

Initial rate determinations

Fromm HJ (1975) *Initial Rate Enzyme Kinetics.* Springer-Verlag, Berlin.
Rudolph FB, Fromm HJ (1979) Plotting methods for analyzing enzyme rate data. In *Methods in Enzymology,* Vol. 63A, DL Purich, Ed. Academic Press, Orlando, FL.

CHAPTER 5

Strategy for the Preparation of Enzymatic Activities from Tissues, Body Fluids, and Single Cells

OVERVIEW

In developing a strategy for the preparation of an enzymatic activity, it is useful to consider two factors. The first is the choice or selection of the biological samples to be used as the starting point for the purification. These samples will clearly differ in terms of their complexity, and this complexity can be used to subdivide the samples into groups.

In the first group are samples that are rather complex not only because they consist of many different cell types but also because they have an extracellular compartment. Such samples include organs, tissues, biological fluids, and microbial cells, together with any other unicellular organisms grown in a culture medium or fermentation broth. For these samples, the initial step is the separation of the cellular from the noncellular compartment. Next, the different cell types within the cellular compartment must be separated, and this set of homogeneous populations of each type of cell becomes the starting point for samples in the second group. With such cells, activities at the cell surface can be directly assayed or the cells can be lysed, providing accessibility to the activities in intracellular organelles and on cytoplasmic fragments.

The samples in the third group are the subcellular fragments liberated by lysis. These include organelles such as mitochondria, as well as those operationally defined as a "membrane fraction" or a fraction containing "soluble components." The initial steps involving samples in this group include the separation of organelles from each other, the separation of insoluble from soluble fractions, and the solubilization of membrane samples or the fractionation and separation of one molecular species from another. Since the strategy developed for purification depends on the choice of starting material, this chapter outlines some of the problems associated with obtaining activities from samples in each group. In addition, we offer solutions to these problems.

There is a second consideration involved in the development of a purification strategy. Derived in part from the first, this consideration relates to the question: To what extent should the activity be purified? The traditional end point of any purification scheme would be a homogeneous protein, given the original demonstration that an enzymatic activity was associated with a single protein molecule. Thus, the question may appear to have only one answer. Several considerations can be used to justify this as the end point of the purification, the most important of which relates to the difficulties associated with the assay of the activity when the preparation is not homogeneous (e.g., when the enzyme remains in a preparation that contains many proteins and many enzymes).

The development of the HPLC method to assay enzyme activities has made it considerably easier to assay a single activity in the presence of others. Thus, attempts to obtain a pure protein during the purification procedure may not be necessary. Since the advent of HPLC to assay enzyme activities, it is possible to stop the purification at a much earlier stage and still assay for a single enzymatic activity. In fact, for some studies, it is even advantageous to assay the activity of interest in the presence of other activities.

Finally, this chapter discusses the use of HPLC itself as an aid in the purification of an enzyme activity. Applications are not restricted to the final stages of a purification. Its use of small sample volumes, its sensitivity, and its speed of separation make HPLC an ideal analytical tool to monitor the efficiency of other steps and procedures that are used during a purification.

5.1 INTRODUCTION

5.1.1 The First Goal: Selection of the Biological Starting Point

Since enzymes are associated with living systems, the development of a strategy for preparation of an activity begins with the selection of a specific biological starting point. This selection may be difficult, since one can start with an organism like an elephant, an organ such as a liver, biological fluids such as blood or saliva, cells that occur naturally such as bacteria or protozoa, or, finally, cultured cells. Preparing an activity from an elephant will present problems quite different from those experienced when preparing that same enzyme from bacteria.

To deal with the array of choices just outlined, samples have been subdivided into three groups. In group I (see Fig. 5.1) are samples that contain both a cellular and an extracellular compartment. The extracellular compartment can contain low molecular weight compounds such as the nutrients found in a fermentation broth, as well as macromolecular materials such as collagen or proteoglycans found in tissues. Also included in this extracellular compartment are fluids, such as tears, saliva, and urine.

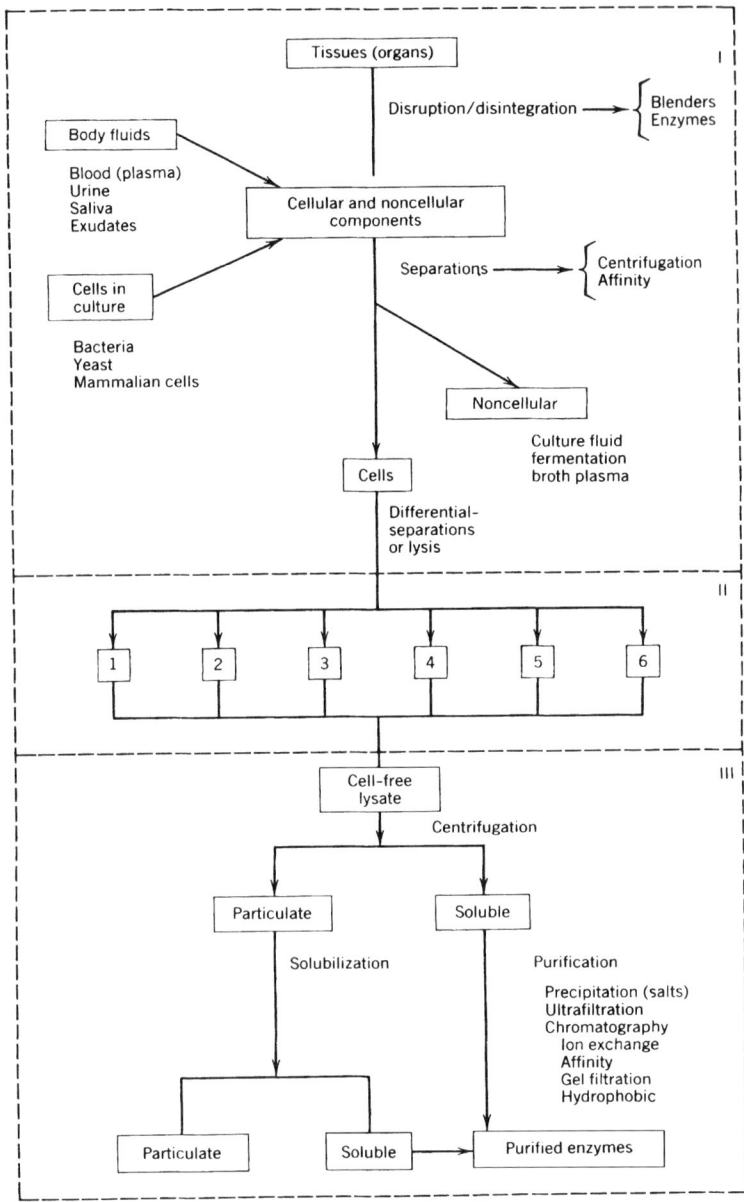

Figure 5.1 Anatomy of a purification scheme: the starting and ending points for three types of purification. (*I*) The starting samples are composed of cells and exracellular materials, the goal of the purification is their separation. (*II*) The source of the enzyme is a heterogeneous population of cells, and the goal is to produce separate homogeneous populations (1–5). (*III*) The source of the enzyme is a cell, and the goal is to isolate a subcellular organelle, fraction, or purified protein.

While samples in group I contain cells, these usually are not of the same phenotype. For example, tissue samples such as skin can contain cells of several types. Therefore, multicellular samples in group I must contain, as part of their purification scheme, a step for the separation of the different cell types.

Samples in group II are composed of homogeneous populations of cells, and samples in group III are formed by the lysis of these cells and can be anything from organelles to soluble purified proteins. It is often possible to start a "purification" scheme with a sample from group III. In this case the strategy for the purification would include steps such as salt precipitation and chromatography.

5.1.2 The Second Goal: Determining the Extent of the Purification, or End Point

One of the most important events in the history of biology was the demonstration of enzymatic activity. Next in importance came the isolation and purification of the activity and the demonstration that catalysis is associated with protein molecules. Thus, advances in enzymology have been intertwined with advances in protein chemistry.

Enzyme isolation and purification as practiced today has progressed from the days when the tools available were few and limited to procedures such as "salting out" with ammonium sulfate, filtration, and centrifugation at rather unimpressive g forces. The procedures of the past involved a significant amount of art, as well as science, with the artistry demonstrated by the last step, which in most schemes usually involved crystallization of the enzyme. To watch a pioneer like Moses Kunitz coax enzyme crystals from an alcohol solution was inspirational. Unfortunately, the technology of videotaping was not around in those days to preserve such events.

Today, the enzymologist has a battery of techniques, including improvements in centrifugation and column chromatography, and enzymes can be purified to homogeneity faster than ever before. But should a pure enzyme be your goal?

Imagine looking under the hood of an automobile and trying to find, somewhere in the engine, a screw 0.016 in. long with a 3/32 in. thread and a round head. One approach to finding such a screw, a minute part of the entire assembly, would be to take the engine apart as quickly as possible to obtain all the screws. These could then be spread out on a table top, and by careful inspection the screw of interest could be located.

Similarly, with an enzyme activity, if your goal is to isolate a special enzyme protein to characterize its size, shape, and amino acid composition, of course it would be wise to disrupt the starting material (organ, cell, or organelle) as quickly and completely as possible to obtain all the proteins, and then spread them out or fractionate them to locate the one of interest.

Returning again to the car engine, suppose we know the screw is present, and now we wish to find its location in the engine. The strategy must be different. For example, it would be wiser to remove each part of the engine carefully and disassemble them separately. In this way, when the screw is found, its location in the engine can be established.

When the question of the localization of an enzyme within the cell had to be answered, purification schemes were developed to take cells apart in a stepwise fashion. For this task a variety of tools were used, including rather commonplace scissors as well as sophisticated centrifugation and chromatographic techniques. And it was only after careful dissection that an understanding emerged of where the enzymes were located.

But how do enzymes function at their respective sites? If the question were asked about the automotive engine screw, one answer might be found by examining that component of the engine to which the screw belonged. For example, if this screw was a part of the carburetor, it might function in regulating the intake of fuel or air; but it might operate in several other ways, as well. In fact, it might not be possible to deduce its function merely by inspection, and to really find out how the screw worked, the intact carburetor would have to be returned to the engine and the engine started.

The same is true for an enzyme. Studies on what might be called its "interenzymatic" function are best carried out while the enzyme is still a part of the organelle or complex to which it belongs. To date such studies have been difficult to perform because of problems in monitoring in a single assay the variety of enzymatic activities that can occur in such complexes. The introduction of HPLC to assay enzymatic activity allows us to consider trying such interenzymatic functional studies, because with this method several activities can be measured simultaneously. Thus, one of the consequences of this advancement in methodology is that we can measure the activity of an enzyme while it remains in a complex and, through such studies, deduce its function within a multienzyme complex.

This advancement in methodology also can affect the strategy employed for the purification of an enzyme, since the goal need not be to isolate the enzyme protein from all other enzyme proteins. Now the goal might be better stated: *Remove from the enzyme preparation only those components or structures that are not necessary for function.*

The scheme in Figure 5.1 can be used to illustrate this point. While this scheme shows three separate starting points, it also shows potential end points. For example, if the sample is multicellular, the purification could be ended after the removal of the extracellular matrix or the medium or broth used to culture cells. Enzyme activities can be measured directly in the multicellular structures or in the extracellular compartment.

Alternatively, the purification can be continued, the cells lysed, and activities assayed after lysis. It is from lysates so produced that organelles such as nuclei and mitochondria are obtained. The disruption of organelles in turn produces soluble enzymes. Ultimately, the end point is determined by the individual study. If questions related to molecular weight, amino acid composition, or catalytic mechanism are to be answered, a pure protein will be required. In contrast, if questions relating one activity to another are of interest, the purification should be ended at the lysate or organelle level.

The sections that follow focus on samples from each of the three groups shown in Figure 5.1. I leave the reader to answer the question: Where do I stop?

5.2 PREPARATION AND ASSAY OF ENZYMATIC ACTIVITIES IN SAMPLES OF TISSUES, ORGANS, AND BIOLOGICAL FLUIDS

5.2.1 Separation of Cellular from Extracellular Compartments

5.2.1.1 Samples Obtained Directly from an Organism Tissues, such as connective tissue, and organs, such as skin or liver, can be thought of as being composed of at least two compartments: the cellular compartment and the extracellular compartment. Since enzymes can be localized in either compartment, one of the first problems is to separate the two compartments.

Techniques should be used that will not damage the cells, since any damage is liable to cause leakage of the contents of the cellular compartment into the extracellular compartment. With tissues or organs, where the noncellular compartment is often a stable fibrillar matrix, a two-step procedure such as that shown schematically in Figure 5.2 is helpful in separating the compartments. Often the matrix is disrupted by cutting or dicing with scissors, shearing

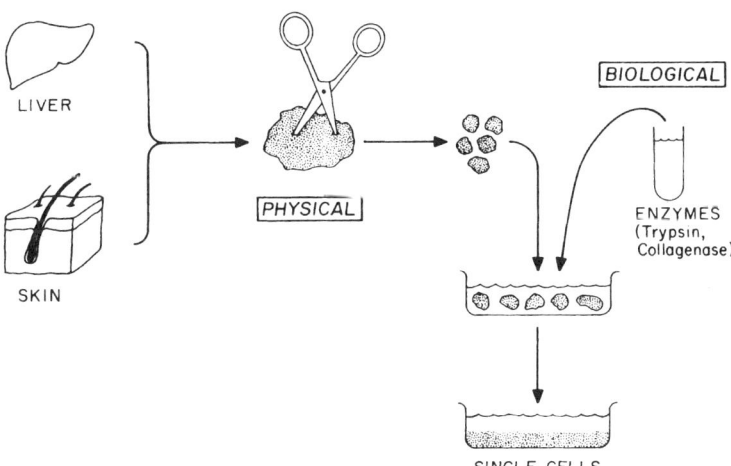

Figure 5.2 Methods used in the disruption of tissues or organs, such as liver and skin. The initial step is usually some physical technique; dicing with scissors is illustrated. The fragments are then treated with an enzyme such as trypsin or collagenase to disrupt the fragments further to obtain single cells.

in a blender, or grinding. With such physical techniques, however, some cellular damage is unavoidable.

The disruption of the matrix can be continued by treating the fragments with purified enzymes, which are often commercially available. These materials are ideal, since they can be chosen for their specificity and also chosen with the composition of the matrix in mind. In samples derived from mammalian tissue, the matrix usually contains collagen, and the enzyme collagenase can be used. Trypsin and other proteolytic activities have also been used with great success. As illustrated in Figure 5.2, the end result of this two-step procedure should be a solution containing intact cells, extracellular fluids, and extracellular components, including some insoluble fragments, some soluble components, and, of course, any enzymes added as reagents.

5.2.1.2 Samples Obtained from Tissue or Organ Culture

Animal tissues and organs can also be grown using a primary culture system by placing the sample on a support, such as an agar surface, a filter, or even a wire mesh, which can be positioned with the sample bathed in a solution of growth medium. A typical arrangement of the latter (Fig. 5.3), consists of a culture

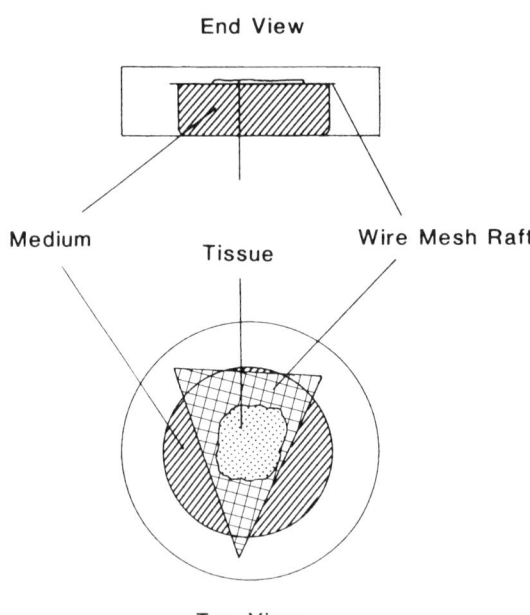

Figure 5.3 Diagram of the apparatus used for the culture of a tissue fragment or an organ. The dish contains a central well (diagonal lines), over which is placed a wire mesh raft to support the tissue. The well is filled with sufficient culture medium to make contact with the tissue. The dish is covered (not shown) and incubated under conditions used to maintain viability of the explant. (From Quintner et al., 1982.)

dish containing a centrally placed well filled with culture fluid. Suspended over the well is a wire screen, which acts as a support for the tissue. The well is filled with culture medium, the tissue is placed on the support, and the dish is covered and incubated.

Samples obtained from such a culture system should be processed by the two-step procedure described above to obtain the individual cells, free from the extracellular compartment. However, note that with cultured samples there is an additional extracellular compartment, which is the culture fluid used to support the growth of the sample. It is not at all uncommon to find enzymatic activities in this fluid. Some of these are normal constituents of the culture medium, while some are a consequence of growth of the sample.

5.2.1.3 Samples Obtained from Biological Fluids

As illustrated in Figure 5.1, biological fluids, which also have two compartments, are placed in group I. Such fluids include blood (often classified as a tissue), urine, semen, tears, and saliva. Many of these fluids contain cells as a normal component, while in others the cells represent a contamination. Fluids like saliva that contact the "outside" and, when collected, often contain microbes, are examples of contaminated materials. Sometimes such microbes are the result of the collection procedure, and their numbers can often be controlled by careful technique. At other times, as is the case with urine, their presence can indicate an underlying disease process. In any case, the study of enzymes from such fluids again requires the separation of the two compartments.

Biological fluids, however, do not contain a fibrillar matrix material, and their separation does not require the two-step procedure described above. Often, centrifugation at a slow speed, such as 5000g for 10 minutes as illustrated in Figure 5.4, will suffice. The pellet produced during this centrifugation should contain most, if not all, of the cellular elements, including any microbes. It is advisable and informative to recover these pellets. Both a sample of the pellet and a sample of the supernatant solution should be examined microscopically for their cellular content. The supernatant solution produced by the centrifugation can be assayed directly for enzymatic activities. However, if excess protein

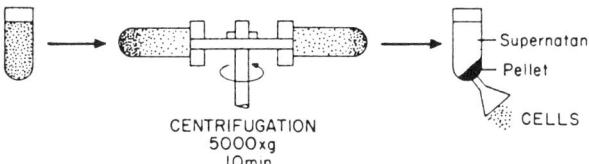

Figure 5.4 Harvesting cells from a culture medium. A sample of the culture medium is transferred to a centrifuge tube and subjected to centrifugation at rather low g forces such as 5000g for a relatively short time, such as 10 minutes. The cells contained in the pellet can be recovered after the supernatant solution has been decanted.

is present in the sample, it will have to be removed from the enzymatic assay samples before HPLC analysis.

5.2.1.4 Samples Obtained from Cell Cultures

Cells of many types are now grown in liquid culture. These include not only mammalian cells but fungi, protozoa, and bacteria. In Figure 5.1 these are placed together under the group I heading because in each case the noncellular elements in the fluid or fermenatation broth should be separated from the cells before analysis. A low speed centrifugation as shown in Figure 5.4 should suffice. The supernatant fraction is collected and assayed for the activity of interest, and the cells set aside for assay or lysis.

5.2.2 Assay of Activities in the Extracellular Compartment

Enzymatic activities will often be found in the extracellular fluid surrounding organs, tissues in biological fluids, or the medium supporting the growth of mammalian cells, bacteria, yeast, or fungi. The enzymatic composition of these fluids can vary considerably, and the assay of enzymatic activities in such fluids presents several major problems.

The first is the presence of proteolytic activities, which must be inhibited early in the procedure to prevent the degradation of other enzyme proteins. Second, the amount of protein present in these fluids is usually in excess of what an HPLC analytical column can handle without becoming clogged. And finally, these fluids often contain many low molecular weight compounds, either those added as nutrients or those present as a result of cellular metabolism. Since such compounds may resemble either the substrate or product, or both, of the enzymatic reaction under study, their presence in the reaction mixture could interfere with the assay. At the very least, such compounds will pass through the analytical column and appear on a chromatogram, confusing the experimental results.

Low molecular weight compounds may be removed, and a variety of methods are available, including dialysis and gel filtration chromatography. The removal of excess protein may be more complicated. It can be dealt with before the assay by further purification of the sample. Alternatively, the excess can remain during the incubation and be removed after the assay by introducing a termination step to precipitate all proteins, which are then removed by filtration. And finally, proteolytic activities can be eliminated by the addition of the "inhibitory cocktail" mentioned below.

A word of caution: Growth media that contain serum contain serum-associated enzymes. The presence of such endogenous activities must be considered in any purification scheme, since they can produce confusing results.

5.2.3 Assay of Activities in the Cellular Compartment

The two-step procedure just discussed serves to disrupt the samples in group I for the purpose of separating the cells from the extracellular fluids. This procedure produces two samples: the solution of extracellular fluids and the cells. In all probability, however, the cells will not all be the same. In fact, depending on the complexity of the starting sample, the cellular population can be quite heterogeneous. Consider two organs, liver and skin, as illustrated schematically in Figure 5.2. Such structures contain several tissue types, including epithelia, connective tissues, and vascular and neural tissues. The preparation of enzymatic activities from such organ samples will require the separation of the different cell types from each other; otherwise, what during the course of a study might appear to be changes in enzymatic composition or activity will in fact be only a reflection of changes in composition of those cells making up the sample. Therefore, for any assay on a complex sample it is important to begin with a preparation of only one cell type. Many methods have been introduced to achieve this separation; some exploit differences in composition of the cells, others utilize differences in cell function, and still others take advantage of differences in composition at the cell surface.

For example, differences in composition of cells are often reflected in differences in density, and therefore cells can often be separated by centrifugation through a solution made of layers of different density—a technique called *buoyant density centrifugation.* At equilibrium the cells can be recovered from the solution at a position that balances their density. Such a procedure is illustrated schematically in Figure 5.5*A*, where the starting tissue is represented as being composed of cells of three different types, and each type is denoted by a different symbol. A centrifuge test tube is filled with a series of sucrose solutions, each with a different density. In the example, three different solutions are used (a discontinuous gradient); the solution of highest density is, of course, placed on the bottom. Sucrose is often used for this purpose. If these cells are layered at the top of the solution, then after centrifugation each of the three cell types will be recovered in a different zone or band within the tube, its position dependent on its density.

Another method entails the addition of chemicals (drugs) that by virtue of their specificity will produce lysis of one or two cell layers, leaving a single cell type unaffected (Fig. 5.5*B*).

Two other techniques, both of which utilize antibodies raised against antigens specific to one of the cell types, have also been employed. In one of these techniques, antibodies can be attached to solid supports, which in turn can be used for column chromatography. In Figure 5.5*C,* the antibody specific for the round cells is shown attached to beads used to pack the column. When a solution of the three cell types is passed through the column, the cells containing the antigen attach to the antibody and are retained, while the other, noninteracting cells pass directly through the column. In this way separation is achieved.

In a variation of the technique above, metallic iron, labeled Fe in Figure 5.5*D,* can be attached to the antibody. The antibody is exposed to the cell, and after its attachment to a specific cell through interaction with the antigen,

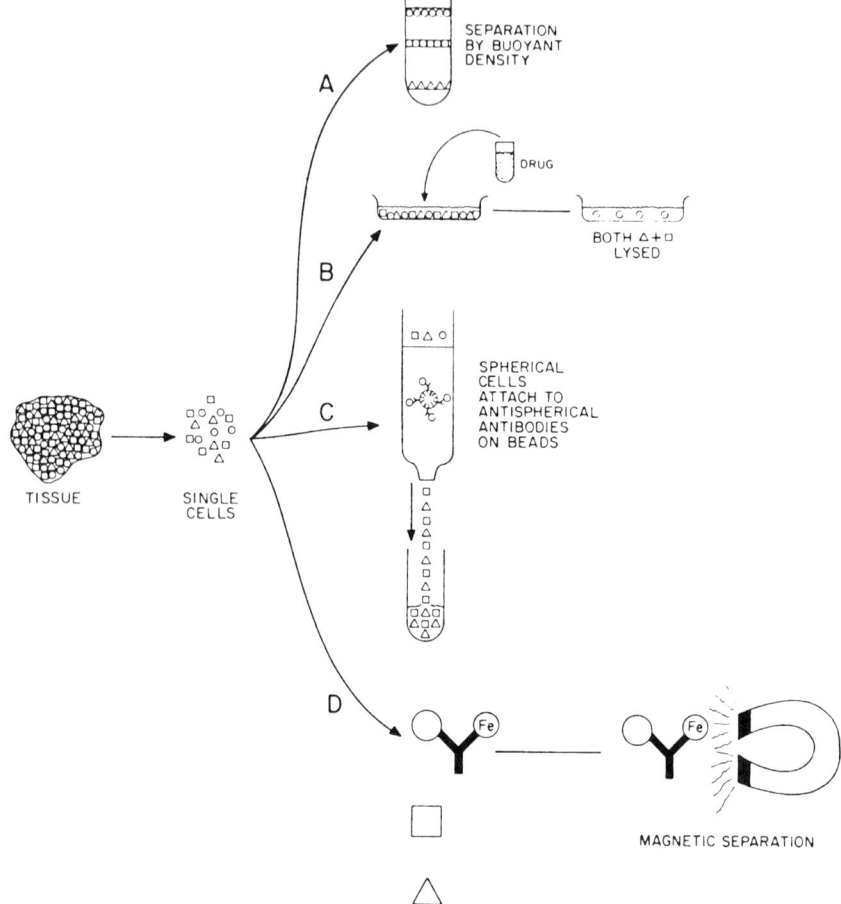

Figure 5.5 Some methods for the separation and isolation of homogeneous cell populations from tissue fragments. The tissue, composed of three cell types (○, □, △), is disrupted and a solution of single cells obtained. The cell types are shown being separated by (A) buoyant density centrifugation, (B) selective cellular lysis, (C) affinity chromatography, and (D) magnetoseparation.

the cells are exposed to a magnetic field that withdraws the "iron-containing" cell from the population.

Even after a homogeneous population of cells has been obtained from an organ or tissue, it is important to remember that when comparing the results of enzyme activity studies obtained on cells from one animal with those from another, differences in the ages of the animals, their genetic backgrounds, or even their nutritional status can alter the enzymatic activities recovered in their cells.

5.3 PREPARATION AND ASSAY OF ACTIVITIES IN INTACT CELLS

In group I, we considered the question of obtaining an activity from a tissue or organ, from a biological fluid, or from cultured cells. The primary task in all these samples was the separation of the extracellular and cellular compartments. Next, the problem of separation of the different cell types within the cellular compartment was considered. In the section that follows, we will open the cell for a look inside. However, let us first consider briefly the surface of the intact cell and the problems associated with the assay of any activities that might be located there.

Setting up an HPLC assay for activities on intact cells requires only one major change: since the reaction mixture will contain cells, the samples for HPLC analysis cannot be injected directly onto the analytical column; thus the reactions must be terminated and any precipitated proteins removed (see Chapter 4). Termination can be accomplished by centrifugation at a low speed or by filtration (see Fig. 4.9B). However, care should be taken to avoid any cell breakage, particularly if the product of the primary or even secondary reactions can also be found as a naturally occurring intracellular component.

Consider an assay for a cell surface ATPase where the reaction product is ADP. A reaction mixture is prepared that contains ATP as the substrate. The addition of the cells will start the reaction, and ADP will be produced directly into the extracellular solution. Lysis of the cell during this assay will release cellular ADP into the incubation medium, altering the results of the assay.

5.4 PREPARATION AND ASSAY OF ACTIVITIES IN SUBCELLULAR SAMPLES

The lysis of cells has its consequences. For example, since most lytic procedures involve disrupting both the plasma membrane and the membranes surrounding internal organelles, opening a cell will result in a loss of boundaries that would otherwise segregate enzymes from degradation. Therefore, following the disruption of most cells, proteolytic activities from some locations such as lysosomes gain access to areas from which they normally would have been excluded. Thus, precautions must be taken if problems arising from proteolytic activities are to be kept at a minimum or, better yet, prevented.

In the absence of any specific information about the nature of such activities, it is often best to mix a cocktail containing several or all of the available proteolytic inhibitors. These include such compounds as 1,10-phenanthroline, soybean trypsin inhibitor, leupeptin, benzamidine, antipain, aprotinin, phenylmethanesulfanyl fluoride, and diisopropylfluorophosphate. As a precaution against any enzyme destruction, such a cocktail is best added to the buffer in which the cellular lysis will be carried out (Fig. 5.6).

Opening the cells can have other consequences. For example, many enzymes require cofactors for catalytic activity. If, as a result of the lysis of cells and organelles, the enzymes become separated from these cofactors, a loss

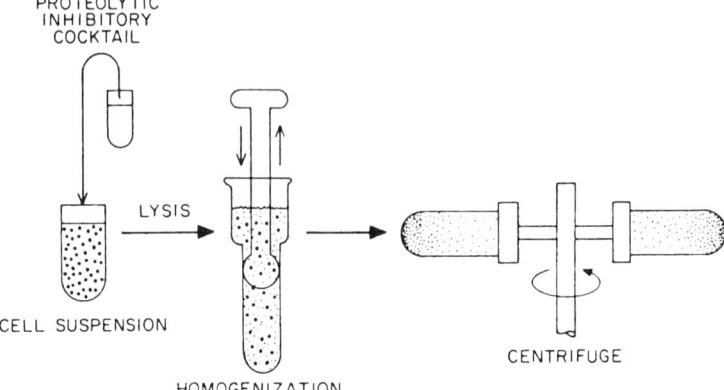

Figure 5.6 Overview of steps involved in the preparation of a cell-free lysate. The cells are resuspended in a buffered solution at a specified cell density. To this suspension is added a "cocktail" containing several proteolytic inhibitors. The cells in the suspension are lysed (here by homogenization). Finally the lysate is subjected to a very low speed centrifugation such as 5000g for 10 minutes to remove unbroken cells.

of activity can result. Also, since the activity of most enzymes is concentration-dependent and since lysis usually results in the dilution of intracellular components, the activity of enzymes present at low intracellular concentrations can be lost. Both these problems are difficult to guard against.

In addition to the foregoing more general concerns are questions concerning the localization of an enzyme activity. The location of an enzyme can determine the type of cell lysis, since it could be more advantageous to lyse the cell completely or in such a manner that the organelles are left intact. For example, some lysis methods such as sonication completely disrupt mitochondria, nuclei, and Golgi systems. If an activity is localized in an organelle such as a mitochondrion, it would seem sensible to adopt a method that leaves these structures intact, to facilitate their separation from the rest of the cellular debris. Thus, for the isolation of mitochondrial enzymes, sonication is not the method of choice for cell lysis.

Methods commonly used for lysis involve such equipment as the French press, the sonicator, and the blender or homogenizer. In Figure 5.6, a homogenizer is represented in sequence with the other steps involved in the lysis process. The choice of the lysis method, while dependent on the localization of the enzyme, is also a function of the type of cell under consideration. For example, when lysing bacterial cells with rigid cell walls, the French press may be one of the few physical methods that works. Alternatively, for disrupting cells that have fragile cell membranes, homogenization as shown in Figure 5.6 is often adequate.

Regardless of the procedure finally adopted, it is wise to measure its success. One of the easiest techniques for accomplishing this assessment is a microscopic examination of samples taken from the lysate after each step of the procedure. Such information is very useful, particularly when one is experimenting with either a new type of cell or a new method for lysis.

5.5 INITIAL PURIFICATION AND ASSAY OF ACTIVITIES IN CELL-FREE LYSATES

While it is possible and often necessary to assay enzymatic activities directly in the lysate, it is often helpful to remove any remaining unbroken cells and any nonbiological debris (e.g., sand grains or glass beads that might have been added to facilitate the breakage). Again, centrifugation at a low speed, such as 5000 rpm, for about 10 minutes should be sufficient.

The supernatant fraction obtained in this step can be centrifuged at 30,000g to produce a second supernatant fraction often referred to as an S-30 fraction. Centrifugation of this S-30 fraction at 100,000g will produce a third supernatant fraction called an S-100 fraction. Each centrifugation removes insoluble material of decreasing size or mass. For example, while the pellet from the S-30 fraction contains many large organelles, such as intact mitochondria, the pellet from the S-100 fraction contains ribosomes, the endoplasmic reticulum, and other smaller membranous structures.

Many enzymes are soluble and will be recovered in the S-100 fraction. In addition, their activity can be measured rather conveniently by adding a sample of the S-100 fraction directly to a reaction mixture. Of course, since the S-100 fraction will contain an excessive amount of extraneous protein, it will be necessary to terminate the reaction and filter the sample prior to injecting it onto the HPLC for analysis.

Note that while the S-100 fraction can be used in this form, it would be best to have it dialyzed to remove unwanted low molecular weight compounds before using it in an assay. In addition to dialysis, a simple salting out can be performed to remove some of the extraneous protein material. Ammonium sulfate is often added for this purpose to remove unwanted proteins or to precipitate the enzyme in question. Any ammonium sulfate should be removed before the sample is used in an assay, because the salt might affect activity. Again, dialysis can be used, or alternatively the sample can be passed through a gel filtration (G-25) column.

If it is necessary to purify the enzyme further, the next step should be one that has a high capacity, that is, one that can process large amounts of protein. Such steps, however, often are not very specific. For example, salting out is a technique of high capacity but low selectivity. This step should be followed by steps of decreasing capacity but greater selectivity. Such techniques include ion-exchange, gel filtration, or affinity chromatography and even HPLC itself. The choice should take into account what is known about the protein, its size and shape, its solubility, and even its substrate specificity. Also, the quantity of protein required should be considered. If only analytical amounts of the enzyme are needed, several methods including HPLC can be included.

5.6 HPLC FOR PURIFICATION OF ENZYMES: A BRIEF BACKGROUND

Early separations of proteins by HPLC relied on controlled-pore glass beads with a coating of 1% polyethylene glycol. In a later modification (3% polyethylene glycol coating on the beads), the technique was used for the separation of plasma proteins. The development of noncompressible ion-exchange supports has allowed various laboratories to separate the isoenzymes of lactate dehydrogenase. Refinement of this anion-exchange support to a microparticulate size (40 μm) has enhanced the resolution of lactate dehydrogenase and creatine kinase isoenzymes as well as the purification of alkaline phosphatase.

Reversed-phase chromatography has also been used to separate proteins. However, the required use of alcoholic gradients or paired-ion reagents with the reversed-phase support should be avoided, to cancel the potential for inactivation of the enzymatic activities.

High pressure gel permeation chromatography (HPGPC) was developed to correspond to gel filtration, where large molecules are partially excluded from the porous coating of the support and thus are eluted before smaller molecules. A silica-based packing with an organochlorosilane chain containing other functional groups has been used to separate plasma proteins on the basis of their molecular size as well as by their ionic, polar, and hydrophobic interactions with the column packing and the mobile phase.

HPLC is useful as an analytical tool in several applications in addition to its role in the purification of an activity. For example, HPLC can be useful in the establishment of gradients. Instruments have been manufactured for use with HPLC that can control the flow and mixing of solvents and thereby generate gradients with a variety of concentrations and "shapes." In addition, since it is possible to carry out separations on the HPLC column in a comparatively short time, a number of these gradients can be applied to an analytical scale column, and the one best suited to the separation established fairly rapidly. Armed with this information, it is a relatively simple matter to carry over these gradient conditions to a non-HPLC ion-exchange column.

The speed with which it is possible to perform an analytical run makes HPLC useful for what might be called a pseudopreparative function. If rather modest amounts of a purified protein are required—for example, to carry out an analysis by polyacrylamide gel electrophoresis or even for antibody production—and if the HPLC column provides adequate separation and purification, it is often possible to produce enough material for such purposes by merely repeating the same run on an analytical column several times. By collecting the appropriate fraction, it is possible to generate sufficient material to advance to the next stage of the purification or to perform the experiments of interest. Again, it is the speed of the separation that makes the approach feasible.

Finally, HPLC can be used as an analytical method to monitor the efficiency of different purification methods. For example, imagine that gel filtration chromatography has just been carried out on a sample of an S-100 fraction. Several peaks are observed. A question usually asked at this point relates to the homogeneity of each of the peaks. HPLC is ideal to answer this question, since it allows each peak to be analyzed in just a few minutes. Also, different columns can be used for the analysis, and therefore the homogeneity of any peak can be verified under a variety of conditions.

As another example, imagine that an S-100 fraction has been prepared, ammonium sulfate has been added, and the fraction that precipitates between 30 and 50% has been obtained. This sample is subjected to affinity and ion-exchange chromatography. Have these procedures been successful in removing extraneous proteins? While it is possible to answer this question by a determination of total protein content, it may be more informative to analyze each of the samples for its constituent proteins.

The speed of the HPLC analysis makes such a determination possible. Figure 5.7 represents the analysis of samples after each of four stages of a typical purification. The analysis was made using gel filtration HPLC, and as shown this could be accomplished in 20 minutes.

Comparison of the profiles reveals that proteins present in the sample prepared by ammonium sulfate fractionation (Fig. 5.7A) were removed following affinity chromatography (Fig. 5.7B) and ion-exchange chromatography (Fig. 5.7C). The profile of the activity following ion-exchange HPLC (Fig. 5.7D) shows considerably less protein than was originally present.

5.7 STRATEGY FOR USE OF HPLC IN THE PURIFICATION OF ACTIVITIES

HPLC has received a great deal of attention as a method for the purification of enzymes, and the results of such studies are rapidly pervading the literature. The reader should consult recent reviews for information concerning the enzyme of interest.

Consistent with our purpose, this volume presents the general principles of how to begin the purification procedure using HPLC. The experience gained over the years, in the purification of enzymes has led to the general approach for purification detailed in this section.

To obtain the separation needed for purification, a few initial runs are carried out to verify solubility and to get some idea of the complexity of the sample. Separation of the proteins can require modification of several variables, including the concentration range of the salt used for the gradient as well as the salt itself. For example, the range of salt concentrations used in the gradient can have a significant effect on the elution profile of a series of

108 STRATEGY FOR THE PREPARATION OF ENZYMATIC ACTIVITIES

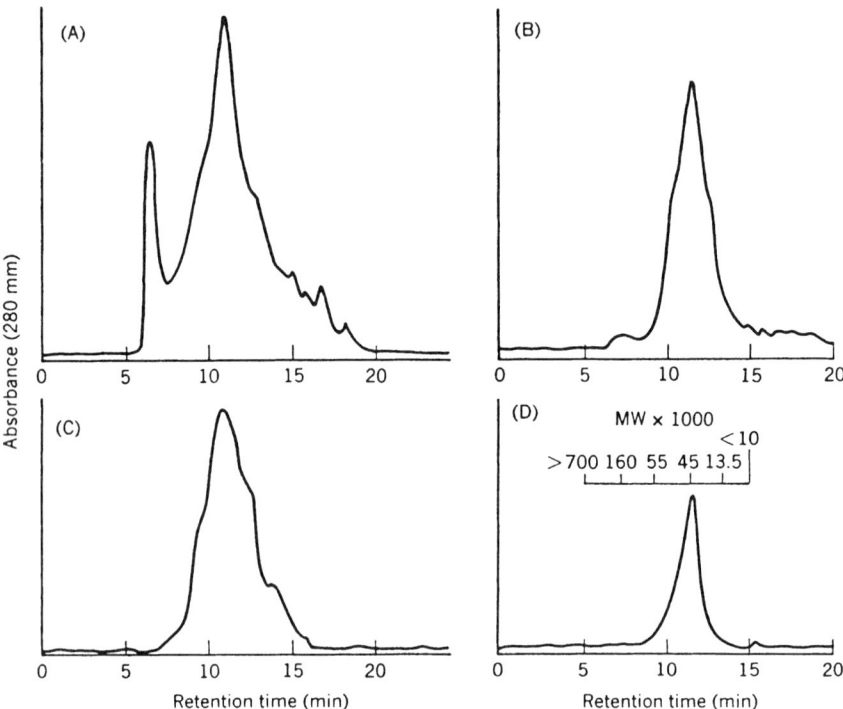

Figure 5.7 The use of HPLC to monitor enzyme purification. These profiles were obtained by gel filtration chromatography during the purification of the enzyme sAMP synthetase. The column was a TSK-250 (BioSil, 7.5 mm × 30 cm), and the mobile phase was 0.1 M potassium phosphate (pH 6.0). The column was monitored at 280 nm. Profiles obtained after (*A*) 30 to 50% ammonium sulfate precipitation, (*B*) affinity chromatography, (*C*) ion-exchange chromatography on DE-52, and (*D*) HPLC ion-exchange chromatography on AX-300.

proteins (Fig. 5.8*A,B*). The pH might also be changed, and the effect of such a change on the separation can be seen by comparing the chromatograms of Figure 5.8(*A,B* vs. *C,D*). During this initial phase, when the purpose is merely to establish conditions for the separation, it is usually not necessary to collect fractions or to assay enzymatic activity. Throughout this phase of the work, the separations can be monitored at 280 or 230 nm in the absence of aromatic amino acids in the protein.

When conditions for the separation have been attained, the enzyme should be located. Another run should be carried out for this purpose, the fractions collected, and the activity of each fraction determined. Figure 5.9 illustrates the results of such an analysis with the absorbance profile obtained at

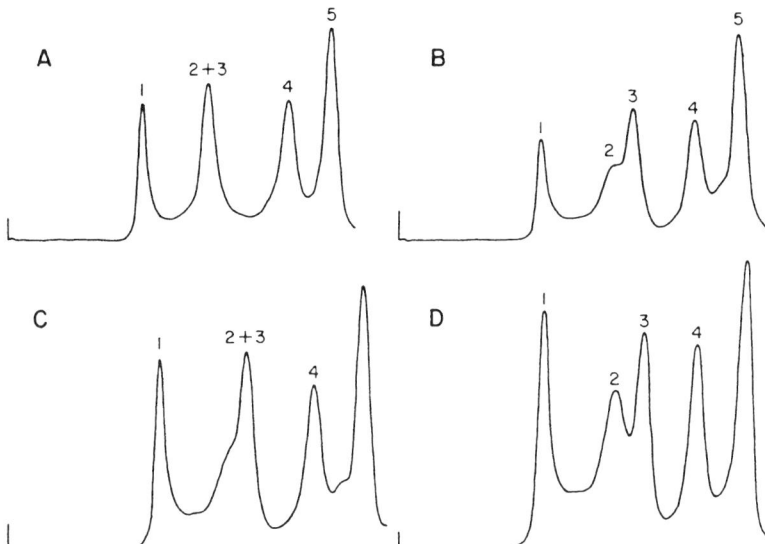

Figure 5.8 Effects of ionic strength and pH on elution profiles obtained by gel filtration HPLC on five proteins: 1, thyroglobulin, 670,000; 2, gamma globulin, 158,000; 3, ovalbumin, 44,000; 4, myoglobulin, 17,000; 5, vitamin D-12, 1350. The column was eluted with sodium acetate (pH 6.8) (A) at 50 mM and (B) at 100 mM. The column was also eluted with (C) 100 mM sodium acetate (pH 6.0) and (D) 150 mM sodium acetate (pH 6.0).

280 nm of the sample as fractionated on an ion-exchange column. The fractions collected were assayed for three different enzymes (Fig. 5.9A); the results are shown in Figure 5.9B–D. Three activities are present, and each activity has been separated into two components. Additional runs may be necessary to obtain sufficient material for subsequent purification. If an HPLC step is introduced early in the purification, several runs will be required to obtain enough sample for additional purifications.

To illustrate the rapidity of HPLC, particularly in comparison with the more conventional techniques, the same sample was separated by conventional ion-exchange chromatography. Figure 5.10 compares the two procedures. These data show that where 14 hours was required for the traditional method, only about 45 minutes is required with HPLC. Therefore, the total time needed to carry out this purification, not counting the time for the enzyme assay, could be as short as 3 to 4 hours. If necessary, the chromatography step could be completely automated. Finally, since each run will use only a fraction of the total volume of the starting material, the entire procedure will be economical.

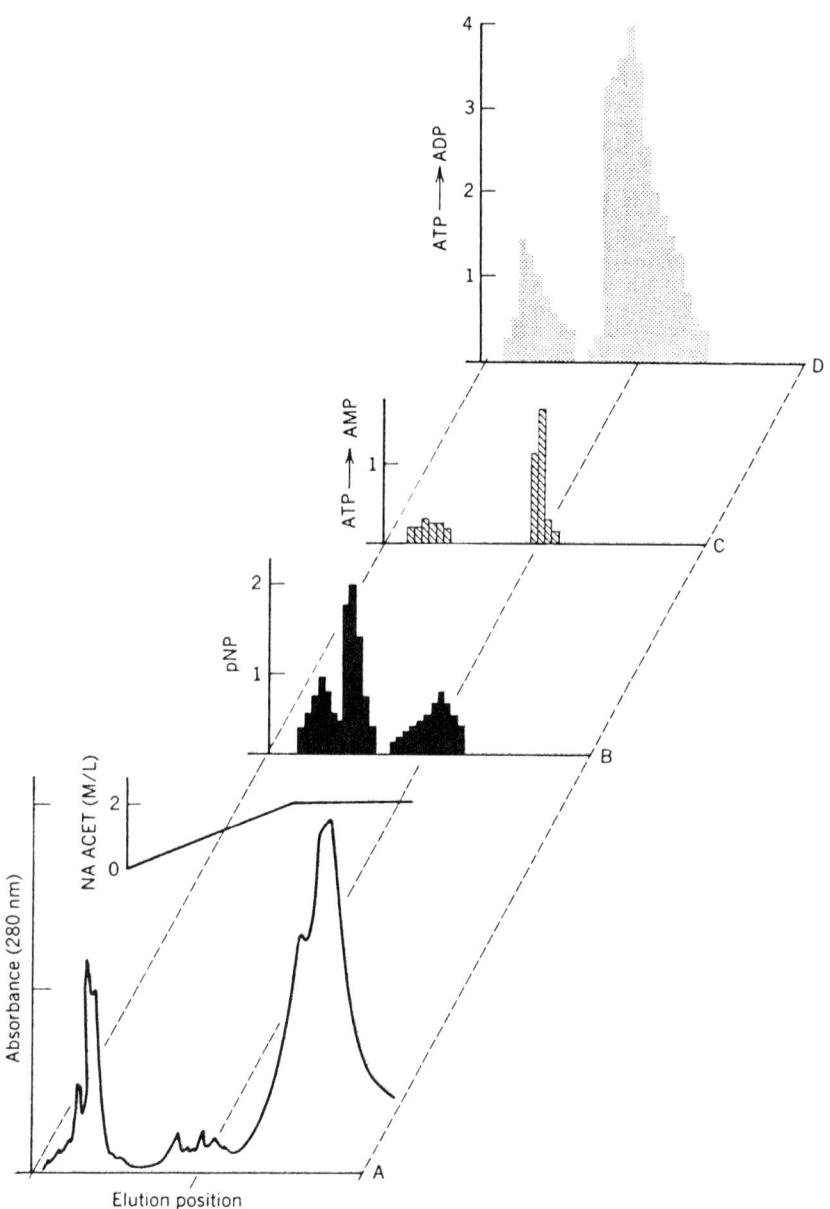

Figure 5.9 Enzymatic activities of fractions following HPLC chromatography. A partially purified preparation was fractionated by ion-exchange HPLC (AX-300) with a mobile phase of 0.1 M potassium phosphate. Proteins were eluted with a gradient of sodium acetate. Column eluent was monitored at 280 nm. Fractions were collected, and each fraction was assayed for three different activities.

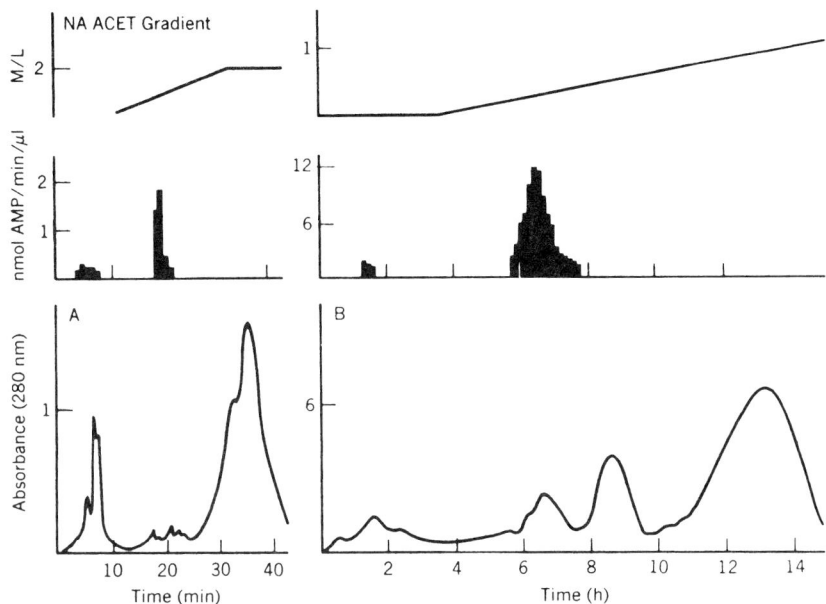

Figure 5.10 Ion-exchange chromatography of a detergent-solubilized membrane fraction. (*A*) Approximately 2 mg of protein from the fraction was injected onto an AX-300 anion-exchange column (250 mm × 4.1 mm) that was equilibrated with a 20 mM sodium acetate buffer (pH 6.3) containing 0.1 mM Z-314. A 20-minute linear salt gradient of 20 mM to 2 M sodium acetate (pH 6.3) with 0.1 mM Z-314 (*top*) was used to elute proteins. The protein profile is shown in the bottom panel. The flow rate was 1 mL/min, and the absorbance was monitored at 280 nm. Fractions of 1 mL were collected and immediately put on ice. For the amount of ATP pyrophosphohydrolase activity contained in each column fraction, the following assay was performed. To 75 μL of each column fraction was added a reaction mixture such that 0.4 mM ATP, 0.4 mM MnCl$_2$, and 50 mM Tris-HCl (pH 7.4) were in 100 μL final volume. The reactions were run at 31°C for 2 hours and terminated at 155°C for 1 minute. An analysis of the reaction components was done by reversed-phase HPLC using μBondapak C$_{18}$ column. The mobile phase was 65 mM KH$_2$PO$_4$ (pH 3.6), 1 mM tetrabutylammonium phosphate, and 2% acetonitrile. Assay tubes samples of 20 μL were analyzed, and the amount of AMP formed by the pyrophosphohydrolase in each column fraction is expressed as nanomoles of AMP per minute microliter. (*B*) Approximately 60 mg of protein from a detergent-solubilized membrane preparation was applied to a DE-52 column (30 cm × 2.5 cm) that had been equilibrated with a 20 mM sodium acetate buffer (pH 6.3) with 0.1 mM Z-314. After washing with 200 mL of buffer, a linear salt gradient (*top*) of 20 mM to 2 M sodium acetate in 2 L (total volume) was used to elute the proteins. Fractions of 15 mL were collected at 4°C, and absorbance was monitored at 280 nm. The protein profile is shown in the bottom panel. For the amount of ATP pyrophosphohydrolase activity contained in each column fraction, the enzymatic assay as described above was performed.

5.8 PROBLEMS RELATED TO THE ASSAY OF ACTIVITIES FOLLOWING THEIR PURIFICATION BY HPLC

While most of the problems in the assay of an activity purified by HPLC are expected and typical of chromatographic work with enzymes, the introduction of this technique into the purification scheme may lead to problems if the fractions obtained from the HPLC purification step are to be measured for enzymatic activity. For example, the salt in each fraction may inhibit any enzymatic activities it contains. Moreover, when ion-exchange HPLC is used the salt concentration will vary in the fractions. Thus it is prudent to study the effects of salt, at the concentration used for elution, on enzyme activity before the chromatography. If the salt is found to be detrimental, it will have to be eliminated or at least reduced in concentration before the chromatography. Removing the salt by dialysis may not be the appropriate way to proceed, however, since the inactivation of enzyme activities is not always reversible.

In addition, the detergents often added to enhance the solubility of proteins can cause problems: They can inhibit activities directly or, if they are removed, the resulting loss of solubility can cause precipitation. However, when HPLC is used for the purification, more often than not the protein concentration of the sample will be low, and the concentration of protein in each of the fractions will be lower still. Therefore, the formation of a precipitate may not be visible, and monitoring the sample at 320 nm, a wavelength useful in monitoring any light scattering, may be the only reliable method of detecting precipitates. As a precaution, it is good practice to dialyze the sample against the mobile phase that is to be used for the elution. When a gradient is used, dialysis against the starting and ending buffers should be carried out as well.

Many detergents, such as Triton X-100, absorb in the ultraviolet range and will therefore interfere with the detection of proteins at 280 nm. The use of nonabsorbing detergents will eliminate this problem. Detergents can also interfere with the operation of ion-exchange columns, and a reduction in detergent concentration may be required for the correct performance of the ion-exchange packing materials.

For example, we found that a 10% zwitterionic detergent was required for complete solubilization of the membrane fraction. However, we also found that the presence of detergent blocked the retention of some proteins on the ion-exchange column and, further, that dialysis of the proteins to remove the detergent resulted in the prompt precipitation of the protein. The problem was solved by trial and error. The protein was dialyzed against detergent solutions of various concentrations until a concentration low enough to permit ion-exchange chromatography but high enough for solubility was found.

Again, it is emphasized that because salt solutions often act to precipitate detergents, as in the precipitation of sodium dodecyl sulfate by potassium, it is necessary to check the solubility of the protein solutions in the detergent against the salt solution at the concentration that will be present at the conclusion of the gradient.

Two other problems often arise following the use of HPLC for purification. The first has to do with the volume of the sample used for assaying activity. Upon successful completion of the ion-exchange step, it is necessary to determine enzymatic activity. These determinations are performed on samples taken from a series collected during the course of the purification. With HPLC purification, however, the volume of each sample collected will probably be no more than a few hundred microliters, and often less. Further, the number of samples is usually small. This situation in HPLC is in contrast to that found in traditional chromatography, where the volume of each sample can be in the milliliter range and the total number of samples or fractions collected can be in the hundreds.

Since to locate the enzyme, it is necessary to assay each of the fractions for activity, the concentration of salts will not be the same in each fraction when gradients are used. And since salts often inhibit activity, false data may be obtained on the distribution of activity across the column if salts remain in the sample.

Finally, the pH at which the purification is carried out may not be suitable for the assay. To solve this type of problem, many investigators, particularly those more familiar with fractions containing volumes such as 10 to 20 mL, usually adjust the sample solution to the assay conditions, either by dilution of an aliquot of the fraction into the buffer used for the enzyme assay or by removing 1 to 2 mL from each and, by dialysis, changing the buffers.

Of course, with the small volumes involved, dialysis is often out of the question. Also, dilution usually consumes most of a given fraction. Therefore, it is necessary to be prepared to carry out the separation step more than once.

Most if not all the problems associated with salt, pH, and the presence of organics can be minimized if not eliminated following concentration by using one of the newly developed microconcentration systems.

5.9 SUMMARY AND CONCLUSIONS

Two considerations dominate the development of a strategy for the purification of enzymatic activities: the choice of the sample to be used as the source of the enzyme and the extent of the purification.

The samples that can be used as a source of starting material can be divided into three groups: multicellular (I), cellular (II), and subcellular (III).

For samples in group I, which includes tissues, organs, biological fluids, and cultured cells, one task is to separate the cellular from the extracellular compartment and another is to obtain a homogeneous population of single cells. These cells constitute the samples of group II. Group III contains samples obtained after cell lysis and includes organelles and purified proteins.

HPLC can be useful for the purification of proteins to homogeneity. It can also be useful as an analytical tool to monitor the purification of proteins using other more preparative procedures.

When HPLC is used to purify enzymes and the enzymes must be located by an analysis of the fractions collected during the separation, the solvents used in the purification may cause problems to develop.

GENERAL REFERENCES

Cell separation techniques

Owen CS (1982) Magnetic cell sorting. In *Cell Separation: Methods and Applications,* Vol. 2, TG Pretlow and TP Pretlow, Eds. Academic Press, Orlando, FL.

Pretlow TG, Pretlow TP (1981) Sedimentation of cells: An overview and discussion of artifacts. In *Cell Separation: Methods and Applications,* Vol. 1 TG Pretlow and TP Pretlow, Eds. Academic Press, New York, 1981.

Quintner MI, Kollar EJ, Rossomando, EF (1982) *Exp. Cell Biol* **50:**222.

Regnier FE (1984) HPLC of membrane proteins. In *Receptor Purification Procedures,* Vol. 2, JC Venter and LC Harrison, Eds. Liss, New York.

Waymouth C (1981) Methods for obtaining cells in suspension from animal tissues. In *Cell Separation: Methods and Applications,* Vol. 1, TG Pretlow and TP Pretlow, Eds. Academic Press, New York.

Protein and enzyme purification by HPLC

Hearn MTW (1994) Reversed-phased high performance liquid chromatography. In *Methods in Enzymology,* Vol. 104, WB Jakoby, Ed. Academic Press, Orlando, FL.

Regnier FE (1984) High performance ion-exchange chromatography. In *Methods in Enzymology,* Vol. 104, WB Jakoby, Ed. Academic Press, Orlando, FL.

Schmuck MN, Gooding KM, Gooding DL (1984) *J Liquid Chromatogr* **7:**2863.

Unger K (1984) High performance size-exclusion chromatography. In *Methods in Enzymology,* Vol. 104, WB Jakoby, Ed. Academic Press, Orlando, FL.

CHAPTER 6

Microdialysis: An in Vivo Method for the Analysis of Body Fluids

with Jan Kehr

OVERVIEW

The technique of in vivo microdialysis is one of the most efficient tools for studying time-dependent changes in biological processes such as metabolism, neurotransmission, and drug clearance. The time courses of biochemical or pharmacological events are traditionally measured by sampling blood, urine, cerebrospinal fluid, or other body fluids. However, the analysis of these products gives an overall and rather delayed set of points of information involving many organs and tissues. Postmortem analysis of dissected tissue or biopsy material provides only a static, one-point-in-time measure of detected substances. Microdialysis allows the continuous monitoring of local chemistry in a very specific tissue structure—for example, frontal cortex of a rat brain or human adipose tissue. Microdialysis is not as invasive as some "older" in vivo perfusion techniques (e.g., push–pull, cortical cup). In addition, it is more general, offers higher specificity than in vivo voltammetry or ion-selective microelectrodes, and is far less expensive than scanning techniques such as positron emission tomography (PET) or nuclear magnetic resonance (NMR) spectrometry. Particularly, the relative simplicity and low cost of microdialysis equipment has contributed to widespread use of this technique, as reflected by more than 2500 published papers (Kehr, 1992, 1993a, 1994), review articles (Benventiste, 1989; Benveniste and Hüttemeier, 1990; Ungerstedt 1991), special journal issues and chapters in books (Hamberger et al., 1982, 1985; Ungerstedt et al., 1982, Ungerstedt, 1984), and a monograph (Robinson and Justice, 1991). Several international conference (Rollema et al., 1991; Louilot et al., 1994) devoted to microdialysis sampling are organized regularly.

Microdialysis is, however, only a sampling technique. This means it provides an investigator with samples that must be chemically analyzed. In fact, the applicability of microdialysis to the study of a particular compound occurring in the extracellular fluid is entirely dependent on the sensitivity of the respective analytical technique. In this context, high performance liquid chromatography (HPLC) plays a key role in terms of sensitivity, reliability, and speed of analysis. In more than 75% of all microdialysis papers, HPLC techniques were used for determinations of monoamines (dopamine, noradrenaline, serotonin) and their metabolites, acetylcholine, amino acids, purines, pyruvate, lactate, and other endogenous or exogenous compounds (Kehr, 1994). In addition, microdialysis is useful for estimations of enzymatic activities based on measurements of substrates or products of enzymatic reactions (Sharp et al., 1986; Westerink et al., 1990). Enzyme inhibitors, cofactors, or any other drugs can be applied either systemically or locally via the microdialysis probe. The enzyme activity to be studied can be present either intra- or extracellularly. The important criterion is that the substrate or product be present in, or diffuse into, the extracellular fluid—the site of microdialysis sampling.

6.1 INTRODUCTION

6.1.1 Principle of In Vivo Microdialysis

Microdialysis is an in vivo bioanalytical sampling technique for monitoring local chemistry in the extracellular fluid of an organ and in body fluids. The sampling principle is quite simple: Molecules diffuse from the body compartment through a semipermeable membrane of the sampling cannula into the perfusion medium flowing inside the probe (Delgado et al., 1972; Ungerstedt and Pycock, 1974). The driving force for this molecular movement is passive diffusion down the concentration gradient between the two compartments. Thus, the technique works in both directions: for recovering substances from a living body, and for delivering substances into this organism.

In practice, a microdialysis probe is implanted into the tissue with tubing connecting it to external components: a perfusion pump and a fraction collector. The probe is perfused continually at low flow rate (0.1–10 μL/min) with an artificial physiological solution. As the perfusate emerges from the probe, fractions are collected and samples of each fraction analyzed.

6.1.2 Extracellular Space

The extracellular space comprises about 18 to 20% of the total tissue volume of most soft organs. This tissue compartment represents an important link between blood capillaries and cell soma for transport of molecules in both directions. The space is filled with the extracellular fluid (ECF) which, in turn, directly communicates with other body fluids such as lymph or cerebrospinal fluid (CSF). The ECF is a glutinous buffered medium consisting of salts and low molecular weight compounds, as well as polysaccharides, glucoproteins, polypeptides, and enzymes forming a complicated extracellular matrix. In the brain, released neurotransmitters can diffuse from the synaptic cleft into the larger extracellular compartment or can be involved in slower signalling over longer distances, so called volume transmission (Bjelke et al., 1994). Because bioanalytical techniques such as homogenization mix cytosol with ECF, it is impossible to use these methods to analyze the ECF contents. In contrast, the microdialysis probe enables "eavesdropping" on intercellular chemical communications.

6.1.3 Microdialysis Probe

The concept of in vivo sampling by microdialysis is based on the construction of a special cannula—a microdialysis probe—which, once implanted in the tissue, will more or less mimic the function of a blood vessel. Just as vessels serve for blood circulation, the probe is perfused with a physiological solution. In the same way that many molecules can pass through the walls of blood capillaries, the probe is provided with a semipermeable membrane allowing free diffusion of low molecular weight compounds. Microdialysis probes are constructed in several different types, among them a concentric design of a perfusion cannula, which is probably the most commonly used in experimental research. The typical examples of such a probe type depicted in Figure 6.1 represent some of the commercially available microdialysis probes. The CMA Model 10 features a "classical" construction of a probe for small and medium-sized laboratory animals, whereas a later modification (CMA 11) allows for microdialysis in very small areas of the brain. For applications in peripheral tissues, where the use of a flexible cannula is desirable, the CMA 20 is recommended.

All three probes shown in Figure 6.1 are similar in design: the inlet and outlet lines consist of two concentric cannulae, where the thinner line is longer and extends into the semipermeable membrane at the tip of the probe. The membrane is glued at its proximal end into the outer shaft of the probe; at its distal end, it is plugged with a glue. Alternatively, it can be glued to the inner cannula (CMA 10), as well, to strengthen the whole construction. Typically, the inner tube serves as the inlet and the outer tube as the outlet for the perfusion medium, but reversal of the flow is also possible. The membrane lengths for most of the commerical probes vary from 1 mm to several centimeters, and the outside diameters range from 0.2 to 0.7 mm.

Several membrane materials are suitable for the construction of microdialysis probes: cuprophane (regenerated cellulose), polycarbonate, polysulfone, polyacrylonitrile. Molecular cutoffs of these membranes are usually in the range of 5000 to 20,000 Da for sampling of small molecules, whereas for large and lipophilic compounds, cutoffs from 20,000 to 100,000 Da are available. Ideally, the membrane should function only by dialysis, not ultrafiltration. This means passage of components only by equilibrating concentration gradients, not by allowing passage of fluid through the membrane due from a high pressure gradient or through large pores in the membrane. Such ultrafiltration phenomena may cause serious disturbances of the tissue homeostasis and reduce dialysis recovery.

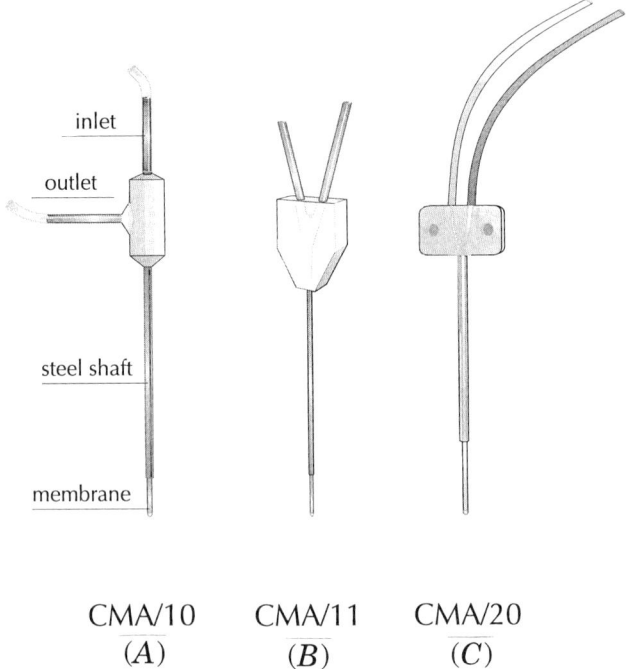

Figure 6.1 Microdialysis probes based on the concentric assembly of the inlet and outlet tubes provide the highest mechanical strength, as well as homogeneous diffusion paths from the surrounding environment; they also cause the lowest tissue damage. The probes differ mainly in the construction of the inlet/outlet lines, as seen when comparing these constructions. The shaft and the capillaries can be either rigid and made of metal (A) or fused silica (B) or flexible and made of polyurethane (C). The outer diameter of the membrane on the CMA/11 probe (B) is 0.24 mm, whereas the membranes of CMA/10 and CMA/20 probes (A and C) are 0.5 mm in diameter.

6.1.4 Dialysis Recovery

Since the first reports on microdialysis in living animals, there have been efforts to estimate "true" (absolute) extracellular concentrations of recovered substances (Zetterström et al., 1983; Tossman et al., 1986). Microdialysis sampling, however, is a dynamic process, and because of a relatively high liquid flow and small membrane area, it does not lead to the complete equilibration of concentrations in the two compartments. Rather, under steady state conditions, only a fraction of any total concentration is recovered. This recovery is referred to as relative or concentration recovery, as opposed to the diffusion flux expressed as absolute or mass recovery. The dependence of recovery on the perfusion flow rate is illustrated in Figure 6.2. As seen, relative recovery will exponentially decrease with increasing flow as the samples become more

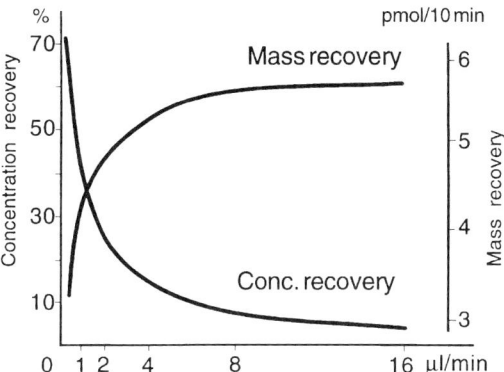

Figure 6.2 In vitro recovery versus flow rate for a typical microdialysis probe (CMA/ 10, 4 mm polycarbonate membrane, cut off 20,000 Da) in a quiescent medium and at ambient temperature. Typical flow rates used from brain microdialysis applications are in the range of 0.1 to 5 μL/min.

and more diluted, while the mass recovered per time unit will increase and reach a plateau.

Relative recovery can be mathematically expressed by Fick's law of diffusion as modified by Jacobson (Jacobson et al., 1985). In this relationship, recovery is the ratio of the concentration in the perfusate to the concentration extracellular. For the mass recovery, an expression similar to a Michaelis–Menten formula for enzymatic reactions was derived (Ekblom et al., 1992). A number of other mathematical models for quantitative microdialysis have been proposed and are reviewed elsewhere (Justice 1993; Kehr, 1993b).

6.2 TECHNICAL ASPECTS OF MICRODIALYSIS

6.2.1 Microdialysis Instrumentation

The microdialysis probe is the heart of the method, as a chromatographic column is the heart of the HPLC instrument. Rigid CMA probes, Models 10, 11, and 12, are used for stereotaxic implantations into the brain, where the probe can be fixed (cemented) to the skull. A flexible probe design (CMA 20) allows the placement of such a catheter into the moving tissues (muscle) or peripheral organs for studies in freely moving animals. The technical difficulties of microdialysis experiments impose requirements for precise liquid delivery, minimized dead volumes, and the capability of handling small sample volumes.

A representative arrangement for microdialysis sampling in anesthetized rats is shown in Figure 6.3. The microinjection pump should include features such as a pulse-free electronically controlled dc motor and should be precali-

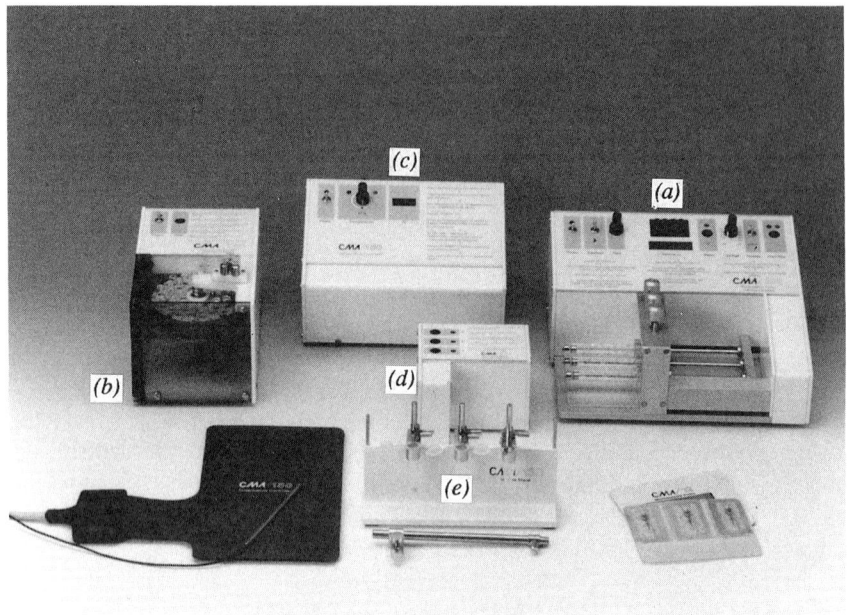

Figure 6.3 Basic setup for microdialysis experiments on small laboratory animals. The system can be used for simultaneous collection from up to three microdialysis probes. The main components are (A) CMA/100 Microinjection Pump, (B) CMA/140 Microfraction Collector, (C) temperature controller, (D) liquid switch for switching between different perfusion fluids, and (E) in vitro stand for storage of microdialysis probes.

brated for precision glass syringes to ensure smooth fluid flows in the range of from 1 nL/min to 1 mL/min. In the arrangement shown, three syringes can be used simultaneously, either for three separate microdialysis probes or for one probe (when switching between different perfusion media is required for stimulations, drug delivery, etc.). Switching between lines must not be allowed to introduce air bubbles into the system.

When not in use, microdialysis probes should be stored wet and protected from mechanical damage. Also, testing the probes by in vitro recovery is easily accomplished if the probes are fixed in an appropriate stand. Fractions can be collected manually, but it is more convenient to use fraction collectors capable of collecting fractions as low as 1 μL.

Several commercial constructions are available for triple (CMA 140, Fig. 6.3B) or dual (CMA 142) probe collecting and with refrigeration (CMA 170) for collecting easily degradable compounds. An alternative is direct coupling of the microdialysis probe's outlet to the high pressure injection value (CMA 160) of an HPLC system (Johnson and Justice, 1983). This enables on-line analysis of samples, with the only delay due to the analysis time of a particular separation method. A rapidly growing interest in clinical applications of microdialysis led to the construction of a special portable microdialysis pump and a flexible microdialysis catheter for use in adipose tissue and noncontracting muscle in humans (Fig. 6.4). Samples are collected manually by using a specially designed microvial.

6.2.2 HPLC Analysis

Essentially, any liquid chromatographic separation mode can be used in conjunction with microdialysis, though the most common are reversed-phase (C_{18} and C_8) materials. Separated compounds are detected most often using electrochemical, fluorescence, and UV detectors. However, some precautions related to the specific characteristics of microdialysis samples should be kept in mind. First, an average total sample volume is usually about 10 μL, and the concentrations of analytes are very low, often close to the limits of detection of most instruments. Therefore, miniaturization of chromatographic systems will be important for continued advances in microdialysis. Microdialysis samples are free from large molecules, lipids, and other anomalies and can therefore be injected directly onto the chromatographic column. Microdialysis sampling is rapid—some 50 samples per day can easily be generated. It is important to optimize the separations for the fastest possible analysis time, often sacrificing the complete resolution of uninterested peaks. The use of automated

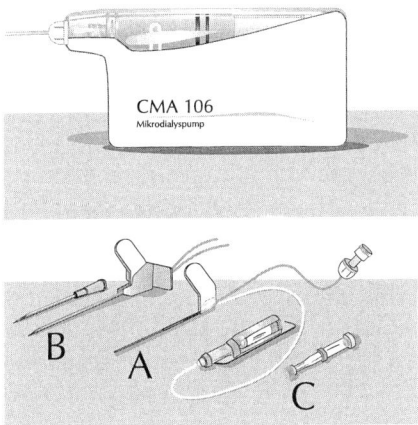

Figure 6.4 The starting setup for the clinical use of microdialysis in human adipose tissue consists of a portable, battery-driven syringe micropump (CMA 106) and a flexible microdialysis catheter (CMA 60). The catheter (*A*) is introduced into the tissue by means of a special introducer (*B*). The needle has a longitudinal channel opening on its side, which allows its removal from the tissue and disconnection from the microdialysis catheter. The sample is collected into a special microvial (*C*).

systems including autosamplers and data acquisition is strongly recommended. A refrigerated autosampler (CMA 200), injecting volumes from 0.1 μL with almost no sample loss, has been built specially for microdialysis samples.

6.2.3 Performing a Microdialysis Experiment on a Rat

Besides the basic apparatus for microdialysis perfusions, fraction collection, and HPLC analysis, several additional instruments and devices are needed, depending on where the microdialysis probe is to be implanted. The most complicated instrumental setup is probably that required for brain dialysis. A stereotaxtic instrument and a stereomicroscope are necessary for precise positioning of microdialysis cannulae into various brain structures. Inhalation anesthesia is preferable and more convenient than injections. However, this type of anesthesia calls for additional equipment, such as air lines, valves, and mixing chamber for halothane or other anesthetic gases, as well as good ventilation of the operation theater.

Once a laboratory animal such as a rat has been anesthetized, it can be fixed into the stereotaxic frame following procedures described in brain stereotaxic atlases (Paxinos and Watson, 1982). The points for fixation are the interaural line and the bottom edge of the upper jaw behind the first incisors. The skull must be exposed to locate the brain coordinates corresponding to the atlas nomenclature. Typically, the bregma (intersection of coronal and sagittal sutures) is chosen as zero "x,y" coordinates. A fine trephine drill should be used to make a hole of some 1 to 2 mm, and all the bone fragments must be carefully removed to avoid damaging the membrane during probe implantation. The surface of the dura mater is usually taken as zero for the "z" coordinate. After the zero coordinates have been marked, a sharp needle is used to cut the dura mater, and the microdialysis probe can be inserted. Some 1 to 3 hours is recommended for stabilization of the outflow of the measured substances. Then three to five basal level fractions are collected before particular stimuli (drugs, ischemia, etc.) are introduced.

There is also the possibility of performing experiments on conscious, free-moving animals. In this case, only the guide cannula is implanted into the anaesthetized animal. Following postoperative recovery (1–7 days), the probe can be inserted directly into the brain through the guide cannula without any need for anesthesia. Chronic microdialysis experiments can be run for up to 3 to 4 days after implantation (Osborne et al., 1991) and still give physiologically relevant data. After 4 to 7 days, a number of tissue reactions such as gliosis and accumulation of polymorphonuclear leukocytes (Benveniste and Diemer, 1987) cause severe alterations in both metabolism and diffusion rate of compounds in the vicinity of the microdialysis probe (Westerink and Tuinte, 1986).

6.2.4 Performing a Microdialysis Experiment on a Human

Although as yet there is no generally approved probe (except in Sweden) for clinicial microdialysis in humans, a number of papers have reported studies on human brain (Hillered et al., 1990; Hillered and Persson, 1991; During, 1991; During et al., 1994), heart (Kannergren et al., 1994), muscle (Rosdahl et al., 1993), skin (Petersen et al., 1992, 1994; Anderson et al., 1994), and adipose tissue (Jansson et al., 1988; Bolinder et al., 1989; Hagström et al., 1990; Meyerhoff et al., 1994). One of the most attractive potential applications is monitoring the subcutaneous glucose levels of diabetic patients (Bolinder et al., 1992, 1993; Pfeiffer, 1994; Sternberg et al., 1994) or newborn children (Korf et al., 1993; DeBoer et al., 1994).

For this application, a sterile microdialysis catheter is implanted percutaneously into the subcutaneous adipose tissue in the periumbilical region using a special stainless steel guide cannula (Rosdahl et al., 1993). The membrane length of 2 to 3 cm and flow rates of 0.1 to 0.5 μL/min should guarantee sufficiently long dialysis time to achieve 100% recovery. Another approach is to use a flat dialysis probe for transcutaneous applications (DeBoer et al., 1993; Korf et al., 1993). Here the skin of newborn babies is first partially removed by stripping with medical tape. Then a microdialysis probe is placed directly onto the exposed skin, usually on the abdomen lateral to the umbilicus (DeBoer et al., 1994).

6.3 APPLICATIONS OF MICRODIALYSIS/HPLC IN ENZYMATIC ANALYSIS

6.3.1 Body Fluids Sampled by Microdialysis

The main target for microdialysis implantations is the ECF of various organs, predominantly the brain. The main objective of these studies is to find chemical correlations, as represented by measured neurotransmitter levels, to pharmaceutical, behavioral, pathological, or other stimuli. However, because of its small size a microdialysis probe allows sampling of fluids from other locations without first removing the fluid. Body fluids other than ECF to which the microdialysis technique has been applied are considered in Sections 6.3.1.1 to 6.3.1.6.

6.3.1.1 Blood Microdialysis in blood can be performed both acutely, using rigid probes and guide cannulae, and chronically by implanting flexible probes with a removable guide–introducer needle. The most common site of implantation is a jugular vein (Hurd et al., 1988; Ekström et al., 1994) or the vena cava (Dubey et al., 1989). The method is attractive for the studies of protein binding (Saisho and Umeda, 1991; Sjöberg et al., 1992; Nakashima et al., 1994), and pharmacokinetic and pharmacodynamic tests (Ståhle, 1991; Ståhle et al., 1993; Wong et al., 1993, Delange et al., 1994). Some advantages are the sampling of free unbound fractions and the possibility of long-term experiments without removing blood or affecting the animal's physiology. Microdialysis is especially useful for bioavailability estimations, since a direct kinetic profile is produced (Wong et al., 1992; Delange et al., 1995).

6.3.1.2 Cerebrospinal Fluid (CSF) Microdialysis allows sampling of CSF (Golden et al., 1993; Malhotra et al., 1994; Togashi et al., 1994) without removal of the fluid and without altering intracranial pressure. CSF can be sampled from the lateral ventricles (Becker et al., 1988) or from the lumbal spinal cord area (Marsala et al., 1995), but most often CFS from the cisterna magna is tested (Knuckey et al., 1991; Matos et al., 1992). To date all studies have been performed in the rat.

6.3.1.3 Vitreous Humor A number of microdialysis studies have been carried on the vitreous fluid of a rabbit eye (Gunnarson et al., 1987; Waga et al., 1991) to study the penetration into this compartment of drugs, including antibiotics (Ben-Nun et al., 1988b) and cytostatics and corticosteroids (Waga and Ehinger, 1995). Polyamide membranes were found to be more suitable than polycarbonate for lipophilic compounds because of their lower carryover and memory effect for these drugs (Waga and Ehinger, 1995). The pathology of the eye was studied using models of experimentally induced ischemia of the retina (Ben-Nun et al., 1988a; Louzadajunior et al., 1992) and laser coagulation (Stempels et al., 1994).

6.3.1.4 Synovial Fluid The compartment of a joint space in the rat knee is filled with the synovial fluid. It has been shown that a microdialysis probe with the 0.5 mm membrane can be used for kinetics and distribution studies of drugs such as bis(5-amidino-2-benzimidazolyl)methane in experimentally induced arthritis (St. Claire and Brouwer, 1992).

6.3.1.5 Perilymph The cochlea of the inner ear contains two compartments: the scala tympani, filled with perilymphatic fluid, and the scala media, filled with endolymphatic fluid. Microdialysis has been performed in the scala tympani of a guinea pig (Laurell et al., 1994). The aim of this study was to measure cisplatin accumulation in this compartment, which has estimated volume of about 15 μL.

6.3.1.6 Bile Microdialysis in the bile duct was performed using a specially constructed shunt probe (Scott and Lunte, 1993). The probe resembles in principle a classical dialysis cartridge with only a single dialysis tube inserted in the outer tubing. The ends of the dialysis tube are inserted into the vessel or bile duct, while the outer tubing is perfused with a Ringer's solution containing sodium taurocholate. Dialysates are collected, and various compounds such as drugs and their metabolites can be determined by HPLC.

6.3.2 Typical Analytes: Small Molecules

Microdialysis is a technique for recovering low molecular weight compounds. The size of the compound recovered is determined by the cutoff parameter of the membranes used for constructing the microdialysis probes. The polycarbonate membrane of 0.5 mm i.d. used for commercial probes manufactured by CMA/Mikrodialysis has a cutoff of 20,000 Da. A 2 mm probe perfused at 2 μL/min with Ringer's solution at ambient temperature will recover some 1 to 2% of a 5000 Da peptide from a quiescent medium. However, more than 10 times higher recovery is achieved for smaller molecules (acids usually give slightly higher recovery than bases) under the same conditions. When a number of different solutes are measured, a typical function of log (in vitro recovery) versus molecular weight can be achieved (Fig. 6.5). Increasing the membrane's cutoff would allow penetration of larger molecules and higher recoveries. For example, using a 40 kDa polyacrylonitrile membrane will improve recoveries for most of the neuropeptides (Maidment et al., 1989). A 100 kDa membrane allows sampling of interleukins, and a recently reported 0.1 μm membrane can even recover enzymes and proteins (Kannergren et al., 1994, 1995; Hamberger, personal communication).

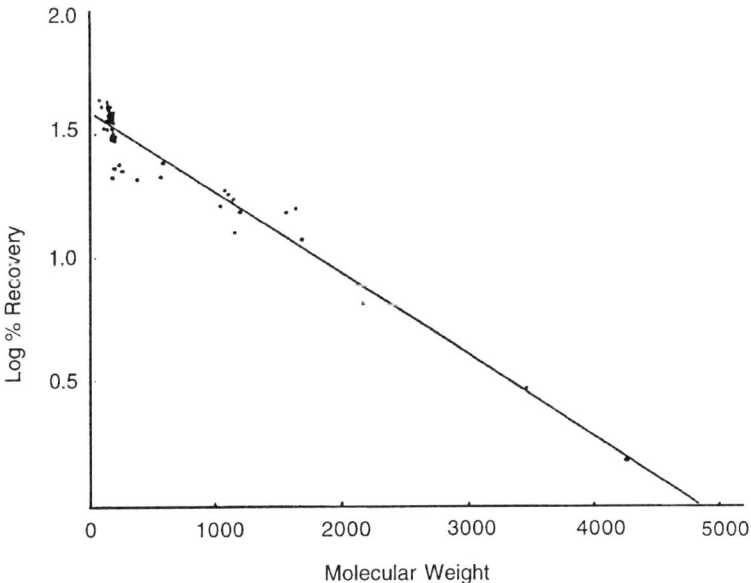

Figure 6.5 The function of in vitro recovery versus molecular weights of 42 tested substances. The CMA 10 microdialysis probe with 5 mm polycarbonate membrane (cut off 20,000 Da) was perfused with Ringer's solution at 2 μL/min.

6.3.3 Estimating Enzymatic Activities

Though it was realized very early that microdialysis can serve as an excellent tool for estimating in vivo enzymatic activities (Sharp et al., 1986), this application remained relatively unexplored until recently (Westerink et al., 1990). Microdialysis offers several advantages over existing in vitro techniques because it affords the opportunity for continuous sampling from organs other than blood and because of its high spatial resolution.

6.3.3.1 Studying Enzymatic Activities in Almost Intact Environments/ Cellular Compartments Westerink et al. (1990) were the first to describe the microdialysis methodology for measuring tyrosine hydroxylase (TH) activity in the rat brain. TH is the rate-limiting enzyme in the synthesis of catecholamines in both the peripheral and central nervous systems. In neurones, TH hydroxylates L-tyrosine to 3,4-dihydroxyphenylalanine (Dopa), which is further transformed to dopamine by aromatic amino acid decarboxylase (AAAD). However, it is impossible to recover measurable extracellular levels of Dopa without prior inhibition of AAAD. Thus, the first step in microdialysis is to start perfusing the dialysis probe with an inhibitor of AAAD such as 3-hydroxybenzylhydrazine (NSD 1015), to achieve detectable levels of Dopa when HPLC is used in conjunction with electrochemical detection (Fig. 6.6). Stabilization of Dopa occurs within 1 to 2 hours, depending on the concentration of NSD 1015 in the perfusion medium (Hashiguti et al., 1993). During this period a corresponding reduction and stabilization of metabolism can be observed as demonstrated for the metabolite 3,4-dihydroxyphenylacetic acid (Dopac). α-Methyl-p-tyrosine, a potent inhibitor of TH activity, caused a rapid decline of Dopa to undetectable levels within 90 minutes of administration (Fig. 6.7).

6.3.3.2 Measuring the Entire Time Course in a Single Experiment To date, microdialysis has been applied mostly to measurements of the rate of biosynthesis (TH activity) in dopaminergic and noradrenergic neurones and to the estimation of tryptophan hydroxylase activity in serotoninergic terminals (Brodkin et al., 1993, Hashiguti et al., 1993). Except for original work of Sharp et al. (1986), only few studies were directed toward evaluation of in vivo degradation of these neurotransmitters by measuring activity of monoamine oxidase (Finberg et al., 1993, Smith and Justice, 1994). In vivo determination of enzymatic activities by microdialysis was proven to be not only a fast and efficient research tool but also more economical and ethically better justified than in vitro preparations. It is possible to obtain both basal (control) levels and the whole time course after stimulation of a single animal. Specific inhibition of the biosynthesis or metabolism of some other neurotransmitter systems (cholinergic, glutamatergic) could be used to increase acetylcholine (Damsma et al., 1987), or to better characterize glutamate (Bakkelund et al., 1993) pools.

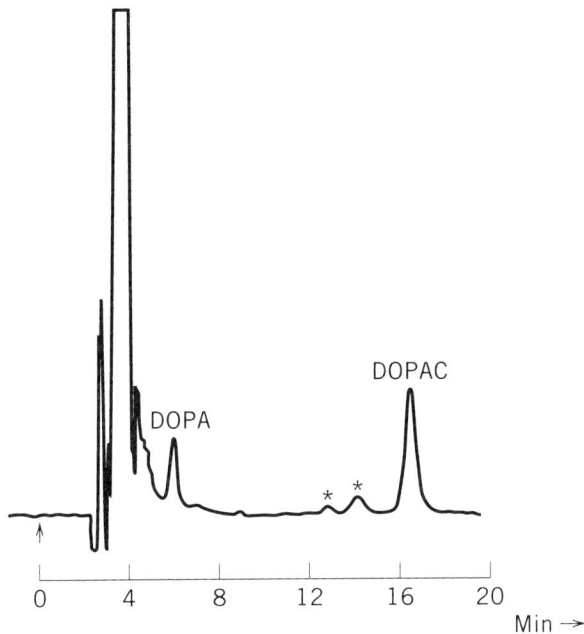

Figure 6.6 Typical chromatogram of a microdialysate sample collected during a 15-minute perfusion. The microdialysis probe was perfused with Ringer's solution containing 0.1 mmol/L NSD 1015. Dopa concentration increases gradually from nondetectable levels and stabilizes after 90 to 120 minutes. The calculated output was 0.45 pmol/min for L-DOPA and 0.7 pmol/min for DOPAC. (From Westerink et al., 1990, with permission).

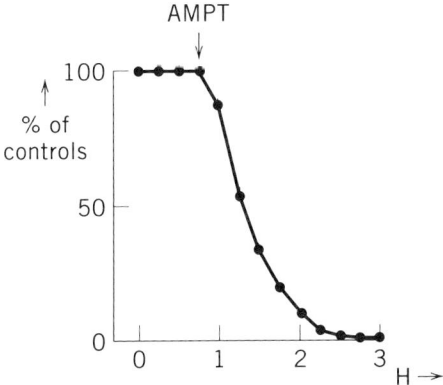

Figure 6.7 Effect of α-methyl-p-tyrosine (AMPT) (250 mg/kg intraperitoneally) on the dialysate levels of L-Dopa. The basal value of L-Dopa was 0.4 ± 0.03 pmol/min (mean $\pm$ SEM, $n = 3$). (From Westerink et al., 1990, with permission).

6.3.3.3 Investigating the Effects of Cofactors and/or Drugs, in Small, Localized Tissue Structures
Once the methodology of measuring an enzyme precursor, intermediate, or a metabolite has been established, it is possible to start testing a number of putative chemical candidates involved in the process of biosynthesis or degradation of neurotransmitters. In the case of TH it was shown that as well as inhibitors of TH activity such as α-methyl-p-tyrosine or the more stable difluoromethyl-Dopa (Robert et al., 1993), the dopamine agonist apomorphine also causes a temporary reduction of Dopa (Westerink et al., 1990). Similar phenomena were observed for noradrenergic terminals in rat hippocampus after administration of clonidine, an α_2-adrenoreceptor agonist (Nisenbaum and Abercrombie, 1992). On the other hand, the dopamine antagonist haloperidol, γ-butyrolactone-blocking dopamine firing, and tetrodotoxin-blocking, voltage-dependent Na channels cause an increase in TH activity in the rat caudatus (Westerink et al., 1990; Carboni et al., 1992). Other typical and atypical neuroleptics, all increasing dopamine biosynthesis rate, were also studied (Guinetdinov et al., 1994). Preloading with phenylalanine but not tyrosine precursors stimulated Dopa formation, and infusion of tryptophan increased 5-hydroxytryptophan accumulation (Westerink and De Vries, 1991). Genetically manipulated mice carrying the human TH gene in the striatum showed comparable activities to the nontransgenic mice (Nakahara et al., 1993).

6.3.3.4 Testing Drugs That Do Not Penetrate the Blood–Brain Barrier
Microdialysis allows direct intracerebral administration of the drugs that normally do not cross the blood–brain barrier (BBB) or are metabolized in the circulation. It was demonstrated earlier (Ozaki et al., 1988; Westerink et al., 1990) that MPP^+, a toxic metabolite of 1-methyl-4-phenyl-1,2,3,6-tetrahydropyridine (MPTP), has a biphasic effect on TH activity when infused via the microdialysis probe for 20 minutes. Since, however, this drug does not penetrate the BBB, it was difficult to investigate its effects in vivo. Microdialysis overcomes this problem and has enabled the testing of a range of other modulators that previously could be studied only on in vitro preparations.

6.3.3.5 Estimating Enzymatic Activities Under Various Physiological/Pathological Stimuli
With continuous microdialysis sampling in conscious animals, it is possible to measure and correlate biochemical parameters, such as enzymatic activities, to behavioral or physiological stimulation. In several studies (Nisenbaum et al., 1991; Nisenbaum and Abercrombie, 1992), increased TH activity and noradrenaline release were measured in rat brain structures and in adrenal glands after exposure to chronic or acute stress. For example, animals chronically stressed by cold showed elevated and more prolonged levels of TH activity (measured as extracellular Dopa accumulation) after a repeated acute tail shock (Nisenbaum and Abercrombie, 1992). Similarly, increased tyrosine hydroxylation was measured in newborn piglets during hypocapnic hypoxia (Tammela et al., 1993) and in the ewe after inhibition of

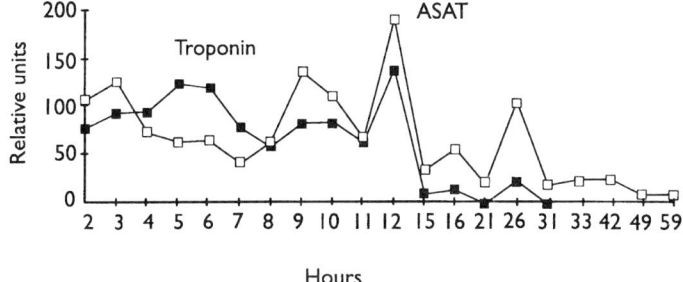

Figure 6.8 Typical profiles of ASAT and troponin relative levels sampled by microdialysis. The microdialysis probe was implanted into the patient's myocardium at time 0 after heart surgery, and fractions were collected during the next 59 hours.

luteinizing hormone secretion by estradiol (Gayrard et al., 1994). Physiological stimuli such as exercise increased dopamine turnover and TH activity in rat striatum (Hattori et al., 1994).

6.3.3.6 Microdialysis Sampling of Enzymes Using the microdialysis probe provided with a large pore-size membrane allows direct sampling of macromolecules from the ECF. Normally, such sampling would be accompanied by problems in the analysis of neurotransmitters and metabolites because the co-recovered enzymes can cause degradation of many small molecules during the transportation at the outlet of the probe or can clog the chromatographic column. However, there is increased interest in using microdialysis for in vivo sampling of large molecules such as proteins (Nyström et al., 1986; Nakamura et al., 1990), interleukins (Woodroofe et al., 1991; Yan et al., 1992), hormones (Jarry et al., 1990; Khandawood et al., 1994), and enzymes (Kennergren et al., 1994, 1995). In the latter case, a special microdialysis probe (Hamberger et al., 1991) with a membrane of 0.1 m pore size was used to monitor aspartate-aminotransferase (ASAT) and troponin during and after heart surgery in humans. Figure 6.8 illustrates typical profiles of ASAT and troponin sampled for more than 2 days in the myocardium of a patient with nonischemic heart disease.

6.4 CONCLUSIONS

The microdialysis technique can be used for continuous in vivo sampling of any body fluid. The main application field of microdialysis is for monitoring the intercellular chemical signaling and transport of molecules in the extracellular fluid. The principle of sampling, based entirely on dialysis (diffusion), ensures that no fluid is removed from or delivered into the sampling area. This principle, together with a fine diameter of the microdialysis cannula, makes it possible to study compartments having dimensions of only few microliters. Relative enzymatic activities can be estimated simply by measuring accumulation or consumption of substrates or metabolites in the particular body fluid.

REFERENCES

Anderson C, Andersson T, Wardell K (1994) Changes in skin circulation after insertion of a microdialysis probe visualized by laser doppler perfusion imaging. *J Invest Dermatol* **102:**807–811.

Bakkelund AH, Fonnum F, Paulsen RE (1993) Evidence using in vivo microdialysis that aminotransferase activities are important in the regulation of the pools of transmitter amino acids. *Neurochem Res* **18:**411–415.

Becker JB, Adams F, Robinson TE (1988) Intraventricular microdialysis: A new method for determining monoamine metabolite concentrations in the cerebrospinal fluid of freely moving rats. *J Neurosci Methods* **24:**259–269.

Ben-Nun J, Alder VA, Cringle SJ, Constable IJ (1988a) A new method for oxygen supply to acute ischemic retina. *Invest Ophthalmol Vis Sci* **29:**298–304.

Ben-Nun J, Cooper RL, Cringle SJ, Constable IJ (1988b) Ocular dialysis. A new technique for in vivo intraocular pharmacokinetic measurements. *Arch Ophthalmol* **106:**254–259.

Benveniste H (1989) Brain microdialysis. *J Neurochem* **52:**1667–1679.

Benveniste H, Diemer NH (1987) Cellular reactions to implantation of a microdialysis tube in the rat hippocampus. *Acta Neuropathol (Berlin)* **74:**234–238.

Benveniste H, Hüttemeier PC (1990) Microdialysis—Theory and application. *Prog Neurobiol* **35:**195–215.

Bjelke B, Strömberg I, O'Connor WT, Andbjer B, Agnati LF, Fuxe K (1994) Evidence for volume transmission in the dopamine denervated neostriatum of the rat after a unilateral nigral 6-OHDA microinjection. Studies with systemic D-amphetamine treatment. *Brain Res* **662:**11–24.

Bolinder J, Hagström E, Ungerstedt U, Arner P (1989) Microdialysis of subcutaneous adipose tissue in vivo for continuous glucose monitoring in man. *Scand J Clin Lab Invest* **49:**465–474.

Bolinder J, Ungerstedt U, Arner P (1992) Microdialysis measurement of the absolute glucose concentration in subcutaneous adipose tissue allowing glucose monitoring in diabetic patients. *Diabetologia* **35:**1177–1180.

Bolinder J, Ungerstedt U, Arner P (1993) Long-term continuous glucose monitoring with microdialysis in ambulatory insulin-dependent diabetic patients. *Lancet* **342:**1080–1085.

Brodkin J, Malyala A, Nash JF (1993) Effect of acute monoamine depletion on 3,4-methylenedioxymethamphetamine-induced neurotoxicity. *Pharmacol Biochem Behav* **45:**647–653.

Carboni E, Tanda G, Di Chiara G (1992) Extracellular striatal concentrations of endogenous 3,4-dihydroxyphenylalanine in the absence of a decarboxylase inhibitor—A dynamic index of dopamine synthesis in vivo. *J Neurochem* **59:**2230–2236.

Damsma G, Westerink BH, de Vries JB, Van den Berg CJ, Horn AS (1987) Measurement of acetylcholine release in freely moving rats by means of automated intracerebral dialysis. *J Neurochem* **48:**1523–1528.

DeBoer J, Plijtergroendijk H, Korf J (1993) Microdialysis probe for transcutaneous monitoring of ethanol and glucose in humans. *J Appl Physiol* **75:**2825–2830.

DeBoer J, Baarsma R, Okken A, Plijtergroendijk H, Korf J (1994) Application of transcutaneous microdialysis and continuous flow analysis for on-line glucose monitoring in newborn infants. *J Lab Clin Med* **124:**210–217.

Delange ECM, Danhof M, Deboer AG, Breimer DD (1994) Critical factors of intracerebral microdialysis as a technique to determine the pharmacokinetics of drugs in rat brain. *Brain Res* **666:**1–8.

Delange ECM, Hesselink MB, Danhof M, Deboer AG, Breimer DD (1995) The use of intracerebral microdialysis to determine changes in blood–brain barrier transport characteristics. *Pharm Res* **12:**129–133.

Delgado JM, DeFeudis FV, Roth RH, Ryugo DK, Mitruka BM (1972) Dialytrode for long-term intracerebral perfusion in awake monkeys. *Arch Int Pharmacodyn Ther* **198:**9–21.

Dubey RK, McAllister CB, Inoue M, Wilkinson GR (1989) Plasma binding and transport of diazepam across the blood–brain barrier. No evidence for in vivo enhanced dissociation. *J Clin Invest* **84:**1155–1159.

During MJ (1991) In vivo neurochemistry of the conscious human brain: Intrahippocampal microdialysis in epilepsy. In *Microdialysis in the neurosciences. Techniques in the Behavioral and Neural Sciences,* Vol. 7, Elsevier Science Publishers, Amsterdam.

During MJ, Fried I, Leone P, Katz A, Spencer DD (1994) Direct measurement of extracellular lactate in the human hippocampus during spontaneous seizures. *J. Neurochem* **62:**2356–2361.

Ekblom M, Gårdmark M, Hammarlund-Udenaes M (1992) Estimation of unbound concentrations of morphine from microdialysate concentrations by use of nonlinear regression analysis in vivo and in vitro during steady state conditions. *Life Sci* **51:**449–460.

Ekström PO, Andersen A, Warren DJ, Giercksky KE, Slordal L (1994) Evaluation of methotrexate tissue exposure by in situ microdialysis in a rat model. *Cancer Chemother Pharmacol* **34:**297–301.

Finberg JPM, Pacak K, Kopin IJ, Goldstein DS (1993) Chronic inhibition of monoamine oxidase type-A increases noradrenaline release in rat frontal cortex. *Naunyn-Schmied Arch Pharmacol* **347:**500–505.

Gardner EL, Chen J, Paredes W (1993) Overview of chemical sampling techniques. *J Neurosci Methods* **48:**173–197.

Gayrard V, Malpaux B, Tillet Y, Thiery JC (1994) Estradiol increases tyrosine hydroxylase activity of the A15 nucleus dopaminergic neurons during long days in the ewe. *Biol Reprod* **50:**1168–1177.

Golden PL, Brouwer KR, Pollack GM (1993) Assessment of valproic acid serum cerebrospinal fluid transport by microdialysis. *Phar Res* **10:**1765–1771.

Guinetdinov RR, Bogdanov MB, Kudrin VS, Rayevsky KS (1994) Remoxipride and raclopride differ from metoclopramide by their effects of striatal dopamine release and biosynthesis in rats. *Neuropharmacology* **33:**215–219.

Gunnarson G, Jakobsson AK, Hamberger A, Sjöstrand J (1987) Free amino acids in the pre-retinal vitreous space. Effect of high potassium and nipecotic acid. *Exp Eye Res* **44:**235–244.

Hagström E, Arner P, Ungerstedt U, Bolinder J (1990) Subcutaneous adipose tissue: A source of lactate production after glucose ingestion in humans. *Am J Physiol* **258:**E888–893.

Hamberger A, Jacobson I, Molin S-O, Nyström B, Sandberg M, Ungerstedt U (1982) Metabolic and transmitter compartments for glutamate. In *Neurotransmitter Interaction and Compartmentation,* HF Bradford, Ed., pp. 359–376. Plenum Publishing, New York.

Hamberger A, Berthold C-H, Jacobson I, Karlsson B, Lehmann A, Nyström B, Sandberg M (1985) In vivo brain dialysis of extracellular nontransmitter and putative transmitter amino acids. In *In Vivo Perfusion and Release of Neuroactive Substances,* A Bayon and R Drucker-Collins, Eds., pp. 119–139. Academic Press, New York.

Hamberger A, Jacobson I, Larsson S, Lönnroth P, Nyström B, Sandberg M (1991) Microdialysis technique for studying brain amino acids in the extracellular fluid: Basic and clinical studies. In *Microdialysis in the Neurosciences. Techniques in the Behavioral and Neural Sciences,* Vol. 7, TE Robinson and JB Justice Jr, Eds., pp. 407–423. Elsevier Science Publishers, Amsterdam.

Hashiguti H, Nakahara D, Maruyama W, Naoi M, Ikeda T (1993) Simultaneous determination of in vivo hydroxylation of tyrosine and tryptophan in rat striatum by microdialysis–HPLC—Relationship between dopamine and serotonin biosynthesis. *J Neural Transm Gen Sect* **93:**213–223.

Hattori S, Naoi M, Nishino H (1994) Striatal dopamine turnover during treadmill running in the rat: Relation to the speed of running. *Brain Res Bull* **35:**41–49.

Hillered L, Persson L, Pontén U, Ungerstedt U (1990) Neurometabolic monitoring of the ischaemic human brain using microdialysis. *Acta Neurochir (Wien)* **102:**91–97.

Hillered L, Persson L (1991) Microdialysis for metabolic monitoring in cerebral ischemia and trauma: Experimental and clinical studies. In *Microdialysis in the Neurosciences. Techniques in the Behavioral and Neural Sceinces,* Vol. 7, TE Robinson and JB Justice Jr, Eds., pp. 389–405. Elsevier Science Publishers, Amsterdam.

Hurd YL, Kehr J, Ungerstedt U (1988) In vivo microdialysis as a technique to monitor drug transport: Correlation of extracellular cocaine levels and dopamine overflow in the rat brain. *J Neurochem* **51:**1314–1316.

Jacobson I, Sandberg M, Hamberger A (1985) Mass transfer in brain dialysis devices—A new method for the estimation of extracellular amino acids concentration. *J Neurosci Methods* **15:**263–268.

Jansson PA, Fowelin J, Smith U, Lönnroth P (1988) Characterization by microdialysis of intracellular glucose level in subcutaneous tissue in humans. *Am J Physiol* **255:**E218–E220.

Jarry H, Einspanier A, Kanngiesser L, Dietrich M, Pitzel L, Holtz W, Wuttke W (1990) Release and effects of oxytocin on estradiol and progesterone secretion in porcine corpora lutea as measured by an in vivo microdialysis system. *Endocrinology* **126:**2350–2358.

Johnson RD, Justice JB (1983) Model studies for brain dialysis. *Brain Res Bull* **10:**567–571.

Justice JB (1993) Quantitative microdialysis of neurotransmitters. *J Neurosci Methods* **48**:263–276.

Kennergren C, Olsson GW, Lönnroth P, Mantovani V, Berggren H, Nyström B, Nyström U, Hamberger A (1994) Microdialysis technique for continuous in vivo surveillence of intramyocardial ischemia. *Scand J Thorac Cardiovasc Surg*, submitted.

Kehr J (1992) *Microdialysis Bibliography* (1–900). CMA/Mikrodialysis AB, Stockholm.

Kehr J (1993a) *Microdialysis Bibliography, Supplement I* (901–1500). CMA/Mikrodialysis AB, Stockholm.

Kehr J (1993b). A survey on quantitative microdialysis—Theoretical models and practical implications. *J Neurosci Methods* **48**:251–261.

Kehr J (1994) *Microdialysis Bibliography, Supplement II* (1501–2200). CMA/Mikrodialysis AB, Stockholm.

Khandawood FS, Gargiulo AR, Dawood MY (1994) Baboon corpus luteum: Autonomous pulsatile progesterone secretion and evidence for an intraluteal oscillator demonstrated by in vitro microretrodialysis. *J Clin Endocrinol Metab* **79**:1790–1796.

Knuckey NW, Fowler AG, Johanson CE, Nashold JR, Epstein MH (1991) Cisterna magna microdialysis of ^{22}Na to evaluate ion transport and cerebrospinal fluid dynamics. *J Neurosurg* **74**:965–971.

Korf J, DeBoer J, Baarsma R, Venema K, Okken A (1993) Monitoring of glucose and lactate using microdialysis: Applications in neonates and rat brain. *Dev Neurosci* **15**:240–246.

Laurell G, Andersson A, Engström B, Ehrson H (1994) Cisplatin in perilymph and blood measured with microdialysis. Case report. CMA/Mikrodialysis AB, Stockholm.

Louilot A, Durkin T, Spampinato U, Cador M (1994) Monitoring molecules in neuroscience. *Proceedings of the Sixth International Conference on In Vivo Methods*, Seignosse, France.

Louzadajunior P, Dias JJ, Santos WF, Lachat JJ, Bradford HF, Coutinhonetto J (1992) Glutamate release in experimental ischaemia of the retina—An approach using microdialysis. *J Neurochem* **59**:358–363.

Maidment NT, Brumbaugh DR, Rudolph VD, Erdelyi E, Evans CJ (1989) Microdialysis of extracellular endogenous opioid peptides from rat brain in vivo. *Neuroscience* **33**:549–557.

Malhotra BK, Lemaire M, Sawchuk RJ (1994) Investigation of the distribution of EAB 515 to cortical ECF and CSF in freely moving rats utilizing microdialysis. *Pharm Res* **11**:1223–1232.

Marsala M, Malmberg AB, Yaksh TL (1995) A chronic spinal dialysis catheter for use in the unanesthetized rat: Methodology and application. *J Neurosci Methods* **62**:43–53.

Matos FF, Rollema H, Basbaum AI (1992) Simultaneous measurement of extracellular morphine and serotonin in brain tissue and CSF by microdialysis in awake rats. *J Neurochem* **58**:1773–1781.

Meyerhoff C, Mennel FJ, Bischof F, Sternberg F, Pfeiffer EF (1994) Combination of microdialysis and glucose sensor for continuous on-line measurement of the subcutaneous glucose concentration: Theory and practical application. *Hormone Metab Res* **26**:538–543.

Nakahara D, Hashiguti H, Kaneda N, Sasaoka T, Nagatsu T (1993) Normalization of tyrosine hydroxylase activity in vivo in the striatum of transgenic mice carrying human tyrosine hydroxylase gene—A microdialysis study. *Neurosci Lett* **158**:44–46.

Nakamura M, Itano T, Yamaguchi F, Mizobuchi M, Tokuda M, Matsui H, Etoh S, Hosokawa K, Ohmoto T, Hatase O (1990) In vivo analysis of extracellular proteins in rat brains with a newly developed intracerebral microdialysis probe. *Acta Med Okayama* **44**:1–8.

Nakashima M, Takeuchi N, Hamada M, Matsuyama K, Ichikawa M, Goto S (1994) In vivo microdialysis for pharmacokinetic investigations: A plasma protein binding study of valproate in rabbits. *Biol Pharm Bull* **17**:1630–1634.

Nisenbaum LK, Abercrombie ED (1992) Enhanced tyrosine hydroxylation in hippocampus of chronically stressed rats upon exposure to a novel stressor. *J Neurochem* **58**:276–281.

Nisenbaum LK, Zigmond MJ, Sved AF, Abercrombie ED (1991) Prior exposure to chronic stress results in enhanced synthesis and release of hippocampal norepinephrine in response to a novel stressor. *J Neurosci* **11**:1478–1484.

Nyström B, Hamberger A, Karlsson J-O (1986) Changes of extracellular proteins in hippocampus during depolarization. *Neurochem Int* **9**:55–59.

Osborne PG, O'Connor WT, Kehr J, Ungerstedt U (1991) In vivo characterisation of extracellular dopamine, GABA and acetylcholine from the dorsolateral striatum of awake freely moving rats by chronic microdialysis. *J Neurosci Methods* **37**:93–102.

Ozaki N, Nakahara D, Mogi M, Harada M, Kiuchi K, Kaneda N, Miura Y, Kasahara Y, Nagatsu T (1988) Inactivation of tyrosine hydroxylase in rat striatum by 1-methyl-4-phenylpyridinium ion (MPP$^+$). *Neurosci Lett* **85**:228–232.

Paxinos G, Watson C (1982) *The Rat Brain in Stereotaxic Coordinates*. Academic Press, New York.

Petersen LJ, Kristensen JK, Bulow J (1992) Microdialysis of the interstitial water space in human skin in vivo—Quantitative measurement of cutaneous glucose concentrations. *J Invest Dermatol* **99**:357–360.

Petersen LJ, Poulsen LK, Sondergaard J, Skov PS (1994) The use of cutaneous microdialysis to measure substance P-induced histamine release in intact human skin in vivo. *J Allergy Clin Immunol* **94**:773–783.

Pfeiffer EF (1994) the "Ulm Zucker Uhr System" and its consequences. *Hormone Metab Res* **26**:510–514.

Robert F, Lambassenas L, Ortemann C, Pujol JF, Renaud B (1993) Microdialysis monitoring of 3,4-dihydroxyphenylalanine accumulation after decarboxylase inhibition—A means of estimate in vivo changes in tyrosine hydroxylase activity of the rat locus ceruleus. *J Neurochem* **60**:721–729.

Robinson TE, Justice JB Jr, Eds. (1991) *Microdialysis in the Neurosciences. Techniques in the Behavioral and Neural Sciences* Vol. 7. Elsevier Science Publishers, Amsterdam.

Rollema H, Westerink BHC, Drijfhout WJ (1991) Monitoring molecules in neuroscience. *Proceedings of the 5th International Conference on In Vivo Methods, Noordwijkerhout*. Meppel: Krips Repro, The Netherlands.

Rosdahl H, Ungerstedt U, Jorfeldt L, Henriksson J (1993) Interstitial glucose and lactate balance in human skeletal muscle and adipose tissue studied by microdialysis. *J Physiol London* **471**:637–657.

Saisho Y, Umeda T (1991) Continuous monitoring of unbound flomoxef levels in rat blood using microdialysis and its new pharmacokinetic analysis. *Chem Pharm Bull (Tokyo)* **39:**808–810.

Scott DO, Lunte CE (1993) In vivo microdialysis sampling in the bile, blood, and liver of rats to study the disposition of phenol. *Pharm Res* **10:**335–342.

Sharp T, Zetterström T, Ungerstedt U (1986) An in vivo study of dopamine release and metabolism in rat brain regions using intracerebral dialysis. *J Neurochem* **47:**113–122.

Sjöberg P, Olofsson IM, Lundqvist T (1992) Validation of different microdialysis methods for the determination of unbound steady-state concentrations of theophylline in blood and brain tissue. *Pharm Res* **9:**1592–1598.

Smith AD, Justice JB (1994) The effect of inhibition of synthesis, release, metabolism and uptake on the microdialysis extraction fraction of dopamine. *J Neurosci Methods* **54:**75–82.

St Claire RL, Brouwer K (1992) Chemical analysis of bis (5-amidino-2-benzimidazolyl)-methane using microdialysis and microcolumn liquid chromatography. Presented at Second International Symposium on Microcolumn Separation Methods.

Stempels N, Tassignon MJ, Sarre S, Nguyenlegros J (1994) Microdialysis measurement of catecholamines in rabbit vitreous humor after retinal laser photocoagulation. *Exp Eye Res* **59:**433–439.

Sternberg F, Meyerhoff C, Mennel FJ, Hoss U, Mayer H, Bischof F, Pfeiffer EF (1994) Calibration problems of subcutaneous glucosensors when applied "in-situ" in man. *Hormone Metab Res* **26:**523–525.

Ståhle L (1991) Drug distribution studies with microdialysis: I. Tissue dependent difference in recovery between caffeine and theophylline. *Life Sci* **49:**1835–1842.

Ståhle L, Guzenda E, Ljungdahl-Ståhle E (1993) Pharmacokinetics and extracellular distribution to blood, brain, and muscle of alovudine (3'-Fluorothymidine) and zidovudine in the rat studied by microdialysis. *J AIDS* **36:**435–439.

Tammela O, Pastuszko A, Lajevardi NS, Delivoriapapadopoulos M, Wilson DF (1993) Activity of tyrosine hydroxylase in the striatum of newborn piglets in response to hypocapnic hypoxia. *J Neurochem* **60:**1399–1406.

Togashi H, Matsumoto M, Yoshioka M, Hirokami M, Tochihara M, Saito H (1994) Acetylcholine measurement of cerebrospinal fluid by in vivo microdialysis in freely moving rats. *Jpn J Pharmacol* **66:**7–74.

Tossman U, Jonsson G, Ungerstedt U (1986) Regional distribution and extracellular levels of amino acids in rat central nervous system. *Acta Physiol Scand* **127:**533–545.

Ungerstedt U (1984) Measurement of neurotransmitter release by intracranial dialysis. In Marsden CA (ed): *Measurement of Neurotransmitter Release In Vivo*. CA Marsden, Ed., pp. 81–105. Wiley, New York.

Ungerstedt U (1991) Microdialysis—Principles and applications for studies in animals and man. *J Intern Med* **230:**365–373.

Ungerstedt U, Pycock C (1974) Functional correlates of dopamine neurotransmission. *Bull Schweiz Akad Med Wiss* **1278:**1–13.

Ungerstedt U, Herrera-Marschitz M, Jungnelius U, Ståhle L, Tossman U, Zetterström T (1982). Dopamine synaptic mechansims reflected in studies combining behavioural recordings and brain dialysis. In *Advances in Dopamine Research,* M Kohsaka et al, Eds, pp. 219–231. Pergamen Press, New York.

Waga J, Ohta A, Ehinger B (1991) Intraocular microdialysis with permanently implanted probes in rabbit. *Acta Ophthalmol (Copenhagen)* **69:**618–624.

Waga J, Ehinger B (1995) Passage of drugs through different intraocular microdialysis membranes. *Graefes Arch Clin Exp Ophthalmol* **233:**31–37.

Westerink BH, Tuinte MH (1986) Chronic use of intracerebral dialysis for the in vivo measurement of 3,4-dihydroxyphenylethylamine and its metabolite 3,4-dihydroxyphenylacetic acid. *J Neurochem* **46:**181–185.

Westerink BH, De Vries JB (1991) Effect of precursor loading on the synthesis rate and release of dopamine and serotonin in the striatum: A microdialysis study in conscious rats. *J Neurochem* **56:**228–233.

Westerink BH, De Vries JB, Duran R (1990) Use of microdialysis for monitoring tyrosine hydroxylase activity in the brain of conscious rats. *J Neurochem* **54:**381–387.

Woodroofe MN, Sarna GS, Wadhwa M, Hayes GM, Loughlin AJ, Tinker A, Cuzner ML (1991) Detection of interleukin-1 and interleukin-6 in adult rat brain, following mechanical injury, by in vivo microdialysis: Evidence of a role for microglia in cytokine production. *J Neuroimmunol* **33:**227–236.

Wong SL, Wang YF, Sawchuk RJ (1992) Analysis of zidovudine distribution to specific regions in rabbit brain using microdialysis. *Pharm Res* **9:**332–338.

Wong SL, Van Belle K, Sawchuk RJ (1993) Distributional transport kinetics of zidovudine between plasma and brain extracellular fluid/cerebrospinal fluid in the rabbit—Investigation of the inhibitory effect of probenecid utilizing microdialysis. *J Pharmacol Exp Ther* **264:**899–909.

Yan HQ, Banos MA, Herregodts P, Hooghe R, Hooghe-Peters EL (1992) Expression of interleukin (IL)-1beta, IL-6 and their respective receptors in the normal rat brain and after injury *Eur J Immunol* **22:**2963–2971.

Zetterström T, Sharp T, Marsden CA, Ungerstedt U (1983) In vivo measurement of dopamine and its metabolites by intracerebral dialysis: Changes after D-amphetamine. *J Neurochem* **41:**1769–1773.

CHAPTER 7

Fundamentals of the Polymerase Chain Reaction and Separation of the Reaction Products

with Kathi J. Ulfelder

OVERVIEW

The analysis of double-stranded DNA (dsDNA) fragments, such as those produced by the polymerase chain reaction* (PCR) and enzymatic digestion, has led to considerable advances in molecular biology. Since its advent in 1985 (Saiki et al., 1985; Mullis and Faloona, 1987), PCR technology has been used to directly detect and quantify viruses, to track inheritance patterns in a family, to diagnose numerous genetic diseases, to identify individuals in forensic applications, and to aid in mapping the human genome.

The classic technique for analyzing the DNA produced by PCR (i.e., slab gel electrophoretic separation, followed by stain or probe detection) has its share of drawbacks. Slab gel electrophoresis is often time-consuming, labor intensive, and difficult to quantitate much less automate. It is, however, accepted as standard methodology for the molecular biology field. Recently, capillary electrophoresis (CE) has been used successfully for the separation of PCR products and DNA restriction fragments with high reproducibility and efficiency (Ulfelder et al., 1992; Landers et al., 1993). The introduction of laser-induced fluorescence detection (LIF) increased the sensitivity of CE (Schwartz and Ulfelder, 1992; Schwartz et al., 1994), a necessary improvement in any competition with autoradiography for low level detection. Given these advances in methodology, CE has become an attractive alternative to the standard method for separation and quantitation of PCR products. Additional information on HPCE is given in Chapter 3.

7.1 INTRODUCTION: POLYMERASE CHAIN REACTION

Since its introduction, the PCR technique (Saiki et al., 1985; Mullis and Faloona, 1987) has been used extensively in molecular biology to amplify specific

---

* PCR is covered by U.S. patents owned by Hoffmann-LaRoche, Inc.

138 FUNDAMENTALS OF THE POLYMERASE CHAIN REACTION

DNA sequences that are present in trace quantities. While the design seems simple—repeated cycles of denaturation, primer binding, and DNA synthesis—this powerful technique can result in 10^6-fold amplification from a single copy of target DNA.

Essentially, all forms of DNA (and RNA) can be amplified by PCR. The amount of DNA template required for amplification depends on the complexity of the genome. Typically, 0.1 to 1 mg of mammalian genomic DNA is required for PCR, while only picogram to nanogram quantities are utilized for bacteria (Sambrook et al., 1989).

A standard PCR reaction mix consists of many components, (Sambrook et al., 1989; Cha and Thilly, 1993). Besides the template, two single-stranded (ss) oligonucleotide "primers," 15 to 30 bases long and having sequences that are complementary to regions on the target, are required to initiate DNA synthesis. The primers are present in excess in the reaction mix—typically between 0.3 and 3 mM of each primer is used. Usually, the ratio of primer to target is kept at least to 10^8 : 1. Too high a ratio creates nonspecific product and primer–dimer; too low a ratio produces very little product. Deoxynucleoside triphosphates (dNTPs) are necessary as the components for primer extension, and are also present in the reaction in excess (37–1.5 mM of each dNTP). The reaction mix is buffered using 50 mM KCl, 10 mM Tris (pH 8.3), and 1.5 mM MgCl$_2$. This buffer is normally optimized for target, template, and DNA polymerase variations.

DNA polymerase is required for synthesis. There are several different polymerases available, each with its own characteristics that will affect its efficacy in the PCR. Since many labs have studied the properties of the different polymerase in current use, much is known about their efficiencies in PCR and rate of base misincorporation (Table 7.1). The thermostable polymerases (*Taq, Vent*) are preferred in PCR over the heat-labile enzymes

TABLE 7.1 Fidelity and Efficiency of DNA Polymerases Used in PCR

Enzyme	Error rate (errors/base)	PCR-induced mutant fraction[a] (%)	Efficiency per cycle (%)	Number of cycles required[b]
Taq	2×10^{-4}	56	88	22
Taq	7.2×10^{-5}	25	36	45 optimized conditions
Klenow	1.3×10^{-4}	41	80	24
T7	3.4×10^{-5}	13	90	22
T4	3×10^{-6}	2	56	32
Vent	4.5×10^{-5}	16	70	26

[a] Fraction of PCR-induced noise following 10^6-fold amplification of a 200-bp target sequence given the error rate.
[b] Number of cycles required to obtain 10^6-fold amplification given the efficiency per cycle.
Adapted with permission from Cha and Thilly, (1993) *PCR Methods Applic.*, 3:S18–S29.

(T4, T7, Klenow) because of their ease of use, especially when automating PCR.

Finally, the reaction mix may contain glycerol for formamide to enhance specificity (Cha et al., 1992; Sarkar et al., 1990), as well as gelatin, Triton X-100, or bovine serum albumin to stabilize the polymerase, and mineral oil to prevent water evaporation during the reaction.

The typical PCR run (Fig. 7.1) consists of three cycles: denaturation (1–2 min at 3 94°C), primer annealing (1–2 min at 50–55°C), and extension (1–2 min at 72°C). This design also requires optimization for each particular PCR. The desired blunt-ended duplex product does not appear until after the third cycle, whereupon it accumulates exponentially in subsequent cycles. The number of cycles required will depend on the efficiency of the reaction per cycle. Once the desired product has reached about 10^{12} copies, PCR efficiency drops significantly, and product stops amassing exponentially. This is the plateau phase; continuing PCR beyond this point often results in contaminating by-products rather than more product (Cha and Thilly, 1993).

7.2 PRINCIPLES OF NUCLEIC ACID SEPARATION

The separation of nucleic acids by CE has become a steadily growing area of interest, especially since the inception of the Human Genome Initiative. This interest originally stemmed from the use of polyacrylamide (PA) or agarose slab gel electrophoresis as the accepted standard for nucleic acid separation (Stellwagen, 1987).

7.2.1 Separation Mechanism

In free solution, the constant linear charge density of DNA molecules affords them a mobility that is independent of molecular weight (Olivera et al., 1964). However, in a support medium such as a slab gel electrophoretic separation by molecular weight occurs. Although the exact mechanism of this molecular sieving effect on nucleic acids is not clear, two theories have been proposed. The first was based on the Ogston model (Ogston, 1958), namely, in a random network of enmeshed fibers, the mobility of a macromolecule will be directly proportional to the volume fraction of pores of a gel it can enter. With increasing gel concentration, the average gel pore size decreases; thus, larger molecules will have difficulty entering the gel pores and therefore show retarded mobility. The second theory is based on the "snakelike" migration of DNA through the pores of gel, known as (biased) reptation (Lumpkin and Zimm, 1982). Depending on the size of the DNA fragment and/or the magnitude of the electric field, nucleic acid species can become aligned parallel to the field and may comigrate in a manner independent of size. Thus, if a DNA fragment is too large, or the field strength too high, no resolution between small or large DNA pieces may be seen.

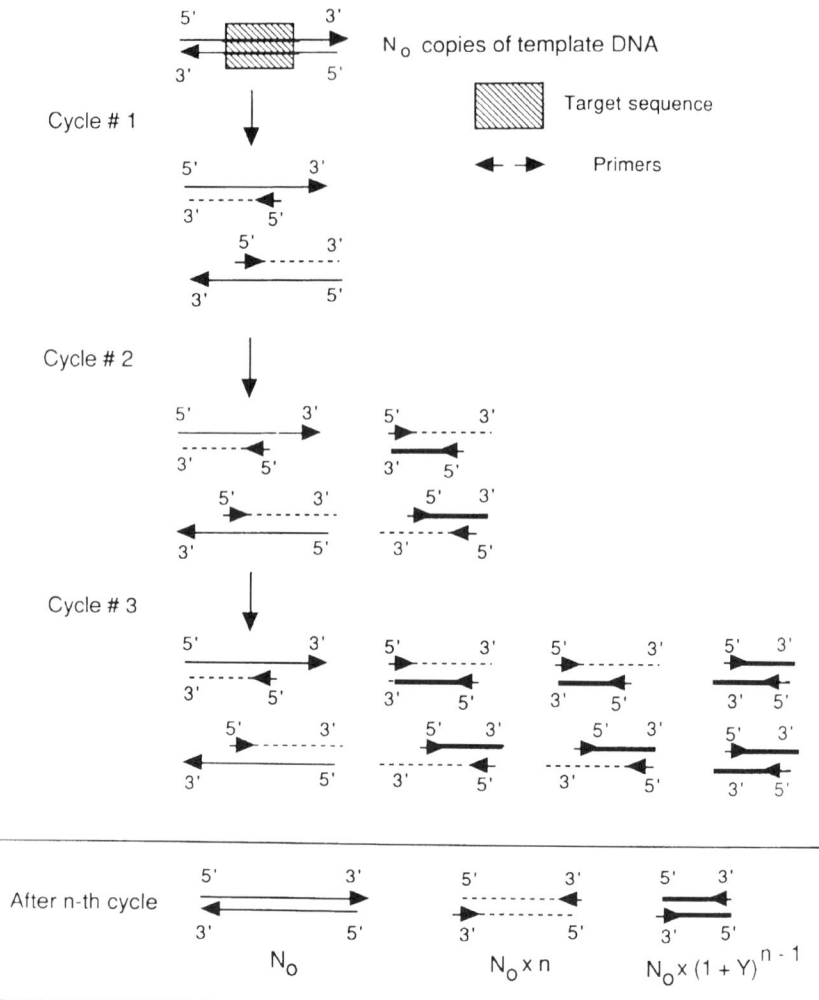

Figure 7.1 Schematic representation of PCR: N_0 copies of duplex template DNA are subjected to n cycles of PCR. During each cycle, duplex DNA is denatured by heating, allowing primers (arrows) to anneal to the targeted sequence (hatched square). In the presence of DNA polymerase and dNTPs, the primers are extended. The desired blunt-ended duplex product (thick bars with arrows) does not appear until after the third cycle, and accumulates exponentially during subsequent cycles. After n cycles of PCR, $N_0 (1 + Y)^{n-1}$ copies of duplex product are present. [Reprinted with permission from Cha and Thilly, *PCR Methods Appl* **3**:S18 (1993).]

7.2.2 Classical Methods of DNA Analysis

In classical electrophoresis, cross-linked PA or agarose slab gels are typically used in a flat bed or cylindrical tube format, where the Joule heating generated in the electrophoresis process can be more readily dissipated. This arrangement allows the use of higher field strengths to improve analyte resolution. Linear PA at low concentrations also provides a molecular sieving mechanism for nucleic acids (Bode, 1977; Tietz et al., 1986).

The many new nucleic acid amplification techniques and methods of determining DNA composition and structure require a final detection step involving PA or agarose electrophoresis. Since slab gel electrophoresis is an established technology, many "cookbook" procedures are available to aid in the techniques of nucleic acid separation (Sambrook) et al., 1989). However, these procedures can be quite time-consuming (> 2 h), technically demanding, labor intensive, and not amenable to automation. Although several samples at a time may be run on a single gel, the gel can be used only once. Reproducibility may be poor, and separated analytes difficult to quantify. Usually, microliters of precious sample are required for analysis; and there is also the problem of waste disposal when toxic substances such as acrylamide and radioactive probes must be used for autoradiography. Because of these problems, CE becomes an attractive alternative to agarose or PA slab gel electrophoresis.

7.3 CE METHODS: PRINCIPLES AND STRATEGIES FOR NUCLEIC ACIDS

7.3.1 CE Principles Related to Nucleic Acids

In a capillary, conditions can be produced that are very similar to a slab gel environment. Using high viscosity buffers or polymerizing a gel inside the capillary creates a matrix not unlike the pores of a slab gel. DNA molecules, negatively charged, migrate toward the anode but must also navigate through the polymer matrix. Smaller molecules will be able to travel with greater ease through this environment and will reach the detector first; the larger the molecule, the more hindered its passage through the matrix. A separation based on molecular mass is thus achieved.

7.3.2 Buffer Systems

The choice of a buffer system becomes the most significant variable in developing a separation of PCR products by CE. Smaller DNA components (such as dNTPs and primers) have been separated by means of a partitioning approach based on affinity for a micelle in solution (Cohen et al., 1987). In separations of larger dsDNA components by CE, a sieving mechanism is required in the capillary to separate species with the same linear charge density. In optimizing a sieving matrix, two approaches have been used: chemical gels and physical gels.

A chemical gel consists of a capillary filled with a cross-linked sieving matrix, such as polyacrylamide. With their well-defined pore structure and size, these gels produce high resolution separations comparable to those obtainable from sequencing gels. A chemical gel, however, cannot withstand high temperatures or high field strengths. Because of this, separations of double-stranded PCR products usually take more than 45 minutes. In addition, electrokinetic sample introduction is the only injection mode feasible, since the high viscosity gel prevents aqueous sample plugs from entering the capillary. As a result, some sample preparation, (e.g., desalting) may be necessary to be sure that enough sample is loaded onto the capillary.

Although chemical gels have been applied to the separation and sequencing of nucleic acids, common molecular biology buffers (Tris–borate–EDTA, Tris–acetate–EDTA) containing organic additives such as low- or zero-crosslinked polyacrylamide, polyethylene glycol, or cellulose, derivatives (e.g., hydroxypropylmethylcellulose, hydroxyethylcellulose) have become increasingly popular for most PCR product separations (Zhu et al., 1989; Heiger et al., 1990; Schwartz et al., 1991; Nathakarnkitkool et al., 1992). These viscous buffers, known as "physical gels," simulate the pore structure of a gel and function as a sieving matrix similar to a chemical gel. The degree of sieving can be controlled by changing the type (e.g., chain length or derivative) and/or concentration of linear polymer additive used, thereby optimizing resolution for a specific DNA size range. These gels are easily replaceable with a simple rinse through the capillary by pressure or vacuum, yielding identical separation conditions from run to run. Moreover, extremely high temperatures (up to 70°C) and high field strengths (1000 V/cm) can be applied to these matrices without damage to the sieving polymer. This ruggedness is beneficial when one is optimizing separation for a specific DNA size range, since manipulation of temperature and/or field strength can improve resolution (Guttman and Cooke, 1991b; Guttman et al., 1992). Most often, a polyacrylamide or polysiloxane-coated capillary is used in conjunction with these buffers to control electroosmotic flow and enhance peak efficiency.

Agarose solutions have also been applied to the separation of DNA fragments (Bocek and Chrambach, 1991a, 1991b, 1992). To take advantage of the light-scattering and -absorbing properties of agarose gels, clear agarose solutions have been used to separate DNA fragments up to 12 kb in length; for this application, molten low-melting agarose is heated and maintained in a coated capillary at 40°C. Alternatively, urea may be added to agarose solutions to prevent gel setup by disrupting hydrogen bond formation.

7.3.3 Intercalators

Intercalators (DNA-binding dyes) have been used for many fluorometric DNA assays, and as a DNA stain in slab gel electrophoresis. Typical dyes include ethidium bromide, a bisintercalator that binds one dye molecule for every 5 base pairs (bp) of DNA; and thiazole orange, a *mono*intercalator having one dye for every 2 bp. A wide variety of dyes has been recently developed (Glazer and Rye, 1992; Benson et al., 1993). Interestingly, addition of these dyes to a CE gel buffer can actually improve dsDNA resolution (Schwartz et al., 1991; Guttman and Cooke, 1991a). These molecules insert themselves between the base pairs of DNA, changing the molecular persistence length, conformation, and charge of the DNA molecules. These modifications result in a change in electrophoretic behavior: Larger DNA molecules move relatively more slowly, allowing the separation time window to widen, thus increasing peak capacity. In some cases, separation is achieved for dsDNA fragments of the same size but different base composition, (Fig. 7.2).

Since DNA intercalators fluoresce strongly when excited by an appropriate light source, use of these dyes presents the opportunity of detecting low levels

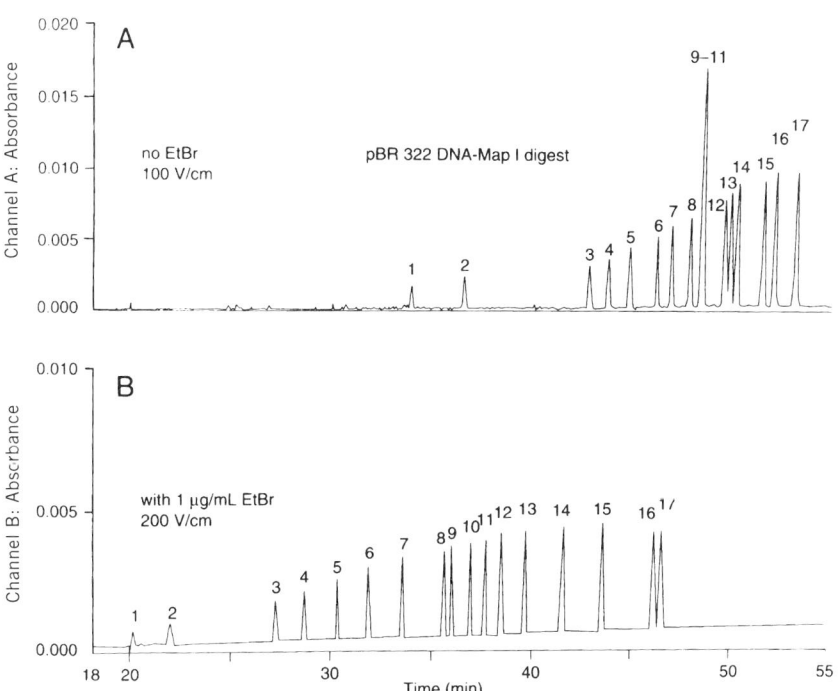

Figure 7.2 Effect of ethidium bromide on the separation of *Msp* I-digested pBR322 DNA. (*A*) No ethidium bromide; 100 V/cm. (*B*) 1 μg/mL ethidium bromide in the gel–buffer system; 200 V/cm. Peaks (in bp): 1, 26; 2, 34; 3, 67; 4, 76; 5, 90; 6, 110; 7, 123; 8, 147; 9, 147; 10, 160; 11, 160; 12, 180; 13, 190; 14, 201; 15, 217; 16, 238; 17, 242. [Reprinted with permission from Guttman and Cooke, *Anal. Chem.* **63:**2038 (1991). Copyright: American Chemical Society.]

of DNA using LIF, without the need for precolumn derivatization of the sample (Schwartz and Ulfelder, 1992).

7.3.4 Typical Instrument Parameters

Capillary electrophoresis of PCR-amplified products is usually performed in the reverse polarity mode (negative potential at the injection end of the capillary). A coated capillary (100 mm i.d., 37–57 cm total length) is filled with a gel buffer system. PCR samples are introduced hydrodynamically or, after desalting, electrokinetically. The PCR sample and a DNA marker of known size may be injected sequentially and allowed to comigrate in the capillary. With a capillary temperature set at 20 to 30°C, separation of PCR products is accomplished at field strengths of 200 to 500 V/cm. Detection is on-line, measuring either UV absorbance at 260 nm, or LIF.

7.3.5 Detection

Although it is clear that CE of nucleic acids with UV detection has many advantages over slab gel techniques—high efficiency peaks (10^7 plates/m) have been demonstrated (Guttman and Cooke, 1991a—a more sensitive approach is again needed to compete with autoradiography for low level detection. The detection limit on the average for double-stranded DNA by UV detection is about 8 ng/mL sampling concentration, (roughly 58 fg on-column). The use of on-column sample stacking (Chien and Burgi, 1992) together with intercalating agents in the buffer system (Ulfelder et al., 1992) can improve sensitivity fivefold. The addition of fluorescent intercalators to the buffer mixture, in conjunction with LIF, has also demonstrated enhanced sensitivity (Schwartz and Ulfelder, 1992).

Under conditions of LIF detection, a fluorogenic intercalator is added to the buffer to identify and quantify the PCR products by means of a laser that excites the resulting DNA–dye complex; the fluorescence emission wavelength of the complex will depend on the fluor chosen. For example, with thiazole orange intercalation, excitation of the complex is accomplished using the 488 nm line of an argon ion laser, with subsequent emission at 530 nm (Haugland, 1992). The DNA–dye complex fluoresces when excited by the appropriate wavelength of a laser, whereas the intercalator alone (as well as non-DNA sample components) will not. Primer and primer–dimer peaks are observed, since they too complex with the dye. However, dNTPs and other PCR reaction species such as *Taq* polymerase are not intercalated and therefore not detected. In all, this mode of LIF detection results in improved resolution of base pairs between the closely migrating fragments, as well as a sensitivity enhanced up to three orders of magnitude compared to UV detection (Schwartz and Ulfelder, 1992; Ulfelder and Shieh, 1993; Zhu et al., 1994).

For example, Figure 7.3 compares spectra from LIF detection of a fluorescent monointercalator and from UV detection. Note that by LIF, the pattern

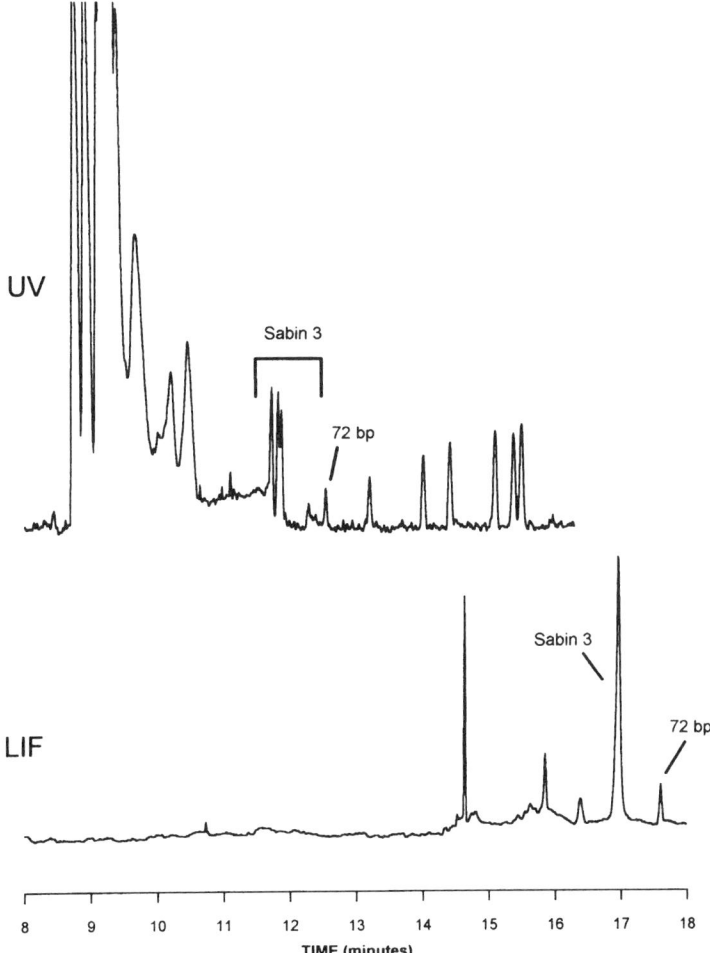

Figure 7.3 UV versus LIF detection of the CE separation of a 53 base pair RT-PCR product from the RNA of the polio virus vaccine, Sabin 3. An *Hae* III digested ϕX174 DNA marker was coinjected with the PCR product for size determination—note the 72 bp fragment. The same Sabin 3 concentration was used for each analysis, whereas the marker total DNA concentration varied from 200 mg/mL for UV analysis, to 20 mg/mL for LIF analysis. Note the unambiguous pattern observed with LIF for the Sabin 3 fragment compared to UV detection of the same fragment. Full scale: UV detection, 0.005 absorbance unit (AU); LIF detection, 10 relative fluorescence units (RFU). [Reproduced with permission from Schwartz et al., *J Capillary Electrophor* **1**:36 (1994). Copyright: ISC Technical Publications, Inc.]

is less ambiguous. Zeptomole (10^{-21} mole) sensitivity on-column has been demonstrated. Alternatively, fluor-labeled primers can be used to create fluorescently labeled PCR products, which are then detected by LIF without the use of intercalators. For subsequent use in PCR leading to fluorescent DNA products, PCR conditions can be optimized to eliminate the need for primer purification.

7.3.6 Data Analysis

In an electropherogram, the area under the peak for a particular DNA fragment can be correlated to the quantity of that fragment. However, one must realize that the peak area of the DNA fragment is related to its residence time in the detector. Slower migrating (large) fragments will remain in the detector window longer than a faster migrating (small) fragment, generating a large peak area that is not representative of the true quantity of DNA passing through the detector. Therefore, the area for a particular peak is "corrected" for detector dwell time by dividing that area by the peak migration time to the detector. The corrected area is then used for quantitation.

Since DNA fragments from the PCR typically are contained in a high salt matrix, their mobility will vary depending on sample salt concentration. Thus, proper identification of these DNA fragments requires the use of an internal standard to normalize analyte velocity. This practice corrects for variance in fragment mobility due to sample matrix differences (i.e., salt content). These internal standards are included for size determination (in bp) as well as a reference for migration time. Candidates for such internal standards include the primer or primer–dimer peaks, since both components are already present in the PCR mixture; alternatively, one or more coinjected standard DNA peak s may be chosen. If any of these fragments are to serve as the internal standard, they must be separated from one another and any PCR product, a precondition that is not easily met when the size of the PCR product is below 60 bp.

7.3.7 Sample Preparation and Injection Considerations

The PCR samples to be analyzed consist of many components, all of which may potentially enter the capillary. Of particular concern are the salts, which will affect sample injection, migration, and peak shape. Since it is preferable not to perform any sample preparation prior to analysis, the method of injection of the PCR product must be examined. In some cases, some sample pretreatment is required to optimize resolution and sensitivity.

With replaceable gels, either an electrokinetic or a hydrodynamic approach to sample introduction is possible. The hydrodynamic injection is generally preferred when quantitation of the PCR product is desired. In this case, the sample, introduced as a plug into the capillary, is exactly the same as that of the sample vial from which it originated. Negative components (Cl^-, dNTPs, primers, primer–dimer, PCR product, DNA template, polymerase) will migrate toward the anodic detector. Template and polymerase may never be detected owing to their size (lower mobility) and concentration on the capillary. With LIF detection, only DNA components can be seen.

With electrokinetic injection, mobility differences between sample components cause a sampling bias (Huang et al., 1988): the higher mobility primers, dNTPs, and salts are preferentially introduced into the capillary over the more slowly moving ds PCR products. The PCR products themselves, having essentially the same mass-to-charge ratio in solution, will not exhibit this sampling bias. Because of differences in salt content from sample to sample, however, variation in load of a sample will affect accurate quantitation, necessitating the use of external and/or internal standards (Butler et al., 1994; Srivatsa et al., 1994). It should be noted that electrokinetic injection often results in more efficient peaks than does pressure injection (Schwartz et al., 1991; Butler et al., 1994). In a matrix of low ionic strength (e.g., after desalting), DNA fragments injected electrokinetically become stacked against the viscous gel buffer, which is higher in ionic strength. This allows a longer injection time, increasing sample loading without sacrificing resolution.

Desalting techniques include ultrafiltration, which simultaneously desalts and concentrates the sample; membrane dialysis (Cooksy, 1992); and simple dilution in low ionic strength buffer or water. The latter method is especially useful when glycerol, Triton X-100, or formamide has been added to the PCR mixture. It should be noted that loss of DNA from adsorption onto the filters in ultrafiltration has been reported (Butler et al., 1994). Thus, as a rule, sample desalting by this method should be used as a last resort.

In some cases, however, the need for high resolution (e.g., 2–4 bp) and sensitivity requires electrokinetic injection of PCR products, and preferably without prior desalting or other sample preparation. Sample stacking techniques can still be applied through the use of a short (3 s) electrokinetic injection of water prior to a long (> 90 s) electrokinetic injection of the PCR product still in its salt matrix. The long sample injection time is required to ensure that enough DNA is loaded in the presence of such high salt. However, the water preinjection also creates a low conductivity focusing zone in which more sample can be loaded onto the capillary, while resolution is maintained. Figure 7.4 demonstrates the increased resolution obtained using this method compared to a pressure injection for a nondesalted PCR sample.

One method of enhancing sample signal in an attempt to achieve sample quantitation cells for the use of longer hydrodynamic injections to increase sample load on the capillary. Resolution and efficiency are greatly compromised, however, by the effects on the matrix of the sample salt (van der Schans et al., 1994). In the example shown in Figure 7.5A, a DNA standard is diluted in high salt. When the DNA is injected for increasing amounts of time, shoulders appear on the peaks. High salt samples cannot take advantage of zone sharpening because their ionic strength is greater than that of the gel buffer. The shoulders on the peaks are caused by DNA in the sample–buffer interface

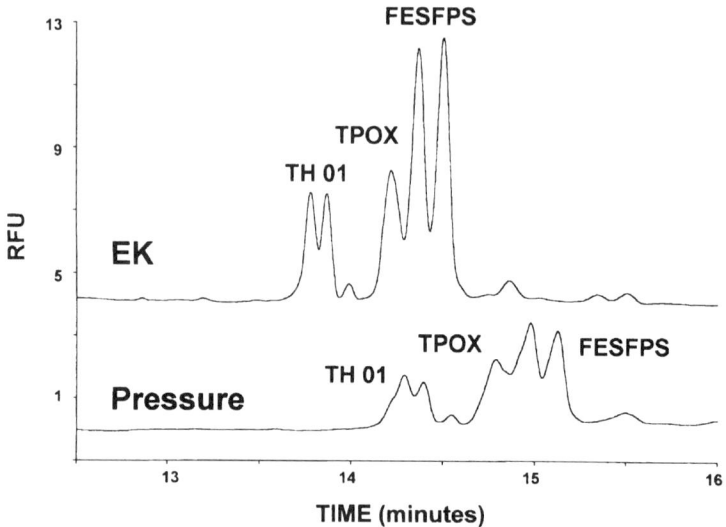

Figure 7.4 Comparison of pressure versus electrokinetic (EK) injection (first water, then sample) for a nondesalted PCR sample on a replaceable gel matrix containing an intercalating dye. Sample: PCR-amplified DNA coding for three short tandem repeat (STR) loci in an individual. Repeat unit is 4 bp. Sample is heterozygous for the loci TH01 and FESFPS, and homozygous for TOPOX. Wavelength, (LIF Ar ion): excitation, 488 nm; emission, 530 nm. Pressure injection: 10 seconds, 0.5 psi. Electrokinetic injection: 3 seconds, 200 V/cm (water), followed by 90 seconds, 200 V/cm (sample). Resolution between the FESFPS peaks is increased from Rs = 0.71 (pressure injection) to Rs = 1.06 (electrokinetic injection).

that diffuse into the buffer zone. This zone will have a higher field strength than the sample plug because of its higher resistance (less salt). DNA fragments that diffuse into this zone migrate faster than the same fragments in the sample plug zone itself. They appear as shoulders on the main fragment peak. The injection prior to the sample of an additional plug of a low resistance solution (e.g., 100 mM Tris–acetate, pH 8.5) results in improved resolution (Fig. 7.5B). The DNA at the sample–buffer interface then decelerate and returns to the sample zone.

An estimation of relative molecular size is possible by comparing an unknown sample's mobility to a standard size curve (log of bp number vs. migration time) generated from the mobility of fragments from a marker diluted in water. However, because of its high salt matrix, PCR product mobility will vary depending on salt concentration. Therefore, as with slab gel analysis, size determination compared to a standard in water is not accurate. Upon injection of the DNA, the salt matrix causes a local decrease in the field strength until the salt has migrated out of the capillary. Lower initial field strength means lower initial velocity for the DNA; the DNA then speeds up when the salt migrates out of the capillary. Size determination, however,

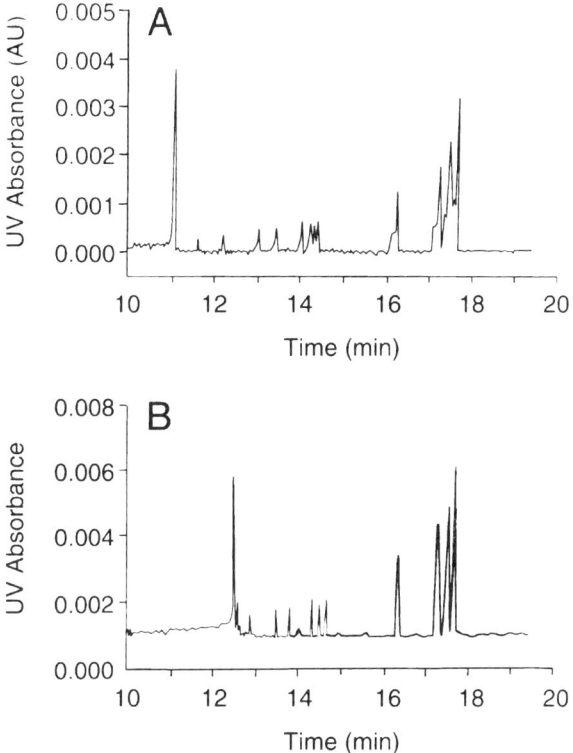

Figure 7.5 (A) Separation of *Hae* III digested ɸX174 RF DNA, 50 μg/mL in 20 mM NaCl. (B) Influence of presample injection of 0.1 M Tris–acetate, pH 8.3. First injection: 10-second pressure injection of Tris–acetate; second injection: 20-second injection of digest. [Reprinted with permission from van der Schans et al., *J Chromatogr A* **680**:511 (1994).]

is now no longer a linear function. To correct this, a double injection technique is employed, whereby an unknown and a standard are injected sequentially and allowed to comigrate in the capillary. Migration times for the coinjected marker fragments generate the molecular size curve, which is then used to determine the relative size of the unknown DNA fragment. Order of injection is important (van der Schans et al., 1994)—since salt from the PCR sample will migrate faster than the DNA components, it can create a leading zone of low field strength. An injection order of sample followed by standard will cause fronting of the PCR product, but sharp peaks for the standard. The opposite is seen for an injection order of standard followed by sample. This effect is demonstrated in Figure 7.6. Alternatively, an internal standard may be included in the sample (e.g., PCR coamplification of a second target sequence, addition of a known quantity of DNA to each sample).

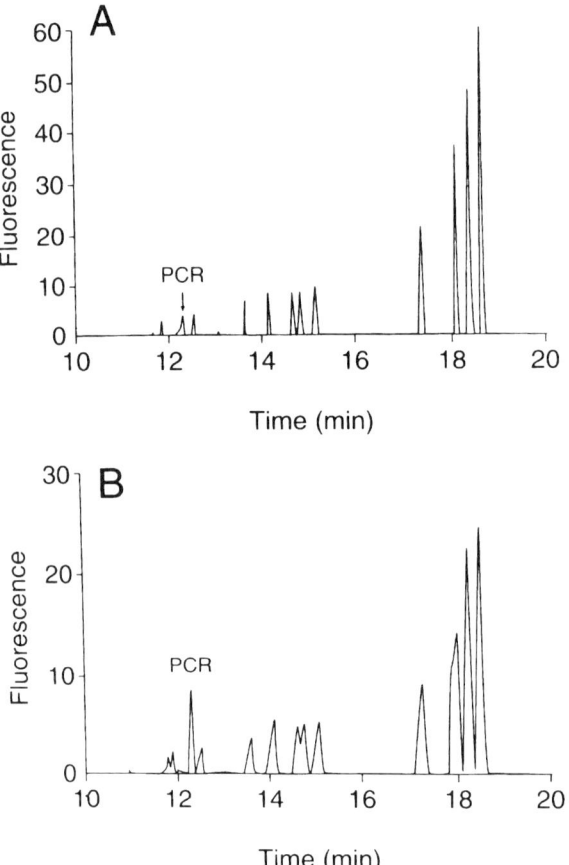

Figure 7.6 (*A*) Electropherogram of PCR sample, then DNA standard. First injection: PCR sample (97 bp); second injection: *Hae* III-digested ϕX174 RF DNA, 10 μg/mL. (*B*) Electropherogram of DNA standard, then PCR sample. First injection: *Hae* III-digested ϕX174 RF DNA, 10 μg/mL; second injection: PCR sample (97 bp). [Reprinted with permission from van der Schans et al., *J Chromatogr A* **680**:511 (1994).]

7.3.8 Artifacts

An injection-related artifact can occur in gel buffers with consecutive electrokinetic injections from the same, low volume (10–200 mL) sample: progressively smaller amounts of sample are introduced into the capillary, resulting in peak heights or areas that decrease with each injection (Schwartz et al., 1995). This effect is due to the migration of cations (e.g., Tris) from the gel buffer into the sample, changing its relative ionic strength. One solution to this problem is to perform an electrokinetic injection from a water vial prior to sample injection. This water injection generates a zone of "ion depletion" (i.e., relatively high resistance and low conductivity) at the cathode end of the capillary. As the PCR sample is subsequently electrokinetically injected into the capillary, very few cations will migrate into the sample vial from the capillary. Hence, the sample solution will stay relatively salt-free, even with subsequent injections from the same vial. This two-step injection procedure not only results in dramatically increased precision (important for quantitative studies), but also increases sample loading. Because a relatively high resistance, low conductivity zone has been introduced into the capillary, the local field strength in this zone is relatively high—therefore, more sample is pulled into the capillary than would have made its way there without the preinjection of water.

A second injection-related artifact can occur in physical gel networks: poor peak shape (i.e., severe tailing), with capillaries improperly cut at the inlet side. An oblique shape at the injection site of the capillary yields a separation efficiency significantly lower than that obtainable with a properly cut, straight-edge capillary (Schwartz et al., 1995). When these jagged-edge capillaries are used, the injection plug length is slightly increased (relative to the straight-edge capillary), resulting in increased extracolumn variance due to the injection plug. Care must be taken to ensure that the capillary ends are properly prepared and maintained.

7.4 PCR APPLICATIONS

7.4.1 Forensic Analysis

Capillary electrophoresis with LIF detection has been applied in the analysis of genetic markers for human identification (McCord et al., 1993; Srinivasan et al., 1993a, 1993b). In the analysis of a locus containing short tandem repeats of 3 bp in the range of 250 or less, a *mono*intercalating dye, YO-PRO-1, was added to the buffer system of a polymer network. To improve separation efficiency, it was also necessary to add ethidium bromide to the run buffer. By means of voltage programming (Guttman et al., 1992), optimal resolution could be obtained for the alleles of *HUMTH01* by a high field separation (405 V/cm) for 5 minutes, followed by a drop in field (to 135 V/cm) for an additional 5 minutes. Separation of all seven alleles, as well as size markers 300 bp and lower, was achieved within 8.3 minutes (Fig. 7.7). Although further reproducibility studies are necessary to ensure that CE with LIF detection is a viable technique for forensic DNA typing, this preliminary work appears to be useful in profiling applications that call for the detection of minute quantities of DNA. (Additional information on the use of PCR and HPCE in forensics, is given in Chapter 8.)

7.4.2 Identification by Hybridization

The Southern method of DNA hybridization (Southern, 1975) is often performed to identify a particular nucleotide sequence in a DNA molecule.

152 FUNDAMENTALS OF THE POLYMERASE CHAIN REACTION

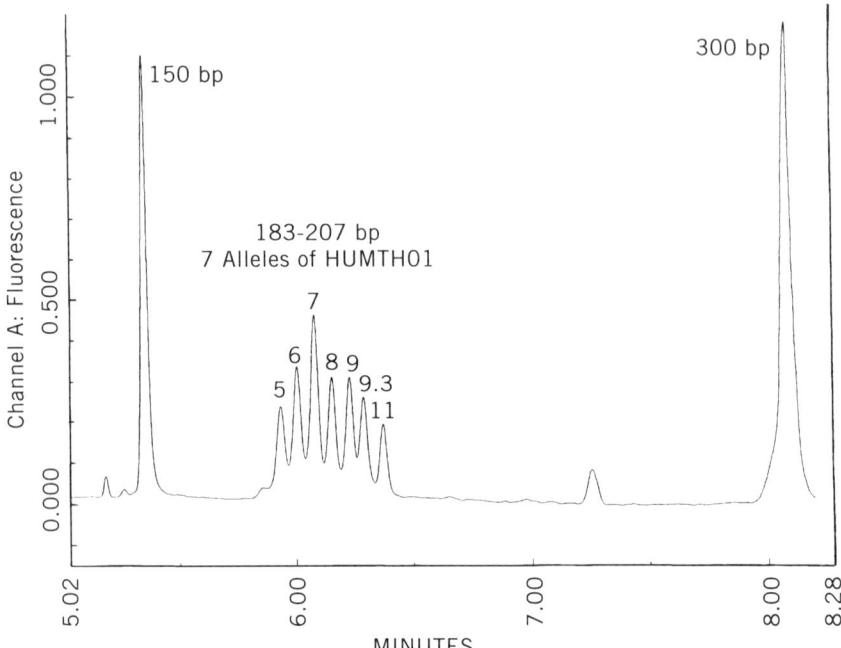

Figure 7.7 High speed forensic typing of *HUMTH01* using voltage programming. The *HUMTH01* allelic ladder, amplified via the PCR, is separated in 7 minutes, demonstrating 4 bp resolution, using a replaceable polymer matrix and LIF detection. The 150 and 300 bp markers allow precise sizing of each allele. Injection: 10 ng/mL *HUMTH01* ladder, injected by 0.5 psi pressure for 30 seconds. Voltage programming: 15 kV for 5.2 minutes; 5 kV for 4.8 minutes. (Electropherogram courtesy of Dr. B. McCord and Dr. J. Butler, FBI Forensic Science Research, Quantico, VA.)

Conventionally, the method entails the gel separation of DNA fragments and their transfer onto a membrane support, with subsequent hybridization using a radiolabeled DNA probe of complementary sequence. Autoradiography of the bound probe can detect gene deletions and rearrangements found in numerous human diseases. Since this method is very labor intensive, DNA hybridization in solution and electrophoretic analysis of the resulting hybrid would be more efficient. The Southern method requires the denaturation of the dsDNA by heat, acid, or alkali (Kornberg, 1980), followed by annealing of the probe.

Given the latest advances in CE methodology and detection, on-line Southern analysis procedures with precolumn DNA hybridization in solution, followed by CE analysis of the resulting hybrid, have been demonstrated (Chen et al., 1991; Bianchi et al., 1994). A fluorescently tagged oligonucleotide of complementary sequence is used as a probe and mixed with a PCR product. The mixture is then heated to 100°C and ramped slowly down to room temperature prior to injection. Capillary gel electrophoresis of the resulting hybrid produces a fluorescing peak with a longer migration time than can be achieved with either the labeled probe or target DNA alone. Moreover, the use of CE for simultaneous hybridization and separation of a PCR product to a fluor-labeled probe was recently reported (Ulfelder et al., 1995). While precolumn hybridization requires heat for dsDNA denaturation, the on-column method uses low pH—acidic enough to denature but not depurinate the dsDNA target. On-column hybridization consists of sequential injections of acid-denatured PCR product and labeled probe, preceded by a high salt injection. This zone of low resistance is required to focus and neutralize the dsDNA, while simultaneously mobilizing the probe into the dsDNA plug. A schematic diagram of injection order is shown in Figure 7.8. Direct identification of DNA fragments by precolumn and on-column hybridization shows much potential in molecular biology studies and in the diagnosis of genetic and infectious diseases currently performed with the Southern blot technique.

7.4.3 Quantitative Analysis

Quantitation of a PCR product is possible through the generation of a standard curve showing linearity between amount of DNA template present prior to amplification and the corrected peak area of resulting PCR product. To develop this standard curve, a serial dilution of template is performed. Alternatively, a ratio of the unknown template's peak area relative to that of an internal standard can be determined (e.g., for competitive PCR). A linear polyacrylamide sieving buffer was used to determine viral burden. The buffer, which contained the intercalating dye EnhanCE, permitted visualization by LIF and quantitation of the products of reverse transcriptase (RT) PCR

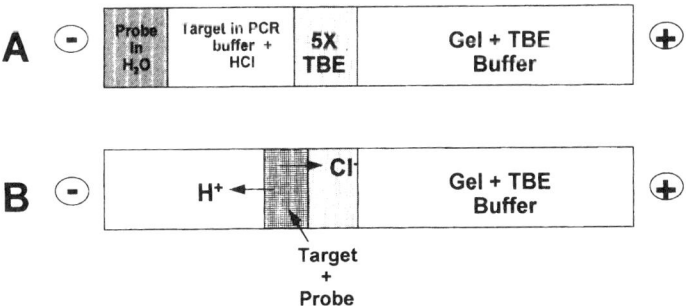

Figure 7.8 Schematic of the injection method for on-column hybridization. (*A*) injection order and (*B*) after 200 V/cm field for 10 seconds. The high ionic strength the 5× Tris-Borate-EDTA (TBE) region causes focusing of the lower ionic strength target and probe, while HCl removal induces renaturation. The probe in water mixes quickly into the target and anneals before the target renatures completely.

generated from the RNA of polio virus (Rossomando et al., 1994). Figure 7.9 shows the LIF results from separation of the 53 bp RT-PCR product derived from the Sabin 3 strain of virus in the polio vaccine. The template amount for each RT—PCR was titrated to create a standard curve. Quantitation was achieved by comparing the corrected peak area for the RT-PCR product generated from the standard curve of known amounts of template RNA (Fig. 7.10).

A second method of PCR product quantitation using CE with fluorescence detection involves the use of fluor-labeled primer to generate a fluorescent PCR product (Ulfelder, 1994). Bacteriophage lambda DNA was used in a PCR, producing a 500 bp fragment with a single fluorescein attached. The labeled PCR product was analyzed by CE using LIF (Fig. 7.11), detecting only the single fluor from each PCR product. With increasing amounts of DNA template, peak height and area of the product also increased up to a point, whereupon the reaction plateaud (Siebert, 1993) (Fig. 7.12). Note the increase in primer and primer–dimer peaks as template availability decreases.

Unfortunately, PCR product size cannot be determined without the use of similarly labeled DNA size standards, since the presence of a fluor label can

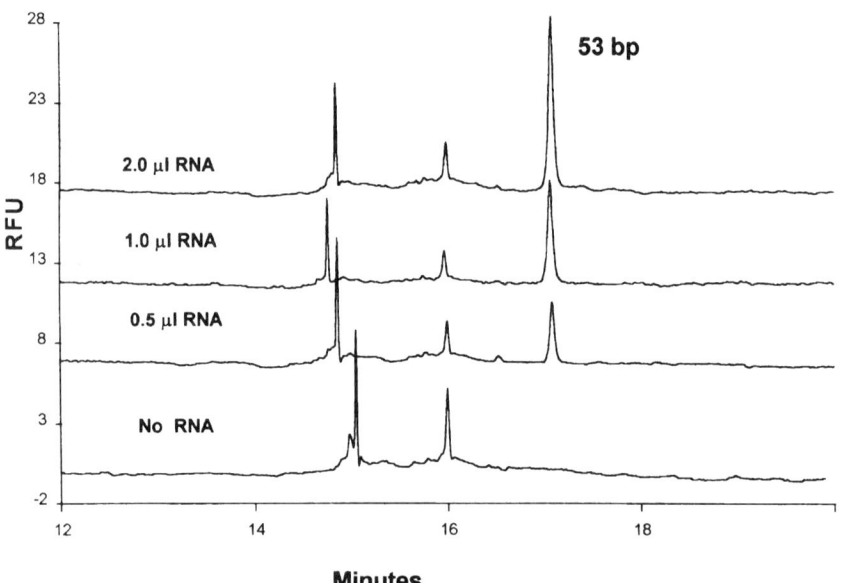

Figure 7.9 CE separation and quantitation of the RT-PCR product (53 bp) from Sabin 3 strain of the polio virus vaccine. Discrete volumes of the template RNA were reverse transcribed and the complementary DNA PCR amplified. The resulting products were analyzed by CE-LIF using a replaceable gel matrix. With increasing amounts of RNA template, peak height and area also increased up to point (> 2.0 μL template), whereupon the reaction plateaued.

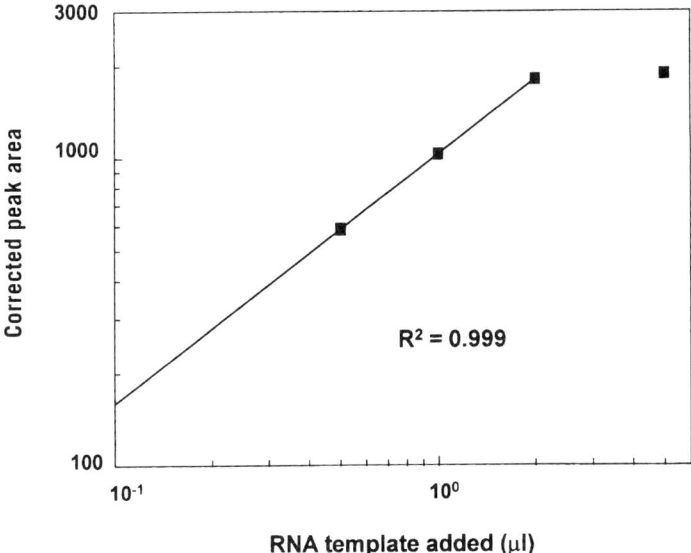

Figure 7.10 Plot of corrected peak area versus RNA template added for RT-PCR. Peak area for the 53 bp Sabin 3 product were corrected for transit time through the detector and plotted as a function of the amount of RNA used in the RT-PCR. A linear relationship is observed up to 2.0 μL RNA added. Inclusion of the 5.0 μL RNA point demonstrates PCR plateauing.

shift electrophoretic mobility. Compared to CE-LIF with the use of intercalators in the buffer, CE-LIF of fluor-labeled PCR products shows more than 10-fold lower detection and quantitation limits. Although the display by a single fluor molecule of a signal greater than that of a molecule intercalated with many fluors would seem unusual there are three possible explanations for this better sensitivity.

First, fluorescence intensity is directly proportional to the product of a fluor's molar extinction coefficient (e) for absorbance and its quantum yield (F) for fluorescence, (see Guilbault, 1990, for a review of fluorescence). Under the conditions used for PCR product separation and detection, both e and F are greater for a single fluorescein molecule than for the intercalator–DNA complex (Haugland, 1992).

Second, although the intercalator has a high quantum yield when complexed to DNA, it still exhibits some low background fluorescence. At high DNA concentrations (> 1 mg/mL), this background appears negligible; however, at very low DNA levels, (producing a signal < 1 relative fluorescent unit full scale), background fluorescence becomes significant for low level quantitation. While the dye concentration in the buffer can be reduced to decrease background, this will affect the signal linearity for the highest DNA concentrations (> 100 mg/mL) because of the insufficient amount of intercalator.

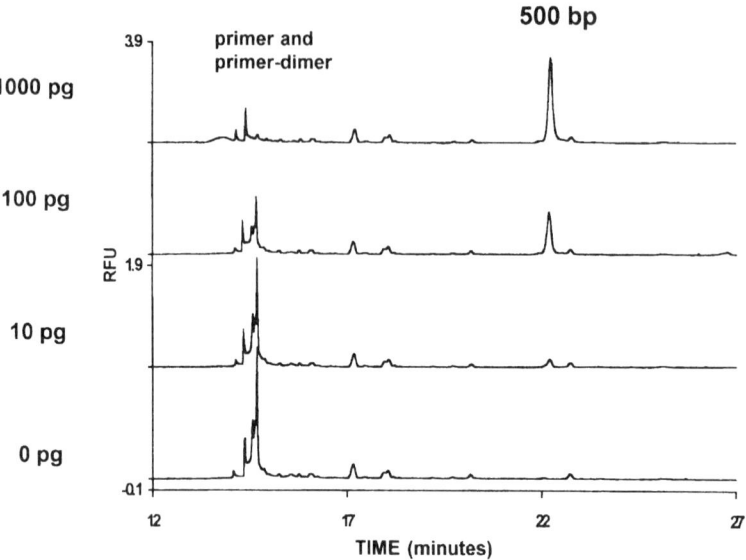

Figure 7.11 CE-LIF separation and quantitation of fluorescein-labeled lambda PCR product. A 500 bp product was generated with titrated amounts of lambda bacteriophage template and primers, one of which was fluorescein-labeled. The fragment was analyzed by CE-LIF, and only the fluor from the PCR product was detected. With increasing amounts of DNA template, peak height and area of the product also increased up to a point, whereupon the PCR plateaued. Note the increase in primer and primer–dimer peaks as template availability decreases. (Reproduced with permission from KJ Ulfelder, *Applications Information Bulletin A-1774*, 1994. Copyright: Beckman Instruments, Inc.)

Third, the presence of excess dye in the buffer may cause self-quenching—the reduction in fluorescence intensity due to interactions between individual fluorophores. Again, although signal reduction is insignificant at high DNA concentrations, it cannot be neglected for very low DNA levels. The dye can be removed from the buffer and the samples prestained *a* with higher affinity intercalator such as TOTO-1 or YOYO-1 (Haugland, 1992), but this approach requires that the sample concentration be known *a priori*, since the correct ratio of DNA bp to dye is critical for signal linearity. For samples of unknown DNA concentration, this is not a viable option. Because of the reduction in signal-to-noise ratio, sensitivity may be compromised for low level DNA, especially compared to single-fluor molecule detection without the use of intercalators.

7.4.4 Quantitative RNA-PCR

For most quantitative applications, it is important that amplification be performed during the exponential phase of PCR—that is, before the reaction

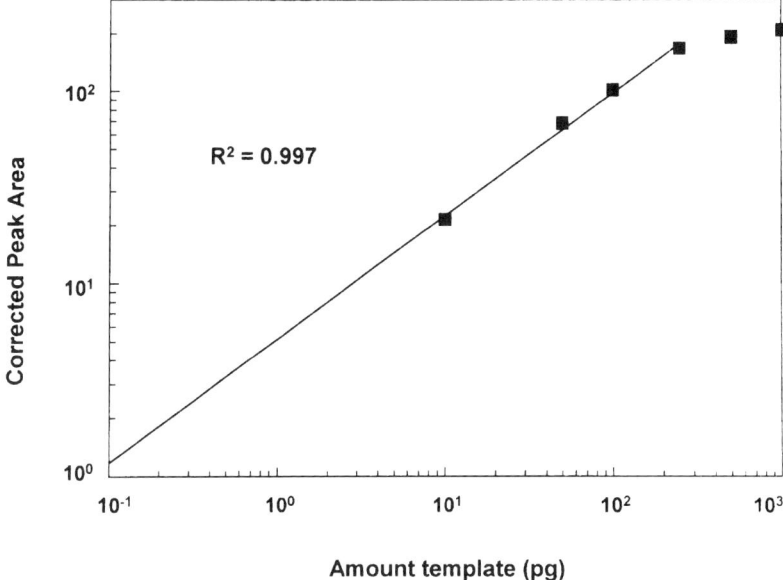

Figure 7.12 Quantitative analysis of labeled lambda bacteriophage PCR product: plot of corrected peak area versus initial DNA template amount. Peak areas for the 500 bp lambda product were corrected for transit time through the detector and plotted as a function of the amount of DNA used in the PCR. A linear relationship is observed up to 250 pg DNA template. Inclusion of the high DNA template data points demonstrates PCR plateauing. (Reproduced with permission from KJ Ulfelder, *Applications Information Bulletin A-1774*, 1994. Copyright: Beckman Instruments, Inc.)

reaches plateau (Cha and Thilly, 1993; Siebert, 1993). In this case, PCR cycle number is optimized by selecting a high level of target and serially diluting; a significant reduction in the number of cycles may be necessary to produce a linear relationship between template and product amounts spanning three orders of magnitude. If low cycle number fails to produce sufficient product for CE-UV detection, LIF may be required.

As an alternative, competitive PCR can be used to quantitate absolute amounts of target without lowering cycle number (Chelly et al., 1988; Piatak, et al., 1993b). In competitive PCR, an unknown amount of target and a known amount of external standard (competitor) compete in a single reaction mixture for the same primers and, therefore, for amplification. The competitor is usually titrated against a constant amount of target. The DNA products resulting from amplification can differ in hybridization properties, restriction enzyme sites, or size. The size difference makes them amenable to separation by electrophoretic methods. The amount of target can be obtained from the competitor concentration when the ratio of their simultaneous amplification products equals one, provided their amplification efficiencies are similar. An important advantage of competitive PCR is that it is possible to obtain useful

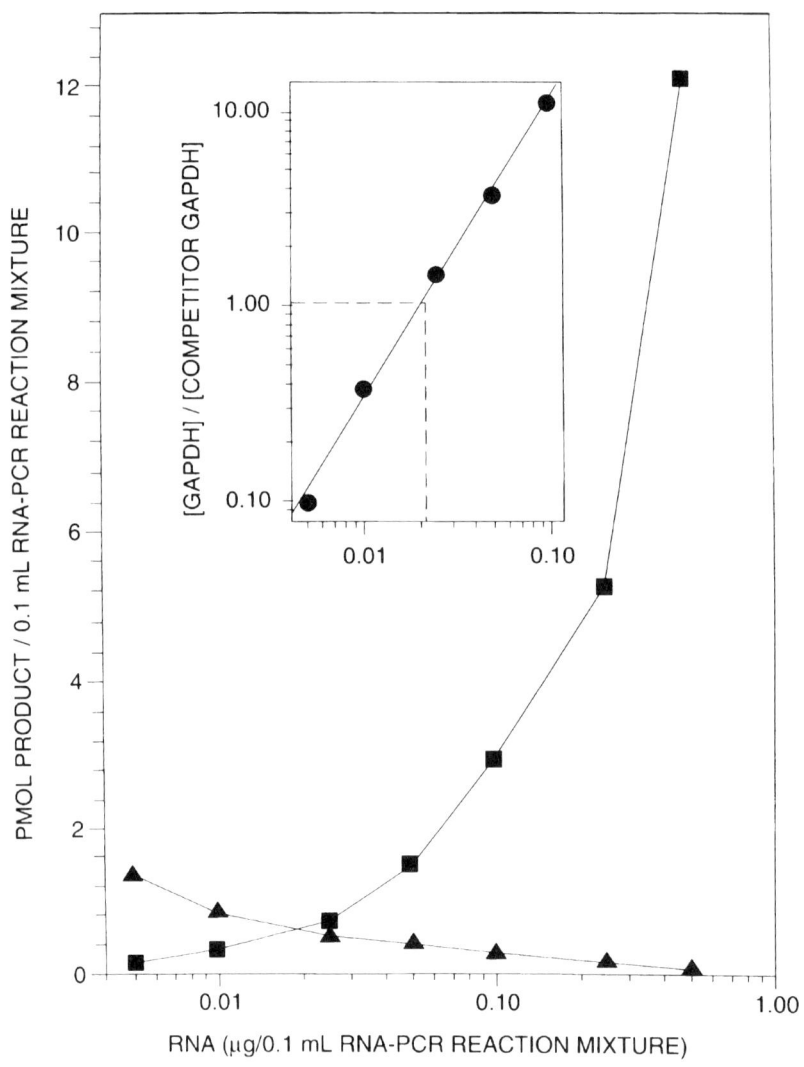

Figure 7.13 Formation of target (■) and competitor (▲) GAPDH during competitive PCR. Equal concentrations were reverse transcribed, mixed at a 1:1 ratio, and diluted. Competitor was kept constant at 0.1 amol/0.1 mL reaction mixture. *Inset:* Replot of the same data showing ratio (target/competitor) as the abscissa. [Reproduced with permission from Fasco et al., *Anal Biochem* **224:**140 (1995).]

data even after the reaction has reached plateau phase, since the ratio of target to competitor remains constant throughout amplification. This makes competitive PCR the preferred method for absolute quantitation of specific DNA or RNA sequences.

Quantitative RNA-PCR has used CE-LIF for detecting variations in gene expression—changes as small as two- to threefold in the amount of cellular messenger RNA coding for a particular protein (Fasco et al., 1995). As a practical application, competitive RNA-PCR was used to study the induction of the cytochrome P450-1A1 gene and relate its expression to that of a noninducible "housekeeping" gene (Piatak et al., 1993a), glyceraldehyde-3-phosphate dehydrogenase (GAPDH). In a variation on the method, the competitor concentration was held constant (at 0.1 amol/0.1 mL reaction mixture), while the target was titrated. Figure 7.13 shows a representative CE-LIF electropherogram for each of the competitive experiments involving P450-1A1 and

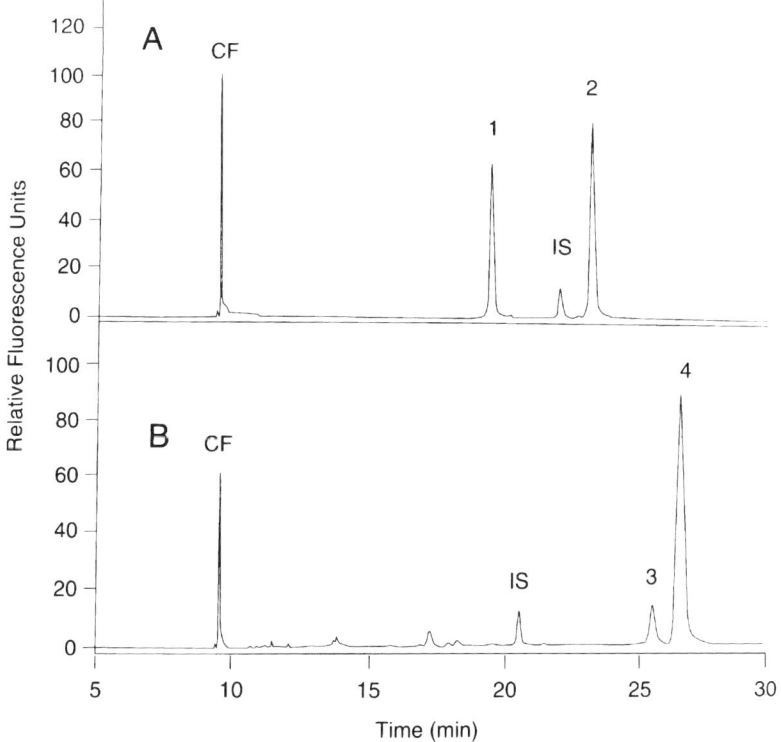

Figure 7.14 (*A*) CE-LIF electropherogram generated during competitive PCR of cytochrome P450-1A1: 1, competitor; 2, target; IS, internal standard; CF, 5-carboxyfluorescein. (*B*) Competitive PCR of GAPDH: 3, competitor; 4, target. [Reproduced with permission from Fasco et al., *Anal Biochem* **224**:140 (1995).]

GAPDH. Formation of product for target and competitor can be plotted versus the initial total mRNA concentration used in reverse transcription (Figure 7.14 for GAPDH alone). Alternatively, a log–log plot of the ratio of target to competitor could be plotted versus the initial total mRNA concentration, (see inset, Figure 7.13). The x value at a ratio of 1 corresponds to the initial mRNA concentration containing 0.1 amol of target GAPDH mRNA: in this case, 0.012 mg/0.1 mL reaction mixture.

7.5 CONCLUSION: FUTURE APPLICATIONS

In the field of molecular biology, new techniques using PCR are constantly evolving which require at the final step electrophoretic analysis. Potential for analysis by capillary electrophoresis has been demonstrated for techniques such as detection of single-stranded conformational polymorphisms (Kuypers et al., 1993), to determine point mutations in a genetic sequence; temperature gradient electrophoresis, to determine changes in DNA melting temperature due to sequence variations (Khrapko et al., 1994); and various hybridization regimes, designed to show DNA–DNA or DNA–protein interactions by an electrophoretic mobility shift of the hybrid (Maschke et al., 1993). In addition, since molecular biology will be increasingly automated in the years to come, one can envision that CE will be an integral part of automated analyzers designed for a variety of diagnostic, biomedical, and forensic research applications.

REFERENCES

Benson SC, Singh P, Glazer AN (1993) Heterodimeric DNA-binding dyes designed for energy transfer: Synthesis and spectroscopic properties. *Nucleic Acids Res* **21:**5727–5735.

Bianchi N, Mischiati C, Feriotto G, Gambari R (1994) Detection of hepatitis C virus by unbalanced polymerase-chain reaction, hybridization to synthetic oligonucleotides and capillary electrophoresis. *Int J Oncol* **4:**903–907.

Bocek P, Chrambach A (1991b) Electrophoretic size separations in liquefied agarose of polystyrene particles and circular DNA. *Electrophoresis* **12:**620–623.

Bocek P, Chrambach A (1992) Capillary electrophoresis in agarose solutions: Extension of size separations to DNA of 12 kb in length. *Electrophoresis* **13:**31–34.

Bocek P, Chrambach A (1991a) Capillary electrophoresis of DNA in agarose solutions at 40°C. *Electrophoresis* **12:**1059–1061.

Bode HJ (1977) The use of liquid polyacrylamide in electrophoresis. *Anal Biochem* **83:**364–371.

Butler JM, McCord BR, Jung JM, Wilson MR, Budowle B, Allen RO (1994) Quantitation of polymerase chain reaction products by capillary electrophoresis using laser fluorescence. *J Chromatogr B* **658:**271–280.

Cha RS, Thilly WG (1993) Specificity efficiency, and fidelity of PCR. *PCR Methods Appl* **3**:S18–S29.

Cha RS, Zarbl H, Keohavong P, Thilly WG (1992) Mismatch amplification mutation assay (MAMA): Application to the c-H-*ras* gene. *PCR Methods Appl* **2**:14–20.

Chelly J, Kaplan J-C, Gautron S, Kahn A (1988) Transcription of the dystrophin gene in human muscle and non-muscle tissues. *Nature* **333**:858–860.

Chen JW, Cohen AS Karger BL (1991) Identification of DNA molecules by precolumn hybridization using capillary electrophoresis. *J Chromatogr* **559**:295–305.

Chien RL, Burgi DS (1992) On-column sample concentration using field amplification in CZE. *Anal Chem* **64**:489A–496A.

Cohen AS, Terabe S, Smith JA, Karger BL (1987) High-performance capillary electrophoretic separation of bases, nucleosides, and oligonucleotides: Retention manipulation via micellar solutions and metal additives. *Anal Chem* **59**:1021–1027.

Cooksy K (1992) Capillary gel electrophoresis of PCR products: Easy sample preparation method. *CE Application Note #2.* J+W Scientific, Folsom, CA.

Fasco MJ, Treanor CP, Spivack S, Figge HL, Kaminsky LS (1995) Quantitative RNA–polymerase chain reaction—DNA analysis by capillary electrophoresis and laser induced fluorescence. *Anal Biochem* **224**:140–147.

Glazer AN, Rye HS (1992) Stable dye–DNA intercalation complexes as reagents for high-sensitivity fluorescence detection. *Nature* **359**:859–861.

Guilbault GG (1990) General aspects of luminescence spectroscopy in Guilbault GG (ed.): *Practical Fluorescence.* Marcel Dekker, Inc., New York, pp. 1–40.

Guttman A, Cooke N (1991a) Capillary gel affinity electrophoresis of DNA fragments. *Anal Chem* **63**:2038–2042.

Guttman A, Cooke N (1991b) Effect of temperature on the separation of DNA restriction fragments in capillary gel electrophoresis. *J Chromatogr* **559**:285–294.

Guttman A, Wanders B, Cooke N (1992) Enhanced separation of DNA restriction fragments by capillary electrophoresis using field strength gradients. *Anal Chem* **64**:2348–2351.

Haugland RP (1992) Nucleic acid stains. In *Moleuclar Probes: Handbook of Fluorescent Probes and Research Chemicals,* KD Laniston, Ed., pp. 221–228. Molecular Probes, Eugene, OR.

Heiger DN, Cohen AS, Karger BL (1990) Separation of DNA restriction fragments by high performance capillary electrophoresis with low and zero crosslinked polyacrylamide using continuous and pulsed electric fields. *J Chromatogr* **516**:33–48.

Huang Z, Gordon MJ, Zare RN (1988) Bias in quantitative capillary zone electrophoresis caused by electrokinetic sample injection. *Anal Chem* **60**:375–377.

Khrapko K, Hanekamp JS, Thilly WG, Belenkii A, Foret F, Karger BL (1994) Constant denaturant capillary electrophoresis (CDCE): A high resolution approach to mutational analysis. *Nucleic Acids Res* **22**:364–369.

Kornberg A (1980) *DNA Replication.* Freeman, San Francisco.

Kuypers AWHM, Willems PMW, van der Schans MJ, Linssen PCM, Wessels HMC, de Bruijn CHMM, Everaerts FM, Mensink EJBM (1993) Detection of point mutations in DNA using capillary electrophoresis in a polymer network. *J Chromatogr* **621**:149–156.

Landers JP, Oda RP, Spelsberg TC, Nolan JA, Ulfelder KJ (1993) Capillary electrophoresis: A powerful micro-analytical technique for biologically-active molecules. *BioTechniques* **14**:98–111.

Lumpkin OJ, Zimm BH (1982) Communications to the editor: Mobility of DNA in gel-electrophoresis. *Biopolymers* **21**:2315–2316.

Maschke HE, Frenz J, Williams M, Hancock WS (1993) Investigation of protein–DNA interaction by mobility shift assays in capillary electrophoresis. Poster T-121, presented at the Fifth International Symposium on High Performance Capillary Electrophoresis, Orlando, FL.

McCord BR, McClure DL, Jung JM (1993) Capillary electrophoresis of polymerase chain reaction-amplified DNA using fluorescence detection with an intercalating dye. *J Chromatogr* **652**:75–82.

Mullis KB, Faloona F (1987) Specific synthesis of DNA in vitro via a polymerase catalyzed chain reaction. *Methods Enzymol* **155**:335–350.

Nathakarnkitkool S, Oefner PJ, Bartsch G, Chin MA, Bonn GK (1992) High resolution capillary electrophoretic analysis of DNA in free solution. *Electrophoresis* **13**:18–31.

Ogston AG (1958) The spaces in a uniform random suspension of fibres. *Trans Faraday Soc* **54**:1754–1757.

Olivera BM, Baine P, Davidson N (1964) Electrophoresis of the nucleic acids. *Biopolymers* **2**:245–257.

Piatak M Jr, Saag MS, Yang LC, Clark SJ, Kappes JC, Luk KC, Hahn BH, Shaw GM, Lifson JD (1993a) High levels of HIV-1 in plasma during all stages of infection determined by competitive PCR. *Science* **259**:1749–1754.

Piatak M Jr, Luk KC, Williams B, Lifson JD (1993b) Quantitative competitive polymerase chain reaction for accurate quantitation of HIV DNA and RNA species. *BioTechniques* **14**:70–80.

Rossomando EF, White L, Ulfelder KJ (1994) Capillary electrophoresis: Separation and quantitation of RT-PCR products from polio virus. *J Chromatogr B* **656**:159–168.

Saiki RK, Scharf SJ, Faloona F, Mullis KB, Horn GT, Erlich HA, Arnheim N (1985) Enzymatic amplification of beta-globin sequences and restriction site analysis for diagnosis of sickle cell anemia. *Science* **230**:1350–1354.

Sambrook J, Fritsch EF, Maniatis T (1989) Analysis and cloning of eukaryotic genomic DNA. In *Molecular Cloning: A Laboratory Manual,* 2nd ed., N Nolan, C Nolan, and M. Ferguson, Eds., pp 9.31–9.62. Cold Spring Harbor Laboratory Press, Plainview, NY.

Sarkar G, Kapelner S, Sommer SS (1990) Formamide can dramatically improve the specificity of PCR. *Nucleic Acids Res* **18**:7465.

Schwartz HE, Guttman A (1995) *Separation of DNA by Capillary Electrophoresis*, Beckman Instruments Primer Series. Vol. VII.

Schwartz HE, Ulfelder KJ (1992) Capillary electrophoresis with laser-induced fluorescence detection of PCR fragments using thiazole orange. *Anal Chem* **64**:1737–1740.

Schwartz HE, Ulfelder KJ, Sunzeri FJ, Busch MP, Brownlee RG (1991) Analysis of DNA restriction fragments and polymerase chain reaction products towards detection of the AIDS (HIV-1) virus in blood. *J Chromatogr* **559**:267–283.

Schwartz HE, Ulfelder KJ, Chen F-TA, Pentoney SL Jr (1994) The utility of laser-induced fluorescence detection in applications of capillary electrophoresis. *J Capillary Electrophor* **1**:36–54.

Schwartz HE, Ulfelder KJ, Guttman A (1995) Injection related artifacts in capillary electrophoresis. Poster P-428, presented at the Seventh International Symposium on High Performance Capillary Electrophoresis, Würzburg, Germany.

Siebert PD (1993) Quantitative RT-PCR. *Methods and Applications Book 3*. Clonetech Laboratories, Inc. Palo Alto, California.

Southern EM (1975) Detection of specific sequences among DNA fragments separated by gel electrophoresis. *J Mol Biol* **98**:503–517.

Srinivasan K, Girard JE, Williams P, Roby RK, Weedn VW, Morris SC, Kline MC, Reeder DJ (1993a) Electrophoretic separations of polymerase chain reaction-amplified DNA fragments in DNA typing using a capillary electrophoresis–laser induced fluorescence system. *J Chromatogr* **652**:83–91.

Srinivasan K, Morris SC, Girard JE, Kline MC, Reeder DJ (1993b) Enhanced detection of PCR products through use of TOTO and YOYO intercalating dyes with laser induced fluorescence–capillary electrophoresis. *Appl Theor Electrophor* **3**:235–239.

Srivatasa GS, Batt M, Schuette J, Carlson RH, Fitchett J, Lee C, Cole DL (1994) Quantitative capillary gel electrophoresis assay of phosphorothioate oligonucleotides in pharmaceutical formulations. *J Chromatogr A* **680**:469–477.

Stellwagen NC (1987) Electrophoresis of DNA in agarose and polyacrylamide gels. In *Advances in Electrophoresis,* A Chrambach, MD Dunn, and BJ Radola, Eds., pp. 177–228. VCH Publishers, New York.

Tietz D, Gottlieb MH, Fawcett JS, Chrambach A (1986) Electrophoresis on uncrosslinked polyacrylamide: Molecular sieving and its potential applications. *Electrophoresis* **7**:217–220.

Ulfelder KJ (1994) Quantitative capillary electrophoretic analysis of PCR products using laser-induced fluorescence detection. Applications Information Bulletin A-1774. Beckman Instruments, Inc., Fullerton, CA.

Ulfelder KJ, Schwartz HE, Hall JM, Sunzeri FJ (1992) Restriction fragment length polymorphism analysis of *ERBB2* oncogene by capillary electrophoresis. *Anal Biochem* **200**:260–267.

Ulfelder KJ, Shieh P (1993) Capillary electrophoretic analysis of PCR products using laser-induced fluorescence detection. Poster T-211 presented at the Fifth International Symposium on High Performance Capillary Electrophoresis, Orlando, FL.

Ulfelder KJ, Dobbs M, Liu M-S (1995) Identification of PCR products by pre-column and on-column hybridization using capillary electrophoresis. Poster P-224, presented at the Seventh International Symposium on High Performance Capillary Electrophoresis, Würzburg, Germany.

Van der Schans MJ, Allen JK, Wanders BJ, Guttman A (1994) Effects of sample matrix and injection plug on dsDNA migration in capillary gel electrophoresis. *J Chromatogr A* **680**:511–516.

Zhu H, Clark SM, Benson SC, Rye HS, Glazer AN (1994) High-sensitivity capillary electrophoresis of double-stranded DNA fragments using monomeric and dimeric fluorescent intercalating dyes. *Anal Chem* **66**:1941–1948.

Zhu M, Hansen DL, Burd S, Gannon F (1989) Factors affecting free zone electrophoresis and isoelectric focusing in capillary electrophoresis. *J Chromatogr* **480**:311–319.

CHAPTER 8

Applications of HPLC/HPCE in Forensics

with Franco Tagliaro, Zdeneck Deyl, and Ivan Mikšík

OVERVIEW

In this book, the focus is on the use of high performance liquid chromatography (HPLC) for the separation and detection of biochemical components produced from enzymatic reactions. But the purpose of the book is also to introduce HPLC and high performance capillary electrophoresis (HPCE) as methods applicable for separation and detection of a variety of components, even those not produced by enzymatic reactions.

For example, HPLC and HPCE have shown great promise in forensics, the application of science to the judicial process. In forensics, HPLC and HPCE have been used to test for illicit drugs, gunshot residues and explosive constituents, pen inks, modified proteins, and nucleic acids. This chapter demonstrates the advantages in the application of these methods, particularly HPCE, to forensics. As the material illustrates, one reason for the success of these methods is that of necessity: forensics deals with minute amounts of sample. The capacity to perform analytical assays under these restrictive conditions is but one factor in the success of HPLC and HPCE for the tasks of forensic work.

This chapter discusses some key papers, with the goal of providing the basic elements of orienting a successful bibliographic search and navigating through a chaotic literature. We begin with an analysis of illicit or controlled drug preparations and proceed to the toxicological analysis of biological samples. Additional details on the HPLC and HPCE methods discussed in this chapter can be found in Chapter 2 and 3, respectively. This chapter also describes the use of the polymerase chain reaction (PCR) in forensics. The fundamentals of PCR are covered in Chapter 7.

8.1 INTRODUCTION

Despite the application of HPCE to analytical disciplines such as chemistry, biochemistry, biotechnology, molecular biology, clinical chemistry, pharmacology, and pharmaceutics, the forensic sciences have so far ignored the technique. This is paradoxical because studies published to date clearly show that HPCE is an ideal complement to the more traditional analytical techniques for solving many of the problems in this area (Northrop et al., 1994).

8.2 FORENSIC TOXICOLOGY

HPCE has been applied to the analysis of drugs and pharmaceuticals (Altria, 1993). In fact, determinations of drugs in pharmaceutical formulations represent one of the most rapidly growing areas for HPCE. In addition, pharmaceutical drug analysis is the starting point for the application of HPCE in forensic toxicology and the forensic sciences (Thormann et al., 1994).

8.3 HPCE ANALYSIS OF ILLICIT DRUG SUBSTANCES

Weinberger and Lurie (1991) first applied HPCE to an analysis of illicit drug substances. These authors used a micellar electrokinetic chromatography (MEKC, discussed in Chapter 3, Section 3.7.2) separation system in 50 μm i.d. bare silica capillaries (length 25–200 cm); the buffer consisted of 8.5 mM phosphate and 8.5 mM borate at pH 8.5 and contained 85 mM sodium dodecyl sulfate (SDS) and 15% acetonitrile. The applied voltage was 25 to 30 kV, and detection was by UV absorption at 210 nm.

These authors reported high efficiency separations of heroin, heroin impurities, degradation products, and adulterants (Fig. 8.1). Also discriminated were acidic and neutral impurities present in heroin seized by law enforcement agencies, as well as in illicit cocaine samples, with resolution of benzoylecgonine, cocaine, *cis*- and *trans*-cinnamoylcocaine. MEKC was also used with a broad spectrum of other compounds of forensic interest, including psilocybin, morphine, phenobarbital, psilocin, codeine, methaqualone, lysergic acid diethylamide (LSD), amphetamine, chlordiazepoxide, methamphetamine, lorazepam, diazepam, fentanyl, phencyclidine hydrochloride (PCP), cannabidiol, and tetrahydrocannabinol (THC), which were all separated with baseline resolution.

The analytical precision of MEKC was characterized by typical relative standard deviations (RSDs) of about 0.5% for migration times and of 4 to 8% for peak areas and peak heights. However, for the peaks with the longest migration times (> 40 min) the analytical precision was worse. This phenomenon was ascribed to the inconsistent evaporation of the organic modifier (acetonitrile) from the buffer reservoirs during the separation. Rapid "aging" of the running buffer, especially in relatively diluted buffers and with volatile organic modifiers, has also been observed. This problem can be overcome by frequent changes of the buffer and selection of electrolyte solutions with high buffering capacity at the working pH. In addition, the calculation of relative

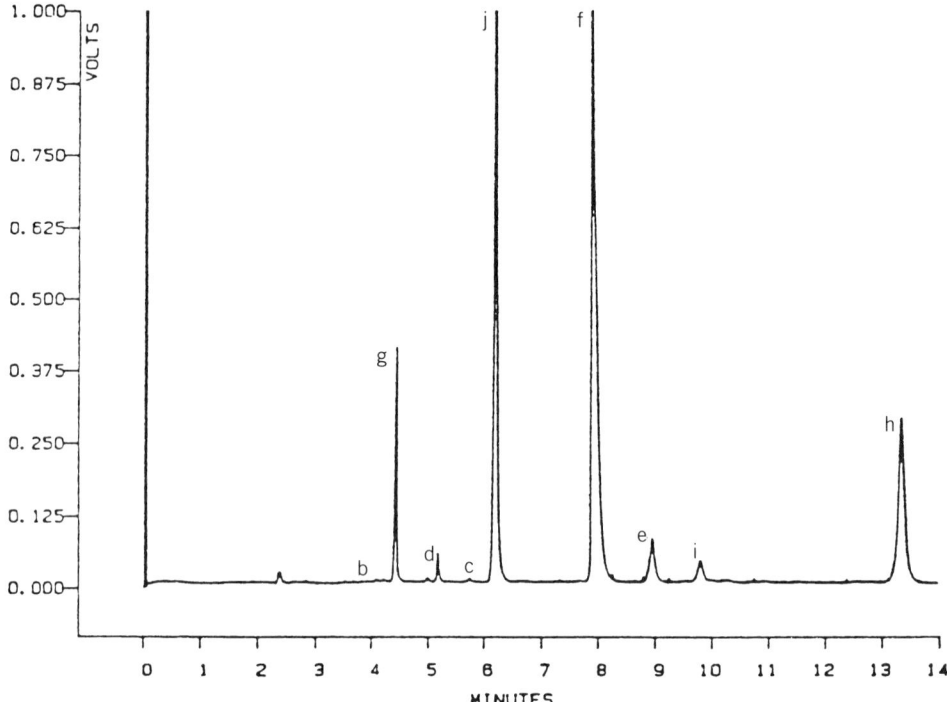

Figure 8.1 MEKC of bulk heroin, heroin impurities, degradation products and adulterants. Conditions: capillary, 25 cm × 50 mm i.d.; voltage, 20 kV; temperature, 40° C; buffer, 85 mM SDS–8.5 mM phosphate–8.5 mM borate–15% acetonitrile, pH 8.5; detector wavelength, 210 nm. *Peaks:* b, morphine; c, 3-MAM; d, 6-MAM; e, acetylcodeine; f, heroin; g, phenobarbital; h, noscapine; i, papaverine; j, methaqualone. (From Weinberger and Lurie, 1991, with permission.)

migration times based on suitable internal standards (IS) can improve the analytical reproducibility, not only in term of migration times but also quantitatively. Improvements in resolution reportedly resulted from the use of capillaries with smaller diameter (25 μm vs. 50 μm i.d.), which gave better heat dissipation and lower diffusion of analytes; sensitivity was reduced, however, because optical path length was reduced. Also, fluorescence detection wavelength was tested, and sensitivity was higher (up to 20 times) than that achieved by UV absorption for fluorescent components of the analyzed mixtures.

Another interesting aspect of this study was the comparison of MEKC and HPLC. While in MEKC the drugs represented in Figure 8.1 migrated in the order morphine (1), phenobarbital (2), 6-monoacetylmoprhine (6-MAM) (3), 3-monoacetyl morphine (3-MAM) (4), methaqualone (5), heroin (6), acetylcodeine (7), papaverine (8), and noscapine (9), in reversed-phase HPLC the elution pattern was substantially different and not correlated with MEKC

with, in order, compounds 1, 4, 3, 7, 6, 2, 9, 8, 5). This finding suggests that one technique should be used to confirm the results of the other. In complex mixtures, about twice as many peaks were observed with MEKC than with HPLC, but HPLC proved more sensitive (up to 80 times). Nevertheless, MEKC proved sensitive enough to detect heroin impurities down to 0.2%.

The successful use of MEKC for the analysis of illicit heroin and cocaine was also reported by Staub and Plaut (1994), who used 50 mM SDS in phosphate–borate buffer (10 and 15 mM, respectively) containing 15% of acetonitrile (pH 7.8). In 25 minutes, these authors achieved separation and determination of paracetamol, caffeine, 6-monoacetylmorphine (6-MAM), acetylcodeine, procaine, papaverine, heroin, and noscapine, although with peaks sometimes skewed. On-line recorded UV spectra of the peaks helped the identification of the individual peaks.

Krogh et al. (1994) adopted a similar analytical approach for the separation of test mixtures of the main alkaloids found in illicit heroin and heroin adulterants (14 compounds), as well as of 7 drugs structurally related to amphetamine. These investigators used a 50 cm $\times$ 50 μm i.d. bare silica capillary and a running buffer of 25 mM SDS in phosphate–borate buffer (10 mM for each salt) pH 9, containing 5% acetonitrile; UV detection occured at 214 nm. Excellent separations were achieved in 15 minutes, with RSDs migration times (relative to crystal violet, the IS) ranging from 0.5 to 1.9% and from 0.89 to 2.23% in within-day and between-day reproducibility tests, respectively. In quantitative studies, standard curves were linear in the range from 0.02 to 0.5 mg/mL, with typical correlation coefficients from 0.997 to 0.999. RSDs were in the range from 2.0 to 4.3% in the analysis of illicit heroin and amphetamine. Reportedly, samples could be injected every 13 minutes, and the fused silica capillary was replaced only after 500 injections to assure good reproducibility. According to the authors, MEKC proved to be a valuable complement to HPLC and GC for the routine analysis of illicit drug preparations.

Trenerry et al. (1994a, 1994b) proposed the use of 50 mM cetyltrimethylammonium bromide (CTAB), instead of SDS, as micellar agent in MEKC. Also, to overcome the reproducibility problems with late peaks observed by Weinberger and Lurie (1991) and ascribed to inconsistent evaporation of acetonitrile from the buffer vials and to capillary wall fouling, Trenerry et al, suggested the following measures: replacing acetonitrile with the less volatile dimethyl sulfoxide (10%), flushing the capillary with the running buffer between runs, and periodically washing the capillary with 0.1 M NaOH and replacing the running buffer with fresh solution.

The use of the cationic micellar agent CTAB (50 mM) in phosphate–borate buffer (10 mM of each salt), pH 8.6, with 10% acetonitrile was preferred because of a faster separation (about 15 min) of heroin and related substances. Because the cationic surfactant, which coats the capillary silica wall with a positively charged layer, reverses the electroosmotic flow (EOF), the voltage (-15 kV) must be applied with a reversed polarity (with the cathode at the injection point). Detection was by UV absorption at 280 nm.

This method was successfully tested with different illicit heroin preparations, and when the results were compared with those obtained with HPLC, a good quantitative correlation was observed. Precision was slightly less in MEKC than in HPLC, but the resolution of complex samples of illicit heroin was better with MEKC.

The same method, with only slight variations (acetonitrile from 10 to 7.5%, detection wavelength from 280 to 230 nm) was adopted for the analysis of illicit cocaine. Benzoylecgonine, cocaine, and *cis*- and *trans*-cinnamoylcocaine, obtained from samples taken from cocaine seizures, were separated. MEKC results correlated well with those from gas chromatography (GC), with similar RSD values, and with HPLC. These authors reported that MEKC methods also proved highly reliable for heroin and cocaine analysis in interlaboratory proficiency tests.

The separation of enantiomers of amphetamine, methamphetamine, ephedrine, pseudoephedrine, norephedrine, and norpseudoephedrine was reported by Lurie (1992) as a means of investigating the synthesis of illicit drug preparations. A derivatization with a chiral reagent, 2,3,4,6-tetra-*O*-acetyl-β-D-glucopyranosyl isothiocyanate, was used, followed by the MEKC separation of the resulting diastereomers. All the enantiomers of the above-mentioned phenethylamines were resolved in a single run, using a bare silica capillary (48 cm $\times$ 50 μm i.d.) and a mixture consisting of 20% methanol and 80% aqueous buffer solution (100 mM SDS, 10 mM phosphate–borate buffer, pH 9.0).

More recently, Lurie et al. (1994) reported the chiral resolution by means of neutral and anionic cyclodextrins (CDs) of a number of basic drugs of forensic interest: amphetamine, methamphetamine, cathinone, methcathinone, cathine, cocaine, propoxyphene, and various α-hydroxyphenethylamines. In this separation, the resolution was optimized by varying the ratio of neutral to anionic CDs. In fact, both types of CD have chiral selectivity, but because of their negative charge, anionic CDs display an electrophoretic countermigration and consequently high retarding effect on analytes.

Chiral resolution was also achieved by means of CE chromatographic techniques, with an enantioselective stationary phase, as reported by Li and Lloyd (1993), who used α_1-acid glycoprotein as stationary phase packed in fused silica capillaries of 50 mm i.d. These authors reported the optimization (by varying pH, electrolyte, and organic modifier concentration in the mobile phase) of the separation of the enantiomers of hexobarbital, pentobarbital, isofosfamide, cyclophosphamide, diisopyramide, metoprolol, oxprenolol, alprenolol, and propranolol.

An excellent review of the applications of HPCE for the analysis of seized preparations of illicit drugs was published by Lurie (1994).

An approach suitable for drug screening alternative to MEKC was proposed by Chee and Wan (1993), who used capillary zone electrophoresis (CZE) with 50 mM phosphate buffer pH 2.35 in a 75 μm i.d. (60 cm long) bare silica capillary. In 11 minutes, they achieved the separation of 17 basic drugs of potential forensic interest: methapyrilene, brompheniramine, amphetamine, methamphetamine, procaine, tetrahydrozoline, phenmetrazine, butacaine, medazepam, lidocaine, codeine, acepromazine, meclizine, diazepam, doxapram, benzocaine, and methaqualone. Detection was by UV absorption at 214 nm.

It was observed that a drug with lower pK_a, hence less positive charge, showed higher migration times. Clear correlation between pK_a values and migration times, however, could not be obtained because of the influence of other factors (according to the authors: molecular size, tendency to interact with the column, and ability to form doubly charged species).

Reproducibility in migration times was characterized by RSD values better than 1%; peak area RSDs ranged from 1.5 to 4.3%. Worse reproducibility was found for analytes with very slow migration (i.e., with pK_a values close to the pH of the background buffer). Notwithstanding an inherent suitability for direct screening of forensic samples, the above-described method was applied only for the analysis of biological fluids. After a simple one-step chloroform–isopropanol (9:1) extraction from plasma and urine samples, previously adjusted to pH 10.5, "clean" electropherograms were obtained, which allowed detection limits of about 0.50 μg/mL of each drug.

Chee and Wan (1993) believe that CZE offers some advantages over MEKC for drug screening, particularly, simple background electrolyte preparation and shorter analysis times. They see the main limitation as the inability to analyze acidic, neutral, and basic drugs together. An additional advantage offered by CZE is that a peculiar separation mechanism, poorly correlated with MEKC, allows the use of this technique in parallel with MEKC for confirmation purposes, with the same instrumental hardware.

MEKC is almost universally adopted for illicit drug analysis not only in pharmaceutical or illicit formulations, but also in biological samples.

Wernly and Thormann (1991) used a phosphate–borate buffer pH 9.1 with 75 mM SDS for the qualitative determination in urine of many drugs of abuse (and their metabolites), including benzoylecgonine, morphine, heroin, 6-monoacetylmorphine, methamphetamine, codeine, amphetamine, cocaine, methadone, methaqualone, and benzodiazepines.

Sample purification and concentration used "double-mechanism" (cation exchange and reversed phase) solid phase extraction cartridges. The extracts from 5 mL of urine were dried, redissolved with 100 μL of running buffer, and injected, to achieve a detection limit of 100 ng/mL. For peak identification, not only the retention times were used, but also the on-line-recorded UV spectra of the peaks, which were compared to computer-stored models.

Because of a sensitivity comparable to that obtainable from nonisotopic immunoassays, as well as low running costs and possibility of automation, Wernly and Thormann proposed MEKC for confirmation testing, following immunometric screenings.

An MEKC system quite similar to that described earlier (50 mM SDS in phosphate–borate buffer, pH 7.8) allowed, also, a high resolution separation of barbiturates, including barbital, allobarbital, phenobarbital, butalbital, thiopental, amobarbital, and pentobarbital (Thormann et al., 1991). Again, on-column multiwavelength detection helped peak identification. Sensitivity was in the order of the low micrograms per milliliter. It is interesting to note that while urine samples needed extraction prior to injection, human serum can be injected directly because some barbiturates, including phenobarbital, elute in an interference-free window of the electropherogram.

Another MEKC separation (75 mM SDS, phosphate–borate buffer, pH 9.1) was reported for the determination of 11-nor-Δ-9-tetrahydrocannabinol-9-carboxylic acid, the major metabolite of Δ-9-tetrahydrocannabinol present in urine (Wernly and Thormann, 1992a). Sample treatment included basic hydrolysis of urine (5 mL), solid phase extraction, and concentration. The resulting sensitivity was 10 to 30 ng/mL (i.e., comparable to the cutoffs of immunoassays). Again, detection was by on-line recording of peak spectra, by means of fast-scanning UV detector.

Wernly et al. (1993) reported also an attempt to use MEKC, without hydrolysis, to determine morphine-3-glucuronide, the major metabolite of morphine (and heroin) in urine. The sensitivity limit (about 1 μg/mL after solid–phase extraction and concentration) was unsatisfactory for confirmation of the results of the usual enzyme immunoassays, but improvements were deemed to be achievable.

Quite recently, Schafroth et al. (1994) determined the major urinary compounds of eight common benzodiazepines (flunitrazepam, diazepam, midazolam, clonazepam, bromazepam, temazepam, oxazepam, and lorazepam). The used MEKC with 75 mM SDS in a phosphate–borate buffer (pH 9.3). After enzymatic hydrolysis and extraction with commercial "double-mechanism" cartridges, the sensitivity was reportedly better than that obtained with the common immunoassay EMIT (enzyme-multiplied immunoassay technique).

The relatively poor sensitivity of HPCE in terms of concentration requires that the sample be concentrated severalfold before injection, if an acceptable degree of sample sensitivity is to be achieved. This concentration is possible provided the biological material is cleansed of all the interfering substances, first of all organic and inorganic ions, that, if present at high concentrations in the injected solution, would alter completely the separation (causing peak distortion and loss of efficiency). Wernly and Thormann (1992b) described a stepwise solid phase extraction for human urine preliminary to MEKC, using commercial "double-mechanism" cartridges exhibiting hydrophobic and ion-exchange interactions. This extraction produced "clean" electropherograms, even injecting extracts from a urine sample concentrated 50 times (Fig. 8.2).

In the same paper, these authors also observed that MEKC with surfactants in plain aqueous buffers failed to resolve the highly hydrophobic analytes (e.g., amphetamine, methamphetamine, methadone, benzodiazepines). Resolution of these compounds could be obtained by adding low percentages of acetonitrile (5–10%), thus bringing about MEKC conditions resembling those reported by Weinberger and Lurie (1991).

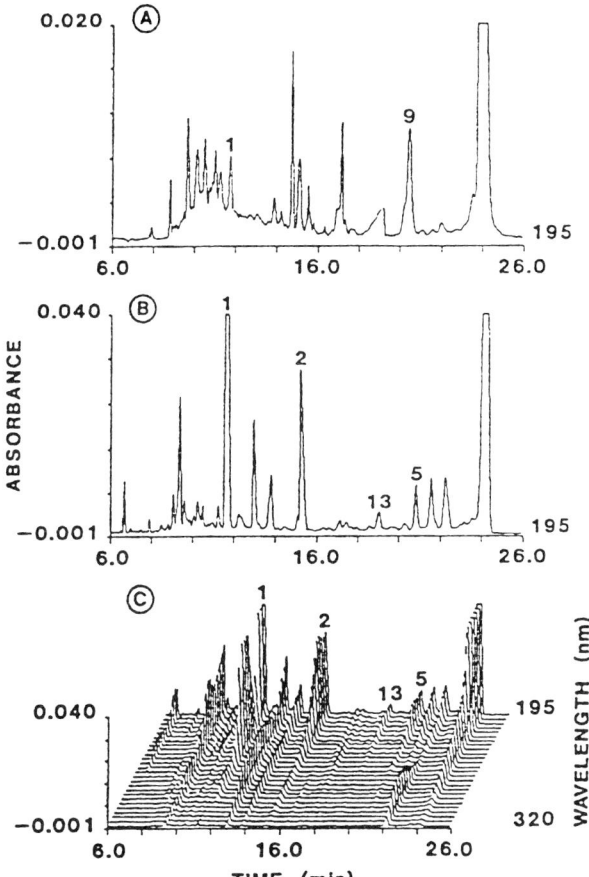

Figure 8.2 Multiwavelength and single-wavelength (195 nm) MEKC electropherograms obtained with the fractions of a two-step extraction of a patient sample, which tested positively for several classes of drugs. *Peaks:* 1, benzoylecgonine; 2, morphine; 5, codeine; 9, methaqualone; 13, 6-MAM.

Thormann et al. (1993) have published an overview of the strategies for using MEKC to monitor drugs in body fluids (serum, urine, saliva); they discuss buffer selection and sample preparation (direct injection, ultrafiltration, solid phase extraction.

The comparative use of MEKC, CZE, and capillary isotachophoresis (CITP) for the determination of drugs in body fluids was reported by Caslavska et al. (1993). Salicylate, paracetamol, and antiepilectics in serum and urine were analyzed with the three techniques. In case of high drug concentrations, body fluids could be injected directly, or with simple dilution (for urine) or ultrafiltration (for serum). Extraction and concentration were necessary for drugs present in the low micrograms per milliliter levels, however. The authors concluded that MEKC and CZE could be applied more easily, whereas CITP required careful selection of buffers and, in general, was less sensitive.

The analysis of hair samples is gaining acceptance in the forensic toxicology environment as a tool for investigating past chronic exposure to illicit drugs (Tagliaro et al., 1993b; Kintz, 1995). The usual analytical strategy, after hair collection, extraction, and extract purification, is based on radioimmunologic screening followed by chromatographic (GC, GC-MS, HPLC) confirmatory analyses.

The use of HPCE for this purpose shows potential advantages related not only to the peculiar separation mechanism, but also to the minimal need of sample, which, for aesthetic reason, is scarce in hair analysis. In recent reports on the analysis of hair samples, Tagliaro et al. (1993a, 1993c) adopted a CZE mode because it appeared to be easier to use and therefore to offer greater possibility for replication by other investigators. The CE background buffer was composed of 50 mM borate, pH 9.2. Simultaneous detection of cocaine and morphine was accomplished at 200 nm wavelength, but a higher selectivity was obtained at the absorbance maxima of each analyte (i.e., 238 nm for cocaine; 214 nm for morphine. Tetracaine and nalorphine were adopted as internal standards for cocaine and morphine, respectively. The separations were highly efficient (up to 350,000 theoretical plates) and repeatable (migration time RSDs: <1% intraday, <3% interdays), and the determination were accurate and precise (intraday RSDs in the range of 3–5%).

Once again, because of the minute (nanoliter) volumes injected and the moderate concentration sensitivity of HPCE with UV detection (in the order of the low micrograms per milliliter), the limit of detection in hair was acceptable (<0.2 ng/mg) only if the hair extracts (from about 100 mg of sample) were reconstituted with 10 to 20 μL of solvent, a procedure too delicate for routine applications. More recently (Tagliaro et al., 1995) reported the use of sample-stacking techniques to allow the injection of about 10 times larger volumes (i.e., about 50 nL), thus permitting reconstitution of the extracts with as much as 100 μL of water, without sacrificing efficiency. The same paper described the application of MEKC to the analysis of hair extracts using a buffer containing 0.1 M SDS in 25 mM borate with 20% methanol. The sensitivity achieved was only slightly less than that obtained with CZE, but the selectivity was much higher because the SDS micelles were able to exploit the additional "reversed-phase-like separation mechanism.

8.4 ANALYSIS OF GUNSHOT RESIDUES AND CONSTITUENTS OF EXPLOSIVES

Forensic investigators analyze gunshot and explosive residues not only to identify persons involved in crimes but also for intelligence purposes (i.e., in analyzing the methods of terrorist and criminal organizations).

Two types of component are important for this purpose: inorganic constituents and organic residues. However, despite a great deal of effort worldwide, using the most sophisticated (and expensive) instrumentation, including neutron activation analysis, mass spectrometry, X-ray and infrared techniques, and a variety of chromatographic techniques, problems persist in the development of methods for routine applications.

HPCE offers the unique possibility of analyzing both organic and inorganic compounds because the separation mechanisms are unique and because a minimal amount of sample is needed for analysis. In an early paper (if not the first) reporting the use of HPCE in this field, Northrop et al. (1991) discussed the performance of MEKC in the separation and determination of some of the constituents of gunshot and explosive materials. Their results demonstrated the possibilities of HPCE for identification of residues. Uncoated silica capillaries 67 cm long with i.d. ranging from 50 to 150 mm were used in experiments aimed at optimizing the separation and at studying the effect of different conditions. The buffer was 2.5 or 5.0 mM in borate, in the pH range 7.8–8.9, and contained SDS concentrations from 10 to 50 mM. The applied potential was 20 kV for the separation, but 5 kV, for 2 seconds, for electrokinetic injections. Detection was by UV absorption. After optimization, the authors decided on 100 mm i.d., 2.5 mM borate, 25 mM SDS, and a detection wavelength of 250 nm (200 nm for nitroglycerin). As many as 11 components of gunshot residues (test mixture), namely, nitroguanidine, nitroglycerin, 2,4-dinitrotoluene (DNT), 2,6-DNT, 3,4-DNT, 2,3-DNT, diphenylamine (DPA), N-nitroso-DPA, 2-nitro-DPA, ethylcentralite, and dibutyl phthalate were all resolved in 10 minutes, with efficiencies between 200,000 and 400,000 theoretical plates and mass detection limits of fractions of nanograms. The results from HPCE were reportedly superior to those from supercritical fluid chromatography.

The same MEKC system was also used for the separation of a test mixture of 15 compounds of interest in high explosive analysis, including nitroguanidine, ethylene glycol dinitrate, diethylene glycol dinitrate, 1,3,5-trinitro-1,3,5-triazacyclohexane (RDX), nitroglycerin, 2,4,6-trinitrotoluene (TNT), pentaerythritol tetranitrate, and picric acid, with excellent resolution except for the overlapping of 1,5- and 1,8-isomers of dinitronaphthalene. Also, separation of all the components (26) of the two sets of standards was attempted with extremely limited coelutions.

In addition, since many compounds have characteristic UV absorption profiles, multiwavelength analysis helped their identification. Because the instrument used was not capable of peak spectra acquisition, these authors had to perform multiple runs of the same sample at different wavelengths and draw absorbance profiles from the comparison of the peak heights in the individual electropherograms. More modern instrumentation fitted with diode array or fast-scanning UV detectors would facilitate analyses of this type.

Application to forensic cases concerned the investigation of spent ammunition casings, reloading powders (IMR 700X, W-W 452AA, HRD, W-W 296,

W-W 748, IMR 4831), and plastic explosives (Semtex, C4, Detasheet, Tovex). Compositional differences among reloading powders were found in products from different manufacturers, but also between powders from the same producers. The reasons for these differences were discussed by the authors.

In a subsequent paper, the same authors (Northrop and MacCrehan, 1992) optimized the sample collection and preparation procedures for the MEKC analysis of gunshot residues. Instead of the widely used swabbing techniques, the authors suggested sample collection with masking adhesive tape (1-inch-square sections). The sample area of interest was, as usual in case of users of revolvers and pistols, the back of the shooting hand along the thumb and forefinger and the webbing between these fingers. The tape "lifts" were stored sealed and refrigerated. Each lift was examined with a binocular stereoscope for gunshot residue particles, which were collected with tweezers and placed in a glass microvial. The particles were added with 50 mL of ethanol and extracted for 30 minutes in an ultrasonic bath. Ethylene glycol (1 mL) was then added to the vial, the solvent was evaporated under a nitrogen stream, and finally the residue was reconstituted with 25 mL of MEKC buffer. Alternatively, sections of the tape lifts were extracted with ethanol, observing the procedure just described.

According to the authors, adhesive tape lifts were clearly superior to swabbing, which was discarded because of unacceptable losses of analytes and high recovery of interferences from the skin. Adhesive tape lifts provided a much cleaner material, without interfering compounds (fats and oils) from the skin. Because of the small sample size required for HPCE, a single particle collected from the skin surface could be subjected to analysis. However, the quality of the adhesive tape was fundamental for the success of the assay; a test of several tapes comparing adhesive character, resistance to extraction solvents, and lack of coeluting interferent peaks showed that masking tape was the most suitable for MEKC analysis.

Another problem the authors addressed was the loss of analytes during evaporation of the extraction solvent. For this purpose, the addition of a small percentage of ethylene glycol, a nonvolatile "keeper," according to the authors, prevented the sample from going to dryness and also avoided adsorption losses. An additional advantage was that ethylene glycol, with a relatively high viscosity, acted to equalize the slight viscosity differences between standard solutions and real samples, which produced some nonreproducibility in migration times. Quantitative results were improved by using an internal standard (β-naphthol). In firing range experiments with subjects who fired different weapons, characteristic gunshot residue constituents were found on each adhesive tape lift on which they were expected to be present, but not on blanks.

HPCE was also applied in the analysis of low explosive ionic residues and compared to ion chromatography (Hargadon and McCord, 1992). Anions and cations left behind from the blast represent useful pieces of evidence from which to infer the type and source of the explosive material used. Although ion chromatography is the ideal tool for this purpose, it suffers from the lack of an adequate complementary technique for peak confirmation. Again, HPCE, which is based on completely different separation and detection principles, gives a unique opportunity to confirm ion analysis results.

To this aim, Hargadon and McCord used CZE (65 cm × 75 mm i.d. capillary) in borate buffer (2 mM borate, 40 mM boric acid) containing 1 mM diethylenetriamine as EOF modifier, at the final pH of 7.8. The applied potential was 20 kV, with reversed polarity. The addition to the running buffer of a dichromate chromophore (1.8 mM) permitted the use of indirect UV detection at 280 nm.

The most important ions present at the blast residue are chloride, nitrite, nitrate, sulfate, sulfide, chlorate, carbonate, hydrogen carbonate, cyanate, thiocyanate, and perchlorate, which are susceptible of CZE determination in a single run. A comparison CZE with ion chromatography showed extensive differences between the relative separation patterns, with a clear advantage of CZE in efficiency and of ion chromatography as far as capacity is concerned. The reproducibility of the elution times in CZE was evaluated over 2 months with the same capillary, with resulting RSD better than 1%, which could be improved by adding bromide as a marker for calculating relative migration times.

It was also observed that peak identity could be checked by comparing the electropherograms recorded at 280 nm with other electrophoretic profiles recorded at 205 nm, a wavelength at which nitrite, nitrate, and thiocyanate absorb, generating positive peaks. Anions that do not absorb also at this wavelength produce negative peaks as at 280 nm. The sensitivity of CZE, with a signal-to-noise ratio of 3, was 0.5 mg/mL, whereas for ion chromatography it was 2 mg/mL; the dynamic range was up to 50 mg/mL for CZE and up to 200 mg/mL for ion chromatography.

The practical consequences of these studies was the discovery that solutions suitable for ion chromatography must be diluted before CZE. This analytical strategy was tested with pipe bomb fragments experimentally prepared with different explosive mixtures (potassium chlorate–Vaseline, black powder, smokeless powder, a mixture of black and smokeless powders) and detonated. Fragments from bombs were extracted with water, filtered, and analyzed. Dual-ion chromatography and CZE analysis were carried out with concordant results.

8.5 ANALYSIS OF PEN INKS

Forensic science ink analysis is one of the most traditional analytical fields. Currently, thin-layer and column chromatography as well as slab gel electrophoresis are used to investigate ink composition, but, recently, CZE has been tentatively applied, with encouraging results (Fanali and Schudel, 1991).

Reversed-polarity separations were carried out in a silica-coated capillary (200 mm × 25 mm i.d.) using 0.1 M ammonium acetate buffer (pH 4.5), with methanol (3:1). Detection was by UV absorbance at 206 nm. Commercially available fiber-tip pen inks were examined, and differences between the relative electrophoretic patterns from different inks were reported. Excellent peak shape and resolution of several components was observed, but quantitation was not attempted. In addition, dye spots separated by thin-layer chromatography were extracted and added to the same ink sample to check by analyte addition the identity of the well-resolved, but individually unknown CZE peaks.

8.6 ANALYSIS OF PROTEINS IN FORENSICS

8.6.1 Separation of Proteins by HPCE

Detailed discussion of the use of HPCE for separation of proteins are available in specialized reviews (Karger, 1989; Rohlicek and Deyl, 1989; Steuer et al., 1990; Mazzeo and Krull, 1991; Deyl et al., 1994a, 1994b; Li, 1994).

In general, three methods may be used for protein separation by HPCE. It is possible by exploiting the charge differences to separate proteins according to their effective charge in acid or alkaline media. Acid buffers are usually preferred for complex protein mixtures, even though the runs may easily take an hour because of the decreased EOF. In alkaline pH the peaks as a rule get sharper and the runs shorter (about 30 min), but fused peaks are frequently observed. The other possibility is to exploit the differences in molecular dimensions and perform protein separations in gel-filled capillaries. Here the separations can be related to molecular mass. The advantage of these separations is an easy interpretation; the disadvantage is that most gel-filled capillaries can withstand only a limited number of runs. This disadvantage has been overcome by using replaceable gels, where the sieving medium is simply added to the run buffer (Cohen et al., 1993; Werner et al., 1993). The third possibility is to add suitable ampholytes to the background electrolyte and to run the capillary under capillary isoelectric focusing (CIEF) conditions. In this operating mode, proteins, after they have been focused, stop, and their zones sharpen. To reach this stage, EOF must be abolished by suitable coating of the capillary (Yao and Regnier, 1993).

Protein in untreated capillaries tend to stick to the inner capillary walls. Several approaches have been used to abolish or at least minimize this problem (McCormick, 1988; Bushey and Jorgenson, 1989; Green and Jorgenson, 1989; Emmer et al., 1991; Kajiwara, 1991). The easiest approach is to perform the separations at very high or very low pH values. In the former case, dissociation of the free amino groups is suppressed and, consequently, interactions between the dissociated silanol groups of the capillary surface and the free amino groups of the protein are minimized. At extremely acid pH, the opposite effect can be introduced; namely, hindrance of dissociation of the silanol groups of the capillary, accompanied by nearly complete dissociation of the free amino groups of the proteins. As a result, the ionic interactions between the protein and the capillary wall are minimized.

Other techniques for suppressing or decreasing the dissociation of both the separated solutes and silanol groups of the capillary include increasing the ionic strength of the run buffer—for example, by adding metal salts (Green and Jorgenson, 1989). However, the increase of conductivity resulting from higher salt concentrations in the buffer requires the use of lower voltages and capillaries of smaller diameter, to provide adequate heat dissipation. Another way of avoiding protein adsorption to the capillary wall is to use zwitterionic buffers, which permit the use of high salt concentrations without excessive heating during the separation (Bushey and Jorgenson, 1989).

Hydrophobic domains in protein molecules may constitute another cause of adsorption to capillary walls. Such interactions can be abolished by using fluorosurfactant buffer additives; however, charge reversal can occur at the surface of the capillary (Emmer el al., 1991). Proteins at a pH below their pI are repelled from the wall. High efficiency separations can be obtained in this arrangement at low ionic strengths. Hydrophobic interaction electrophoresis can be performed in the presence of amphophilic polymers (e.g., stearoyl dextran), together with ethylene glycol. These additives change the degree of the hydrophobic interaction between the protein and the capillary wall, improving resolution (Kajiwara, 1991).

Alternatively, protein sticking may be prevented by modification of the capillary surface. Successful separations can be obtained with capillary columns coated with glycidoxy-propyltrimethoxysilane and polyethylene glycol (Bruin et al., 1989). This approach is limited to separations at pH 3 to 5 because at higher pH values the coatings becomes unstable. Other coatings are represented by polyethyleneimine (Towns and Regnier, 1990) and treatment with Tween and Brij (Towns and Regnier, 1991). These coatings can be used successfully over a wide pH range, typically 3 to 15. The deactivation of the capillary surface through aryl pentafluoro modifiers represents another possibility (Swedberg, 1990; Maa et al., 1991). Also C_8 and C_{18} modifications have been successfully used (Dougherty et al., 1991). Modification of the inner surface of the capillary with vinyl-bonded polyacrylamide (Cobb et al., 1990) not only eliminates sorption of the separated proteins but eliminates the EOF as well. Similar effects were observed with polymethylglutamate-coated capillaries (Bentrop et al., 1991). Hydrophilic coatings with hydroxylated polyether functionalities represent another way of successful shielding of the capillary wall (Nashabeh and El Rassi, 1991).

In using this approach to prevent sticking, the problem of coating stability must be considered. Therefore the procedure referred to as dynamic coating was introduced (Wu and Regnier, 1991). A number of commercially available coated capillaries are currently marketed.

Protein sorption to the capillary wall is not an issue with gel-filled capillaries because the wall is shielded by the gel filling. Gel-filled capillaries can be either prepared in the lab or purchased. An easy and inexpensive approach is the dynamic filling of the capillary with a diluted gel solution (typically 4% linear polyacrylamide) (Wu and Regnier, 1992).

Whereas in free capillaries the addition of SDS to protein samples before separation results in a loss of resolution, with gel-filled capillaries the runs can be performed in the presence of SDS, and separations result from the sieving effect of the gel.

8.6.2 Separation of Protein Mixtures by Two-Dimensional Techniques

Another broad field is the separation of complex protein mixtures. While separations using monodimensional techniques have had only limited success, applications of two-dimensional techniques are emerging (Giddings, 1984; Davis and Giddings, 1985; Giddings, 1987).

The first attempt at a two-dimensional approach involved the coupling of HPCE with HPLC. Because coupling of CZE with HPLC results in incompatibility of the time constants of the two separation steps, HPLC (gel permeation chromatography) should always be the first separation step, followed by the electrokinetic separation. Lemmo and Jorgenson (1993) developed such a system for two-dimensional separations, combining HPLC gel chromatography and CZE, and it exhibits practical applicability. The size exclusion was carried out in a 1 m × 250 mm microcolumn. The effluent from this size-exclusion chromatography microcolumn filled a sample loop on a computer-controlled six-port valve. The contents of the loop were transferred past the grounded end of the electrophoresis equipment capillary for electromigration injection. Detection was by UV absorbance at 214 nm. The success of this arrangement was based on the extremely fast separations in the HPCE section. The overall electrophoresis separation should take 4 minutes or less, to allow for sufficiently fast sampling of the gel column effluent.

Another problem for two-dimensional separations occurs when the first dimension involves silica-based size-exclusion chromatography of proteins, performed at high salt concentration, often of the order of 0.5 M. Such salt concentrations are not compatible with electrophoresis systems, which in most cases are run in an environment that includes 50 mM buffer. To solve the problem, Zorbax GF 450, which has relatively low excess surface charge, was proposed as a suitable packing material for the size-exclusion step (Lemmo and Jorgenson, 1993). This addition permitted the size-exclusion separation step to be carried out at a low salt concentration, which was compatible with subsequent electrophoresis. However, this change limited the method to the separation of proteins with pI values less than 8.23. When the pI was higher, there was adsorption of proteins both to the size-exclusion column packing and to the capillary wall of the electrophoresis equipment, distorting the separations.

8.7 ANALYSIS OF NONENZYMATICALLY MODIFIED PROTEINS

In the analysis of proteins in forensic applications, the chemical modifications that occur to proteins, posttranslationally and nonenzymatically, are of primary importance. These chemical changes are a result of chemical reactions between side chains of the protein and reactive groups of metabolites and/or exogenous toxicants, including drugs present in extracellular fluid such as serum. The analytical accessibility of these modified proteins depends on their rate of turnover. For example, those with a slow turnover rate will be long-lived, and such problems will be much more easy to identify than those with faster turnover rates.

This section discusses two families of structural proteins, namely, extracellular keratins and collagens. Each example represents a family of closely related chemical entities, which means dealing with complex protein mixtures. We consider two types of chemical modifications: modifications arising from increased levels of glucose in body fluids (obviously related to diabetes) and those derived from acetaldehyde, the first metabolic product of ethanol ingestion. As noted, a common feature of the latter nonenzymatic modification reactions is the presence of an aldehydic functionality in the nonprotein reactant (Jelínková et al., 1995; Deyl and Mikšik, 1995).

8.7.1 Nonenzymatic Modifications to Keratins by Ethanol

It has been reported that in the aldehydic and/or glucose modification of proteins, only some lysine residues of the protein molecule are modified (Harding et al., 1989; Monnier et al., 1992; Vlassara et al., 1994). The reasons for these dissimilar results appear to reflect the structural differences in the amino acid residues surrounding the lysine residues. The first step in the glucose-mediated reaction is the formation of an aldimine followed by a rearrangement of the reaction product to an Amadori product and a set of further reactions, as demonstrated in Figure 8.3. As a consequence of these reactions, there is a decrease in the number of free amino groups in the protein and an increase in the effective negative charge of the modified protein. Thus, provided a sufficient number of modifications have occurred along the polypeptide chain, these modified proteins will migrate more swiftly to the anode. However, modification of the positively charged lysine residues will change the hydrophobic properties of the protein as well (Reiser et al., 1992).

An example of the effects of aldehydic changes was obtained by monitoring the effect of ethanol ingestion on the keratin taken from rat hair (Jelínková et al., 1995). In this study CE profiles of hair keratins taken from rats fed 10% ethanol and from rats fed water were compared by single-dimension

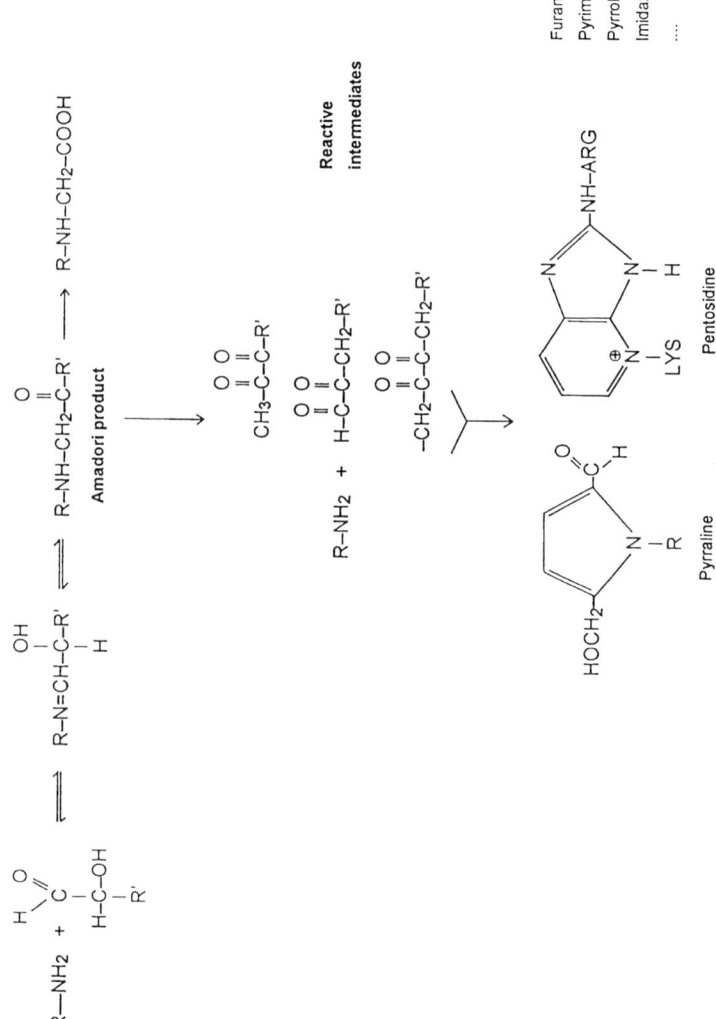

Figure 8.3 Scheme of the Maillard reaction.

8.7 ANALYSIS OF NONENZYMATICALLY MODIFIED PROTEINS 181

electrophoresis (Fig. 8.4). The peaks can be placed in two groups: the peaks in the first group, with a retention time of about 5 minutes (two in the controls and three in the ethanol-treated animals) correspond to the fast-moving (narrow) zone shown in panel A_1. The peaks in the second group, moving with the retention time of 7.5 to 10 minutes, and additional peaks moving with retentions of 15 to 18 minutes, correspond to the other (broad) zone seen panel A_1.

In contrast to the results with single-dimension electrophoresis are the findings reported from two-dimensional gel separations (lower panels of Fig. 8.4). Two groups of spots are observed: a group of keratin proteins moving fast in the cathodic direction and a group distributed along a digonal. The literature suggests the two fast-moving peaks in the controls and the three in the ethanol-fed rats are low-sulfur keratins, while those in the second group, the diagonal smear, are high-sulfur keratin proteins.

This conclusion can be tested by passing the keratin proteins through a cation-exchange cartridge that retains only low-sulfur proteins. The retained fraction (released by washing the cartridge with 0.1 M triethylamine followed by overnight dialysis of the filtrate) corresponds to the first two (or three in alcohol-treated animals) fractions appearing during the capillary electrophoresis runs (see Figure 8.5A,C). The unretained proteins correspond to the remaining, more anodically moving peaks. Indeed, amino acid analysis of the unretained fraction indicated a higher proportion of sulfur-containing amino acids in the unretained fraction, justifying the conclusion that the proteins in group 1 are low-sulfur and those in group 2 are high-sulfur proteins.

As mentioned above, the profiles in Figure 8.4 also indicate the presence of an additional low-sulfur component (arrow) in the alcohol-fed animals. This additional component has an apparent relative molecular mass of 70,000.

Further investigation of the high-sulfur proteins by two-dimensional electrophoresis under alkaline conditions was not possible because of smearing in the high-sulfur protein zone. Therefore two-dimensional electrophoresis separations were run at pH (Fig. 8.6). These separations showed two spots in the 25,000–35,000 relative molecular mass region. In CZE separations (Fig. 8.6A,B) run at pH 3.5, two distinct peaks appeared between 10 and 20 minutes running time in samples obtained from the alcohol-treated animals (indicated by arrows). These peaks were absent in control rats. When the fraction retained in the ion-exchange cartridge was run under similar conditions, the results seen in Figure 8.6C were obtained. By comparing the two sets of results, it is possible to identify the peaks moving within 12 minutes and less as low sulfur. Also, peaks moving more rapidly to the anode represent the high-sulfur fraction. This type of analysis leads to the conclusion that of three additional peaks of the alcohol-treated animals, two belong to the high-sulfur proteins, while the third fraction may be classified as low in sulfur.

Further characterization of the remaining fraction in the low-sulfur keratin group in the ethanol-treated animals included analysis of the trypsin-released peptides by plasma desorption mass spectrometry. These results indicated the

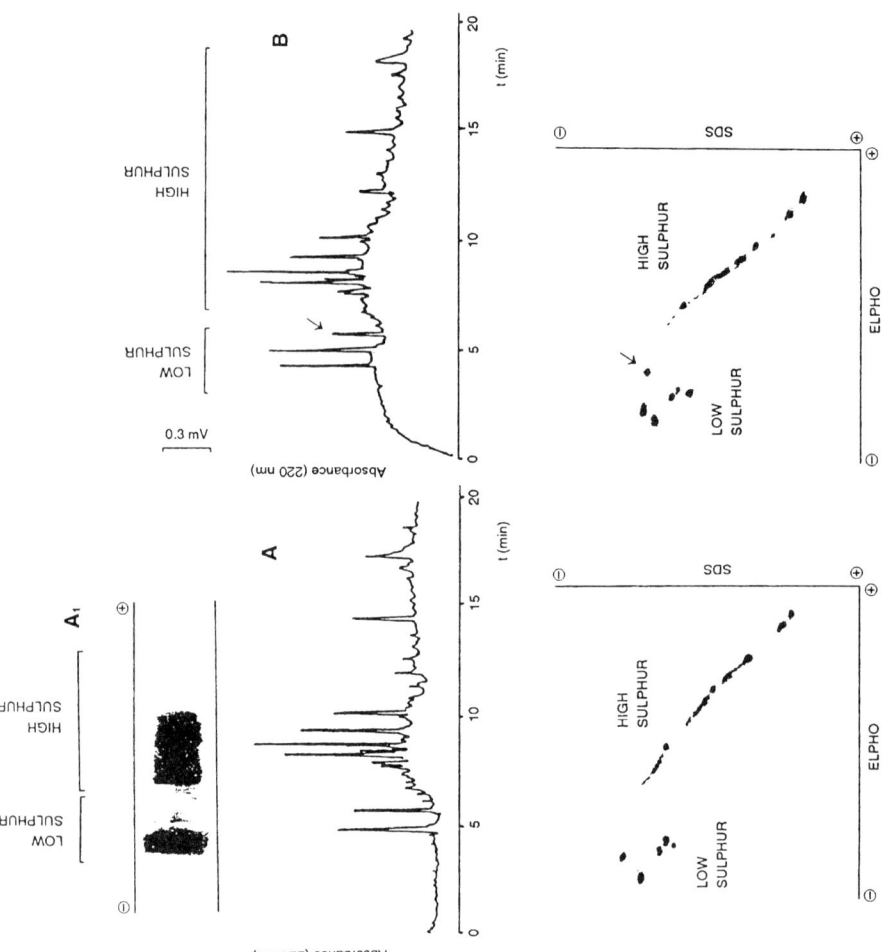

presence of a peptide differing in its molecular mass by three acetyl residues from another peptide found in controls. This finding suggests that three lysine residues were modified through the reaction with acetaldehyde.

The foregoing analysis makes clear that capillary electrophoresis can reveal differences in keratin proteins obtained from animals treated with ethanol compared to controls. However, it is also apparent that because of the complexity of the protein matrix, other methods must be used as well to identify the modified fraction.

8.7.2 Nonenzymatic Modifications to Collagen by Glucose

Another example of the application of capillary electrophoresis for revealing posttranslationally modified proteins is the separation of collagen polypeptide chains after modification by glucose. Deyl and Mikšík (1995) performed separations using 4% polyacrylamide gel-filled capillaries. As shown in Figure 8.7, SDS electrophoresis in capillaries containing linear polyacrylamide offers good separations of collagen's polypeptide chains, as well as any polymers, separations that are at least comparable with those resulting from slab gel electrophoresis. Moreover, CZE separations offer the possibility of quantitation, by measuring the area of individual peaks, a difficult procedure with slab gels. Also, CE in non-cross-linked gels offers the possibility of separating polymers, (i.e., of components having molecular mass $\geq 300,000$). This capability is of considerable importance, because no method offering sufficient selectivity is available for analyzing such polymers. Even in diluted polyacrylamide slab gels, such high relative molecular mass fractions either stick to the starting line or move as a smear between the γ fraction (the collagen α-chain trimer) and the start. Other methods, such as gel permeation low pressure column chromatography, do not provide sufficient selectivity.

Incubation of a collagen type I sample with glucose produces the results seen in Figure 8.8. Such profiles exhibit a twofold difference from untreated collagen electropherograms: the α peaks are split into two each, and the peaks of γ-chain polymers and higher are increased. Both these effects can be ascribed to the nonenzymatic reaction of the ε amino groups of lysine residues with glucose. Reiser (1991) reported that such reactions are nonspecific, affecting several lysine residues in the collagen molecule and, upon prolonged incubation, leading to progressive insolubilization of the collagen

---

Figure 8.4 Comparison of capillary electrophoretic (ELPHO) profiles (pH 9.2) of hair keratin samples obtained from control rats (A) and alcohol-treated rats (B). The dilution of extracts was 50 mL to 5 mL. Position of the peak present in alcohol-loaded animals but absent in controls is indicated by an arrow. Assignment of "high-sulfur" and "low-sulfur" proteins is based on the results shown in Figure 8.5. (From Jelínková et al., 1995, with permission.)

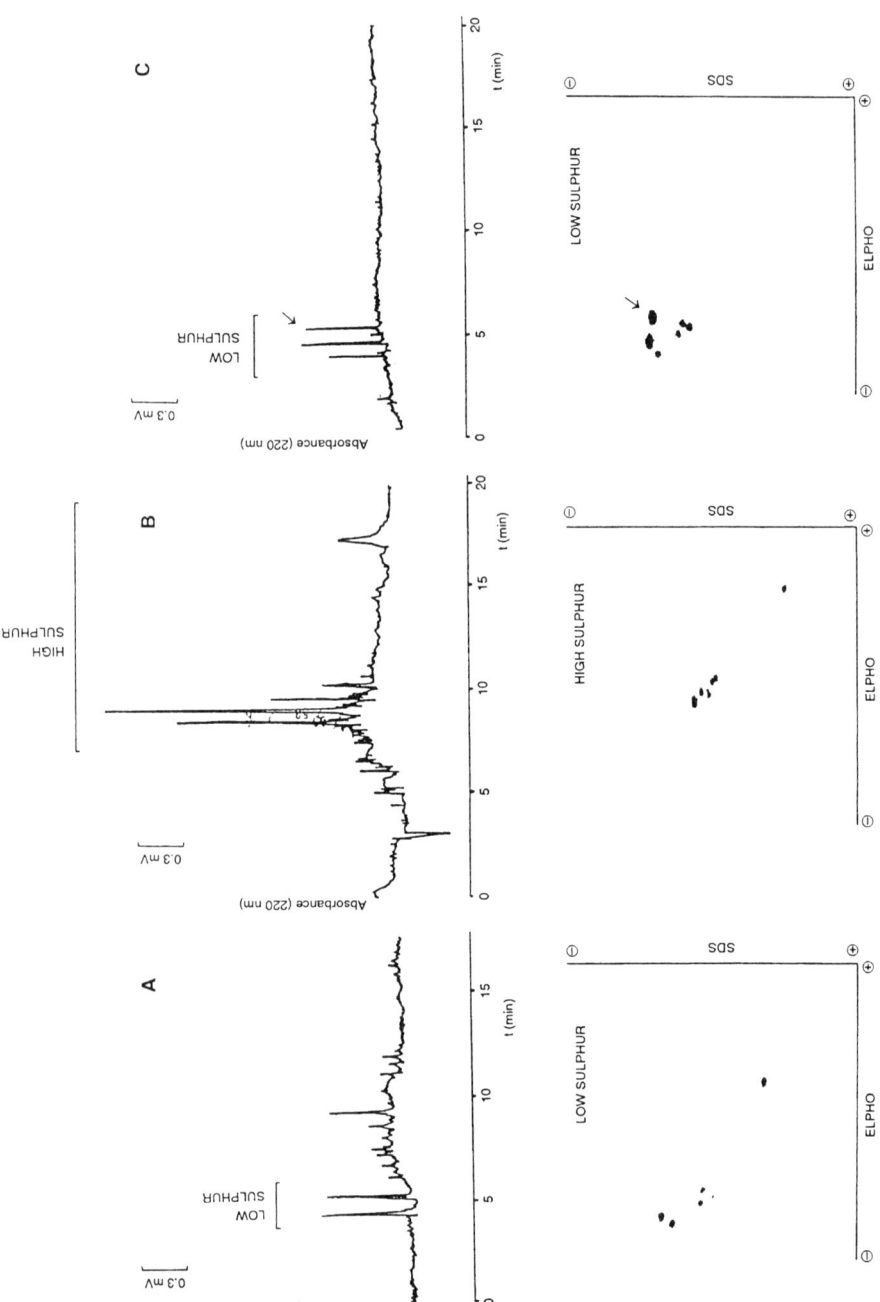

proteins. Based on these observations, it is reasonable to assume that splitting of the two α-chain peaks into four, following incubation with glucose, may reflect nonspecific binding of glucose to several sites of collagen α chains, resulting in a decrease of positive charges of these polypeptide chains. Because these separations are performed with excess SDS, splitting of the α peaks can hardly be attributed to the change in effective charge of individual polypeptide chains. It is more probable that glycation alters the hydrophobic properties of collagen α chains and modifies the amount of SDS bound to the protein molecule. Decreased binding of SDS to proteins, which possess a high proportion of sugars (though O-glycosidically bonded), has been well documented. The occurrence of polymers with molecular mass higher than γ-collagen molecules, as well as the increased proportion of γ-chain polymers in the overall profile, can be ascribed to further polymerization of constituting α chains, which is known to lead, under in vitro conditions, to complete insolubilization of collagen proteins.

8.8 ENZYMATIC ACTIVITY ASSAY BY CAPILLARY ELECTROPHORESIS

CZE offers the potential of ultramicroanalyses of enzymatic activities in biological samples. This application could be useful in forensic cases when only trace amounts of biological specimens are available—for example, in the detection of amylase activity from saliva extracted from stamps or envelopes.

The idea for microassays was developed by Bao and Regnier (1992) and demonstrated in their work with glucose-6-phosphate dehydrogenase. Their approach begins with the familiar mechanism of enzyme-catalyzed reactions, in which a substrate S reacts with an enzyme E to form an enzyme–substrate complex ES. This complex dissociates to form a product P and the enzyme. Because E, S, and ES are different molecular entities, they may have different net charges and may exhibit different electrophoretic mobilities. Bao and Regnier (1992) take as a hypothetical example an enzyme with a net charge $+10$, a substrate with a net charge -2, and a product with a net charge -1. Given the relative magnitudes of these charges, the mobility of the enzyme

---

Figure 8.5 CZE profiles (pH 9.2) of the retained (A) and unretained (B) fractions obtained after passing the crude extract through Bio-Rad AG 50W-X8 filter, and of proteins retained on the cation-exchange cartridge from the alcohol-treated rat hair preparation (C). High-sulfur proteins were recovered in the filtrate, while low-sulfur proteins were released by 0.1 M triethylamine. Comparison with the two-dimensional gel separations is visualized under each CZE run. The position of the fraction of low-sulfur proteins occurring in alcohol-treated animals but absent in controls is indicated by an arrow. (From Jelínková et al., 1995, with permission.)

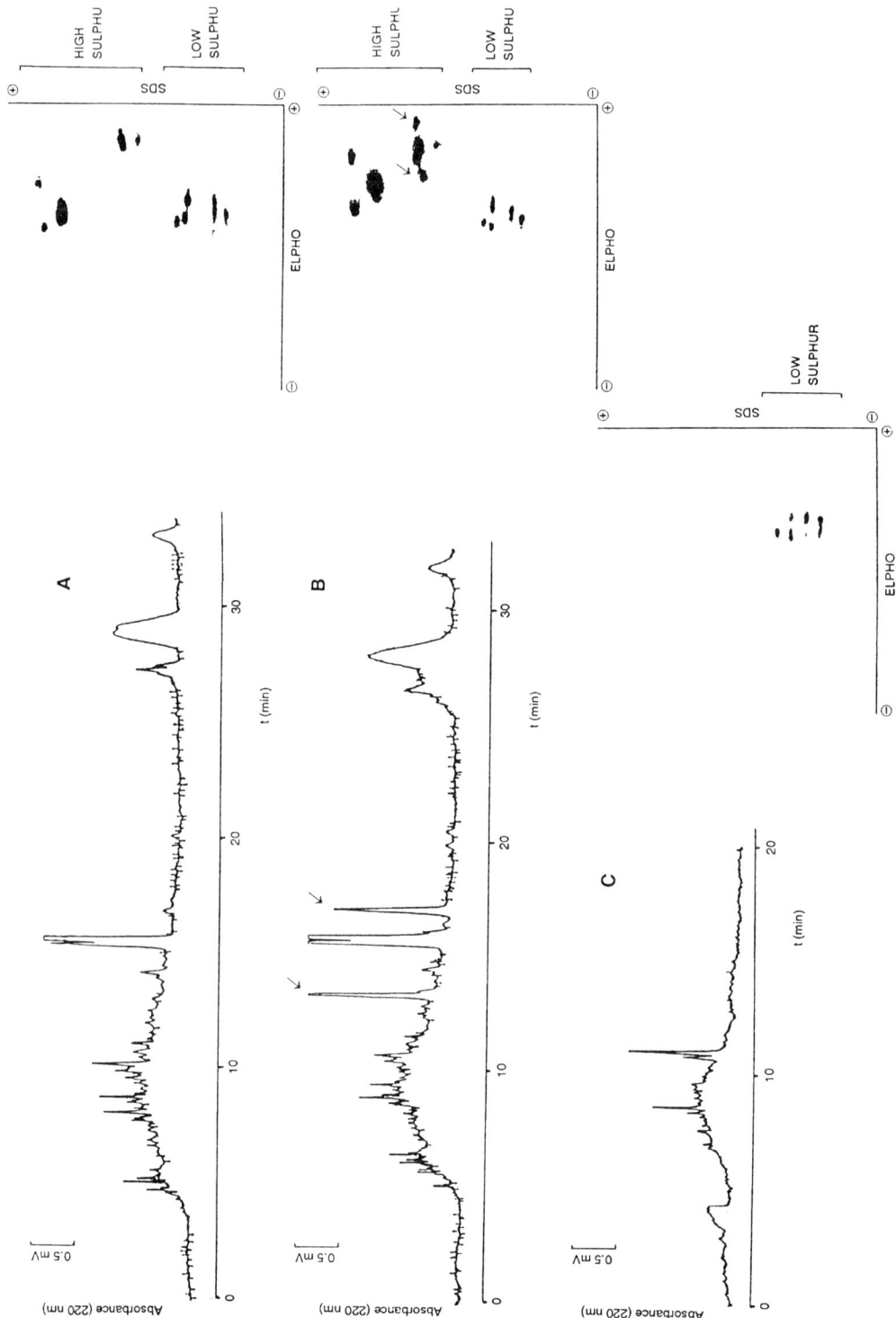

would be expected to be greater than that of either the substrate or the product, and we can predict that migration will occur in the opposite direction. By applying the electric potential across the capillary, it is possible to design a CZE system in such a way that the band of the enzyme will overrun the band of the substrate. Using this example, Bao and Regnier (1992) introduced the idea of electrophoretic mixing. They argue that there is no dilution in such a system; moreover, turbulence is not required and, consequently, spreading of the zone will be minimal. The only band-spreading force will be diffusion. All this, however, occurs in the absence of the EOF. Finally, Bao and Regnier emphasize that such a mixing can be quite fast, of the order of seconds.

When the substrate and the enzyme are electrophoretically mixed, most of the enzyme will appear in the enzyme–substrate complex. Returning to the example used by Bao and Regnier, the net charge of the enzyme–substrate complex will be $10 - 2 = 8$. Since enzyme activity measurements are performed under substrate-saturating conditions, all the enzyme will be bound to substrate, and the mobility of the enzyme should be the mobility of the ES complex.

Following ES formation, the complex is incubated to initiate the reaction. To stop the flow, the potential is brought to zero. With neither electrophoretic nor electrokinetic transport taking place, all the components of the mixture will remain within a single, rather narrow zone. Another method for incubation is to run the mixture under constant potential. In this method the reactants and products will be separated and mixed continuously. At first glance it might seem that in this method an assay would be precluded because the enzyme is continuously being separated from the product. This, however, is not the case, as in practice the enzymatic reaction occurs at a rate orders of magnitude faster than that of the separation. Bao and Regnier (1992) calculated that product formation could be 100 to 100,000 times greater than the amount of the enzyme, depending on the turnover number of the enzyme.

Finally, when the product is formed in the system and separated from the parent compound in some eletrophoretic mode, it will be detected. Clearly, at this stage, the product has to be transported to the detector by a combination of electrokinetic and electrophoretic processes.

A practical arrangement could be visualized as follows: A surface-deactivated capillary is filled with an enzyme and a saturating concentration of the substrate. Buffer and other cofactors are added, as well. The enzyme is introduced as in any other capillary electrophoretic system (e.g., either

---

Figure 8.6 CZE profiles and two-dimensional electrophoresis of rat hair extracts under acidic conditions. CZE run at pH 3.5, two-dimensional separation at pH 3. (*A*) Controls, (*B*) alcohol-treated rats, 7 weeks, (*C*) fraction of control animals retained on the Bio-Rad AG 50W-X8 filter and released with 0.1 *M* triethylamine corresponding to "low-sulfur" proteins. (From Jelínková et al., 1995, with permission.)

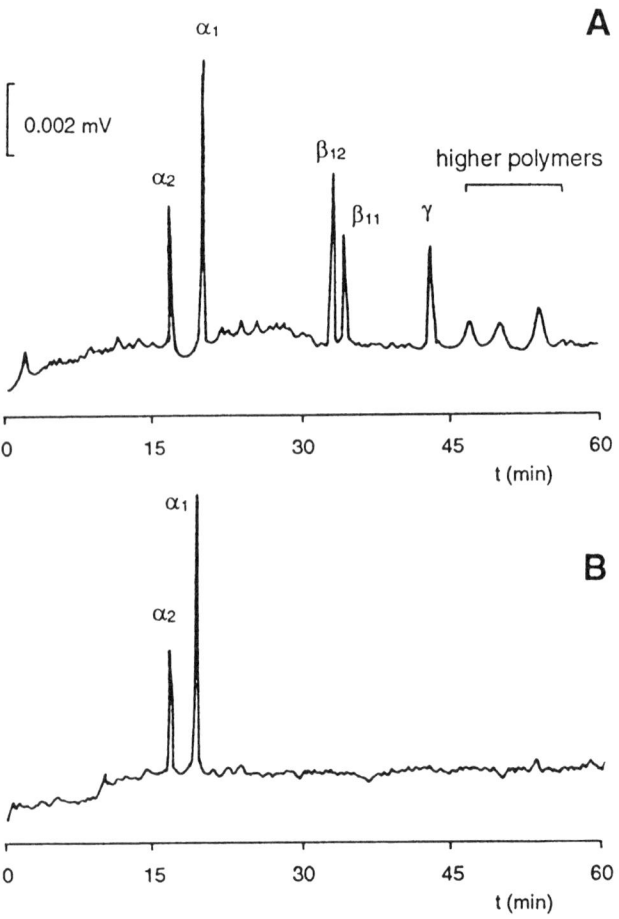
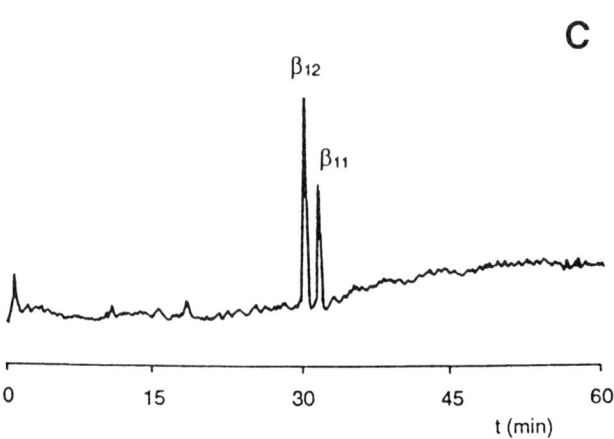

electrokinetically or hydrodynamically). When the potential is switched on, the components will mix and the reaction will be started. With few exceptions, the product and the substrate–enzyme complex will be transported at different velocities. Product formation will continue until the enzyme leaves the system. The theoretical elution profile will be as shown in Figure 8.9. Here it is assumed that the enzyme–substrate complex moves with a higher velocity than the product (Fig. 8.9A). The first product detected at point A corresponds to the enzyme as it migrates past the detector. The spike at position B is that of the product formed during the short interval between introducing the enzyme into the system and switching on the potential. If the transport velocity of the product is higher than that of the enzyme–substrate complex, the electropherogram will be reversed, as shown in Figure 8.9B. At point A, we observe the artificial spike caused by the existence of the interval between introducing the enzyme and switching on the potential. From the position of the artificial peak on the electropherogram, it is possible to determine the relative migration velocities of the enzyme–substrate complex and the product. The level of the plateau indicated as C in Figure 8.9A,B is directly proportional to enzyme concentration at constant potential. Figure 8.9C depicts the situation that may occur with an isoenzyme mixture and a common product.

Switching the system to zero potential for a fixed period of time within the time window bracketing the passage of the enzyme by the detector would allow for product accumulation. When the power is switched on again, the enzyme will be separated from the product and, if the product has suitable detection properties, it will appear as a peak on the enzyme plateau. This approach is sometimes referred to as "parked reaction." A practical application featuring the activity assay of D-glucose-6-phosphate:NADPH oxidoreductase is shown in Figure 8.10.

A less complex approach can be used for enzyme activity estimation; namely, fractions can be collected and enzymatic activity measured in these fractions. The practical realization involves a normal run before fraction collection to detect the migration time of the peak, which should be collected. The time T_c when the peak is going to leave the capillary can be estimated as retention time multiplied by the length of the capillary divided by the capillary length to the detector. When fractions are to be collected, the potential is switched off just before T_c. A new vial is placed at the end of the separation

---

Figure 8.7 Capillary gel electrophoresis profiles of collagen chains and chain polymers in 4% polyacrylamide; 75 mm i.d. capillary, 45 cm long (35 cm to the detector). Tris–glycine buffer, pH 8.8, 10 kV. Detection by UV absorbance at 220 nm. (*A*) Complete sample. (*B*) α-Region proteins sampled from a preceding slab gel run. (*C*) As in (*A*), but β-region excised from the slab gel. The appearance of the gel separation is shown at top (cathode on the right side of the gel). (From Deyl and Mikšik, 1995, with permission.)

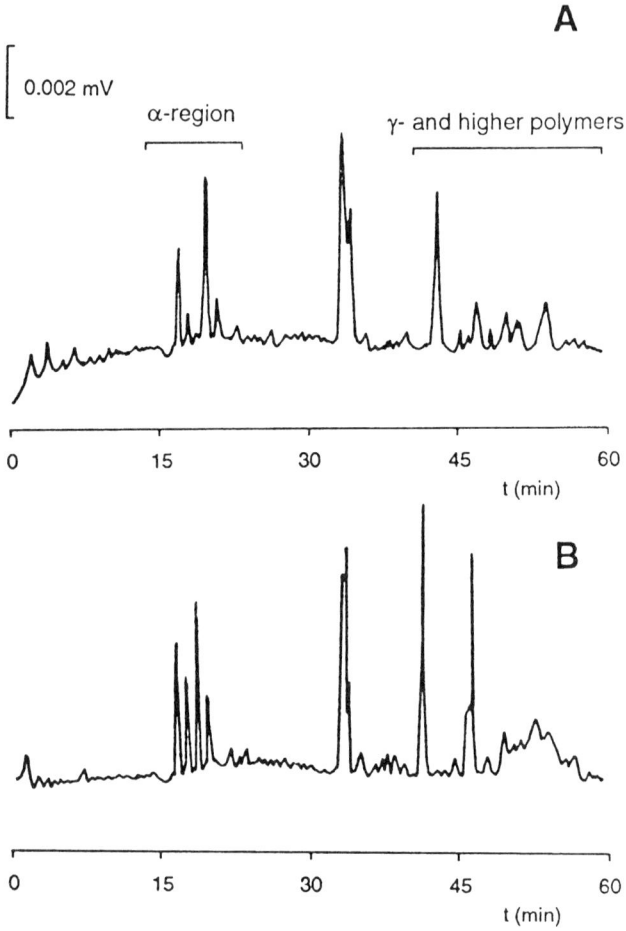

Figure 8.8 Capillary gel electrophoresis profiles of collagen α-chains and chain polymers in 4% polyacrylamide; after incubation for 4 days (*A*) and 7 days (*B*) at 30°C. Other conditions as in Figure 8.7. Note the splitting of peaks in the α-region and the increase of γ and higher chain polymers. Unincubated profile (control) is shown in Figure 8.7*A*. (From Deyl and Mikšik, 1995, with permission.)

capillary, the potential is switched on again, and the peak is forced to move to the ample collection vial. In practice in this step the potential was set to 7.5 kV, which made the enzyme peak to move into the fraction collection vial within 0.6 minute. The substrate is added to the vial of the collected peak, the vial is incubated for a defined period of time, and the product of enzymatic reaction may be assayed (e.g., in a second capillary electrophoresis run). This approach was used for assaying proteolytic activity in fermentation broth; but with slight modifications, the principle is applicable to any other enzyme source (Mulholland et al., 1993). In this particular case, to obtain as much

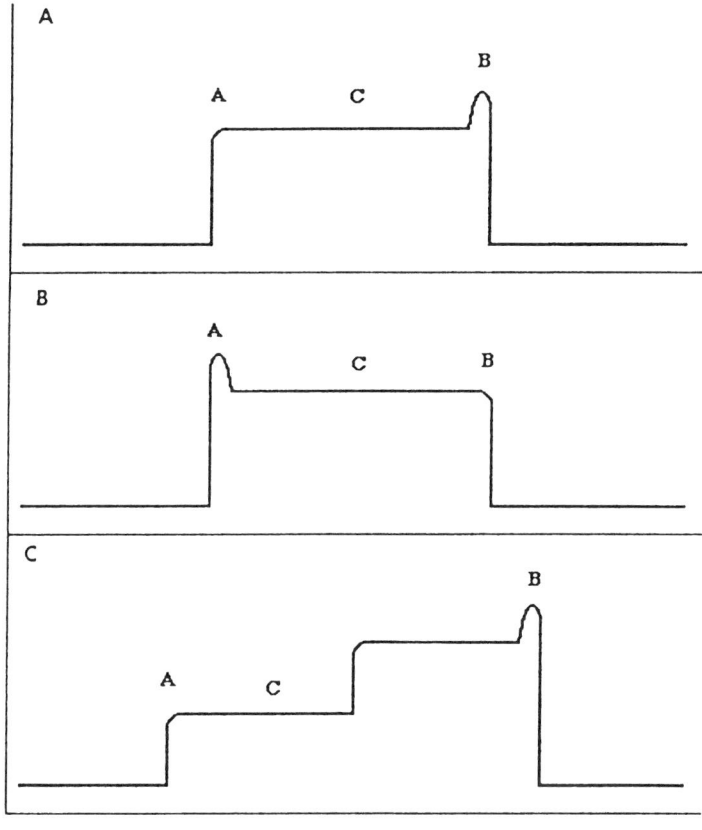

Figure 8.9 Predicted models showing various electropherograms in a capillary electrophoretic enzyme assay. The moving velocities are ESP (*A*) and PES (*B*); (*C*) multiple-isoenzyme form, with the moving velocity of the common product smaller than those of these isoenzymes. (From Bao and Regnier, 1992, with permission.)

material as possible, undiluted fermentation broth was injected directly into the capillary. Also, the volume of the injected sample was as much as 30 nL, which, on the other hand, decreases separation efficiency. For separation, 33 mM phosphate buffer (pH 9.5) was used, and the separation was carried out at 15 kV.

Finally, if the product of the enzymatic reaction is known, the activity of the enzyme can be assayed on the basis of quantitating the product "off line." Thus, for instance, glutathione peroxidase activity can be measured by quantitating the oxidized and reduced forms of glutathione by capillary electrophoresis, as demonstrated by Pascual et al. (1992). The electrophoretic separation buffer used was 100 mM tetraborate (pH 8.2), containing 100 mM SDS.

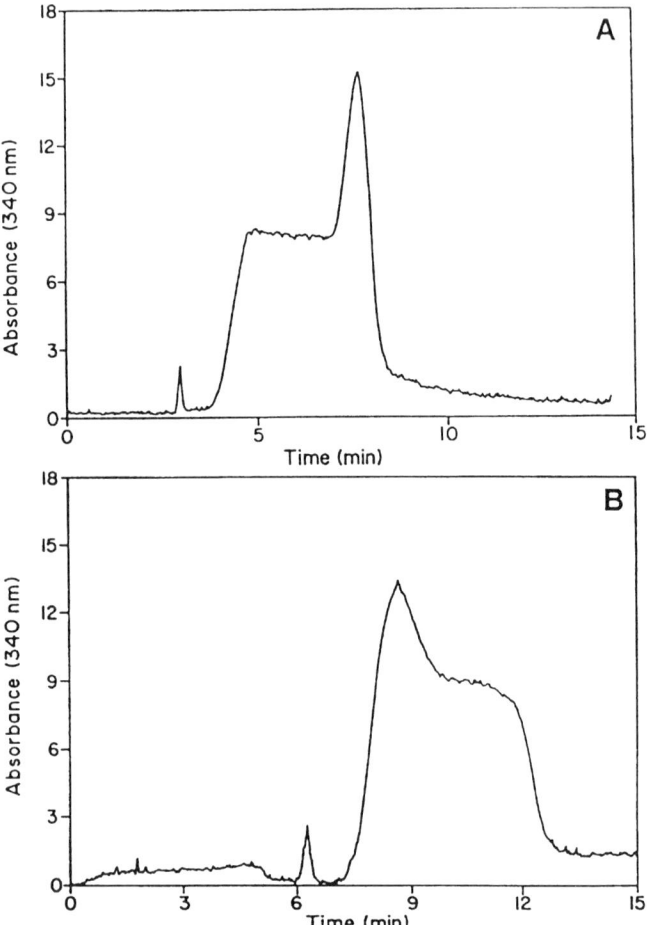

Figure 8.10 Electropherograms showing accumulated peaks resulting from the parked reaction at different running times. (*A*) NADPH accumulated at the beginning before electrophoresis. (*B*) NADPH accumulated just before glucose-6-phosphate oxidoreductase passed the detection window. (From Bao and Regnier 1992, with permission.)

8.9 PROTEIN–DRUG BINDING ASSAYS

It is well known that drugs bind to plasma proteins, particularly to serum albumin and α-acid glycoprotein, and that only the unbound, or free, fraction is responsible for any pharmacological effect. For protein–drug binding studies size-exclusion chromatography in one of three variants—namely, the Hummel–Dreyer method (1962), the vacancy peak method (Sebille, et al., 1979), and frontal analysis (Cooper and Wood, 1968)—is the traditional method of choice. Kraak et al. (1992) tested all three variants in a CZE version and showed that the frontal analysis procedure represents the preferred approach.

Size-exclusion chromatography methods exploit the difference in the exclusion of the drug and drug–protein complex from the column packing. In CE, separation of bound and unbound drug is based on charge and size differences. Because the protein molecule is much larger and carries more charge than a drug molecule, it is reasonable to assume that if the drug is bound to the protein, neither the protein's charge nor its molecular mass will be significantly altered. Consequently, both the drug and the drug–protein complex will have the same electrophoretic mobility, and methods similar to those developed for size-exclusion chromatography can be used with capillary electrophoresis, provided the protein and the unbound drug have different electrophoretic mobilities.

When the Hummel–Dreyer method is applied in CZE, the capillary is filled with the background electrolyte containing the drug, which causes a large detector background. Then a small sample containing the drug, buffer, and the protein is injected. The total concentration of the drug in the sample is the same as in the background electrolyte, but a part of it is bound to the protein. If potential is switched on, the protein–drug complex starts to move to the cathode, leaving a local deficiency in the drug concentration in the direction to the anode. This deficiency causes a negative peak, which moves with the mobility of the drug, and its size corresponds to the amount of the bound drug. During the migration, the protein–drug complex is in equilibrium with the free drug in the buffer. Thus, the protein–drug complex will give a positive peak. As stressed by Kraak et al. (1992), if at the detection wavelength the absorbance of the binding protein is zero, and if the molar absorptivities of the drug and drug–protein complex are the same, then the areas of the positive and negative peaks must be equal. This situation, however, rarely occurs in practice (Fig. 8.11).

To apply the vacancy peak method, the capillary is filled with the background buffer containing the protein as well as the drug. The situation is similar to the Hummel–Dreyer method described above, and the background signal in the detector is quite high. Next a small plug of the buffer only is applied and the power is switched on. Assuming that the mobilities of the protein and the protein–drug complex are higher than the mobility of the drug itself, the following effects are seen at the front and rear edge of the buffer plug:

- At the front edge the drug is migrating more slowly than the protein and therefore it stays behind.
- At the rear end the protein migrates faster than the plug.

This process progresses until both fronts reach one another. Then, since the whole process is reversible, in the middle of the plug the protein starts to

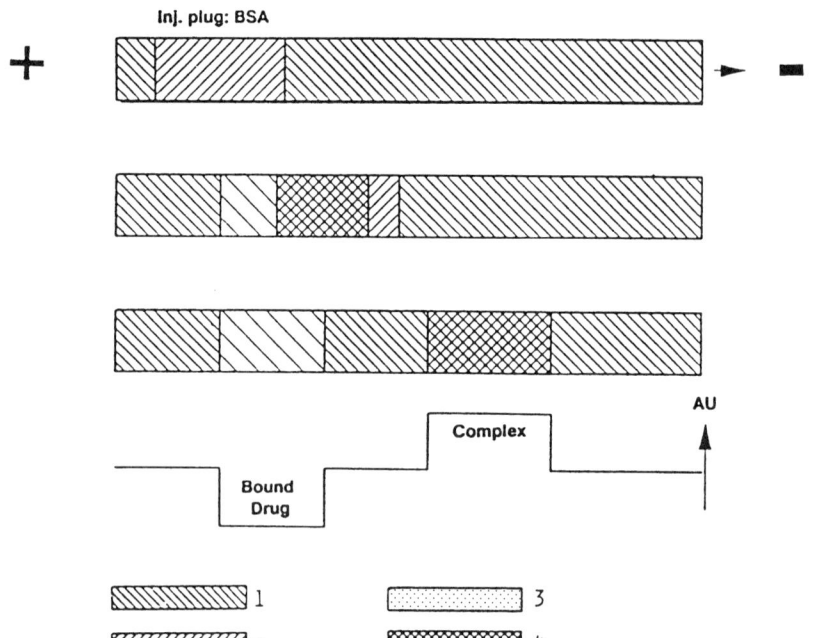

Figure 8.11 Schematic representation of the Hummel–Dreyer capillary zone electrophoretic method. Zone identification: 1, drug; 2, bovine serum albumin; 3, buffer; 4, BSA–drug complex. (From Kraak et al., 1992, with permission.)

absorb drug molecules again. Reabsorption continues to equilibrium (i.e., the drug is reconcentrated to its original level). From this moment, a steady state is reached and two negative bands appear on the electropherogram; the first corresponds to the bound drug, the second represents the free drug zone (Fig. 8.12).

The last approach, also applicable to CZE, is analogous to frontal analysis. Here the capillary at the beginning of the experiment is filled with the background electrolyte only, and a very large sample drug containing buffer, protein, and the drug is injected into the capillary. Because of the differences in electrophoretic mobility between the drug and the protein, the free drug starts to leak out at the rear edge of the plug. A plateau corresponding to the free drug concentration is formed. At the front edge of the plug another plateau is created, representing the free protein. Thus, the result of this analysis is a three-step profile in which the free protein plateau is followed by the zone of protein–drug complex and the profile is terminated by a plateau corresponding to free drug. The height of the drug plateau is taken as a measure of free drug concentration. As stressed by Kraak et al. (1992), depending on the detection wavelength used, the first plateau is frequently not seen by the detector (Fig. 8.13).

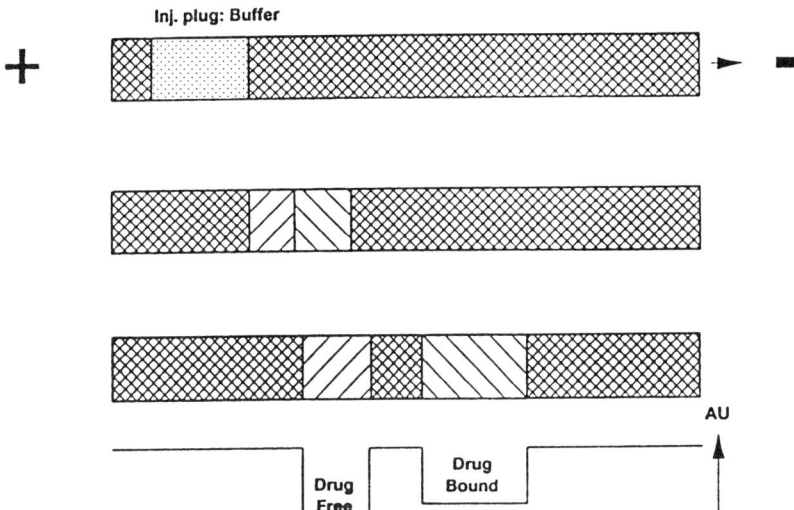

Figure 8.12 Schematic representation of the vacancy peak capillary zone electrophoresis method. Zone identification as in Figure 8.11. (From Kraak et al., 1992, with permission.)

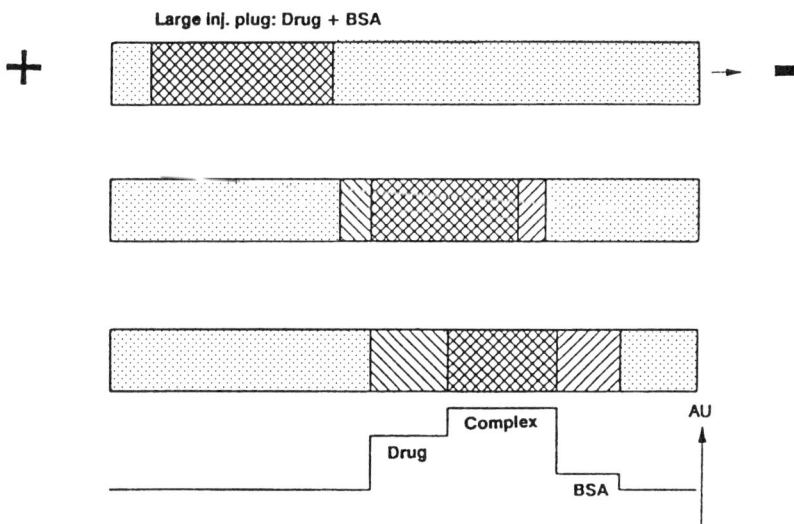

Figure 8.13 Schematic representation of the frontal analysis zone electrophoretic method. Zone identification as in Figure 8.11. (From Kraak et al., 1992, with permission.)

Thus the three approaches to drug binding assays are represented according to the schemes in Figures 8.11 to 8.13. Kraak et al. (1992), testing binding of warfarin by bovine serum albumin, used 0.067 potassium phosphate buffer at pH 7.4 as background electrolyte. For the practical evaluation of data by the Hummel–Dreyer method, the simplified approach presented by Pinkerton and Koeplinger (1990) was applied. In this approach two injections at a given warfarin concentration were applied: the sample consisting of buffer and protein and the blank buffer. From the peak areas the bound drug [Db] concentration can be calculated.

8.10 NUCLEIC ACIDS AND THEIR CONSTITUENTS

Capillary electrophoretic separation of nucleic acids has been reviewed (Cohen et al., 1987a; Kuhr, 1990; Gebauer and Thormann, 1991). In nucleotide and nucleoside analysis, MEKC has been the method of choice, using SDS (Row et al. 1987; Cohen et al., 1987b; Kasper et al., 1988), dodecyltrimethylammonium bromide, or hexadecyltriethylammonium bromide (Liu et al., 1989). Other applications concerning chemically modified nucleotides, nucleosides, and nucleobases can be found in papers by Lecoq et al. (1991) and Thormann et al. (1992).

Special techniques were developed for the analysis of nucleotide phosphates. For the separation of these compounds, either dynamic pH gradient (Sustacek et al., 1989) or linear polyacrylamide-coated capillaries (Takigiku and Schneider, 1991) were used. In the latter case separations were done in the reversed-polarity mode (i.e., from cathode to anode).

In forensic toxicology, these methods are potentially useful for detecting modified nucleotides and nucleosides in the middle of a large excess of the unmodified species present in biological materials. Knowledge of these properties may be useful for investigations into direct interactions of exogenous (e.g., environmental) toxicants with the DNA matrix.

However, by far the most interesting application of CE to DNA analysis is in the field of criminal identification and in paternity tests. The analysis of DNA polymorphisms is based on exponential amplification of individual loci, followed by analysis of local differences in length/sequence. Presently, length polymorphisms are by far the most important class of genetic variability studied for forensic as well as for genetic purposes. Consequently, electrophoresis on agarose or polyacrylamide slab gels is the easiest procedure adopted for typing genotype differences. Generally, short (<1000 bpg: amplified fragment length polymorphisms, AmpFLPs) or very short (down to 200 bp: short tandem repeats, STRs) gene segments are targeted, mostly implying length differences as little as 1%. Since there is a constant mass–charge ratio in nucleic acid stretches, reliable separation of these small differences requires very long gels with high sieving effect and denaturing conditions.

Recent experience with CE has shown that this analytical technique may compete with the polyacrylamide and agarose gel electrophoresis procedures in giving fast, reproducible, and reliable DNA strand separation.

In both cases, characterization of the genetic pattern is based on tiny differences in length of nucleotide sequences, which have the peculiarity of a constant mass–charge ratio for any length. As in traditional slab gel electrophoresis, in CE the general separation strategy is based on sieving gels (CGE).

To achieve length polymorphism separation, gel-filled columns have been used, typically with 2 to 6%T and 3 to 6%C polyacrylamide, as reported by Cohen et al. (1988), Paulus and Ohms (1990), Paulus et al. (1990), Lux et al. (1990), Baba et al. (1991a, 1991b), and Wang et al. (1991). It is worth mentioning that gel-filled capillaries are commercially available, ensuring effective DNA fragment separation (Guttman and Cooke, 1991a, 1991b; Guttman et al., 1992). Unfortunately gel-filled columns have a short lifetime, owing to rapid gel degradation, which is believed to be caused by localized overheating, resulting from the high voltages generally applied in CE separations (Guttman and Cooke, 1991b). Another problem reported in the literature is the irreversible binding of DNA strands to gel-filled columns, which reduces separation efficiency over time (this is particularly true for high molecular mass DNA).

As an alternative to gel-filled columns, nongel sieving media based on high viscosity background buffer containing a water-soluble linear polymer, such as methylcellulose or linear polyacrylamide, have been introduced (Zhu et al., 1989; Grossman and Sloane, 1991; Schwartz et al., 1991; MacCrehan et al., 1992; Nathakarnkitkool et al., 1992). In these media DNA fragments are separated when they become entangled with the polymer network inside the capillary (Grossman and Sloane, 1991; Chiari et al., 1993). The main advantage of such an arrangement is that the nongel sieving media are stable and can be renewed with each run, since fresh sieving buffer is pumped into the column prior to every analysis and the old filling of the capillary is pumped out. Another advantage of this technique is the absence of problems with remnants in the capillary from the preceding run, since the material that has not passed the capillary column during the run is simply pumped out in the postwash step.

However, a price is paid in return for these advantages: nongel sieving systems usually have a lower resolving power compared to gel-filled capillaries. The resolving power, however, depends on the nature of the nongel sieving medium, and a recently introduced polymeric hydroxyethylcellulose shows a separation power close to that of gel-filled capillaries (Nathakarnkitkool et al., 1992).

Nongel sieving media have been applied to problems of forensic interest by McCord et al. (1993a), who used the following procedure for buffer preparation: 0.1 mM EDTA was added to 100 mM Trisma base and 100 mM boric acid; pH was adjusted to 8.7 with cesium hydroxide. Hydroxyethylcellulose was dissolved in this buffer at a concentration of 0.5% (w/v), and the solution was filtered through a 0.45 mm cellulose acetate filter. Prior to analysis, ethidium bromide was added to a concentration of 1.27 mM. Phenylmethyl-coated silica columns (70 cm × 100 mm i.d., phase thickness 50 mm) were used. The authors (McCord et al., 1993a) recommend a 2-minute rinse with methanol followed by a 6-minute rinse with buffer. The methanol rinses help to keep the column bubble-free by wetting the phenylmethyl coating and also reduce the buildup of impurities on the column.

A few specific problems arising from the association of CE and PCR-based polymorphisms should be stressed. First, the presence of Cl$^-$ (from MgCl) in the PCR mixture has a detrimental effect on the electrokinetic injection of DNA molecules, since the chloride ions inject preferentially, owing to their lower mass–charge ratio. Therefore, the PCR samples have to be desalted (according to McCord et al., 1993a, float dialysis is the procedure of choice). Second, large quantities of excess primers and other low molecular mass material must be eliminated. This is done by ultrafiltration, before float dialysis. Furthermore, optimum polymer concentration is a crucial point. A concentration of 0.5 to 0.75% has a positive effect on resolution of smaller fragments, likely reflecting a reduction in effective pore size, although other interactions cannot be excluded (Grossman and Sloane, 1991). According to McCord et al. (1993a), DNA fragments in the 150–400 bp range may require even higher concentrations of the polymer, with concomitant negative effect on the resolution of higher molecular mass fragments. Besides, the upper limit of polymer concentration is determined by the viscosity of the arising solution, since highly concentrated polymer solutions may not be pumped into the tiny separation capillary.

Increasing ethidium bromide concentration from 0.635 to 6.35 mM reportedly improves resolution, but higher concentrations have deleterious effect on the column performance (particularly if left for longer periods in the column). Adjusting the pH of the buffer with CsOH is another means of improving separation efficiency, probably because ion pairs form between DNA fragments and cesium ions. This phenomenon is believed to decrease the migration rate of DNA fragments, improving their separation.

In spite of all precautions, a lifetime of phenylmethyl-coated silica columns is limited. Reportedly they last for 2 weeks, after which period they should be replaced to prevent loss of resolution.

In forensic biology, CE of PCR-amplified DNA has already been used to solve classical problems of genetic typing. It is usually carried out in loci that contain a variable number of tandem repeat sequences (VNTRs). The length of a repeat unit can be as short as 2 bp, and at a given locus each individual has a pair of such sequences (alleles). McCord et al. (1993a) attempted to use nongel sieving CE technique for the analysis of two genetic markers, namely D1S80 and SE 33.

D1S80 maps chromosome 1 and has a tandem repeat unit of 16 base pairs, which confers on this locus a large repeat unit polymorphism. McCord et al. (1993a) reported the successful resolution of a mixture of standard D1S80 alleles, spanning from 403 to 1069 bp; they used 0.635 mM ethidium bromide in the background buffer containing 0.5% hydroxyethylcellulose. In the hands of these authors, alleles could be separated with a gradual loss of intensity as molecular mass increased.

In SE 33, the other example of a forensic genetic marker amenable to effective CE separation, alleles differing by as little as 4 bp, in a molecular range of 230 to 350 bp, exist, with potentially 2 bp repeats (half-alleles). This similarity poses much higher analytical demands on the separation power of CE. Notably, CE carried out in 1.27 mM ethidium bromide and 0.5% hydroxyethylcellulose was capable of resolving an allelic test mixture. This example shows, much more than D1S80, the potential of CE in forensic biology.

In a further paper, McCord et al. (1993b), using a similar CE nongel sieving system, reported the application of LIF detection of PCR-amplified DNA. Since DNA has weak native fluorescence, PCR products must either be tagged by means of fluorescent primers or stained with fluorescent dyes. The use of fluorescent dyes intercalating DNA reported yields more sensitivity than PCR tagging (a greater number of fluorescent molecules can be introduced); in addition, DNA intercalation (usually with ethidium bromide) is known to improve also the separation, by increasing in the rigidity of the fragments. The dye YO-PRO-1 was adopted as visualizing and intercalating agent because of its strong binding constant to DNA and low intrinsic fluorescence, if not bound to DNA. The application of this method (nongel sieving buffer CE separation with LIF detection) to the analysis of PCR-amplified DNA fragments from 120 to 400 bp in size showed an excellent sensitivity of about 500 pg/mL of DNA, with resolution up to 6 bp. For better resolution (≤4 bp), it was necessary to add ethidium bromide to the buffer as an additional intercalating agent. With this procedure, alleles from a number of VNTRs of interest in genetic typing were successfully investigated.

The application of CE-LIF using intercalating dyes TOTO-1 and YOYO-1 (with a sample prestaining procedure) and separation systems employing either polymer network (0.5% methylcellulose) or cross-linked polyacrylamide gel (3%T, 3%C) was also reported by Srinivasan et al. (1993). Excellent sensitivity (about 500 pg/mL) was reported; besides, it was found that the polymer network capillaries have a broader DNA size range of effectiveness and provide higher resolution and efficiency than gel-filled capillaries.

Specific reports on applications of this technique for forensic purposes are still scarce. The apparent reason for this slow penetration of CE into forensic genetics is twofold. First, the analysis of genetic markers by CE is not a simple technology, and it must be performed in a specialized laboratory. Second, the field of DNA fragment separation by CE is still being developed methodologically and is not an established technique as compared, for example, to slab gel electrophoresis. Nevertheless, this technique should become widespread in forensic biology in the relatively near future.

8.11 CONCLUSION

This chapter introduced the application of HPLC and HPCE to forensics, the use of science in the judicial process. We explored how these separation techniques are used in testing for illicit drugs, gunshot residues and explosive constituents, pen inks, modified proteins, and nucleic acids.

The chapter illustrated that the success of HPLC/CE in forensics is due in part to the ability of these techniques to deal with minute amounts of sample and to perform analytical assays under restrictive conditions.

Key papers were reviewed to provide the reader with a "road map" to the literature that covers this area. The chapter concluded with a brief discussion of the role of the polymerase chain reaction in forensics.

REFERENCES

Altria KD (1993) Capillary electrophoresis for pharmaceutical research and development. *LC-GC Int* **6**:616–619.

Baba Y, Matsuura T, Wakamoto K, Tsuhako M (1991a) Comparison of high-performance liquid chromatography with capillary gel electrophoresis in single base resolution of polynucleotides. *J Chromatgr* **558**:273–284.

Baba Y, Tsuhako M, Enomoto S, Chin AM, Dubrow RS (1991b) High performance separations of nucleic acids using capillary electrophoresis and high-performance liquid chromatography. *J High Resolut Chromatogr* **14**:204–206.

Bao J, Regnier FE (1992) Ultramicro enzyme assays in a capillary electrophoretic system. *J Chromatogr* **608**:217–224.

Bentrop D, Kohr J, Engelhardt H (1991) Poly(methylglutamate)-coated surface in HPLC and CE. *Chromatographia* **32**:177–178.

Bruin GJM, Chang JP, Kuhlman RH, Zegers K, Kraak JL, Poppe H (1989) Capillary zone electrophoretic separation of proteins in polyethylene glycol-modified capillaries. *J Chromatogr* **471**:429–436.

Bushey MM, Jorgenson JW (1989) Capillary electrophoresis of proteins in buffers containing high concentrations of zwitterionic salts. *J Chromatogr* **480**:301–310.

Caslavska J, Lienhard S, Thormann W (1993) Comparative use of three electrokinetic capillary methods for the determination of drugs in body fluids. *J Chromatogr* **638**:335–342.

Chee Gl, Wan TSM (1993): Reproducible and high-speed separation of basic drugs by capillary zone electrophoresis. *J Chromatogr* **612**:172–177.

Chiari M, Nesi M, Righetti PG (1993) Movement of DNA fragments during capillary zone electrophoresis in liquid polyacrylamide. *J Chromatogr* **652**:31–39.

Cobb KA, Dolnik V, Novotny M (1990) Electrophoretic separations of proteins in capillaries with hydrolytically stable surface structures. *Anal Chem* **62**:2478–2483.

Cohen AS, Paulus A, Karger BL (1987a) High-performance capillary electrophoresis using open tubes and gels. *Chromatographia* **24**:15–23.

Cohen AS, Terabe S, Smith JA, Karger BL (1987b) High-performance capillary electrophoretic separation of bases, nucleosides, and nucleotides: Retention manipulation via micellar solutions and metal additives. *Anal Chem* **59**:1021–1027.

Cohen AS, Najarian DR, Paulus A, Guttman A, Smith JA, Karger BL (1988) Rapid separation and purification of oligonucleotides by high-performance capillary gel electrophoresis. *Proc Natl Acad Sci USA* **85**:9660–9663.

Cohen AS, Smisek DL, Keohavong P (1993) Capillary gel electrophoresis of biopolymers. *Tends Anal Chem* **12**:195–202.

Cooper PF, Wood GC (1968) Protein-binding of small molecules: New gel filtration method. *J Pharm Pharmacol* **20** (Suppl):150–156.

Davis JM, Giddings JC (1985) Statistical method for estimation of number of components from single complex chromatograms: Application to experimental chromatography runs. *Anal Chem* **57**:2178–2182.

Deyl Z, Mikšik I (1995) Separation of collagen type I chain polymers by electrophoresis in non-crosslinked polyacrylamide filled capillaries *J Chromatogr A* **698**:369–373.

Deyl Z, Tagliaro F, Mikšik I (1994a) Capillary electrophoretic techniques: A new tool in clinical chemistry. *Eur J Lab med* **1**:161–171.

Deyl Z, Tagliaro F, Mikšik I (1994b) Biomedical applications of capillary electrophoresis. *J Chromatogr B* **656**:3–27.

Dougherty AM, Woolley CL, Williams DL, Swaile DF, Cole RO, Sepaniak MJ (1991) State phases for capillary electrophoresis. *J Liquid Chromatogr* **14**:907–921.

Emmer A, Jansson M, Roeraade J (1991) Improved capillary zone electrophoretic separation of basic proteins, using a fluorosurfactant buffer additive. *J Chromatogr* **547**:544–550.

Fanali S, Schudel M (1991) Some separations of black and red water-soluble fiber-tip pen inks by capillary zone electrophoresis and thin-layer chromatography. *J Forensic Sci* **36**:1192–1197.

Gebauer P, Thormann W (1991) Isotachophoresis of proteins in uncoated open-tubular fused-silica capillaries. *J Chromatogr* **558**: 423–429.

Giddings JC (1984) Two dimensional separations: Concept and promise. *Anal Chem* **56**:1258A.

Giddings JC (1987) Concepts and comparisons in multidimensional separation. *J High Resolut Chromatogr, Chromatogr Commun* **10**:319–323.

Green JS, Jorgenson JW (1989) Minimizing adsorption of proteins on fused silica in capillary zone electrophoresis by the addition of alkali metal salts to the buffers. *J Chromatogr* **478**:63–70.

Grossman PD, Sloane DS (1991) Experimental and theoretical studies of DNA separations by capillary electrophoresis in entangled polymer solutions. *Biopolymers* **31**:1221–1228.

Guttman A, Cooke N (1991a) Denaturing capillary gel electrophoresis. *Int Biotech Lab* **9**(4):10.

Guttman A, Cooke N (1991b) Effect of temperature on the separation of DNA restriction fragments in capillary gel electrophoresis. *J Chromatogr* **559**:285–194.

Guttman A, Arai A, Magyar K (1992) Influence of pH on the migration properties of oligonucleotides in capillary gel electrophoresis. *J Chromatogr* **608**:175–179.

Harding JJ, Beswick HT, Ajiboye R, Huby R, Blakytny R, Rixon KC (1989) Non-enzymic post-translational modification of proteins in aging. *Mech Ageing Dev* **50**:7–16.

Hargadon KA, McCord BR (1992) Explosive residue analysis by capillary electrophoresis and ion chromatography. *J Chromatogr* **602**:241–247.

Hummel JP, Dreyer WJ (1962) Measurement of protein-binding phenomena by gel filtration. *Biochim Biophys Acta* **63**:530–532.

Jelínková D, Deyl Z, Mikšik I, Tagliaro F (1995) Capillary electrophoresis of hair proteins modified by alcohol intake in laboratory rats. *J Chromatogr A* **709**:111–119.

Kajiwara H (1991) Application of high-performance capillary electrophoresis to the analysis of conformation and interaction of metal-binding proteins. *J Chromatogr* **559**:345–356.

Karger B (1989) High-performance capillary electrophoresis. *Nature* **339**:641–642.

Kasper TJ, Melera M, Gozel P, Brownlee RG (1988) Separation and detection of DNA by capillary electrophoresis. *J Chromatogr* **458**:303–312.

Kintz P, Ed. (1995) Hair analysis in forensic toxicology. Hair analysis II. Special Issue. *Forensic Sci Int* **70**:1–222.

Kraak JC, Busch S, Poppe H (1992) Study of protein–drug binding using capillary zone electrophoresis. *J Chromatogr* **608**:257–264.

Krogh M, Brekke S, Tonnesen F, Rasmussen KE (1994) Analysis of drug seizures of heroin and amphetamine by capillary electrophoresis. *J Chromatogr* **674**:235–240.

Kuhr WG (1990) Capillary electrophoresis. *Anal Chem* **62**:403R–414R.

Lecoq AF, Leuratti C, Marafante E, Di Biase S (1991) Analysis of nucleic acid derivatives by micellar electrokinetic capillary chromatography. *J High Resolut Chromatogr* **14**:667–672.

Lemmo AV, Jorgenson JW (1993) Two-dimensional protein separation by microcolumn size-exclusion chromatography–capillary zone electrophoresis. *J Chromatogr* **633**:213–220.

Li SFY (1994) *Capillary Electrophoresis, Principles, Practice and Applications*, 2nd ed. Elsevier, Amsterdam.

Li S, Lloyd DK (1993) Direct chiral separations by capillary electrophoresis using capillaries packed with an α_1 acid glycoprotein chiral stationary phase. *Anal Chem* **65**:3684–3690.

Liu J, Banks JF Jr, Novotny M (1989) High speed micellar electrokinetic capillary chromatography of the common phosphorylated nucleotides. *J Microcol Sep* **1**:136–141.

Lurie IS (1992) Micellar electrokinetic capillary chromatography of the enantiomers of amphetamine, methamphetamine and their hydroxyphenethylamine precursors. *J Chromatogr* **605**:269–275.

Lurie IS (1994) Analysis of seized drugs by capillary electrophoresis. In *Analysis of Addictive and Misused Drugs*, JA Adamovics, Ed., pp. 151–219. Dekker, New York.

Lurie IS, Klein RFX, Dal Cason TA, LeBelle MJ, Brenneisen R, Weinberger RE (1994) Chiral resolution of cationic drugs of forensic interest by capillary electrophoresis with mixtures of neutral and anionic cyclodextrins. *Anal Chem* **66**:4019–4026.

Lux JA, Yin H-F, Schomburg G (1990) A simple method for the production of gel-filled capillaries for capillary gel electrophoresis. *J High Resolut Chromatogr* **13**:436–437.

Maa Y-F, Hyver KJ, Swedberg SA (1991) Impact of wall modifications on protein elution in high performance capillary zone electrophoresis. *J High Resolut Chromatogr* **14**:65–67.

MacCrehan WA, Rasmussen HT, Northrop DM (1992) Size-selective capillary electrophoresis (SSCE) separation of DNA fragments. *J Liquid Chromatogr* **15**:1063–1080.

Mazzeo JR, Krull IS (1991) Coated capillaries and additives for the separation of proteins by capillary zone electrophoresis and capillary isoelectric focusing. *BioTechniques* **10**:638–645.

McCord BR, Jung JM, Holleran E (1993a) High resolution capillary electrophoresis of forensic DNA using a non-gel sieving buffer. J Liquid Chromatogr **16**:1963–1981.

McCord BR, McClure DL, Jung JM (1993b) Capillary electrophoresis of polymerase chain reaction–amplified DNA using fluorescence detection with an intercalating dye. *J Chromatogr* **652**:75–82.

McCormick RM (1988) Capillary zone electrophoretic separations of peptides and proteins using low pH buffers in modified silica capillaries. *Anal Chem* **60**:2322–2328.

Monnier VM, Sell DR, Miyata S, Nagaraj RH, Odetti P, Lapolla A (1992) Advanced Maillard reaction products as markers for tissue damage in diabetes and uremia: Relevance to diabetic nephropathy. *Acta Diabetol* **29**:130–135.

Mulholland F, Movahedi S, Haque GR, Kasumi T (1993) Monitoring tripeptidase activity using capillary electrophoresis. Comparison with the ninhydrin assay. *J Chromatogr* **636**:63–68.

Nashabeh W, El Rassi Z (1991) Capillary zone electrophoresis of proteins with hydrophilic fused-silica capillaries. *J Chromatogr* **559**:367–383.

Nathakarnkitkool S, Oefner PJ, Bartsch G, Chin MJ, Bonn GK (1992) High-resolution capillary electrophoretic analysis of DNA in free solution. *Electrophoresis* **13**:18–31.

Northorp DM, MacCrehan WA (1992) Sample collection, preparation, and quantitation in the micellar electrokinetic capillary electrophoresis of gunshot residues. *J Liquid Chromatogr* **15**:1041–1063.

Northrop DM, Martire DE, MacCrehan WA (1991) Separation and identification of organic gunshot and explosive constituents by micellar electrokinetic capillary electrophoresis. *Anal Chem* **63**:1038–1042.

Northrop DM, McCord BR, Butler JM (1994) Forensic applications of capillary electrophoresis. *J Capillary Electrophor* **1**:158–168.

Pascual P, Martinez-Lara E, Bárcena JA, López-Barea J, Toribio F (1992) Direct assay of glutathione peroxidase using high-performance capillary electrophoresis. *J Chromatogr* **581**:49–56.

Paulus A, Ohms JI (1990) Analysis of oligonucleotides by capillary gel electrophoresis. *J Chromatogr* **507**:113–123.

Paulus A, Gassman E, Field MJ (1990) Calibration of polyacrylamide gel columns for the separation of oligonucleotides by capillary electrophoresis. *Electrophoresis* **11**:702–708.

Pinkerton TC, Koeplinger KA (1990) Determination of warfarin–human serum albumin protein binding parameters by an improved Hummel–Dreyer high performance liquid chromatography method using internal surface reversed-phase columns. *Anal Chem* **62**:2114–2122.

Reiser KM (1991) Nonezymatic glycation of collagen in aging and diabetes. *Proc Soc Exp Biol Med* **196**:17–29.

Reiser KM, Amigable MA, Last JA (1992) Nonenzymatic glycation of the type I collagen. *J Biol Chem* **267**:24207–24216.

Rohlicek V, Deyl Z (1989) Simple apparatus for capillary zone electrophoresis and its application to protein analysis. *J Chromatogr* **494**:87–99.

Row KH, Griest WH, Maskarinec MP (1987) Separation of modified nucleic acid constituents by micellar electrokinetic capillary chromatography. *J Chromatogr* **409**:193–203.

Schafroth M, Thormann W, Alleman D (1994) Micellar electrokinetic capillary chromatography of benzodiazepines in human urine. *Electrophoresis* **15**:72–78.

Schwartz HE, Ulfelder KJ, Sunzeri FJ, Busch MP, Brownlee RG (1991) Analysis of DNA restriction fragments and polymerase chain reaction products towards detection of the AIDS (HIV-1) virus in blood. *J Chromatogr* **559**:267–283.

Sebille B, Thuaud N, Tillement JP (1979) Equilibrium saturation chromatographic method for studying the binding of ligands to human serum albumin by high-performance liquid chromatography. Influence of fatty acids and sodium dodecyl sulfate on warfarin–human serum albumin analog. *J Chromatogr* **180**:103–110.

Srinivasan K, Girard JE, Williams P, Roby RK, Weedn VW, Morris SC, Kline MC, Reeder DJ (1993) Electrophoretic separations of polymerase chain reaction–amplified DNA fragments in DNA typing using a capillary electrophoresis–laser induced fluorescence system. *J Chromatogr* **652**:83–91.

Staub C, Plaut O (1994) High performance capillary electrophoresis. A new tool in forensic toxicology? In *Proceedings of the 31st International Meeting of TIAFT, Leipzig '93*, RK Mueller, Ed., pp. 452–458. MOLIAN press, Leipzig.

Steuer W, Grant I, Erni F (1990) Comparison of high-performance liquid chromatography, supercritical fluid chromatography and capillary zone electrophoresis in drug analysis. *J Chromatogr* **507**:125–140.

Sustacek V, Foret F, Bocek P (1989) Simple method for generation of a dynamic pH gradient in capillary zone electrophoresis. *J Chromatogr* **480**:271–276.

Swedberg SA (1990) Characterization of protein behavior in high-performance capillary electrophoresis using a novel capillary system. *Anal Biochem* **185**:51–56.

Tagliaro F, Antonioli C, Moretto S, Archetti S, Ghielmi S, Marigo M (1993a) High-sensitivity low-cost methods for determination of cocaine in hair: High-performance liquid chromatography and capillary electrophoresis. *Forensic Sci Int* **63**:227–238.

Tagliaro F, Chiarotti M, Deyl Z, Eds. (1993b) Hair analysis as a diagnostic tool for drugs of abuse investigation. Special Issue. *Forensic Sci Int* **63**:1–314.

Tagliaro F, Poiesi C, Aiello R, Dorizzi R, Ghielmi S, Marigo M (1993c) Capillary electrophoresis for the investigation of drugs in hair: Determination of cocaine and morphine. *J Chromatogr* **638**:303–309.

Tagliaro F, Smyth WF, Turrina S, Deyl Z, Marigo M (1995) Capillary electrophoresis: A new tool in forensic toxicology. Applications and prospects in hair analysis for illicit drugs. *Forensic Sci Int* **70**:93–104.

Takigiku R, Schneider RE (1991) Reproducibility and quantitation of separation for ribonucleoside triphosphates and deoxyribonucleoside triphosphates by capillary zone electrophoresis. *J Chromatogr* **559**:247–256.

Thormann W, Meier P, Marcolli C, Binder F (1991) Analysis of barbiturates in human serum and urine by high-performance capillary electrophoresis–micellar electrokinetic capillary chromatography with on-column multi-wavelength detection. *J Chromatogr* **545**:445–460.

Thormann W, Minger A, Molteni S, Caslavska J, Gebauer P (1992) Determination of substituted purine in body fluids by micellar electrokinetic capillary chromatography with direct sample injection. *J Chromatogr* **593**:275–288.

Thormann W, Lienhard S, Wernly P (1993) Strategies for the monitoring of drugs in body fluids by micellar electrokinetic capillary chromatography. *J Chromatogr* **636**:137–148.

Thormann W, Molteni S, Caslavska J, Schmutz A (1994) Clinical and forensic applications of capillary electrophoresis. *Electrophoresis* **15**:3–12.

Towns JK, Regnier FE (1990) Polyethyleneimine-bonded phases in the separation of proteins by capillary electrophoresis. *J Chromatogr* **516**:69–78.

Towns JK, Regnier FE (1991) Capillary electrophoretic separations of proteins using nonionic surfactant coatings. *Anal Chem* **63**:1126–1132.

Trenerry VC, Wells RJ, Robertson J (1994a) The analysis of illicit heroin seizures by capillary zone electrophoresis. *J Chromatogr Sci* **32**:1–16.

Trenerry VC, Robertson J, Wells RJ (1994b) The determination of cocaine and related substances by micellar electrokinetic capillary chromatography. *Electrophoresis* **15**:103–108.

Vlassara H, Bucala R, Striker L (1994) Pathogenic effects of advanced glycosylation: Biochemical, biologic, and clinical implications for diabetes and aging. *Lab Invest* **70**:138–151.

Wang T, Bruin GJ, Kraak JC, Poppe H (1991) Preparation of polyacrylamide gel-filled fused-silica capillaries by photopolymerization with riboflavin as the initiator. *Anal Chem* **63**:2207–2208.

Weinberger R, Lurie IS (1991) Micellar electrokinetic capillary chromatography of illicit drug substances. *Anal Chem* **63**:823–827.

Werner WE, Demorest DM, Stevens J, Wiktorowicz JE (1993) Size-dependent separation of proteins denatured in SDS by capillary electrophoresis using a replaceable sieving matrix. *Anal Biochem* **212**:253–258.

Wernly P, Thormann W (1991) Analysis of illicit drugs in human urine by micellar electrokinetic capillary chromatography with on-column fast scanning polychrome absorption detection. *Anal Chem* **63**:2878–2882.

Wernly P, Thormann W (1992a) Confirmation testing of 11-nor-delta-9-tetrahydrocannabinol-9-carboxylic acid in urine with micellar electrokinetic capillary chromatography. *J Chromatogr* **608**:251–256.

Wernly P, Thormann W (1992b) Drug of abuse confirmation in human urine using stepwise solid-phase extraction and micellar electrokinetic capillary chromatography. *Anal Chem* **64**:2155–2159.

Wernly P, Thormann W, Bourquin D, Brenneisen R (1993) Determination of morphine-3-glucuronide in human urine by capillary zone electrophoresis and micellar electrokinetic capillary chromatography. *J Chromatogr* **616**:305–310.

Wu D, Regnier FE (1992) Sodium dodecyl sulfate–capillary gel electrophoresis of proteins using non-cross-linked polyacrylamide. *J Chromatogr* **608**:349–356.

Yao X-W, Regnier FE (1993) Polymer- and surfactant-coated capillaries for isoelectric focusing. *J Chromatogr* **632**:185–193.

Zhu M, Hansen DL, Burd S, Gannoni F (1989) Factors affecting free zone electrophoresis and isoelectric focusing in capillary electrophoresis. *J Chromatogr* **480**:311–319.

CHAPTER 9

Survey of Enzymatic Activities Assayed by the HPLC Method

with David Lambeth

OVERVIEW

The use of HPLC to assay enzymatic activities has continued to grow in popularity since publication of the first edition of this book in 1987. Papers describing methods of assay for approximately 150 different enzymes were reviewed in preparation for writing this chapter.

There are several reasons for the development of HPLC-based methods for assaying enzymatic activities. First is the desirability of accurately assessing enzymatic activity in preparations as near the biological state as possible. This often requires the use of turbid preparations, and the product to be measured must be separated from other components of the reaction mixture. However, the use of crude preparations often carries the price of needing to correct for secondary reactions. HPLC allows secondary reactions to also be measured.

Second, the resolving power of HPLC greatly exceeds that of other separation methods. An enzymatic assay that was developed to use thin-layer or paper chromatography may readily be improved by switching to HPLC. Furthermore, HPLC is automated.

Third, the hazards of handling and disposing of radioisotopes have made it desirable to find alternative methods of assay. When the assay involves fluorescent or intensely absorbing substrates or products, the sensitivity of HPLC-based assays rivals that of radiochemical assays.

Fourth, column eluates can be monitored by several means, including UV-visible absorbance, fluorescence, and electrochemical or radiochemical detection. Postcolumn mixers and reactors can readily be incorporated into the eluate stream, thus allowing modification of eluted compounds to increase sensitivity of detection. And finally, the resolving power and sensitivity of HPLC allows monitoring of subtle changes in molecules, especially macromolecules.

Each assay is presented according to the scheme used throughout this book. The primary reaction is introduced, followed by the methods used for separation, including stationary phase, mobile phase, and the elution protocol. The method of detection is also described.

The enzymatic assay is then described, including buffers and pH, the method for initiating the reaction, and the process used for termination. Next, the methods used in the preparation of the sample for HPLC analysis are described, including centrifugation, filtration, or any type of purification preceding injection into the HPLC system. For many of the assays, time span and range of protein concentration for which the reaction is linear are also indicated.

Finally, the source of the enzyme activity is mentioned, including disclosure of purification procedures that were employed.

Adoption of a published procedure for an HPLC-based enzyme assay is not always a straightforward matter. For example, most separations are obtained by reversed-phase chromatography on a C_{18} column. The chromatographic behavior of some compounds is sensitive to the amount of unreacted silanols present in the matrix. The extent to which silanols are "end-capped" varies from manufacturer to manufacturer, and even between lots from the same vendor. Second, the composition of the mobile phase may not be subscribed sufficiently. For example, the starting materials for preparing a phosphate buffer may not be clearly stated, and it may not be indicated whether pH adjustment was made before of after addition of organic solvent. Third, it is advisable to validate an assay whenever the enzyme source is different from the one described in the published method. It is especially important to determine the time span and protein range in which product formation is linear with time. Deviations from expected behavior can be due to competing (secondary) reactions as well as differences in properties of the enzyme assayed.

The general reference section contains additional citations that are relevant to the enzymes described, to provide readers with a more extensive survey of the HPLC assay methods that have been developed for these activities.

9.1 INTRODUCTION

It is clear from the preceding chapters, that for the HPLC method to be used to assay an enzymatic activity, both a stationary support phase and a mobile phase must be available for the separation of the substrate and the product.

Of course, the converse is also true. If a system for separation of two compounds that have a substrate–product relationship has been developed, it may be used as the basis for developing an enzymatic assay.

9.2 CATECHOLAMINE METABOLISM

9.2.1 Tyrosine Hydroxylase (Haavik and Flatmark, 1980; Naoi et al., 1988; Mandai et al., 1992)

Tyrosine hydroxylase is a monooxygenase that catalyzes the conversion of L-tyrosine to Dopa. The activity is found in peripheral and cholinergic neurons and chromaffin cells of the adrenal medulla. HPLC methods have been developed for the assay of this activity.

In one method, the Dopa formed during the reaction was partially purified by ion-exchange and aluminum oxide chromatography and the amount present quantified by reversed-phase HPLC (ODS column). The mobile phase consisted of 0.1 M potassium phosphate buffer at pH 3.5. The column was eluted isocratically and the eluent monitored by means of an electrochemical detector.

The volume of the reaction mixture used with the HPLC assay was only about one-fifth the volume usually required with other assay methods, resulting in a considerable saving in reagents. The reaction was terminated with perchloric acid, the pH of the solution was returned to about 8 with potassium carbonate, and the sample was clarified by centrifugation. First the supernatant solution was purified using the double-column chromatographic procedure mentioned above, and then the samples were injected onto the HPLC column and analyzed for Dopa.

With the electrochemical detector, it is possible to quantitate the amount of Dopa present. The presence of pterins, cofactors required for activity, can also be detected.

Also, as noted, the HPLC method eliminates the need to know the concentration of tyrosine in the tissue. Such information would be required in radiochemical assays of activities, since any unlabeled substrate, in this case tyrosine, would reduce the specific activity of the labeled tyrosine.

In the second method, there was no prepurification: Dopa was measured directly. The separation of Dopa and tyrosine was carried out on a cationic ion-exchange HPLC (sulfonated fluorocarbon polymer) protected with a precolumn packed with silica pellets. The column was eluted with a mobile phase of 10 mM acetate buffer (pH 3.70) with 1% (v/v) propanol. The separation of tyrosine and Dopa is shown in Figure 9.1A. Because of its unique natural fluorescence, the Dopa can be monitored with a fluorescence detector without interference from endogenous substances. An excitation wavelength of 281 nm was used with the emission of 314 nm.

The reaction mixture contained the substrate L-tyrosine. Benzyloxyamine was added to inhibit any secondary reactions catalyzed by the enzyme aromatic L-amino acid decarboxylase, an activity often present in these samples. Chromatograms of samples taken during an incubation are shown in Figure 9.1B, a zero-time control with a single peak of tyrosine, and Figure 9.1C, after 20 minutes of incubation showing the peak of Dopa formed as a result of enzymatic activity.

The samples, prepared from bovine adrenal glands, were homogenized, and the homogenate was purified further by centrifugation in 0.2 M sucrose.

The assay described by Naoi et al. (1988) is sensitive enough to eliminate the need for the purification and concentration steps required in other procedures.

L-Dopa was separated on a Cosmosil 5 C_{18} column (4.6 mm × 250 mm). The mobile phase contained 90 mM sodium acetate, 35 mM citric acid, 130 μM disodium EDTA, and 230 μM sodium n-octanesulfonate in 10.5%

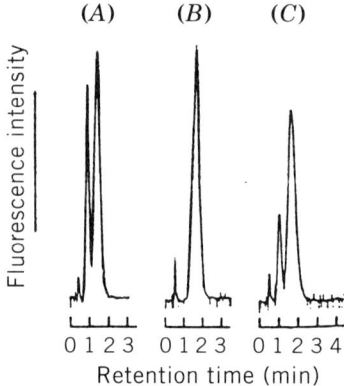

Figure 9.1 (A) Chromatogram of 384 pmol of Dopa (1.03 min) and 1.92 nmol of L-tyrosine (1.55 min). (B) and (C) Chromatograms of an acidic ethanol extract of an incubation mixture in the assay of tyrosine 3-monooxygenase activity of bovine adrenal medulla microsomes. (B) Zero-time control with a single peak of L-tyrosine (1.55 min). (C) The formation of Dopa (1.03 min) following a reaction period of 20 minutes. Volumes of 20 µL of the diluted (twice) incubation mixture were injected into the liquid chromatograph. Detection wavelengths for excitation and emission were 281 and 314 nm, respectively. (From Haavik and Flatmark, 1980.)

methanol. The sample was eluted at a flow rate of 0.8 mL/min at room temperature. Detection and quantitation was by use of a Coulochem electrochemical detector. The voltage of a conditioning cell was set at +250 mV, and those of the first and second electrodes of an analytical cell were set at +50 and −300 mV, respectively. The output of the latter electrode was monitored. Quantitation of L-Dopa was carried out by comparison of the peak area with that of standards.

The reaction mixture contained in a total volume of 100 µL: 200 µM L-tyrosine, 100 mM sodium acetate buffer (pH 6.0), 1 mM (6R)-L-*erythro*-5,6,7,8-tetrahydrobiopterine or (6RS)-methyl-5,6,7,8-tetrahydropterin, 10 µg of catalase, and 1 mM NSD-1055 (inhibitor of aromatic L-amino acid decarboxylase). The stock pterin solution were made up to be 10 mM in 1 M mercaptoethanol. Assays of brain homogenates also included 2 mM ferrous ammonium sulfate. The amount of the source of tyrosine hydroxylase was brain homogenate (500–700 µg protein) or PC12h cells (50 µg protein). Reactions were started by adding tyrosine and the pterin cofactor. After incubation at 37°C for 10 minutes, the reaction was terminated by addition of 100 µL of 0.1 M perchloric acid, containing 0.4 mM sodium metabisulfite and 0.1 mM EDTA. The resulting mixture was allowed to stand on ice for 10 minutes before centrifugation and analysis of an aliquot of the supernate by HPLC. The reaction was linear from 5 to 200 µg of PC12h cell protein, and with time up to 10 minutes.

Brain samples were homogenized with 10 volumes of 10 mM phosphate buffer (pH 7.4). The homogenates were centrifuged through a Centricut centrifuge tube, which retained molecules of molecular mass less than 20,000. The retentate was assayed. A clonal rat pheochromocytoma cell line, PC12h, was suspended in 10 mM potassium phosphate buffer (pH 7.4) and homogenized by sonication.

In the assay described by Mandai et al. (1992), the product, L-Dopa, was separated on a TSK-ODS-120T column (4.6 mm × 250 mm). The mobile phase contained 10.5% methanol, 90 mM sodium acetate, 35 mM citric acid, 0.13 mM disodium EDTA, and 0.23 mM sodium n-octanesulfonate. The effluent was monitored by fluorescence using excitation and emission wavelengths of 281 and 314 nm, respectively.

The standard assay contained retinal homogenate (0.2–3 mg protein), 0.05 mM tyrosine, 1 mM (6R,S)-5,6,7,8-tetrahydro-L-biopterin, 3.5 mM NADH, 0.02 unit of dihydroxypteridine reductase, 15 μg catalase, 40 mM sodium acetate buffer (pH 6.0), and 0.1 mM o-benzylhydroxylamine (inhibitor of Dopa decarboxylase) in a final volume of 100 μL. After incubation of the mixture at 37°C for 5 to 20 minutes, the reaction as stopped by adding 100 μL of 0.5 M perchloric acid containing 0.4 mM sodium metabisulfite and 0.1 mM disodium EDTA. The supernate obtained after centrifugation was used for HPLC analysis.

The retina of bovine eye was homogenized with a Teflon homogenizer in one volume of Dulbecco's phosphate-buffered saline (PBS) containing protease inhibitors (4 μg/mL of p-amidinophenylmethanesulfonylfluoride and 10 μg/mL of leupeptin). The supernate obtained by centrifugation was passed through a Sephadex G-25 column before being used for assays.

9.2.2 5-Hydroxytryptophan Decarboxylase (Rahman et al., 1980)

Aromatic L-amino acid decarboxylase catalyzes the decarboxylation of L-5-hydroxytryptophan (L-5-HTP) to serotonin (5-HT). In the assay, L-5-HTP was used as the substrate and the formation of 5-HT was measured.

The separation of reactant from product was carried out on a reversed-phase column eluted isocratically with an elution buffer of 0.1 M potassium phosphate containing 10% methanol at pH 3.2. N-Methyldopamine (N-M-DA) was added to each reaction mixture as an internal standard. The eluent was monitored with an electrochemical detector. The separation of these three compounds is shown in Figure 9.2A.

The reaction mixture contained L-5-HTP as substrate, pyridoxyl phosphate as a cofactor, pargylcine HCl, and the enzyme. The reaction was terminated by the addition of trichloroacetic acid (TCA), and after addition of the internal standard the reaction mixture was clarified by centrifugation. The sample was prepurified on Amberlite, and the 5-HT eluted and injected onto the HPLC column for quantitation. The results obtained the following the incubation of 5-HTP with the homogenate are shown in Figure 9.2B. The formation of

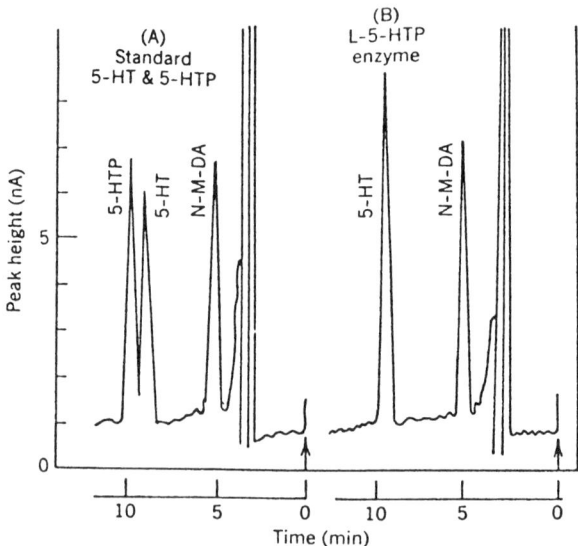

Figure 9.2 HPLC elution pattern of L-5-HTP decarboxylase incubation mixtures with homogenate of rat cerebral cortex as enzyme. The standard incubation mixture contained 5 mg of rat cerebral cortex. (*A*) Standard mixture of 50 μL containing 17.5 pmol each of L-5-HTP, 5-HT, and *N*-methyldopamine (*N*-M-DA). (*B*) Experimental incubation with L-5-HTP; 250 pmol of *N*-M-DA was added to each sample after incubation. (From Rahman et al., 1980.)

the reaction produce 5-HT is indicated by the peak of this material on the chromatogram. The rate of product formation is shown in Figure 9.3.

The enzyme was prepared from rat and human cerebral cortex. Cortical samples were homogenized in a sucrose solution, and the homogenate was used directly as the enzyme source.

9.2.3 Dopa Decarboxylase(L-Aromatic Amino Acid Decarboxylase) (Nagatsu et al., 1979; D'Erme et al., 1980)

In the presence of the cofactor pyridoxyl phosphate, Dopa decarboxylase catalyzes the decarboxylation of L-dopa to dopamine. This enzyme has been shown to be the same protein as 5-hydroxytryptophan decarboxylase, and both are referred to by the name aromatic L-amino acid decarboxylase (AADC).

In this assay, the substrate, Dopa, and the reaction product were separated by reversed-phase HPLC and eluted isocratically with 0.1 M potassium phosphate buffer at pH 3.0. The eluent was monitored with an electrochemical detector. The separation obtained with this procedure is shown in Figure 9.4*B*, together with results obtained after incubation of L-Dopa with the enzyme from rat cerebral cortex for 20 minutes at 37°C (Fig. 9.4*A*). Using a calibration curve of the type shown in Figure 9.5, it was possible to show that 1.55 nmol

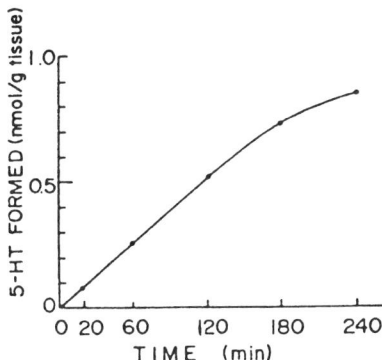

Figure 9.3 Rate of 5-HT formation using the homogenate of rat cerebral cortex as enzyme at 37°C. The incubation mixture contained 1 mg of rat cerebral cortex. (From Rahman et al., 1980.)

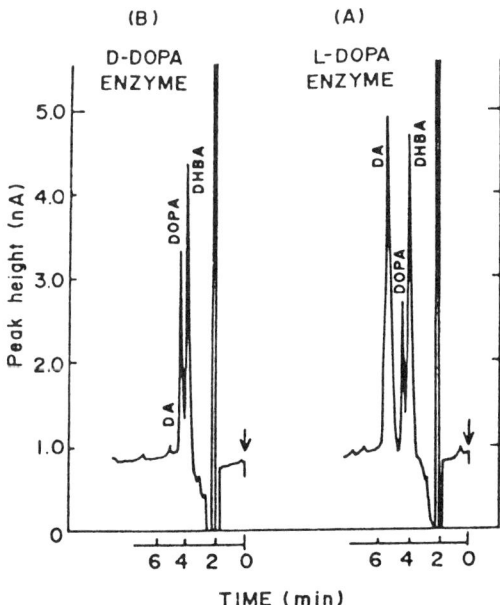

Figure 9.4 Elution pattern of AADC incubation mixtures with the homogenate of rat cerebral cortex as enzyme from HPLC. The standard incubation mixture contained 0.5 mg of rat cerebral cortex. (*A*) Experimental incubation with L-dopa. (*B*) Blank incubation with D-dopa. Dihydroxybenzylamine (100 pmol) was added to each sample after incubation. DHBA, dihydroxybenzylamine: DA, dopamine. (From Nagatsu et al., 1979.)

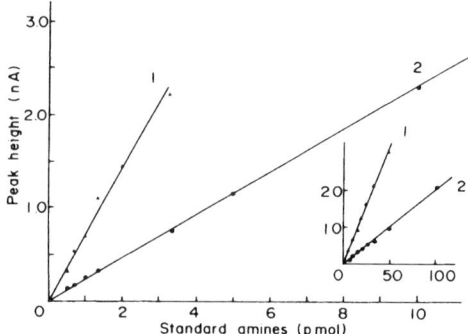

Figure 9.5 Standard curves of Dopa and dihydroxybenzylamine by HPLC with voltammetric detection for the peak height. One hundred microliters of a sample containing various amounts (500 fmol to 100 pmol) of dopamine (1) and hydroxybenzylamine (2) was injected into the column and was detected by a voltammetric detector. (From Nagatsu et al., 1979.)

of dopamine was formed. Figure 9.6 shows the rate of dopamine formation with the homogenate.

In another report the compounds were separated with a mobile phase composed of methanol–water–glacial acetic acid (25:75:1), and the eluent was monitored directly at 280 nm. The separation of Dopa and dopamine using these conditions is shown in Figure 9.7. The reaction mixture contained either Dopa or α-methyldopa, an analog, as the substrate. The dopamine formed after 5 minutes at 25°C with the homogenate is shown in Figure 9.8. Again, quantitation was achieved through the use of a calibration curve (see Figure 9.9). The reaction was started by the addition of the homogenate and was terminated with 12% TCA. These solutions were clarified by centrifuga-

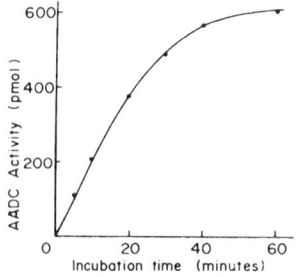

Figure 9.6 The rate of dopamine formation using the homogenate of rat cerebral cortex as enzyme at 37°C. The incubation system contained 0.5 mg of rat cerebral cortex. (From Nagatsu et al., 1979.)

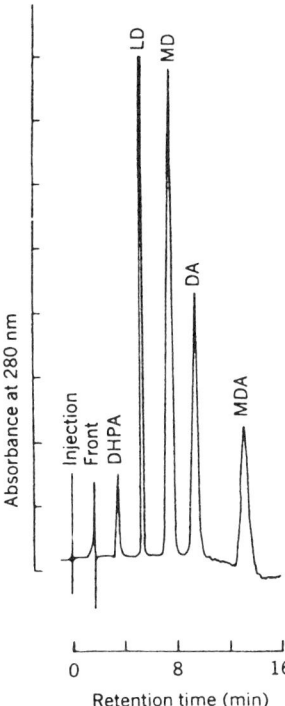

Figure 9.7 Chromatogram obtained by HPLC from a mixture of 3,4-dihydroxyphenylacetone (DHPA), L-Dopa (LD), α-methyldopa (MD), dopamine (DA), and α-methyldopamine (MDA) under the conditions described in the text. (From d'Erme et al., 1980.)

tion, the supernatant components were prepurified using Amberlite, and samples were injected onto the HPLC column for analysis.

Samples were obtained from rat cerebral cortex. They were homogenized, and samples of the homogenate were used directly as the source of the activity.

9.2.4 Dopamine β-Hydroxylase (Feilchenfeld et al., 1982; Lee et al., 1987)

Dopamine β-hydroxylase is a monoxygenase that catalyzes the hydroxylation of dopamine to form norepinephrine. This enzyme is localized in the chromaffin granules of the adrenal medulla and in the storage vesicles of central and peripheral catecholaminergic neurons. Since these compounds are unstable, this activity is often assayed by following the formation of octopamine from tyramine. For example, in the assay developed by Feilchenfeld et al. (1982), the reactant tyramine was separated from the product octopamine by reversed-phase, ion-paired HPLC (μBondapak C_{18} using a mobile phase of 17% (v/v)

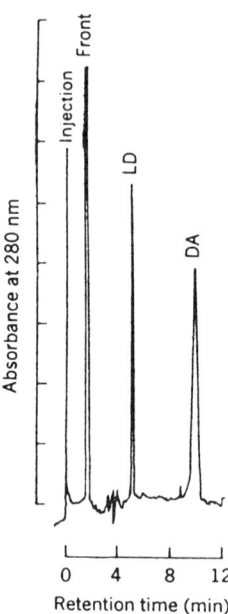

Figure 9.8 Chromatogram obtained by HPLC of the acidic supernatant of the Dopa decarboxylase–catalyzed reaction, using L-Dopa as substrate, after 5 minutes at 25°C in the presence of 300 enzyme units. (From D'Erme et al., 1980.)

methanol in water containing 10 mM acetic acid, 10 mM 1-heptanesulfonic acid (an ion-pairing reagent), and 12 mM tetrabutylammonium phosphate. Figure 9.10A illustrates the separation of tyramine and octopamine from each other and from other components of the reaction mixture. Peaks were detected by absorbance measurements at 280 nm.

The reaction mixture contained the substrate tyramine hydrochloride (1 mM), sodium fumarate, ascorbic acid, catalase, and an acetate buffer at pH 5.0. The reaction was started by the addition of the activity, and samples were removed at intervals as short as 6 minutes and injected directly onto the HPLC column for analysis. Figure 9.10A shows the chromatogram obtained before the addition of the activity. The tyramine peak is observed. (Note that the detector sensitivity was changed midway through the elution from ×1 to ×100, an example of the "sensitivity-shift" technique described Section 4.2.5.) Figure 9.10B shows a chromatogram after approximately 14 minutes of incubation. A new peak, octopamine, is clearly visible. The concentration of octopamine was determined from the area of the peak on the chromatogram, and when these data were plotted as a function of reaction time, the data in Figure 9.11 were obtained. The rate of product formation is seen to be linear for about 12 and 9 minutes, respectively, for two concentrations of enzyme obtained from a commercial source.

In contrast, in the assay described by Lee et al. (1987), epinephrine and norepinephrine were separated as the 1,2-diphenylethylenediamine deriva-

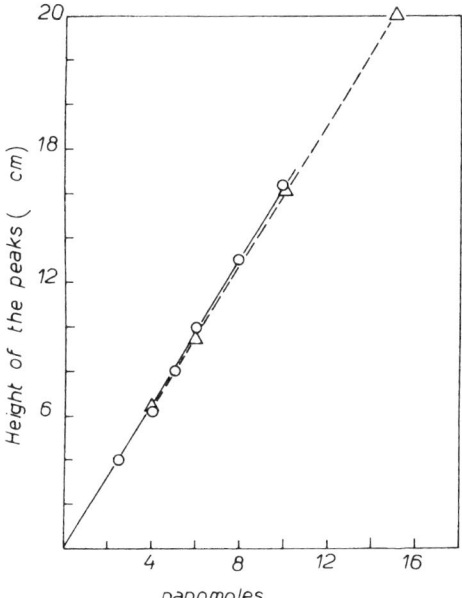

Figure 9.9 Quantitation of dopamine and Dopa by HPLC. The heights of the chromatographic peaks are reported as a function of the amounts of dopamine (○) or of dopa (△), respectively. (From D'Erme et al., 1980.)

tives at ambient temperature on a TSK gel ODS-80TM column (4.6 mm × 150 mm). The mobile phase was composed of acetonitrile–methanol–0.1 M acetate buffer, pH 5.0 (4:2:5, v/v/v), and used at a flow rate of 1 mL/min. Detection was by fluorescence, using excitation and emission wavelengths of 350 and 475 nm, respectively.

The reaction mixture contained 20 µL of enzyme preparation, 90 µL of 0.5 M acetate buffer (pH 5.0), 20 µL of 0.3 M N-ethylmaleimide, 20 µL of

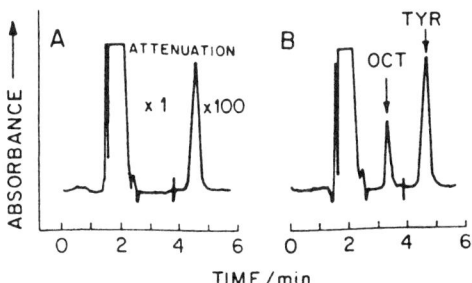

Figure 9.10 HPLC of the assay mixture (A) prior to and (B) 13.65 minutes after addition of dopamine β-hydroxylidase to the assay stock sample. OCT, octopamine (3.2 min); TYR, tyramine (4.5 min). (From Feilchenfeld et al., 1982.)

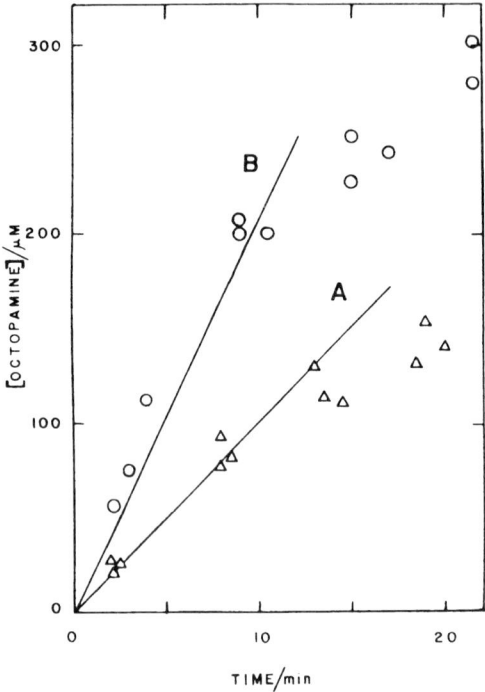

Figure 9.11 Time course of production of octopamine with varying DBH concentrations: 0.01 and 0.02 mg/mL for line A (△) and line B (○), respectively. Lines represent the least-squares fits of the initial data. (From Feilchenfeld et al., 1982.)

10 μM copper(II) sulfate, 10 μL of 10 mg/mL catalase, 10 μL of 60 mM (for assay of human plasma) or 250 mM (for rat serum) L-ascorbic acid, and 10 μL of 0.3 M sodium fumarate. The reaction was started by the addition of 20 μL of 80 mM dopamine. After 15 to 30 minutes at 37°C, the reaction was stopped by the addition of 700 μL of 0.5 M trichloroacetic acid, and 100 μL of 4 μM epinephrine was added as an internal standard. After centrifugation, 100 μL of the supernate was added to a Toyopak IC-SP M cartridge. The cartridge was washed in succession with 10 mL of water, 2 mL of 0.2 M sodium phosphate buffer (pH 6.0), 2 mL of water, and 2 mL of 50% acetonitrile. The adsorbed amines were eluted with 2 mL of a 1:1 mixture of acetonitrile and 0.6 M KCl. To the eluate were added 200 μL of 1,2-diphenylethylenediamine solution (212 mg in 0.1 M HCl) and 10 μL of 0.3 M potassium hexacyanoferrate(III). The mixture was allowed to stand at 37°C for 40 minutes before analysis of 100 μL by HPLC. The amount of dopamine formed was linear up to 15 μL of human plasma, and with time up to 30 minutes.

The assay was validated using rat serum and human plasma from heparinized blood of healthy volunteers.

9.2.5 Catechol O-Methyltransferase (Pennings and Van Kempen, 1979; Smit et al., (1990)

Catechol *O*-methyltransferase plays an important role in the catabolism of catecholamine neurotransmitters such as dopamine, norepinephrine, and epinephrine, and inactivation of catechol estrogens and catechol xenobiotics. Several different methods have been developed.

In one, *S*-adenosyl-L-methionine was the methyl donor, norepinephrine was the substrate, while products of the reaction, nor*met*anephrine and nor*para*nephrine, were converted to the more stable and more easily obtained compounds vanillin and isovanillin, respectively, by periodate oxidation. The oxidation also allowed for the extraction of the incubation mixture with organic solvents such as ethyl acetate, affording a more complete deproteinization.

In this study (not shown), the compounds vanillin and isovanillin together with *p*-hydroxyacetanilide, added as an internal standard, were separated by reversed-phase HPLC (LiChrosorb) with a methanol–50 mM phosphate buffer (pH 7.2) (3:7, v/v) as the mobile phase. The compounds were eluted isocratically and the eluent monitored by an electrochemical detector.

In another study, the substrate was 3,4-dihydroxybenzoic acid, and the reaction products were 3-methoxy-4-hydroxybenzoic acid and 3-hydroxy-4-methoxybenzoic acid. The substrate and the two products were separated by HPLC on a reversed-phase column (LiChrosorb) with a mobile phase of 0.05 M acetic acid in methanol–water (1:4, v/v), pH 3.2 Figure 9.12 shows the separation obtained under these conditions.

The reaction mixture was contained in a volume of 1 mL Tris-HCl buffer (pH 7.9), *S*-adenosylmethionine, $MgCl_2$, 3,4-dihydroxybenzoic acid, and dithiothreitol. Reactions were started by the addition of the activity and terminated

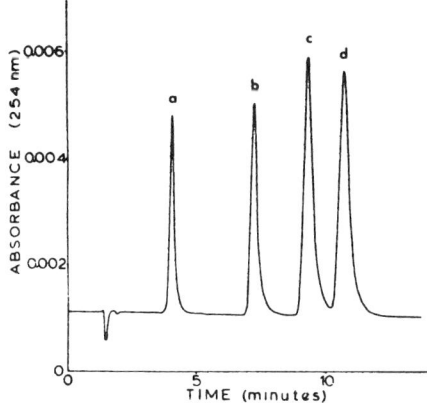

Figure 9.12 Optimal separation of DHBA (a), 4-hydroxybenzoic acid (b), MHBA (c), and HMBA (d) on a LiChrosorb 5 RP 18 column. (From Pennings and Van Kempen, 1979.)

after 10 minutes by placing tubes in a boiling water bath for 2 minutes. After cooling and centrifugation, the supernatant solution was loaded onto a DEAE-Sephadex column, and O-methylated products were eluted with 0.075 mM HCl. Samples were analyzed, and Figure 9.13 shows the chromatogram of a sample. Peaks representing the unreacted substrate, an internal standard, and the reaction products are observed.

Samples obtained from rat liver were homogenized and an S-100 solution prepared. Samples of this S-100 were the source of the transferase.

In the assay described by Smit et al. (1990), 5,6-dihydroxyindole-2-carboxylic acid was the substrate. Two products are formed, 5-hydroxy-6-methoxyindole-2-carboxylic acid and 6-hydroxy-5-methoxyindole-2-carboxylic acid.

The substrate and two products are separated on a μBondapak TM C_{18} column (3.9 mm $\times$ 300 mm). The mobile phase consisted of 0.05 M acetate buffer (pH 4.7), 10 mM $Na_2S_2O_5$, 1 mM EDTA, and 25% methanol. The column effluent was monitored by fluorimetry with excitation, and emission wavelengths being 295 and above 340 nm, respectively.

The enzyme assay contained 0.1 mM 5,6-dihydroxyindole-2-carboxylic acid in 30 mM Tris-HCl buffer (pH 7.8), 2.5 mM $MgCl_2$, 1 mM S-adenosylmethionine (as methyl donor), 0.5% Triton X-100, 0.5 mM EGTA, and 5 mM dithiothreitol in a total volume of 250 μL. The reaction was stopped with 25 μL of 4 M perchloric acid. Following centrifugation, the supernate was analyzed by HPLC. Product formation was linear with time for 60 minutes.

The source of enzyme was normal human melanocytes grown in cultured media. Homogenates were prepared by subjecting washed cells to sonication or homogenization in a Potter tube. In their discussion Smit et al. make a

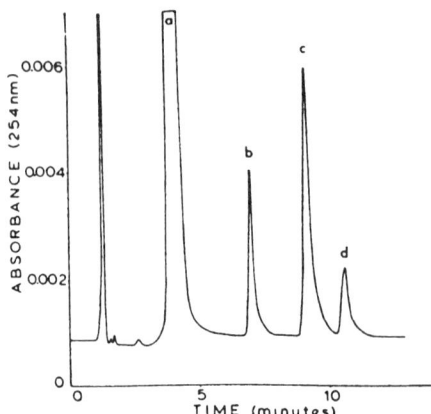

Figure 9.13 The determination of m- and p-O-methylated products (c,d) on a LiChrosorb 5 RP 18 column. Also shown are 4, hydroxybenzoic acid (a) and 3-methoxy-4-hydroxybenzoic acid (MHBA) (b). (From Pennings and Van Kempen, 1979.)

number of interesting comments on why HPLC assays can be more useful than other methods.

9.2.6 Phenylethanolamine N-Methyltransferase (Trocewicz et al., 1982; Beaudouin et al., 1993)

In the assay by Trocewicz et al. (1982), the enzyme phenylthanolamine N-methyltransferase catalyzes the conversion of noradrenaline (NA) to adrenaline (AD).

The assay, which measured only the amount of AD formed, used ion-paired, reversed-phase HPLC chromatography. The separation was carried out on a C_{18} (Nucleosil) column with a mobile phase of 0.1 M sodium phosphate buffer (pH 2.3–3.5) containing 5 mM sodium pentanesulfonate as the counterion to form ion pairs with the catecholamines, and 0.5% (v/v) acetonitrile. The separation of NA from AD is shown in Figure 9.14.

The reaction mixture (250 μL) contained pargylcine to inhibit secondary reactions catalyzed by monoamine oxidase activity, S-adenosylmethionine as the donor, and NA as the acceptor. Dihydroxybenzylamine (DHBA) was added as an internal standard. The reaction was terminated with perchloric acid containing EDTA. The pH was then adjusted to 8.5 with Tris-HCl, and the mixture was centrifuged. The clear supernatant was passed through an aluminum oxide column. The adsorbed AD was eluted, and a sample injected onto the HPLC column for analysis and quantitation. An electrochemical detector was used for detection. The chromatogram obtained after incubation with the homogenate is shown in Figure 9.14A, where the AD peak is clearly seen. In contrast, a similar incubation but without enzyme (Fig. 9.14B) showed no AD peak. The rate of AD formation is shown in Figure 9.15.

Samples prepared from rat brain was homogenized and used directly. Care was taken during the dissection to keep the samples on ice.

In the assay described by Beaudouin et al. (1993), the phenylethanolamine N-methyltransferase catalyzes the transfer of a methyl group from S-adenosyl-L-methionine to noradrenaline to form adrenaline and S-adenosyl-L-homocysteine as the final step of adrenaline biosynthesis. Adrenaline is mainly synthesized in the adrenal medulla.

The reactants and products were separated on an MOS Hypersil column (4.6 millimeters × 200 mm, 5 μm). The mobile phase was composed of a 90:10 mixture of solvent A, consisting of 0.1 M sodium acetate, 0.02 M citric acid, 0.93 mM sodium octanesulfonate, and 0.12 mM disodium EDTA (pH 4.6), and solvent B, methanol UV detection was used, with the optimal wavelength being 258 nm for the adenoxyl derivatives and 279 nm for adrenaline and noradrenaline. Quantitation was normally based on the S-adenosyl-L-homocysteine formed.

The reaction mixture contained 880 μL of 0.1 M phosphate buffer (pH 7.5), 10 μL of 1 mM noradrenaline, 10 μL of 1 mM S-adenosyl-L-methionine, and 100 μL of enzyme preparation in a final volume of 1 mL. After 10

222 SURVEY OF ENZYMATIC ACTIVITIES ASSAYED BY THE HPLC METHOD

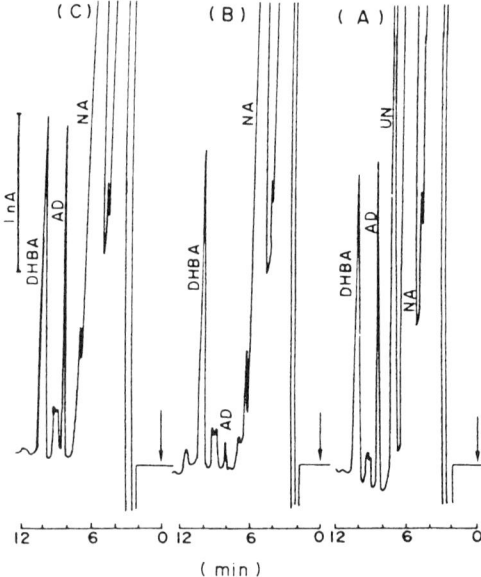

Figure 9.14 Typical elution pattern of phenylethanolamine N-methyltransferase incubation mixtures with the homogenate of rat pons plus medulla oblongata as enzyme. The incubation mixture contained 10 mg of rat pons plus medulla oblongata as enzyme and 16 µM noradrenaline (NA) and 18 µM S-adenosyl-L-methionine (SAM) as substrates. (A) Experimental incubation with homogenate of 10 mg of rat pons plus medulla oblongata. (B) Blank incubation without enzyme. (C) Another blank incubation, to which was added 15 pmol of adrenaline (AD) the reaction had been stopped. Formation of 16. 6 pmol of AD from NA during 60 minutes of incubation at 37°C was calculated from a calibration curve. DHBA, dihydroxybenzylamine (internal standard); UN, unknown peak. (From Trocewicz et al., 1982.)

incubation in the dark at 37°C, the reaction was stopped by the addition of 25 µL of glacial acetic acid. A 200 µL aliquot of the reaction mixture was analyzed directly by HPLC.

Enzyme preparations assayed were a partially purified enzyme from bovine adrenal medulla obtained commerically, and the supernate derived from homogenates of rat adrenal medulla.

9.2.7 Monoamine Oxidases A and B (Freeman et al., 1993)

Monoamine oxidases catalyze oxidative deamination of many primary, secondary, and tertiary amines. They have a wide tissue distribution including brain, liver, and intestine. A variety of endogenous amines, such as catecholamines, and pharmacological substances are metabolized. The products of primary amines are the corresponding aldehydes, ammonia, and hydrogen peroxide.

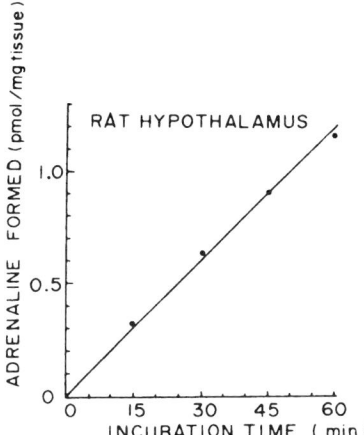

Figure 9.15 The rate of adrenaline formation using a homogenate of rat hypothalamus as enzymes at 37°C. Standard incubation system containing 10 mg of tissue was used. (From Trocewicz et al., 1982.)

Two distinct isoforms of monoamine oxidase, the A and B isoforms, have been identified.

The HPLC method described here allows the assay of both isoforms in the same run by using 5-hydroxytryptamine and 3-methoxy-4-hydroxybenzylamine as highly selective substrates for the A and B isoforms, respectively. The product of 5-hydroxytryptamine that is quantitated is 5-hydroxyindole-2-carboxylic acid, which is obtained by including aldehyde dehydrogenase in the assay mixture to oxidize the intermediate aldehyde. The product of monoamine oxidase B is 3-methoxy-4-hydroxybenzaldehyde. The internal standards used are 5-hydroxyindole-2-carboxylic acid and 3,4-dihydroxybenzoic acid.

The enzyme assay components including internal standards are resolved within 3 minutes on a 4.6 mm × 3.3 cm, 3 µm reversed-phase ODS column (Perkin-Elmer). The mobile phase contained 0.100 M citric acid, 0.50 mM tetrapentylammonium chloride (an ion-pairing agent), 50 µM EDTA, and 12.5% acetonitrile (v/v) titrated to pH 4.70 with NaOH. Eluted compounds were detected with an electrochemical detector. The separation obtained is shown in Figure 9.16.

The reaction mixture included phosphate buffer, alcohol dehydrogenase, NAD^+, substrates and internal standards, and mercaptoethanol in a final volume of 190 µL. The enzyme preparation is preincubated with all components of the assay system except substrates, which are added to initiate the reaction. The reaction is terminated by adding 100 µL of the incubation mixture of 300 µL of reaction-terminating solution containing acetate–perchlorate buffer and ascorbic acid. Samples could be stored up to 2 weeks at −80°C before thawing and filtering through 0.45 µm nitrocellulose filters. Five-microliter aliquots were injected into the HPLC.

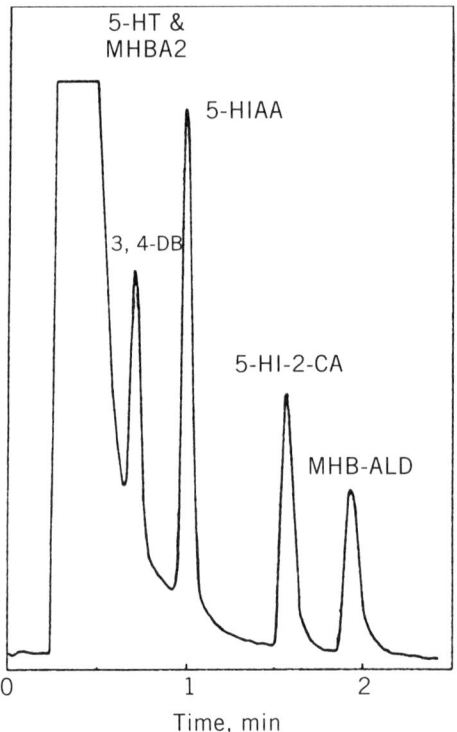

Figure 9.16 Typical chromatogram resulting from whole mouse brain homogenate using dual-assay conditions. (From Freeman et al., 1993.)

Fresh whole mouse brains were homogenized in a ground-glass homogenizer and subjected to ultrasonication before being used in the assays.

9.2.8 Arylsulfatase (Bradley and Manowitz, 1988)

Arylsulfatase cleaves sulfate esters. Cerebroside sulfate is its natural substrate. The enzyme is also active toward p-nitrocatechol sulfate, which is the basis for this assay.

The p-nitrocatechol formed was separated at 30°C with a reversed-phase C_{18} column (4 millimeters × 125 mm) from EM Science. The mobile phase was prepared by mixing 13 g of monochloroacetic acid, 4.99 g of NaOH, 0.177 g of sodium octylsulfate, 130 mL of acetonitrile, 10 mL of tetrahydrofuran, and approximately 620 of mL water. After the pH had been adjusted to 3.3 with approximately 220 mL of 8 M acetic acid, the mixture was diluted to 1 liter with water. The electrochemical detector was set at an applied potential of 0.750 V.

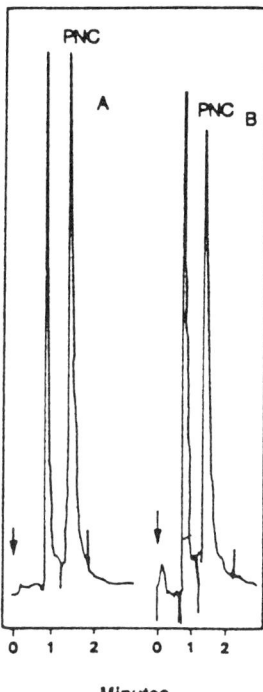

Figure 9.17 Typical HPLC elution pattern from (A) an aqueous solution containing *p*-nitrocatechol (PNC), 3.87 pmol per injection, and (B) the ethanol-deproteinized extract of an ASA incubation mixture containing a dialyzed human leukocyte and platelet lysate. The arrows denote time of injection. (From Bradley and Manowitz, 1988.)

The reaction mixture was prepared by mixing 800 μL of substrate solution containing 0.01 M *p*-nitrocatechol sulfate (dipotassium salt), 1.71 M NaCl, and 0.5 M acetate buffer (pH 5.0) with 200 μL of either leukocyte and platelet lysate or saliva. After incubation for an hour at 37°C, a 50 μL aliquot was mixed with 100 μL of bovine serum albumin (13 mg/mL) and 1800 μL of cold 95% ethanol. After centrifugation to remove proteins, 20 μL was injected into the HPLC system. The reaction was linear with up to 200 μL of lysate.

The assay as described was used with leukocyte and platelet lysate, but it is suitable with slight modification for determination of activity in saliva and amniotic fluid. A typical HPLC elution pattern is shown in Figure 9.17.

9.2.9 Monoamine Oxidase and Phenol Sulfotransferase (Sim and Hsu, 1990)

Monoamine oxidase catalyzes the conversion of dopamine to 3,4-dihydroxyphenylacetic acid and tyramine to 4-hydroxyphenylacetic acid. Phenol sulfotransferase transfers sulfate from 3'-phosphoadenosine-5'-phosphosulfate to dopamine, 3,4-dihydroxyphenylacetic acid, and phenol. This assay is sufficiently sensitive to measure the activity of either enzyme in 5 mg of rat brain and liver.

Radiolabeled products were separated from substrates by chromatography on a Merck C_{28} column (5 μm). The mobile phase contained 0.1 M sodium acetate, 0.1 M citric acid, 0.1 mM sodium octylsulfate, 0.15 mM EDTA, and 0.2 mM dibutylamine in 10% methanol (v/v). The pH was 4 for the monoamine oxidase assay and 3.7 for phenol sulfotransferase. A flow-through radioisotope detector was used to quantitate the amount of radioactivity in the eluted peaks.

The assay for monoamine oxidase contained in a final volume of 100 μL to 30 μL of homogenate and 50 μM [7-^{14}C] dopamine (0.93 mCi/mmol) or [7-^{14}C] tyramine (3.11 mCi/mmol) in 0.5 M phosphate buffer (pH 7.4). The assay for phenol sulfotransferase was also initiated by adding 30 μL of homogenate. The mixture contained 1.7 μM [^{35}S] 3'-phosphoadenosine-5'-phosphosulfate (1.51 Ci/mmol) and 50 μM dopamine, 3,4-dihydroxyphenylacetic acid, or phenol in 10 mM phosphate buffer (pH 6.4). After various incubation periods, the activity of either enzyme was stopped by addition of 30 μL of 2 N HCl. The resulting mixtures were centrifuged or filtered before analysis by HPLC.

Half of a rat brain hypothalamus (8.5–10 mg), the anterior pituitary (5–7 mg), or minced liver (4 mg) was homogenized by sonicating in 100 μL of 0.5 M phosphate buffer (pH 7.4) for the assay of monoamine oxidase, or in pH 6.4 buffer for assay of phenol sulfotransferase.

Figure 9.18 shows a representative chromatogram of the substrates and products of liver monoamine oxidase activity. Figure 9.19 illustrates results with phenol sulfotransferase activity.

9.2.10 N-Acetyltransferase (Martin and Downer, 1989)

N-Acetyltransferase uses acetyl–CoA to acetylate the amino moiety of arylalkylamines. In mammalian pineal gland, this enzyme catalyzes the production of N-acetyl-5-hydroxytryptamine, which is the precursor of melatonin. It is also involved in the inactivation of monoaminergic neurotransmitters in insects.

Substrates (p-octopamine, dopamine, or 5-hydroxytryptamine) were separated from their N-acetylated products on an Ultrasphere I.P. C_{18} column (4.6 mm × 250 mm, 5 μm). The mobile phase was comprised of 75 mM monobasic sodium phosphate, 1 μM EDTA, 0.35 mM 1-octanesulfonate (sodium salt), 11% methanol, and 4% acetonitrile. Coulometric detection was used. Detection of p-octopamine and its N-acetylated product was achieved at a potential of +0.75 V, while dopamine, 5-hydroxytryptamine and their products were detected at +0.50 V.

Microassays were used in which the final assay volume was 10 μL. Microvolumes were dispensed by using a 100 μL Hamilton syringe and a repeating dispenser. Reactions were carried out in 96-well polystyrene microplates. A typical well contained 2 μL of sodium phosphate buffer, 2 μL of water, and

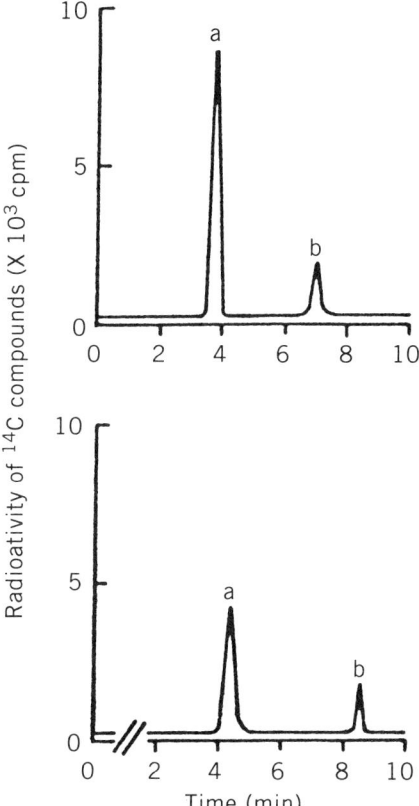

Figure 9.18 Chromatograms of the substrates and products of liver monoamine oxidase activity. Upper tracing: a, dopamine, b, Dopac. Lower tracing: a tyramine, b, Hopac. The retention time for dopamine, Dopac, tyramine, and Hopac was 4.0, 7.0, 6.0, and 10.7 minutes, respectively. (From Sim and Hsu, 1990.)

4 μL of tissue extract. The reaction was started with the addition of 2 μL of a solution containing both the monoamine substrate and acetyl–CoA. The final assay contained 200 mM KCl, 20 mM sodium phosphate buffer (pH 7.0), 0.52 mM dithiothreitol, 1.0 mM monoamine substrate, and 2.5 mM acetyl–CoA. The plate was covered with Parafilm during the incubation at 30°C. The reaction was terminated by the addition of 100 μL of ice-cold 0.1 M perchloric acid. After the acidified mixture had been allowed to stand on ice for 10 minutes, 15 μL was diluted further with 200 μL of 0.1 M perchloric acid. A 5 μL aliquot was used for HPLC analysis. Assays were linear to 60 minutes and with 2.5 to 15 μg of protein.

Brains from adult male American cockroaches were homogenized either by sonication or by glass–Teflon pestle in 100 μL of ice-cold 0.5 M potassium

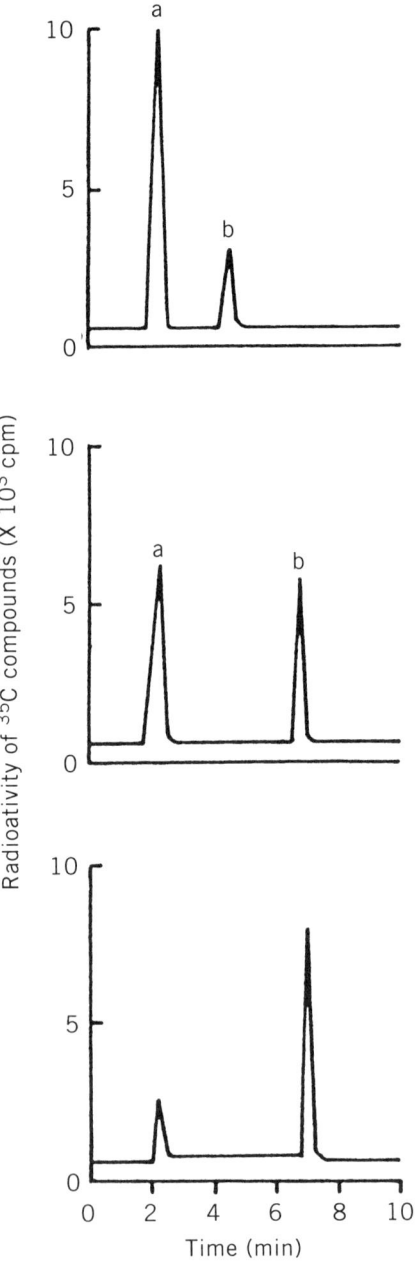

Figure 9.19 Chromatograms of the cosubstrate (PAPS) and products of liver phenol sulfotransferase activity. Upper tracing: a, PAPS, b, sulfate-conjugated dopamine. Middle tracing: a, PAPS, b, sufate-conjugated Dopac. Lower tracing: a, PAPS, b, sulfate-conjugated phenol. The retention time for sulfate-conjugated dopamine, Dopac, and phenol was 4.5, 6.8, and 7.0 minutes, respectively. (From Sim and Hsu, 1990.)

chloride and 1.3 mM dithiothreitol. The homogenates were centrifuged at 40,000 g for 20 minutes and the resulting supernates were assayed for activity.

9.2.11 Acetyl–CoA/Arylamine N-Acetyltransferase (Manneus et al., 1990; Thomas et al., 1990)

The N-acetylation of different aryl- and arylalkylamines, using acetyl–CoA as the acetyl donor, proceeds when catalyzed by acetyl–Co A : arylamine N-acetyltransferase. A natural substrate is serotonin, which is converted to N-acetylserotonin in the pineal gland and other regions of the brain.

Serotonin and N-acetylserotonin are separated on a Spherisorb ODS 2 column (3.2 mm × 150 mm). The mobile phase was a 3:1 mixture of 0.1 M sodium acetate (pH 4.75) and methanol. The column effluent was monitored at 275 nm. Greater sensitivity can be obtained by using fluorescence or electrochemical detection.

The enzyme assay mixture contained in a final volume of 200 μL: 20 μmol glycine–KOH buffer (pH 9.5), 0.30 μmol serotonin–HCl, 0.12 μmol acetyl–CoA, and 50 μL of enzyme solution with a maximum activity of 100 mU/mL. After 5 minutes of incubation at 35°C, the reaction was stopped by diluting 10-fold with 0.1 M perchloric acid. After filtering, a 25 μL aliquot of the filtrate was injected onto the HPLC column.

Acetyl–CoA : arylamine N-acetyltransferase was purified from pigeon liver by using steps involving protamine sulfate, ammonium sulfate fractionation, and affinity chromatography on immobilized amethopterin. Figure 9.20 shows a representative chromatogram. In another study, Thomas et al. (1990) used tryptamine as the substrate. In addition, whereas the study by Manneus et al. (1990) appeared to require pure enzyme, Thomas's study did not.

9.3 PROTEINASE

9.3.1 Vertebrate Collagenase (Gray and Saneii, 1982)

In both the α_1 and α_2 chains of type I collagen, degradation of the protein begins with cleavage of the Gly-Ile or the Gly-Leu bond by collagenase.

In the assay developed for this activity, the collagen was replaced by the peptide dinitrophenyl (DNP)-Pro-Gln-Gly-Ile-Ala-Gly-Gln-D-Arg. During the course of the reaction, this substrate was cleaved into the two products DNP-Pro-Gln-Gly and Ile-Ala-Gly-Gln-D-Arg. The substrates and the two products were separated on a reversed-phase column (Varian MCH-10) eluted as follows: Initially the column was equilibrated with a mobile phase consisting of 0.1% H_3PO_4–CH_3CN (80:20, v:v), which was followed by a mobile phase in which the proportion of CH_3CN was increased linearly up to 40% by volume. This gradient resulted in the elution of both the DNP–tripeptide reaction product and the unreacted substrate, as shown in Figure 9.21. The eluent was

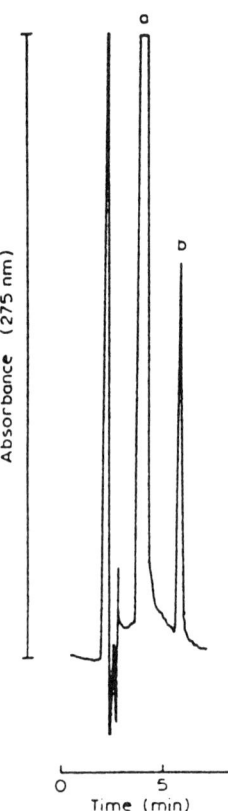

Figure 9.20 Chromatogram illustrating N-acetylserotonin formation by the enzyme. *Peak:* a, serotonin hydrochloride; b, N-acetylserotonin. (From Manneus et al., 1990).

monitored at 206 nm, although monitoring for the DNP derivative at 365 nm was also feasible.

The reaction mixture included Tris-HCl buffer, NaCl, $CaCl_2$, and the DNP–octapeptide substrate. The reaction, initiated by the addition of enzyme, was incubated at 37°C. The total volume of the reaction mixture varied between 10 and 25 μL depending on the number of samples to be analyzed. Successive aliquots of 3 to 10 μL were removed with a syringe at timed intervals and applied directly to the column without further treatment. [The column was protected by a guard column packed with pellicular packing (V/dac Reverse Phase Pellicular Packing).] The results of incubation of the substrate with tadpole collagenase are shown in Figure 9.22. Chromatograms for three time points are shown. In Figure 9.23, the rate of hydrolysis is shown as a function of enzyme concentration. Because the gradient had to be reversed prior to injection of a new sample, the interval between samples was 18 to 20 minutes.

The enzyme vertebrate collagenase was partially purified from a lyophilized tissue culture medium of back skin from tadpoles. The medium was harvested,

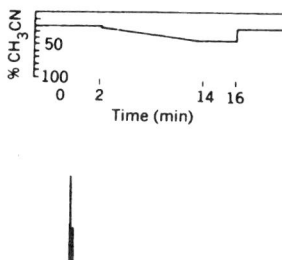
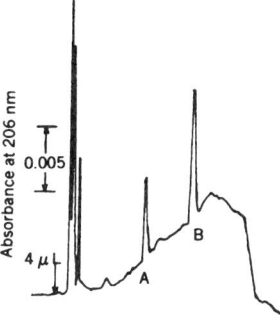

Figure 9.21 Separation of DNP-Pro-Gln-Gly (A) from DNP-Pro-Gln-Gly-Ile-Ala-Gly-Gln-D-Arg (B) using reversed-phased HPLC. An aliquot of 4.0 μL was injected at the arrow. The initial solvent was 0.1% H_3PO_4/CH_3CN (80:20). The column was eluted with this solvent for 2.0 minutes, following which the CH_3CN was increased to 40% linearly over a 12-minute time period. The flow rate was 2.0 mL/min. Upper graph shows the gradient used. (From Gray and Saneii, 1982.)

and the protein precipitated by 40% ammonium sulfate collected by centrifugation. The precipitate was recovered and purified further on G-200 Sephadex.

9.3.2 Dipeptidyl Carboxypeptidase (Angiotensin I Converting Enzyme, EC 3.4.15) (Baranowski et al., 1982; Doig and Smiley, 1993)

Angiotensin I converting enzyme is a dipeptidyl carboxypeptidase that cleaves angiotensin I to angiotensin II and the dipeptide histidyl-leucine (His-Leu). The enzyme is bound to the membrane of lung arterial endothelium and is involved in the renin–angiotensin system that regulates blood pressure and fluid balance.

In the assay reported by Baranowski et al. (1982), the rate of formation of angiotensin II and the dipeptide His-Leu from the substrate angiotensin I was followed.

The reactant was separated from the product as the fluorescamine derivatives on a reversed-phase HPLC column (Partisil ODS) and eluted isocratically using a two-solvent system. To elute the dipeptide, the solvent was 60% acetonitrile in water diluted 9:1 with 1 M acetic acid, at a final pH of 3.5 (Fig. 9.24). To elute the angiotensin compounds, 38% acetonitrile in 1 M ammonium acetate (pH 4.0) was used (Fig. 9.25). The eluent was monitored with a fluorometer.

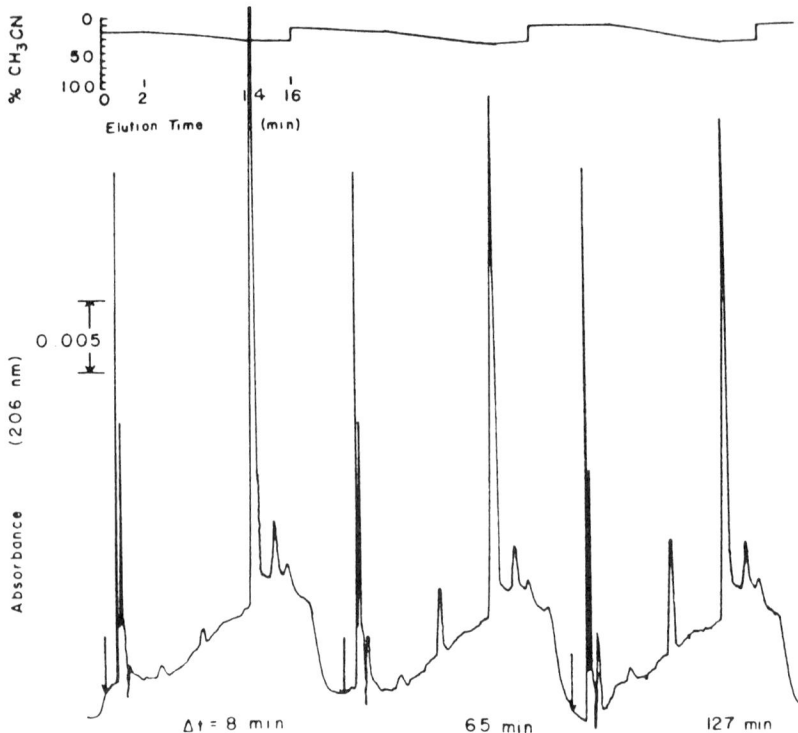

Figure 9.22 Separation of hydrolysis products of the action of tadpole collagenase on DNP-Pro-Gln-Gly-Ile-Ala-Gly-Gln-D-Arg by reversed-phase HPLC. Upper tracing shows the gradient employed. The reaction was initiated at zero time by the addition of 4 μg of enzyme. At the indicated times, a 4.0 μL aliquot of the reaction mixture was injected onto the HPLC column and the peptides separated. (From Gray and Saneii, 1982.)

The reaction mixture contained angiotensin I, a phosphate buffer at pH 8.0, NaCl (chloride is required for activity), and the enzyme. Incubations were at 37°C for 30 minutes and were terminated by treatment with a boiling water bath for 5 minutes. Centrifugation was used to remove precipitated protein. Fluorescamine in acetone was added to the supernatant solution, and samples were injected for analysis. The enzymatically formed fragments were separated on HPLC as shown in Figures 9.24 and 9.25; the rate of His-Leu formation is shown in Figure 9.26.

The dipeptidyl carboxypeptidase was prepared from rat lungs. A microsomal fraction was prepared and extracted with detergent (sodium deoxycholate) and clarified by centrifugation. The supernatant solution was dialyzed against sodium phosphate, and the dialysate was stored frozen.

In the assay described by Doig and Smiley (1993), the enzyme was assayed by using a synthetic peptide, hippuryl–His-Leu, as substrate and measuring

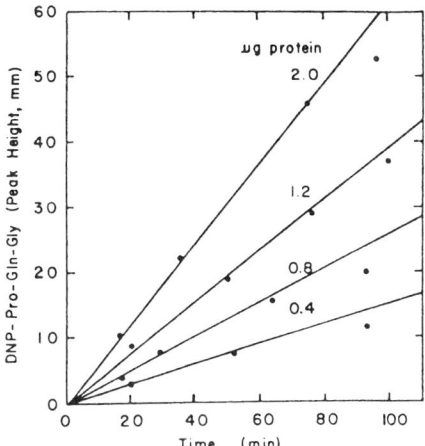

Figure 9.23 Dependence of rate of hydrolysis of DNP-Pro-Gln-Gly-Ile-Ala-Gly-Gln-D-Arg on collagenase concentration. Substrate concentration was 1.8 mM at pH 7.6 and 37°C. (From Gray and Saneii, 1982.)

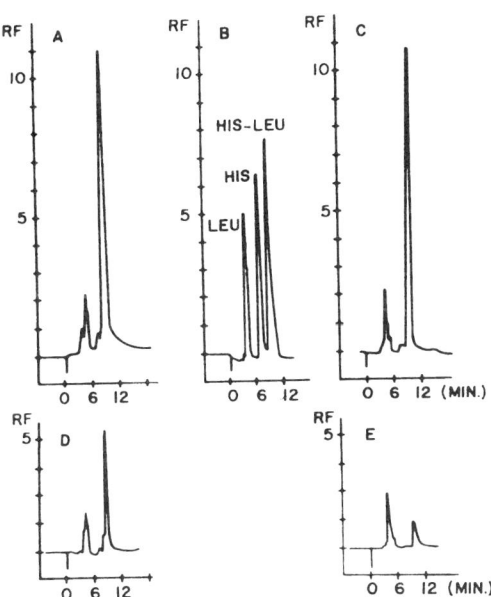

Figure 9.24 HPLC separation of a standard mixture of fluorescamine derivatives of histidine, leucine, and histidyl-leucine (B). HPLC separation of enzymatically formed peptide fragments in the presence (A, C) and absence, (D,E) of sodium chloride in the incubation mixture. (From Baranowski et al., 1982.)

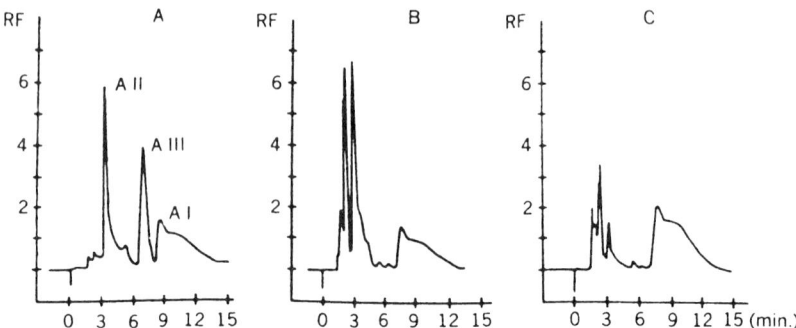

Figure 9.25 HPLC separation of a standard mixture of the fluorescamine derivatives of angiotensins I, II, and III (*A*). HPLC separation of enzymatically formed peptide fragments in the presence (*B*) and absence (*C*) of sodium chloride in the incubation mixture. (From Baranowski et al., 1982.)

the hippuric acid produced by hydrolysis. The substrate and product were separated on a Supelco HISEP SHP column (250 mm × 4.6 mm) that was preceded by a 20 mm × 4.6 mm HISEP column. The shielded hydrophobic phase (SHP) column in this design excludes macromolecules by incorporating a hydrophilic outer surface. Because proteins are excluded and elute at the void volume, the need for sample cleanup is eliminated. The mobile phase contained 180 mM ammonium acetate–acetonitrile (95:5, v/v). The absorbance of the eluate was monitored at 254 nm.

The enzyme assay contained 5 mM hippuryl–His-Leu, 100 mM phosphate buffer (pH 8.3), and 276 mM NaCl in a final volume of 650 μL. When rat lung extracts were the enzyme source, the assay mixture was maintained at

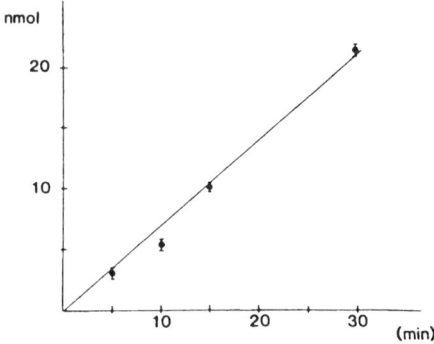

Figure 9.26 Linear realtionship observed between the amount of His-Leu generated by angiotensin-converting enzyme and incubation time. Number of determinations are 3, 5, 5, and 5 for 5, 10, 15, and 30 minutes, respectively. (From Baranowski et al., 1982.)

37°C for 15 to 60 minutes before centrifuging for 2 minutes and injecting 20 μL on the column. To assay the enzyme in serum 20 μL aliquots were removed from the assay mixture and injected directly.

Lung tissue from rat was homogenized with a Teflon tissue grinder in 20 volumes of ice-cold 50 mM phosphate buffer (pH 8.3). The homogenate was centrifuged at 1000g for 10 minutes. The supernate was recentrifuged at 30,000g for 60 minutes, and the resulting pellet was resuspended and used as the source of angiotensin-converting enzyme.

9.3.3 Luteinizing Hormone–Releasing Hormone Peptidase (Advis et al., 1982)

The luteinizing hormone–releasing hormone peptidase is one of many peptidases that catalyze the hydrolysis of neuropeptides.

The assay used for this activity was based on the separation of the substrate, luteinizing hormone–releasing hormone (LHRH) from its degradation product, a pentapeptide. A reversed-phase HPLC (C_{18}) column was used, and LHRH was eluted isocratically in an aqueous mobile phase composed of 42% acetonitrile containing 0.7 mM tetraethylammonium phosphate. To analyze for degradation products, a mobile phase of sodium phosphate–H_3PO_4 (pH 2.5, both at 0.1 M) was used to equilibrate the C_{18} column. The reaction fragments were eluted with an exponential gradient of 0 to 30% acetonitrile. Several fragments generated by the reaction, include $LHRH_{1-5}$, $LHRH_{6-10}$, and $LHRH_{1-3}$, and all were separated and collected.

The LHRH peptidase activity was assayed with LHRH in a phosphate buffer (pH 7.2). The reaction was started by addition of LHRH and terminated by heating at 110°C for 6 minutes in a heating apparatus (Reacti-Thermi, Pierce). The sample was clarified by centrifugation and stored prior to activity determinations. The chromatograms obtained at zero time and after 2-hour incubation, are shown in Figure 9.27. The hypothalamus from the female rat was sampled using a "punch" technique. Samples were homogenized in ice-cold phosphate buffer at pH 7.4 and the lysate was centrifuged for 5 minute at 10,000g at room temperature. The supernatant fraction was used as the source of peptidase activity and stored at −80°C.

9.3.4 Papain Esterase (Chen et al., 1982)

Papain is a proteolytic enzyme from plants. In the HPLC assay developed to measure its activity, the water-soluble N-benzyl-L-arginine ethyl ester (BAEE) is used as the substrate.

The separation of the substrate from the reaction product benzoylarginine was carried out by reverse phase HPLC on a μBondapak CN column with a mobile phase of 0.05 M ammonium acetate–methanol (85:15). The column was eluted isocratically, and the eluent was monitored at 254 nm.

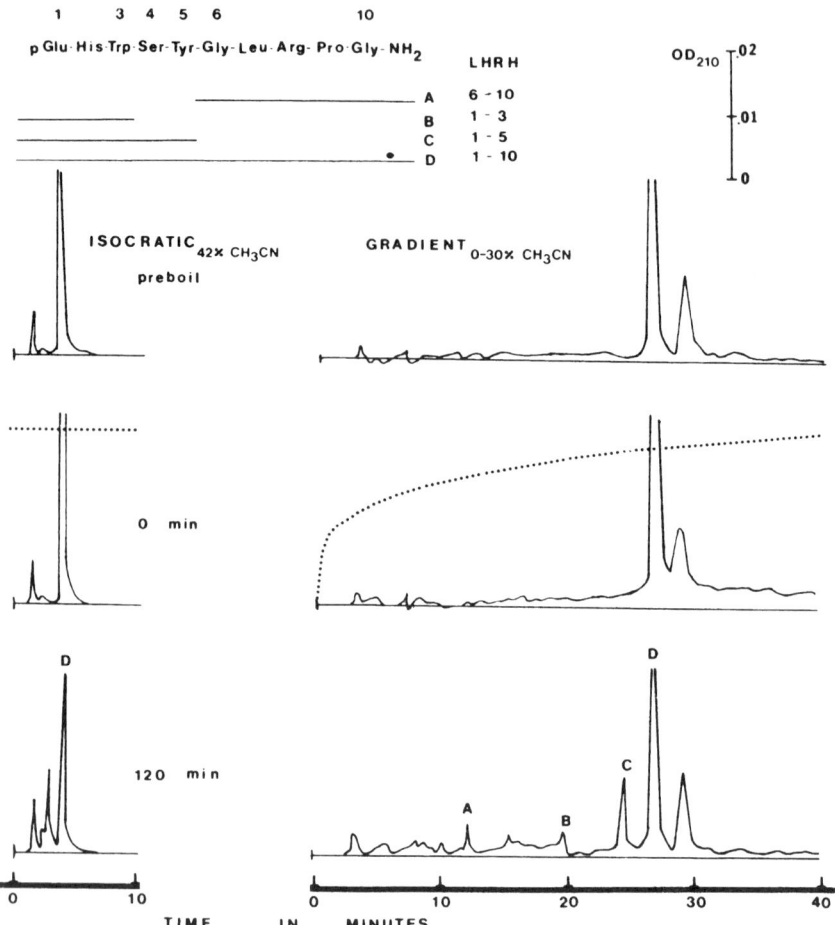

Figure 9.27 Isocratic and gradient HPLC analysis of median eminence (ME) supernatant fraction after incubation with LHRH. Total peptidase activity was assessed in ME and was analyzed by HPLC. Upper HPLC tracing represents a preboiled control, the middle tracing a zero-time incubation, and the lower a 2-hour preincubation. The peptide fragments are indicated in the upper panel by letter (A–D) and the corresponding peaks they represent are shown by the same letter. (From Advis et al., 1982.)

The reaction mixture contained the substrate BAEE, mercaptoethanol, EDTA, and sodium chloride. The reaction was started by the addition of papain, and the immediate adjustment of the incubation mixture to pH 6.2. After 5 minutes, the reaction was terminated by the addition of 30% acetic acid. Precipitated material was removed by filtration, using a 0.45 μm Millipore filter, and the filtrate was analyzed by HPLC. The results of an assay are

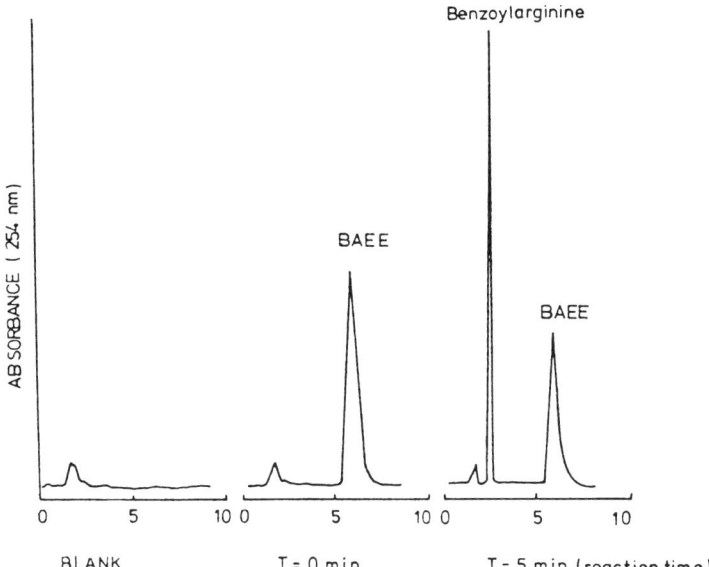

Figure 9.28 HPLC profile of papain-catalyzed BAEE hydrolysis product. Conditions: column, µBondapak CN; eluent, 0.05 M ammonium acetate–methanol (85:15); flow rate, 2.0 mL/min; detector, 254 nm at ambient temperature (From Chen et al., 1982.)

shown in Figure 9.28, where the appearance of the reaction product benzoylarginine is evidence of enzymatic activity.

The papain was obtained from commercial sources.

9.3.5 Plasma Carboxypeptidase N (Kininase I, Bradykinin-Destroying Enzyme, EC 3.4.12.7) (Marceau et al., 1983)

Plasma carboxypeptidase N degrades and therefore inactivates bradykinin. This activity may have a role in the regulation of inflammatory peptides. The HPLC method developed for its assay uses the dipeptide hippuryllysine (Hip-Lys) as the substrate and measures activity by measuring the release of hippuric acid.

The separation of substrate from the product was carried out by reversed-phase HPLC (C_{18} µBondapak) using a mobile phase of a 1:4 mixture of methanol and 0.001 M K_2HPO_4 and H_3PO_4 (pH 3.0). The column was protected by a precolumn packed with Corasil. The column was eluted isocratically, and detection was at 230 nm. The separations obtained under these conditions are shown in Figure 9.29.

The reaction mixture contained Hepes [N-(2-hydroxyethyl)piperazine-N^1-2-ethanesulfonic acid] buffer (pH 7.75) with NaCl and the substrate Hip-Lys in a volume of 500 µL. The plasma was added to start the reaction, and the

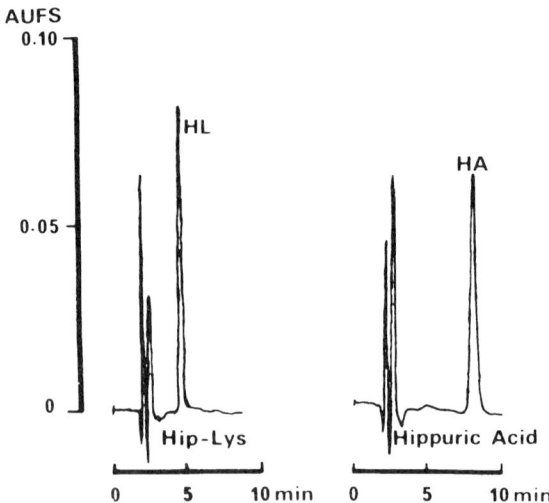

Figure 9.29 HPLC traces of standard solutions (32 μg/mL) of Hip-Lys (HL) and hippuric acid (HA). (From Marceau et al., 1983.)

reaction was terminated by the addition of absolute ethanol. The mixture was clarified by centrifugation, and the supernatant solution applied to the HPLC for analysis. The result of a 15-minute incubation is shown in Figure 9.30.

The carboxypeptidase was prepared from plasma.

9.3.6 Dipeptidase (Horiuchi et al., 1982)

In a study of dipeptidase by Horiuchi et al. (1982), the tripeptide hippurylhistidylleucine (Hip-His-Leu) was used as a substrate, and the assay involved measuring the amount of hippuric acid released by the enzyme.

The substrate was separated from the product by reversed-phase HPLC (Nucleosil 7 C_{18}) using a mobile phase of methanol–10 mM KH_2PO_4 (1:1) adjusted to pH 3.0 with phosphoric acid. Detection was at 228 nm. The separation of the components of the reaction mixture is shown in Figure 9.31.

The reaction mixture contained a phosphate buffer at pH 8.3 with NaCl and the substrate Hip-His-Leu. The reaction was incubated with the dipeptidase preparation for 30 minutes at 37°C and terminated with 3% metaphosphoric acid. The mixture was centrifuged, and a sample of the supernatant solution was injected onto the HPLC column for analysis. The results of an assay are shown in Figure 9.31. The appearance of the hippuric acid is taken as evidence of enzymatic activity. The rate of product formation is shown in Figure 9.32.

Peptidase preparations obtained from rat blood, lung, and kidney were chopped and homogenized, and clarified by centrifugation at 20,000g for 20 minutes. The supernatant solution served as the source of peptidase activity.

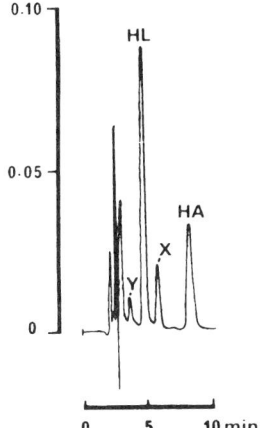

Figure 9.30 HPLC traces of solutions of Hip-Lys containing 75% human plasma incubated for 15 minutes. The substrate concentration was 1.02 mM. X and Y represent unknown substances from plasma. (From Marceau et al., 1983.)

9.3.7 Aminopeptidase (Mousa and Couri, 1983)

Aminopeptidases are involved in the metabolism of opioid peptides including enkephalins and β-endorphins. An HPLC method was developed to measure the hydrolysis of these compounds by measuring the formation of tyrosylglycylglycine, tyrosylglycine, and tyrosine.

The separation was accomplished by reversed phase HPLC (Ultrasphere ODS column) with a mobile phase of 50 mM sodium phosphate buffer (pH 2.1) with acetonitrile and methanol (90:5:5). The column was eluted isocratically, and an electrochemical detector was used. The separation of the standards is shown in Figure 9.33.

The reaction mixture for peptidase activity contained the substrate methionine enkephalin with Tris-HCl (pH 7.4) as buffer. The reaction was started by the addition of serum peptidase, and after various intervals the reaction was terminated by adding 1 M HCl.

The aminopeptidase was obtained from either serum, rat brain, or synaptosomal membrane.

9.3.8 Enkephalinases A and B (Ohno et al., 1988)

Enkephalinase A (dipeptidyl carboxypeptidase) and enkephalinase B (dipeptidyl aminopeptidase) are involved in the degradation of enkephalin, which has an opiate-like activity. Enkephalinase A cleaves methionine enkephalin (Tyr-Gly-Gly-Phe-Met) after the second glycine, whereas enkephalinase B cleaves after the first glycine. A postcolumn derivatization system is used to detect N-terminal tyrosine-containing peptides.

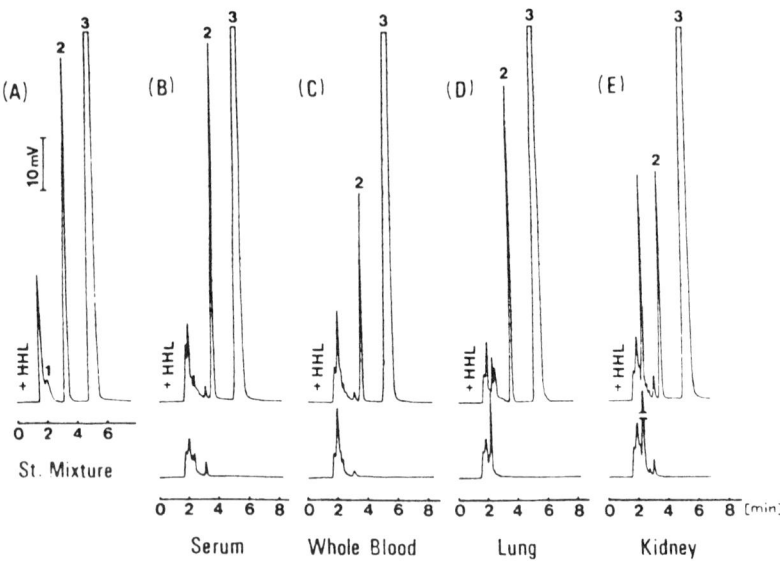

Figure 9.31 Chromatograms obtained from various samples incubated with (upper diagram) or without (lower diagram) Hip-His-Leu (HHL). (*A*) Standard mixture of 2.7 nmol His-Leu, 2.7 nmol hippuric acid, and 100 nmol Hip-His-Leu. (*B*) A 50 μL aliquot of serum or (*C*) whole blood was incubated with or without 5 mM Hip-His-Leu. After 30 minutes, 0.75 mL of 3% *m*-phosphoric acid was added and centrifuged. (*D*) Lung or (*E*) kidney was homogenized in 5 volumes of chilled Tris-HCl buffer containing 0.5% Nonidet-P40, and centrifuged. The supernatant was incubated with or without 5 mM Hip-His-Leu. In the case of lung, the supernatant was diluted 20 times with the buffer prior to incubation with Hip-His-Leu. *Peaks:* 1, His-Leu; 2, hippuric acid; 3, Hip-His-Leu. (From Horiuchi et al., 1982.)

Substrate and product peptides were separated on a reversed-phase column (200 mm × 4 mm) packed with TSK gel ODS-120T (5 μm). The mobile phase was formed from solvents A and B formed from acetonitrile–sodium phosphate buffer (0.3 M, pH 2.3)–water mixed in ratios of 1:20:79 and 3:1:1 (v/v/v), respectively. For detection of N-terminal tyrosines (other tyrosines do not react), the eluate stream was mixed first with 50 μm $CoCl_2$ and 20 mM NH_2OH, and then with 0.3 M borate (pH 11.4). The resulting mixture was passed through a reaction coil maintained at 75°C before monitoring at 435 nm (emission) and 335 nm (excitation). Calibration graphs for the products were linear up to at least 5 nmol per assay.

The enzyme reaction ws conducted by adding 30 μL of enzyme sample to a mixture containing 50 μL of 0.5 M Tris-HCl (pH 7.4), 50 μL of 1 mM bestatin, 20 μL of 40 mM captopril, and 50 μL of 100 μM methionine enkephalin. Bestatin and captopril are included to inhibit aminopeptidase and angiotensin-coverting enzyme, respectively. Incubation at 37°C proceeded for 30 minutes, and then the reaction was terminated by heating at 100°C for 2

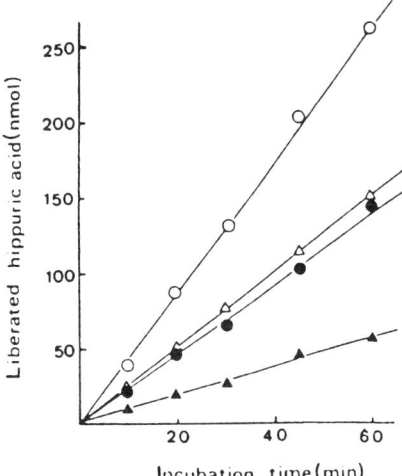

Figure 9.32 Dependence on incubation time of hippuric acid formation from Hip-His-Leu catalyzed by serum (○), whole blood (●), lung (△), and kidney (▲). A 50 μL aliquot of sample was used. Each point represents the mean of three determinations. (From Horiuchi et al., 1982.)

minutes. After 25 minutes of centrifuging at 2000g, a 50 μL portion of the supernate was injected into the column.

Enzyme preparations were obtained by removing brains from anesthetized rats and separating each brain into the striatum, cortex, pituitary, hypothalamus, hippocampus, and amygdala regions. A portion of each region (ca. 100 mg) was homogenized with 2 mL of water. The homogenate was centrifuged at 800g for 10 minutes and the pellet was extracted with 5 mL of Triton X-100 solution (1 mg/mL). After standing for an hour, the mixture was centrifuged at 2000g for 25 minutes and the supernate was used in the enzyme reaction.

9.3.9 Rhinovirus 3c Protease (Hopkins et al., 1991)

A number of proteinase assays are based on a fluorescein-labeled peptide. Hopkins et al. (1991) used a 14-mer synthesized with and without fluorescein as a label to assay rhinovirus 3c protease. The label did not affect the rate of hydrolysis. The sequence of the peptide containing fluorescein was M-E-A-L-F-Q-G-P-L-Q-Y-K-E(fluorescein)-L, and cleavage occurred after the first six residues.

The technique known as high speed reversed-phase HPLC was applied, and the fluorescein-labeled substrate and its product were separated within 2.5 minutes on C_{18} column (4.6 × 33 mm) with a 76:24 (v/v) mixture of (50 mM sodium phosphate and 50 mM sodium acetate at pH 7) and acetonitrile. The column was run at ambient temperature at 3 mL/min, and 75 μL

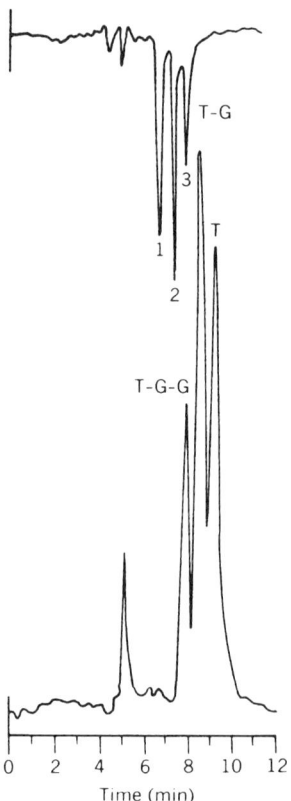

Figure 9.33 Representative chromatogram of tyrosine (T), tyrosylglycine (T-G) and tyrosylglycylglycine (T-G-G) standards. Upper (HPLC-UV) and [HPLC–electrochemical detection (ED)] chromatograms shows the separation of 55 ng of T-G-G (1), 70 ng of T-G (2), and 35 ng of T (3). Analytes were detected at 205 nm (0.1 A) and + 1.25 V oxidation potential (100 nA). (From Mousa and Couri, 1983.)

of sample was injected. Analysis time could be reduced to 12 seconds by applying the techniques of ultrahigh speed HPLC. In this case, an M-pel C_8 column (4.6 mm × 15 mm, 2 μm) was eluted at 4 mL/min and 80°C after injection of 1 μL of sample. A fluorescence detector with excitation at 441 nm and detection at 521 nm was used. Exemplary separations are shown in Figure 9.34.

The reaction mixture contained 100 mM Hepes, 300 mM KCl, 1 mM EDTA, and 6 mM dithiothreitol at pH 7.5 and 37°C. Substrate concentration was typically 50 μM, and the reaction was started by adding 0.13 μM enzyme. The reaction was terminated by removing 75 μL aliquots and adding them to 63 μL of 2% (v/v) trifluoroacetic acid.

The enzyme preparation used was more than 90% pure and was obtained by expressing the gene in *E. coli*.

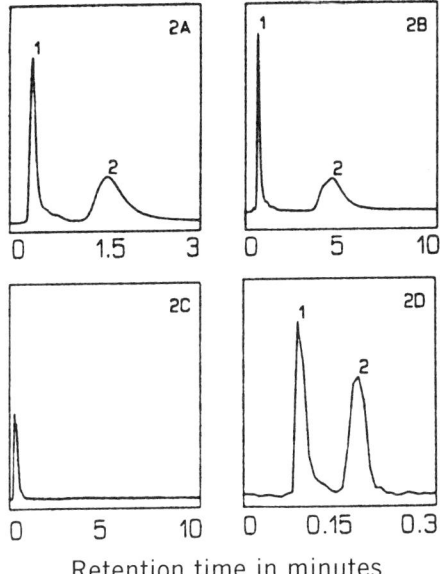

Retention time in minutes

Figure 9.34 High speed and ultrahigh speed chromatograms of separations of substrate and product from assays of rhinovirus 3c protease. (*A*) high-speed HPLC separations of the fluorescein-labeled product, peak 1, and labeled substrate, peak 2, which were eluted from a C_{18} column at 3 mL/min by a 76:24 (v/v) mixture of buffer (50 mM sodium phosphate and 50 mM sodium acetate in pH 7) and acetonitrile. The peptides were injected in a total volume of 75 μL and chromatography was performed at ambient temperature. (*B* and *C*) Same peaks as above except buffer/acetonitrile ratio was 77:23 (v/v). (*D*) Ultrahigh-speed chromatogram using an M-pel C_8 column at 80°C eluted at 4 mL/min after injection of 1 μL of sample. (From Hopkins et al., 1991.)

An assay utilizing many of the same principles described above has been described for HIV-protease (Tamburini et al., 1990; Betageri et al., 1993).

9.3.10 Stromelysin (Harrison et al., 1989)

Stromelysin, which degrades proteoglycan and other proteins of the matrix of cartilage, is a metalloendoproteinase secreted by mammalian fibroblasts. The assay described here uses substance P (Arg-Pro-Lys-Pro-Gln-Gln-Phe-Gly-Leu-Met-NH$_2$) in an automated, semicontinuous assay that is well suited for kinetic analyses. Cleavage occurs specifically between positions 6 and 7 of the peptide.

The products are separated isocratically on a Whatman RAC II Partisil 5 μm C_8 column (10 cm × 4.6 mm) using 0.1% TFA (pH 2) and acetonitrile in a ratio of 67:33 (v/v). The flow rate was 1.0 mL/min, and 20 μL aliquots were injected. The eluate was monitored at 215 nm.

A typical reaction contained 750 µL of a buffered solution of substrate in a 1.5 mL glass autosampler vial. The enzyme was added in 250 µL aliquots to give final concentration of 2 µg/mL. After mixing, the vial was placed in the thermostated compartment of the autosampler. Tight temperature control at 25°C (not available on many autosamplers) is crucial to obtaining good data. Samples were automatically withdrawn by the autosampler and injected onto the column. Quenching of the enzymatic reaction occurs immediately upon mixing with the mobile phase.

Stromelysin was isolated as the zymogen from the culture media of interleukin-1-stimulated human gingival fibroblasts. The zymogen is activated by incubating with trypsin, which in turn is inactivated by adding soybean trypsin inhibitor.

9.3.11 Dipeptidyl Peptidase IV/Amino Peptidase-P (Harada et al., 1988)

Dipeptidyl peptidase IV hydrolyzes Xaa-Pro-Yaa- peptides to Xaa-Pro and Yaa-, while amino peptidase-P cleaves on the amino side of Pro. These two enzymes are found in the microvillous fraction of rabbit kidney. Catalysis is nonspecific with respect to Xaa and Yaa, except the Yaa position cannot be occupied by Pro or Hyp. Harada et al. have published several papers on assaying cleavage of proline-containing peptides found in structural proteins.

Gly-Pro-Ala was used as a substrate for dipeptidyl peptidase IV and was separated from Gly-Pro on a Zorbax ODS (15 cm × 4.6 mm) column by means of a mobile phase containing 10 mM KH_2PO_4(pH 2.5) and 1.0 mM 1-octanesulfonate as an ion-pairing agent. The substrate used for amino peptidase-P was Gly-Pro-Hyp, which was separated from Pro-Hyp by means of a mobile phase containing a 15:85 (v/v) mixture of (10 mM KH_2PO_4) (pH 2.5) and 1-octanesulfonate) and acetonitrile. The eluate was monitored at 210 nm.

The reaction mixture contained in a final volume of 500 µL 0.03 mmol Tris-HCl (pH 8.0), 0.3 µmol of peptide substrate, and the appropriate amount of microvillous fraction, which was prepared from rabbit kidney. In the aminopeptidase-P assay, 1.0 µmol of manganese chloride was also included. Reactions were run for 30 minutes at 37°C and were stopped by adding 400 µL of 10% perchloric acid. After centrifugation, 20 µL aliquots were injected. The amount of product formed was linear with amount of enzyme added out to 250 µg and 1000 µg of microvillus protein for dipeptidyl peptidase IV and amino peptidase-P, respectively.

Figure 9.35 shows a typical chromatogram and Figure 9.36 the dependence of protein concentration and product formation.

9.3.12 Carboxypeptidase N (Grimwood et al., 1988)

Carboxypeptidase N (kininase I) cleaves C-terminal lysyl and argininyl residues from peptides. The assay described here is about 1000-fold more sensitive

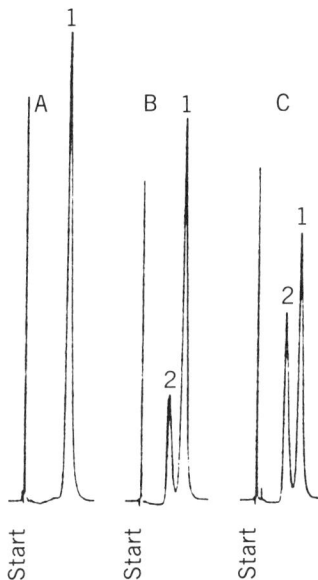

Figure 9.35 Typical chromatographic patterns of the hydrolysis of Gly-Pro-Ala with different concentrations (A, 0 μg; B, 100 μg; C, 200 μg) of rabbit kidney microvillous fractions. Peak 1 (A), Gly-Pro-Ala (3.3 nmol, retention time 6.4 min) decreased; peak 2 (B,C), Gly-Pro (retention time 5.0 min) increased at each protein concentration. Mobile phase, 10.0 mM potassium dihydrogen phosphate containing 1.0 mM, 1-octanesulfonate (pH 2.5). (From Harada et al., 1988.)

than a spectrophotometric assay. Furylacryloyl-Ala-Lys or furylacryloyl-Ala-Arg is used as substrate. As little as 51 pg of purified carboxypeptidase N can be detected.

The enzyme is assayed by quantifying by HPLC the furylacryloyl-Ala produced from either of the two substrates. The substrate and product are separated on a C_{18} RP column (4 μm, Waters) by using 0.1% TFA in 80:20 water–acetonitrile. The eluate is monitored at 280 nm. The formation of product is linear with time and proportional to enzyme concentration when hydrolysis does not exceed 20 to 25%.

The standard assay used by Grimwood et al. (1988) consisted of 10 μL 2.0 mM substrate in 0.05 M Hepes (pH 7.75) and 10 μL of enzyme. The reactions proceeded at 37°C and were terminated with 10 μL 0.4 M H_3PO_4. Samples of 20 μL were injected. The lysine-containing peptide is cleaved three times more rapidly than the argininyl peptide. A guard column containing the same chromatographic medium was replaced every 2 weeks.

Applicability of the assay was shown for human carboxypeptidase N purified from plasma as well as enzyme in unpurified plasma and conditioned medium from a human hepatoma cell line (Hep G2).

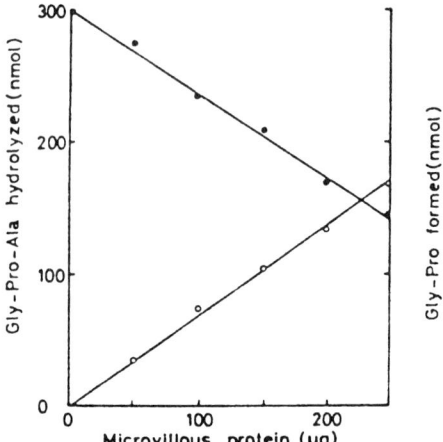

Figure 9.36 Dependence of microvillus protein concentration of Gly-Pro formation from Gly-Pro-Ala catalyzed dipeptidyl peptidase IV. Each value of the sample concentration represents the mean of three different determinations. Hydrolysis of Gly-Pro-Ala (●) and formation of Gly-Pro (○) were stoichiometrically related. (From Harada et al., 1988.)

9.3.13 Renin (Nakamura-Imajo et al., 1992)

The activities of renin in plasma and serum are measured in the diagnosis of hypertension. The natural substrate for renin is angiotensin. Several synthetic peptides have been used in renin assays. Recently, Nakamura-Imajo et al. (1992) used N-(2-pyridyl)glycine (pg) as a fluorescent tag on a nonapeptide.

Renin cleaves the substrate, pg-His-Pro-Phe-His-Leu-Val-Ile-His-β-Ala-OH between Leu and Val. The pyridyl-containing product and substrate are separated on a Wakosil 5 C_{18} HG column (4.6 mm × 250 mm). With purified renin where extraneous proteases are absent, a short (4.6 millimeters × 10 mm) column could be used. The columns were equilibrated with 50 mM ammonium acetate buffer (pH 6.0) containing 10% acetonitrile. The peptides were eluted by using a 10 to 30% acetonitrile gradient. Chromatography was carried out at 40°C. Pyridyl-containing peptides were detected fluorimetrically by using excitation and emission wavelengths of 300 and 360 nm, respectively.

The assay was conducted by adding 10 μL of serum or plasma to 100 μL containing 500 μM substrate in 100 mM Mops-NaOH (pH 7.0), 1 mM phenylmethylsulfonyl fluoride, and 3 mM o-phenanthroline. The reaction mixture was incubated at 37°C for an hour, and then 60 μL was directly injected into the column.

The assay is suitable for assaying enzyme in serum and plasma, and also recombinant human prorenin expressed in Chinese hamster ovary cells. Prorenin was activated by trypsin digestion.

9.4 AMINO ACID AND PEPTIDE METABOLISM

9.4.1 Ornithine Aminotransferase (O'Donnell et al., 1978)

Ornithine aminotransferase (OAT) is a mitochondrial matrix enzyme that catalyzes the primary reaction

L-ornithine + α-ketoglutarate $\rightarrow$ glutamic $-$ γ-semialdehyde + glutamate

However, following its formation, the semialdehyde undergoes a spontaneous cyclization and is converted to a Δ'-pyrroline-5-carboxylic acid (P5C), a proline precursor.

The assay developed for this activity involves the reaction of the P5C with o-aminobenzaldehyde (OAB) to form the reaction product dihydroquinozolinium (DHQ). The DHQ and unreacted OAB were separated by reversed-phase HPLC (LiChrosorb C18). The column was eluted isocratically with a mobile phase of 1 part methanol to 2 parts water, and the separation shown in Figure 9.37 obtained. The reaction mixture contained L-ornithine (35 mM), α-ketoglutarate, potassium phosphate (pH 7.4), and pryidoxyl phosphate in a total volume of 2 mL. The reaction was started by the addition of the homogenate and terminated by the addition of 1 mL of 3 N HCl containing the OAB. Precipitated protein was removed by centrifugation (3000 rpm), and samples of the supernatant solution (10 μL) were injected for analysis.

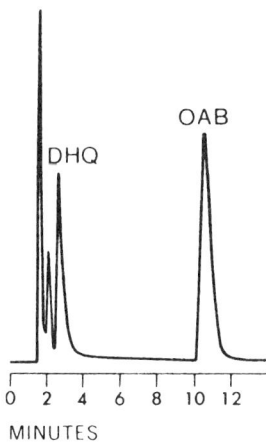

Figure 9.37 Isocratic reversed-phase HPLC chromatogram showing separation and detection of dihydroquinozolinium (DHQ) and o-aminobenzaldehyde (OAB). The column was LiChrosorb C_{18}, and 4.6 mm $\times$ 250 mm, and the solvent system was methanol–water (1:2), pumped at 1.5 mL/mm. Detection was at 254 nm with an Altex 310 chromatograph. (From O'Donnell et al., 1978.)

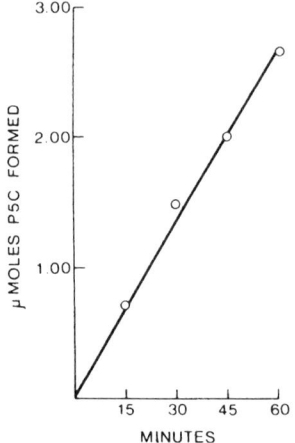

Figure 9.38 OTA activity as determined by HPLC plotted against time. The reaction was linear for 60 minutes. (From O'Donnell et al., 1978.)

Quantitation was by means of peak height rather than peak area. The OTA activity was determined over a 60-minute incubation, and, as shown in Figure 9.38, it was linear during this interval as well as proportional to the amount of homogenate added (Fig. 9.39).

The homogenate was prepared from the liver of an adult female rat in 20% sucrose containing potassium phosphate (pH 7.4). The homogenate was centrifuged at 1000 rpm for 15 minutes and stored at $-20°C$.

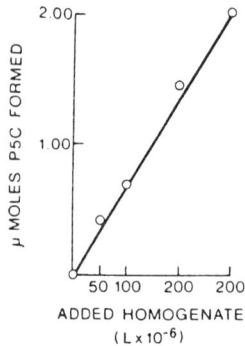

Figure 9.39 OTA activity determined by HPLC plotted against added homogenate. The reaction was linear with homogenate. (From O'Donnell et al., 1978.)

9.4.2 Glutamine Synthetase, Glutamate Synthetase, and Glutamate Dehydrogenase (Martin et al., 1982; Marques et al., 1989)

Depending on the organism, assimilation of ammonia takes place via the activities of glutamine synthetase, glutamate synthetase, and/or glutamate dehydrogenase. The assays described, which measure glutamate and/or glutamine formation and disappearance, were developed for use in situ and in vitro.

In the HPLC assay method described by Martin et al. (1982), glutamic acid and glutamine are separated by reversed-phase HPLC after derivatization with o-phthaldialdehyde/2-mercaptoethanol (OPA). The separation was carried out on a C_{18} column (μBondapak) using a mobile phase of 20 mM sodium phosphate buffer (pH 6.8) and methanol (64:36 v/v). The column was eluted isocratically and monitored at 340 nm. The separation, carried out at room temperature, is shown in Figure 9.40. The reaction mixtures were prepared according to published procedures. The reactions were terminated using either chloride or TCA. Precipitates were removed by centrifugation, and the supernatant solution was used for the assay. Figure 9.41, which shows chromatographic profiles obtained after incubation for 15 minutes, also indicates the formation of glutamine (gln) and glutamate (glu).

The enzymes were prepared by published procedures either from root nodules or from rice leaves.

In the assay described by Marques et al. (1989), the o-phthaldialdehyde derivatives of glutamine and glutamate were separated at 45°C on a μBondapak C_{18} or a Novapak C_{18} column (3.9 mm $\times$ 150 mm). The mobile phase was 20 mM sodium phosphate buffer (pH 6.5) containing 22% (v/v) methanol and 2% (v/v) tetrahydrofuran. Fluorescence detection was used, with excitation and emission at 338 and 425 nm, respectively.

Glutamine synthetase activity was determined by following glutamine formation. The reaction contained 50 mM Hepes–NaOH (50 mM). The reaction was started by addition of ATP and stopped after 15 minutes of incubation at 30°C by addition of 0.6 mL of 1 N HCl.

Glutamate synthetase activity was determined by following glutamate formation. The reaction mixture contained in a final volume of 0.9 mL: 45 μmol potassium phosphate buffer (pH 7.0), 5 μmol L-glutamine, 1 μmol 2-oxoglutarate, 5 μmol aminooxyacetate, 10 nmol *Synechococcus* ferredoxin, and enzyme. The reaction was started by adding 0.8 mg of sodium dithionite freshly dissolved in 0.1 mL of 0.12 M NaHCO$_3$, and stopped after 15 minutes at 30°C by the addition of 0.6 mL of 1 N HCl.

Glutamate dehydrogenase aminating activity was determined as glutamate formation in a reaction mixture that contained in 1 mL: 85 μmol Tris-HCl buffer (pH 8.0), 10 μmol 2-oxoglutarate, 50 μmol NH$_4$Cl, 0.2 μmol NADPH, and enzyme. The reaction was started by adding NH$_4$Cl, and stopped after 15 minutes of incubation at 30°C by the addition of 0.6 mL of 1 N HCl.

With each assay described, samples were processed by taking 0.4 mL from the reaction mixture, centrifuging, and diluting an aliquot of the supernate

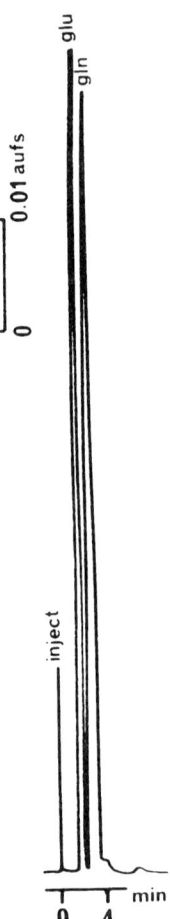

Figure 9.40 Chromatograms of *o*-phthaldialdehyde, glutamic acid, and glutamine standards. Sample contained 5 nmol of glutamic acid and 10 nmol of glutamine. (From Martin et al., 1982.)

25-fold with 50 mM potassium phosphate buffer (pH 7.5). For each assay, 50 µL of diluted sample was mixed with 150 µL of a derivatizing solution prepared as follows: 27 mg *o*-phthaldialdehyde were dissolved in a mixture containing 0.5 mL of methanol, 4.5 mL of 0.4 M sodium borate buffer (pH 10.0), and 20 µL of 2-mercaptoethanol. After a 90-second incubation at 22°C, 40 µL of sample was injected into the HPLC system.

The sources of enzyme assayed was *Synechococcus* PCC 6301 and PCC 6803. For in situ assays, cells were permeabilized by mixed alkyltrimethylammonium bromide or toluene.

Figures 9.42 and 9.43 show chromatograms of enzyme assays.

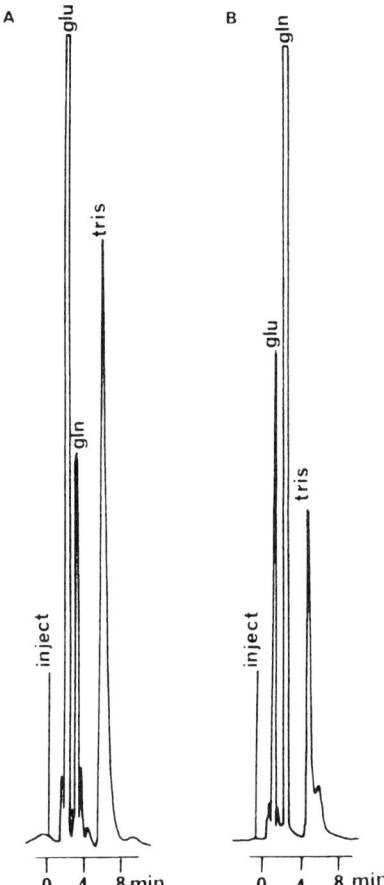

Figure 9.41 Chromatograms of enzyme assay media. (*A*) Elution profile of the assay medium of glutamine synthetase. Alder root nodule enzyme plus assay mixture was incubated for 15 minutes. (*B*) Elution profile of the assay medium of glutamate synthetase. Rice leaves enzyme plus assay mixture was incubated for 15 minutes. (From Martin et al., 1982.)

9.4.3 Asparagine Synthetase (Unnithan et al., 1984)

Asparagine synthetase catalyzes the reaction

$$\text{L-Asp} + \text{L-Gln} + \text{ATP} \rightarrow \text{L-Asn} + \text{L-Glu} + \text{AMP} + \text{PP}_i$$

In this assay the asparagine, aspartate, glutamine, and glutamate are separated by reversed-phase HPLC (C_{18}) using a mobile phase of 70% sodium acetate buffer (pH 5.9), and 30% methanol as shown in Figure 9.44.

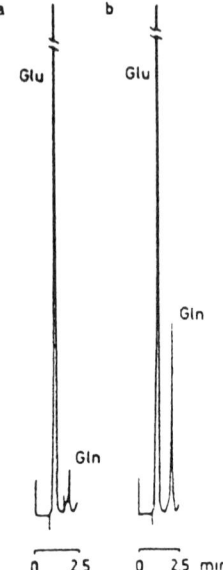

Figure 9.42 Chromatograms corresponding to a glutamine synthetase in situ assay. (*a*) Sample taken at zero time. (*b*) Sample taken after completion of the assay. (From Marques et al., 1989.)

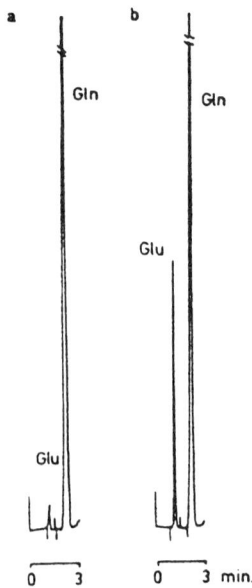

Figure 9.43 Chromatograms corresponding to a glutamate synthetase in situ assay. (*a*) Sample taken at zero time. (*b*) Sample taken after completion of the assay. (From Marques et al., 1989.)

9.4 AMINO ACID AND PEPTIDE METABOLISM   253

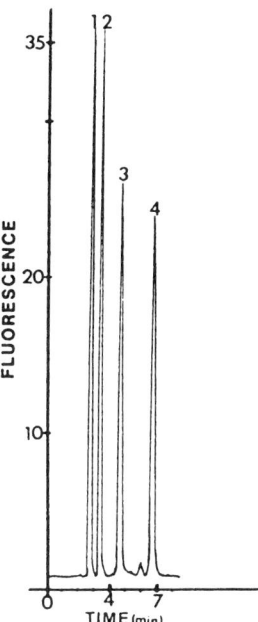

Figure 9.44 Chromatogram of assay of asparagine synthetase. *Peaks:* 1, *o*-phthaldialdehyde aspartate; 2, glutamate; 3, asparagine; 4, glutamine. A mixture of 0.025 mM of each amino acid was made and 20 μL injected. (From Unnithan et al., 1984.)

The reaction mixture contained Tris-HCl, MgCl, glutamine, ATP, and aspartate. The reaction was started by adding the reaction mixture cocktail to the enzyme preparation. The solution was incubated at 37°C, and at intervals the reaction was terminated by transferring samples to boiling sodium acetate buffer. The solutions were cooled and centrifuged, and the amino acids derivatized with *o*-phthaldialdehyde (OPA). Figure 9.45 illustrates the formation of reaction products by comparing the profiles obtained at zero time and after 30 minutes of incubation. The formation of the two reaction products, asparagine and glutamate, is clearly seen. Figure 9.46 shows the time course of glutamate formation.

The enzyme was prepared from bovine pancreas by published procedures.

9.4.4 Tryptophanase (Krstulovic and Matzura, 1979)

Tryptophanase catalyzes the conversion of tryptophan to indole and acetic acid. Pyridoxal phosphate is a required cofactor. The HPLC method developed to assay this activity involves the separation of the tryptophan from the indole.

The separation was carried out by reversed-phase HPLC (C_{18} μBondapak) using a mobile phase of anhydrous methanol–water (1:1, v/v). The column was

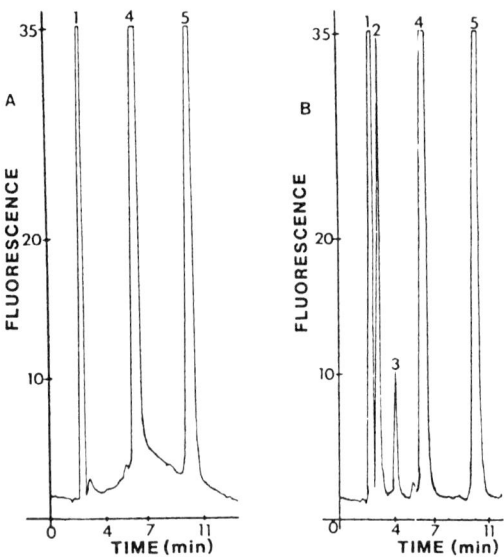

Figure 9.45 Chromatography of enzyme assay media. *Peaks:* 1, aspartate; 2, glutamate; 3, asparagine; 4, glutamine; 5, Tris-HCl buffer. Elution profile of the assay medium incubated for (A) zero time and (B) 30 minutes. (From Unnithan et al., 1984.)

eluted isocratically and detected fluorometrically with excitation and emission wavelengths of 285 and 320 nm, respectively. The separation obtained is shown in Figure 9.47.

The reaction mixture contained potassium dihydrogen phosphate buffered to pH 7.0 and bacterial cells. The cells were sonicated and preincubated with pyridoxal phosphate, and the reaction was started by the addition of

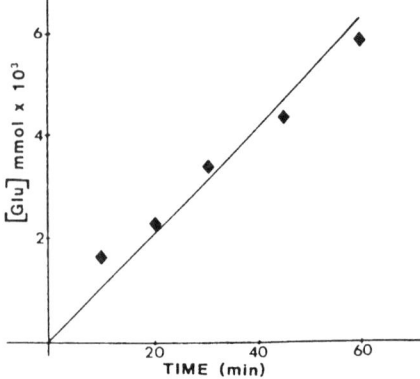

Figure 9.46 L-Glutamate formation with respect to incubation time at 37°C due to the glutaminase activity. (From Unnithan et al., 1984.)

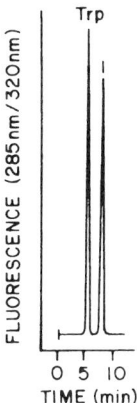

Figure 9.47 Separation of a synthetic mixture of tryptophan (Trp) and indole (I). Chromatographic conditions: column, C_{18} μBondapak; eluent, anhydrous methanol–water (1:1 v/v); flow rate, 1.0 mL/min; temperature, ambient; detection, fluorescence, 285 nm excitation, 320 nm emission cutoff filter. (From Krstulovic and Matzura, 1979.)

tryptophan. At intervals, reactions were terminated with TCA and the solution clarified by centrifugation, filtered, and then analyzed. Several chromatograms showing the time-dependent increase in indole formation are illustrated in Figure 9.48.

The tryptophanase was obtained from *E. coli*.

9.4.5 Dihydroxyacid Dehydratase (Smyk-Randall and Brown, 1987)

Dihydroxyacid dehydratase is involved in the biosynthesis of valine and isoleucine. L-α,β-dihydroxyisovaleric acid and L-α,β-dihydroxy-β-methylvaleric acid are dehydrated to form the α-keto acid precursors of valine and isoleucine.

The 2,4-dinitrophenylhydrazone derivatives of α-ketoisovaleric acid and α-keto-β-methylvaleric acid are separated from 2,4-dinitrophenylhydrazine by chromatography at room temperature on a Zorbax C_{18} column (4.6 mm × 250 mm). Solvent A was 25% acetonitrile in water containing 0.1% triethylamine (v/v) and adjusted to pH 4.5 with acetic acid. Solvent B was acetonitrile. A linear gradient from 20 to 50% B was made within 20 minutes. The effluent was monitored at 254 nm.

The reaction mixture contained in a total volume of 1 mL: 50 mM Tris-HCl buffer (pH 8.0), 100 mM MgCl$_2$, 3.8 mM L-α,β-dihydroxyisovaleric acid (sodium salt), and crude extract containing 0.5 mg of protein. The reaction mixture was incubated for various times at 37°C, and the reaction was terminated by transferring 100 μL of the assay mixture to a 200 μL centrifuge tube containing 20 μL of 50% trichloroacetic acid. Precipitated proteins were removed by centrifugation. Then 50 μL of the supernate was mixed with

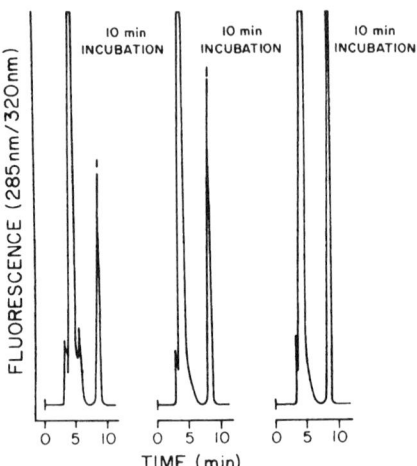

Figure 9.48 Chromatograms of the *E. coli* extracts incubated with tryptophan for 10, 20, and 40 minutes, respectively. Samples were deproteinated with TCA, and the extracts neutralized with solid Tris. Chromatographic conditions same as in Figure 9.47. Volume injected, 5 μL; attenuation, 0.5 μA. (From Krstulovic and Matzura, 1979.)

30 μL of 0.2% 2,4-dinitrophenylhydrazine in 2 M HCl (w/v) in a 2.5 mL polypropylene centrifuge tube. After 30 minutes at room temperature, samples were diluted with 60% acetonitrile in water to a final volume of 1 mL. After filtration, 20 μL was injected onto an HPLC column.

Enzyme activity was measured in cell-free extracts prepared from *E. coli* strain K-12.

9.4.6 Glutaminyl Cyclase (Consalvo et al., 1988)

Several proteins and bioactive peptides contain an N-terminal pyroglutamic acid residue produced by cyclization of an N-terminal glutaminyl residue. This assay for glutaminyl cyclase follows the reaction for a model peptide.

The substrate, Gln-Leu-Tyr-Glu-Asn-Lys-ε-(dansyl)-OH, is separated from the product containing the N-terminal pyroglutamyl residue by chromatography on a Hypersil ODS column (4.6 mm × 100 mm). The mobile phase contained 100 mM sodium acetate (pH 6.5) in 24% acetonitrile. A fluorescence detector was used with excitation at 352 to 360 nm and emission at 482 nm. Fluorescence detection is about 100 times more sensitive than monitoring absorbance at 340 nm.

The enzymatic reaction was carried out in 50 mM Tris-HCl (pH 8.0) and was initiated by adding either the protein extract (up to 0.5 mg/mL) or peptide substrate (1–98 μM). Aliquots of 10 to 20 μL were removed at various times, and the reaction was quenched by the addition of 10 μL of 10 mM phenanthroline. Aliquots of 10 μL were then analyzed by HPLC.

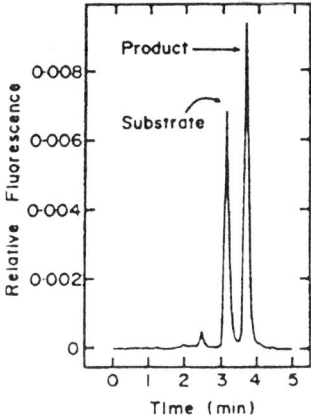

Figure 9.49 HPLC separation of a reaction mixture containing the substrate Gln-Leu-Tyr-Glu-Asn-Lys-ε-(Dns)-OH and the enzymatically generated product <Glu-Leu-Tyr-Glu-Asn-Lys-ε-(Dns)-OH. Partial conversion of the substrate to product was achieved by incubation of Gln-Leu-Tyr-Glu-Asn-Lys-ε-(Dns)-OH (43 μM) with crude bovine pituitary homogenate (0.5 mg/mL) in 50 mM Tris-HCl, pH 8, for 10 minutes. Peak heights correspond to the injection of 80 pmol of substrate and 130 pmol of product. (From Consalvo et al., 1988.)

The source of glutaminyl cyclase was bovine pituitaries. Extracts were partially purified by ammonium sulfate fractionation and chromatography on Sephacryl S-300.

Figure 9.49 shows representative chromatograms.

9.4.7 Leucine 2,3-Aminomutase (Aberhart, 1988)

Leucine 2,3-aminomutase interconverts the α and β forms of leucine. Similar mutases are known for other amino acids including lysine, tyrosine, and serine. The resolving power of HPLC is useful in separating and quantitating the α and β derivatives of these amino acids.

The o-phthaldehyde derivatives of the amino acids were separated on a Waters Resolve C_{18} column (3.9 mm × 150 mm). For separating α- and β-leucine, a mobile phase containing 20% acetonitrile and 80% 0.1 M potassium phosphate (pH 7.5) was used. The effluent was monitored by a fluorometer with the excitation and emission wavelengths set at 365 and 418 nm, respectively.

Leucine 2,3-aminomutase was assayed in incubation mixtures containing 100 μL of 1 M triethanolamine–HCl buffer (pH 8.5), 30 μL of water, 50 μL of 10 μM flavin adenine dinucleotide, 50 μL of 10 mM coenzyme A, 50 μL of 10 mM nicotinamide adenine dinucleotide, 50 μL of 10 mM pyridoxal-5′-phosphate, 50 μL of 360 mM glutamate (pH 8.5, with KOH), 200 μL of

100 mM DL-β-leucine, and 100 μL of enzyme (about 10 mg protein). Adenosine cobalamin (50 μL of 100 μM) was added last, and mixtures were incubated in the dark at 37°C for 1 hour. Then 50 μL of 10% Na$_2$WO$_4$ and 50 μL of 6 N HCl were added. After centrifugation, 10 μL samples containing 0.01 to 0.50 μmol of amino acid were mixed with 200 μL of o-phthaldehyde reagent and allowed to react for 1 minute at room temperature. An aliquot of 20 μL was injected into the HPLC system. The o-phthaldehyde reagent was prepared fresh daily by mixing 100 μL of o-phthaldehyde solution (350 mg/5 mL absolute ethanol) with 10 mL of 0.1 M borate buffer (pH 10.4) and then adding 20 μL of 2-mercaptoethanol.

Rat liver was homogenized in 0.2 M potassium phosphate buffer (pH 7.0) containing 0.8% Triton X-100. The supernate obtained by centrifuging the homogenate at 40 000g for 30 minutes at 0°C was used to assay leucine 2,3-aminomutase.

Figure 9.50 shows typical chromatograms.

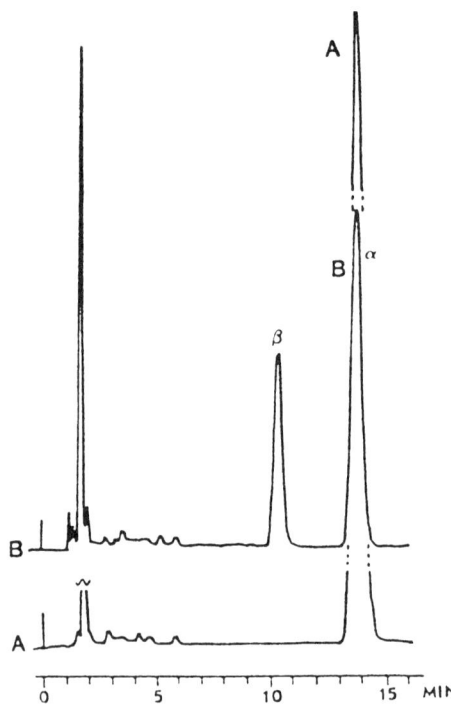

Figure 9.50 Typical lysine 2,3-aminomutase run. (A) Initial incubation mixture (t = 0), (B) Analysis after incubation for 135 minutes. (For quantitation, peak areas are determined; the β peak area is multiplied by 1.85, and total, (1.85 $\times$ β) + α), normalized to 100%. (From Aberhart, 1988.)

9.4.8 Diaminopimelate Epimerase and Decarboxylase (Weir et al., 1989)

Diaminopimelate decarboxylase catalyzes the final step in lysine biosynthesis in bacteria. The epimerase catalyzes the interconversion of the LL- and *meso* isomers of diaminopimelate. Because these enzymes are absent in mammals, they are considered to be potential targets for antimicrobial agents.

The *o*-phthaldialdehyde derivatives of LL-diaminopimelate, *meso*-diaminopimelate, norvaline (internal standard), and lysine were separated on a Spherisorb C_{18} column (4.5 mm × 250 mm). A linear gradient from 100% solvent A (30 methanol, 70% 50 mM sodium acetate buffer (pH 5.9) to 30% solvent A and 70% methanol was imposed over 35 minutes. Detection was by fluorescence, with excitation and emission wavelength of 340 and 455 nm, respectively.

The normal incubation mixture contained in 0.1 M potassium phosphate (pH 7.0) in a final volume of 1.5 mL: 15 μmol recrystallized *meso*-diaminopimelate, 0.1 μmol pyridoxal 5-phosphate, 3.75 μmol norvaline, and desalted cell-free extract (about 0.5 mg protein). The reaction was carried out at 37°C and was started by adding the extract. Aliquots of 500 μL were taken immediately before and after the 15-minute incubation period and transferred to 10 mL Pyrex tubes held in a boiling water bath, each containing 1.0 mL sodium acetate buffer (1.2 M, pH 5.2). After heating for 5 minutes, the tubes were centrifuged and an aliquot of the supernate was diluted 20-fold with water. The derivatization reaction was carried out by mixing 200 μL of the diluted supernate with 200 μL of a sodium dodecyl sulfate solution (2%, w/v, in 400 mM sodium borate buffer, pH 9.5), and adding 400 μL of the *o*-phthaldialdehyde/2-mercaptoethanol derivatizing solution. After exactly 1 minute of derivatization, an aliquot was injected onto the HPLC column.

The source of enzyme was cell-free extracts prepared from a lysine-overproducing strain of *Bacillus subtilis* NCIB 3610.

Figure 9.51 shows representative chromatograms.

9.4.9 Lysine–Ketoglutarate Reductase (Davis, 1989)

Ketoglutarate reductase catalyzes the formation of saccharopine from lysine, which is the first step in a major pathway for lysine catabolism. NADPH and α-ketoglutarate are cofactors in the reaction.

The *o*-phthaldialdehyde derivatives of lysine and saccharopine were separated on a Ultrasphere-XL ODS column (4.6 mm × 70 mm, 3 μm). The column was equilibrated using methanol–0.1 M sodium acetate (pH 6.7) in a 16:84 ratio. The flow rate was 1.6 mL/min. After injection, methanol was increased linearly to 20% in 0.5 minute. A further increase to 20% methanol occurred at 11.5–12 minutes. A return to 16% methanol occurred between 19 and 19.5 minutes, and the column was equilibrated for 8.5 minutes prior to

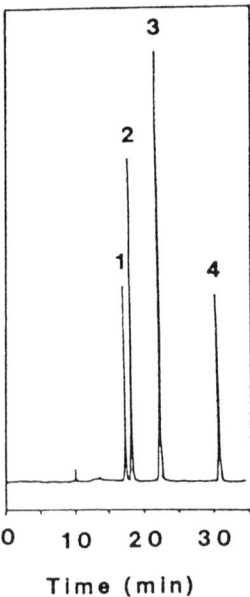

Figure 9.51 HPLC analysis of an enzyme reaction mixture after 30 minutes of incubation. *Peaks:* 1, o-phthaldialdehyde derivatives of LL-DAP; 2, *meso*-DAP; 3, norvaline; 4, lysine. (From Weir et al. 1989.)

the next injection. An electrochemical detector was used with an applied potential of +0.50 V; the working electrode was glassy carbon.

The assay mixture contained 2 mM NADPH, 3 mM L-lysine, 5 mM α-ketoglutarate, and 0.5 mL of homogenate in a volume of 0.9 mL. Upon addition of lysine, the tubes were flushed with nitrogen, sealed, and incubated for 30 minutes at 37°C. The reaction was terminated by the addition of 0.25 mL of 12% perchloric acid. Following addition of 0.1 mL of α-amino-β-guanidinopropionic acid as an internal standard, the tube was held on ice for 30 minutes before centrifuging. The supernate was transferred to a tube containing 0.25 mL of cold 2 M potassium bicarbonate. After 30 minutes of incubation on ice and centrifugation, the supernates were filtered (0.45 μm) and a 50 μL aliquot was transferred to an autosampler vial. Then 50 μL of o-phthaldialdehyde mixture [50 mg of o-phthaldialdehyde, 1.25 mL of methanol, 11.25 mL of 0.4 M sodium borate (pH 9.5), and 50 μL of mercaptoethanol] was added. The resulting solutions were allowed to react for 1 minute before 50 μL was injected into the HPLC system.

The homogenate used as enzyme source was prepared from the livers of pregnant rats that had been homogenized in 0.1 M potassium phosphate buffer (pH 7.4).

9.4.10 γ-L-Glutamylcyclotransferase (Fasolato and Galzigna, 1988)

γ-L-Glutamylcyclotransferase catalyzes the conversion of some γ-glutamylamino acids into 5-oxo-L-proline and free amino acid. The γ-glutamyl cycle may operate in the transport of amino acids.

The substrate, γ-glutamylglutamine, and 5-oxo-L-proline are separated on a C_{18} SIL-X-10 column (4.6 mm × 250 mm, 10 μm). The mobile phase was 0.025 M ammonium phosphate (pH 3.5) and the flow rate was 1 mL/min. The effluent was monitored at 212 nm.

The reaction mixture contained 0.1 mL of 0.002 M L-glutamylglutamine in 0.08 M Tris-HCl buffer (pH 8.0). The reaction was started by adding 0.1 mL of the preparation of synaptosomes. Incubation was continued at 37°C for 30 to 120 minutes, and the reaction was stopped by denaturation at 96°C for 2 minutes. The tubes were centrifuged and 10 μL of the supernate was used for analysis by HPLC.

The source of enzyme was synaptosomes isolated from rat brain. The synaptosomes were suspended in 0.005 M Tris-HCl (pH 8.6) to give a concentration of about 2 mg/mL.

9.4.11 γ-Glutamylcysteine Synthetase and Glutathione Synthetase (Nardi et al., 1990)

The γ-glutamylcysteine synthetase (GC-s) reaction involves the reaction of L-glutamate and L-cysteine and is driven by the hydrolysis of ATP. The glutathione synthetase reaction uses ATP hydrolysis to drive the formation of glutathione for γ-glutamylcysteine and glycine.

The monobromobimane derivatives of cysteine, γ-glutamylcysteine, and glutathione were separated on a Merck Supersphere RP-18 column (4.0 mm × 250 mm). The mobile phase was methanol–water (18:82) containing 0.25% acetic acid adjusted to pH 3.8 with NaOH. The flow rate was 1.0 mL/min. Detection was by fluorescence, excitation and emission wavelengths 370 and 485 nm, respectively.

The incubation mixture for assay of γ-glutamylcysteine synthetase contained in a final volume of 0.3 mL: 0.1 M Tris-HCl (pH 8.2), 6 mM ATP, 50 mM KCl, 6 mM dithiothreitol, 20 mM $MgCl_2$, 3 mM L-cysteine, and 15 mM L-glutamic acid. The mixture was incubated at 37°C for 15 minutes to ensure the complete reduction of cysteine. The reaction was initiated by addition of hemolysate. Samples of 20 μL were withdrawn at various times between 10 and 30 minutes and added to 50 μL of 50 mM N-ethylmorpholine (pH 8.4) and 10 μL of monobromobimane in acetonitrile. After incubation for 15 minutes in the dark at room temperature, the reaction was stopped by addition of 80 μL of 10% sulfosalicylic acid. The volume was made up to 500 μL with water before centrifugation and analysis of the resulting supernate by HPLC.

The incubation mixture for glutathione synthetase was the same as above, except the substrates with 3 mM γ-glutamylcysteine and 30 mM glycine. The reaction mixtures were processed as described above.

The source of enzyme activities was human blood prevented from clotting by EDTA. Washed erythrocytes were hemolyzed in cold distilled water.

9.4.12 Glutamic Acid Decarboxylase (Chakraborty et al., 1991)

Glutamic acid decarboxylase catalyzes the decarboxylation of L-glutamic acid to form the neurotransmitter γ-aminobutyric acid.

The o-phthaldialdehyde derivatives of glutamate, γ-aminobutyric acid (GABA), and δ-aminovaleric acid (DAVA: internal standard) were separated on an Altex Ultrasphere ODS column (4.6 mm × 250 mm, 5 μm). The mobile phase contained 0.2 M sodium acetate (pH 3.8), 100 mg/mL EDTA, and 40% acetonitrile (v/v). Detection was by fluorescence, with excitation and detection wavelengths of 330 and 440 nm, respectively.

Glutamic acid decarboxylase was assayed in a final volume of 100 μL containing 40 μL of 200 mM potassium phosphate buffer (pH 6.8), 10 μL of 5 mM L-glutamic acid, 5 μL of 0.2 mM pyridoxal 5'-phosphate, 40 μL of homogenate (2 mg protein/mL), and 1 μg of gabaquline (inhibitor of γ-aminobutyric acid degradation). The reaction was stopped by the addition of 10 μL of 100% trichloroacetic acid. After centrifugation, 5 μL of standard δ-aminovaleric acid solution and 90 μL of o-phthaldehyde solution (2 mg/mL 0.4 M borate buffer, pH 10.4) was added and the mixture was allowed to react for 3 to 5 minutes before injection of 20 μL into the HPLC system. The reaction was linear for 20 minutes.

The source of enzyme was a homogenate of rat brain prepared by sonication in 0.01 M 3-N-morpholinopropanesulfonic acid (pH 7.4). The homogenate was diluted 1:1 with buffer containing 4 mM pyridoxal phosphate, 0.6 M tetraethylamine, and 2 mM 2-aminoethyl isothiomonium. The final suspension was treated with 0.4% Triton X-100 (final concentration), and the supernate obtained by centrifugation was used for assays.

9.4.13 Histamine N-Methyltransferase (Fukuda et al., 1991)

Histamine N-methyltransferase catalyzes the transfer of a methyl group from S-adenosyl-L-methionine to histamine to form N-methylhistamine. This assay is suitable for following activity during enzyme purification.

Histamine and N-methylhistamine were separated on a weak cation exchange (TSK CM2SW) column (4 mm × 150 mm). The mobile phase was prepared by dissolving 5.25 g of citric acid and 10 g of imidazole in 880 mL of water, and then mixing the resulting solution with 200 mL of acetonitrile. The separated compounds were derivatized at room temperature with o-phthaldialdehyde by mixing that reagent with the column eluate at a flow rate of 1.0 mL/min. The o-phthaldialdehyde reagent was prepared by dissolving 2.47 g of boric acid and 0.12 g of Brij 35 in 100 mL water, adjusting the pH to 10.5 with 5 M KOH, and then adding 60 mg of o-phthaldialdehyde in 5 mL of methanol together with 100 μL of 2-mercaptoethanol. The reaction coil was made of Teflon tubing (1 m × 0.5 mm), and the resulting products were detected by fluorescence. The ranges of excitation and emission wavelengths were 230 to 400 and 410 to 800 nm, respectively.

The reaction mixture in a final volume of 1 mL contained 0.7 mL of 0.1 M phosphate buffer (pH 7.4) containing 0.1 mM pargyline and 0.1 mM aminoguanidine, 0.1 mL of 1.0 mM histamine, 0.1 mL of 2.5 mM S-adenosyl-L-methionine, and 0.1 mL of enzyme preparation. After incubation at 37°C, the reaction was terminated by adding 50 μL of 60% perchloric acid. After centrifugation, the supernate was adjusted to pH 6.5 with 5 M KOH after adding 20 μL of 5% bromothymol blue as a pH indicator. The precipitated potassium perchlorate was removed by centrifugation and the supernate was applied to a column of Amberlite CG-50 (4 mm × 20 mm), which had been equilibrated with 0.2 M sodium phosphate buffer (pH 6.5). The column was washed with 3 mL of 5 mM Na$_2$EDTA (pH 6.5) and with 3 mL of 0.1 M hydrochloric acid. Histamine and N-methylhistamine were eluted with 1.0 mL of 0.5 M HCl, and the eluate was applied to the HPLC column.

The source of enzyme was a dialyzed, high speed supernate obtained after fresh rat kidney had been homogenized in 4 volumes of ice-cold 0.05 M sodium phosphate buffer (pH 7.4) containing 1 mM dithiothreitol and 1% polyethylene glycol (average molecular mass 300).

9.4.14 Amino Acid Decarboxylase (Kochhar et al., 1989)

The assay described for amino acid decarboxylase can be used to quantitate the substrates and products associated with the decarboxylation of arginine, aspartate, 2,6-diaminopimelate, histidine, glutamate, lysine, and ornithine.

Kochhar et al. (1989) characterized an assay for glutamate decarboxylase activity. Glutamate and 4-aminobutyrate were separated on a Nucleosil C$_8$ column. The mobile phase was 13 mM trifluoroacetate and 1 mM 1-octanesulfonate. Detection was by postcolumn derivatization with o-phthaldialdehyde reagent (1 mL/min) mixed with the column eluate (also 1 mL/min). The Teflon reaction coil (3 m × 0.3 mm) was kept at room temperature. The o-phthaldialdehyde reagent was prepared by dissolving 800 mg of o-phthaldialdehyde in 20 mL of ethanol plus 2.5 mL of 2-mercaptoethanol and mixing with 980 mL of 0.4 M sodium borate (pH 9.7) and 3 mL Brij 35. The fluorometer was set to give excitation at 350 nm and emission was measured at 450 nm.

The reaction mixture contained 5 mM L-glutamate and 0.5 to 10 mU of enzyme in 100 mM sodium phosphate (pH 7.2) containing 0.1 mM pyridoxal 5'-phosphate and 1 mM S-2-aminoethylisothiouronium bromide. The reaction was started by adding 5 to 10 μL of enzyme solution to give a final volume of 50 μL. Aliquots of 10 μL were removed at 0, 10, and 20 minutes and mixed with 10 μL of prechilled 0.2 M perchloric acid. After centrifugation, 10 μL of the supernate was mixed with 190 μL of the HPLC mobile phase; HPLC analysis took 40 µL. Assays were linear for 40 minutes and with protein added up to 200 µg/mL.

The assay was used for both a partially purified preparation of glutamate decarboxylase from *E. coli* and rat brain homogenates. Chromatography conditions for the other substrate–product mixtures listed above were also described.

9.4.15 Aromatic L-Amino Acid Decarboxylase (Zuo and Yu, 1991)

Decarboxylation of L-3,4-dihydroxyphenylalanine to dopamine, and of 5-hydroxytryptophan to serotonin, is catalyzed by aromatic L-amino acid decarboxylase. A single enzyme may be responsible for both activities. This assay permits simultaneous determination of both activities.

The substrates and products just noted were separated on an Ultrasphere I.P. C_{18} column (4.6 mm × 250 mm, 5 µm). The mobile phase contained 75 mM sodium phosphate (pH 2.75), 1 mM sodium octylsulfate, 500 µM EDTA, and 13% (v/v) acetonitrile. Quantitation was by electrochemical detection of the products using 0.75 V versus an Ag/AgCl reference electrode.

The standard incubation mixture contained 0.2 M phosphate buffer (pH 7.5), 0.02 mM pyridoxal phosphate, 0.1 mM pargyline, 0.2 mM L-dihydroxyphenylalanine, 0.1 mM 5-hydroxytryptophan, and enzyme in a total volume of 200 µL. After incubation at 37°C for 30 minutes, the reactions were terminated by addition of 800 µL of chilled 0.1 M perchloric acid containing 0.1 mM sodium metabisulfite and 0.2 mM EDTA. After centrifugation, 10 µL aliquots were used for HPLC analysis.

Tissue homogenates in 5 volumes of chilled 0.05 M phosphate buffer (pH 7.4) were prepared from the liver, kidney, adrenal, brain, heart, lung, and small and large intestine of rats.

9.4.16 D-Amino Oxidase (Biondi et al., 1991)

D-Amino oxidase is a peroxisomal enzyme that catalyzes the oxidative deamination of D-amino acids to give the corresponding α-keto acids, ammonia, and hydrogen peroxide. In this assay, α-ketovaleric acid from D-valine was quantitated.

The quinoxalinol derivatives of α-ketovaleric acid and ketovaleric acid (internal standard) were separated on a LiChrospher 100 RP-8 column (4 mm × 250 mm). The mobile phase was a 60:40 ratio of 0.35 M ammonium acetate and acetonitrile. The eluate was monitored by fluorescence with the excitation and emission wavelengths set at 340 and 420 nm, respectively.

The reaction mixture contained in 1.0 mL: 20 µmol of D-Val, 10 nmol of FAD, 10 µg of catalase, and 2 nmol of ketovaleric acid in 0.1 M pyrophosphate buffer (pH 8.5). The enzymatic reaction was started by adding 10 to 100 µL of beef kidney homogenate. At different times, 200 µL aliquots were withdrawn and added to 100 µL of ice-cold 6 M HCl. After centrifugation, a

200 μL aliquot of o-phenylenediamine solution (0.44 M in 2 M HCl containing 5 μL/mL 2-mercaptoethanol) was added. The resulting mixture was flushed with nitrogen and then heated for 15 minutes at 80°C in a sand bath. After cooling, 0.3 mL of 6 M sodium acetate was added and 20 μL aliquots were injected into the HPLC. The assay was linear for up to 20 min and with up to 10 μg of a commercial preparation of D-amino acid oxidase.

Sources of enzyme were partially purified type II D-amino acid oxidase from porcine kidney, and the supernate from a beef kidney homogenate prepared in 2.5 volumes of 0.1 M pyrophosphate buffer (pH 8.5) containing 10 μM FAD. The supernate obtained by centrifugation was used.

9.4.17 Threonine/Serine Dehydratase (Singh et al., 1993)

Threonine dehydratase catalyzes the formation of 2-ketobutyrate from threonine, and pyruvate from serine. This assay can be used to examine the reaction in the presence of both substrates.

The quinoxalone derivatives of pyruvate and 2-ketobutyrate were separated on a μBondapak C_{18} reversed-phase column (3.9 mm × 300 mm). The mobile phase was a mixture of acetonitrile, methanol, and 40 mM phosphate buffer (pH 7) (5:8:12, v/v/v) run at a flow rate of 1 mL/min. Detection could be carried out spectrophotometrically at 254 nm, or fluorometrically with excitation and emission wavelengths of 340 and 389 nm, respectively.

The standard assay contained 100 mM Tris-HCl (pH 9), 100 mM KCl, 20 mM threonine and/or serine, and the enzyme. After incubation at 37°C for various times, the reaction was terminated by addition of 3 N HCl. The reaction mixture was centrifuged and 250 μL of the supernate was mixed with 250 μL of o-phenylenediamine solution (10 mg/mL in 3 N HCl). The tubes were tightly capped and heated in a boiling water bath for 30 minutes. The derivatized keto acids were extracted with ethyl acetate. An aliquot of the ethyl acetate layer was dried under nitrogen. The residue was taken up in 1 mL of the running solvent for HPLC. Assays were linear for 30 minutes and with amount of protein added.

Extracts prepared from the leaves of seedling tomatoes were used as the source of enzyme.

9.4.18 Tryptophan Dioxygenase (Holmes, 1988; Yim et al., 1987)

Tryptophan dioxygenase catalyzes the first step in the oxidative metabolism of L-tryptophan. L-Kynurenine is the product.

L-Kynurenine was separated from 3-hydroxykynurenine (observed in some assays) by chromatography on a Beckman Ultrasphere ODS column (4.6 mm × 150 mm). The mobile phase (1.5 m/min) was a 5:235 mixture of acetonitrile and 0.1 M ammonium acetate buffer (pH 4.65). The column effluent was monitored at 365 nm.

The reaction mixture in a 25 mL Erlenmeyer flask contained 1 mL of 0.2 M sodium phosphate (pH 7.0) and 0.075 M ascorbate, 100 μL of 42 μM hematin HCl (prepared in 0.01 M NaOH), 100 μL of water, and 700 μL of rat liver homogenate. After a 10-minute preincubation at 30°C, the reaction was started by adding 100 μL of 40 mM L-tryptophan. At various times between 0 and 80 minutes, aliquots were removed and added to tubes containing 0.1 volume of 2.4 M perchloric acid. After chilling and centrifuging, then storage at room temperature for 2 hours to promote transformation of any formylkyneurine present to kyneurine, 25 μL samples of each supernate were assayed.

Liver tissue from rat was homogenized in 16 volumes of ice-cold normal saline by sonication. The whole homogenates were used in assay.

Figure 9.52 shows chromatograms of the reaction assay. Another report by Yim et al. (1987) describes this activity from fruit flies.

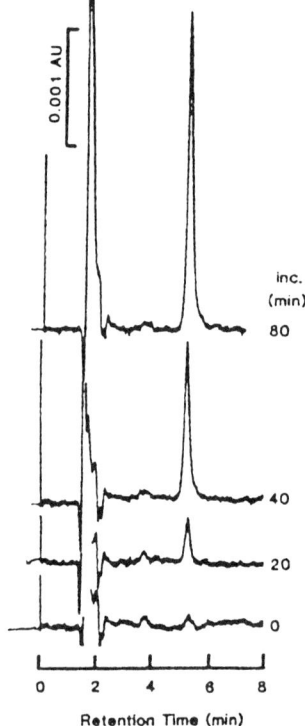

Figure 9.52 HPLC analysis of kynurenine in perchloric acid supernatants of tryptophan dioxygenase incubation mixtures. A sample of rat liver homogenate (containing 39 mg wet wt of tissue) was assayed for tryptophan dioxygenase. Aliquots of the assay mixture were removed at zero time and after 20, 40, and 80 minutes of incubation at 30°C, and quenched by the addition of perchloric acid. Perchloric acid supernatants (25 μL) were analyzed for kynurenine by HPLC. (From Holmes, 1988.)

9.4.19 Tryptophan 2,3-Dioxygenase (Seifert, 1993)

Tryptophan 2,3-dioxygenase catalyzes the conversion of L-tryptophan to N-formyl-L-kynurenine. The product is further converted, both spontaneously and enzymatically, to L-kynurenine. This assay measures both products.

L-Kynurenine and N-formyl-L-kynurenine are separated by chromatography on Perkin-Elmer CR Pecosphere column (4.6 mm × 30 mm, 3 μm). The mobile phase contained 1 mM phosphate buffer (pH 2.4) in 5% acetonitrile. The flow rate was 0.5 mL/min, and 5 μL was injected for analysis. Both compounds were detected at 254 nm and quantified by comparison with standard calibration curves.

The reaction mixture contained 10 to 15 pieces of liver slices (average weight of 4 ± 1 mg/slice) suspended in 1 mL of 15 mM tryptophan. The reaction was allowed to proceed at 37°C for 2 hours before being terminated by immersing the reaction vials into an ice bath and transferring the contents into 1 mL of precooled (−5°C) methanol. The mixture was allowed to stand on ice for 30 minutes, after which it was centrifuged to remove protein. The supernate was filtered (0.2 μm) before analysis by HPLC. Formation of product was linear up to 80 mg of liver slices and 4 hours of incubation.

Livers from mice were manually sectioned into 1 to 1.5 mm cubes. In contrast to liver homogenates or purified enzyme, liver slices did not require exogenous methemoglobin or ascorbic acid for activation of tryptophan dioxygenase.

9.4.20 Kynureninase (Ubbink et al., 1991)

Kynureninase is involved in the oxidative metabolism of tryptophan. It catalyzes the conversion of L-kynurenine to anthranilic acid. The enzyme also converts L-3-hydroxykyneurenine to 3-hydroxyanthranilic acid. The latter compound has a high fluorescence, which is the basis for detection in this assay.

The reaction mixture was separated on a Whatman Partisphere C_{18} column (4.6 mm × 150 mm, 5 μm). The mobile phase at a flow rate of 1 mL/min was 0.1 M potassium phosphate buffer (pH 5.8) containing 1% acetonitrile. The fluorescence detector was set for excitation and emission wavelengths of 322 and 414 nm, respectively.

Homogenates of lymphocytes were preincubated at either 15 or 30°C for 10 minutes before the reaction was started, by addition of 3-hydroxykynurenine to a final concentration of 1 mM. When the assay was performed at 30°C, the presence of 4 μM pyridoxal-5′-phosphate was needed to obtain a linear reaction. The reaction was terminated after 5 minutes by the addition of 150 μL of 10% trichloroacetic acid. HPLC analysis was carried out on 30 μL of the supernate obtained after centrifugation.

Lymphocytes were isolated from human blood collected with EDTA as anticoagulant. Packed lymphocytes (10 μL) were resuspended in 225 μL of ice-cold sodium 5,5-diethylbarbiturate–HCL buffer (pH 8.4). After addition

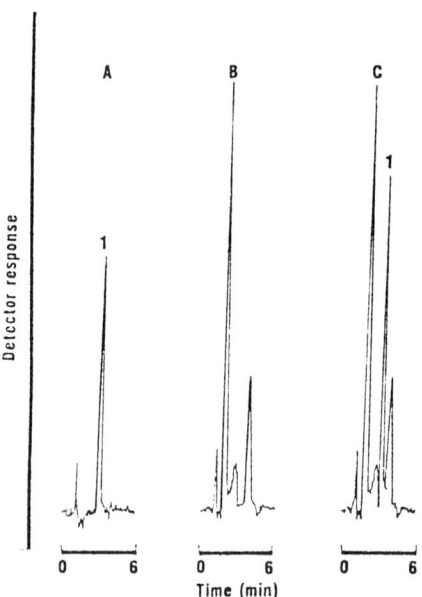

Figure 9-53 Determination of lymphocyte kynureninase activity levels using HPLC. Enzyme activity is measured by quantification of formation of the product, 3-hydroxyanthranilic acid (3-HAA). (*A*) 3-HAA standard (12.0 nmol/L). (*B*), Lymphocyte homogenate blank. (*C*) Lymphocyte 3-HAA production after 5 min of incubation in presence of 3-hydroxy-kynurenine. Peaks: 1, 3-HAA; unmarked peaks are unidentified components. (From Ubbink et al., 1991.)

of 25 μL of 10% Triton X-100, the suspension was sonicated to produce the homogenate.

Figure 9.53 shows chromatograms.

9.4.21 Kynurenine 3-Monoxygenase (Wiseman and Nichols, 1990)

Kynurenine 3-monoxygenase is present in the outer membrane of rat liver mitochondria, where it catalyzes the oxidation of L-kynurenine to 3-hydroxy-L-kynurenine. This assay follows the disappearance of the substrate, L-[G-^{3}H]kynurenine.

Kynurenine was separated from L-3-hydroxykynurenine by chromatography on a Waters μBondapak C_{18} column (8 mm × 100 mm, 10-μm), using a flow rate of 3 mL/min. The mobile phase was 0.02 M sodium acetate (pH 5.5) containing 2% methanol. Radioactivity was quantitated with a flow-through detector, using scintillation fluid at 3 mL/min.

The reaction mixture contained 0.1 M Tris acetate (pH 8.0), 10 mM KCl, 1 mM EDTA, 3 mM glucose-6-phosphate, 1 U/mL glucose-6-phosphate dehydrogenase, 50 to 250 μM L-$[G$-^{3}H]kynurenine (0.04 μCi/mL), 100 μM NADPH, and enzyme (0.25–0.5 mg/mL). After 30 minutes of incubation at 37°C, assays were quenched with the addition of 5 mL of acetic acid per milliliter and heating for 2 minutes at 95°C. After centrifugation, a 0.4 mL aliquot was analyzed by chromatography. The assay was linear with time up to 120 minutes and with protein concentration up to 1 mg/mL.

Rat liver mitochondria were isolated in 0.25 M sucrose by standard procedures.

9.4.22 N^5-Methyltetrahydrofolate-homocysteine Methyltransferase (Garras et al., 1991; Lee et al., 1992; Goeger and Gauther, 1993)

The enzyme also known as methionine synthetase catalyzes the conversion of homocysteine to methionine using a folate derivative as the methyl donor. The assay of Garras et al. (1991) is based on quantitation of the o-phthaldehyde derivative of methionine.

The o-phthaldialdehyde derivatives of methionine and norvaline (internal standard) are separated on an ODS Hypersil column (4.6 mm × 250 mm, 5 μm). The flow rate was 2 mL/min and a linear 30 to 55% methanol gradient in 50 mM phosphate buffer (pH 5.0) was developed in 12 minutes. The methanol concentration was then changed linearly from 55 to 80% between 12 and 15 minutes. The fluorescence detector was set for excitation and emission wavelengths of 336 and 450 nm, respectively.

The incubation mixture contained in a final volume of 100 μL: 400 μM DL-homocysteine, 500 μM ($\pm$)-L-N^5-methyltetrahydrofolate, 50 μM cyanocobalamin, 300 μM S-adenosylmethionine, 125 mM 2-mercaptoethanol, 20 μM L-norvaline, 50 mM potassium phosphate buffer (pH 7.4), and 50 μL of liver or cell extract. The incubation mixture was immediately flushed with nitrogen and overlayered with 50 μL of bis(3,5,5-trimethylcyclohexyl)-phthalate. The incubation, carried out at 37°C in the dark, was stopped by the addition of 10 μL of 4 N perchloric acid. The acid was then neutralized by addition of 10 μL of 4 N KOH containing 3.3 M potassium bicarbonate. After centrifugation, 90 μL of supernate was mixed with 175 μL of o-phthaldialdehyde reagent (prepared by mixing 1 mL of 56 mM o-phthaldialdehyde in methanol with 9 mL of 0.1 M sodium borate buffer, pH 9.5, then adding 40 μL of 2-mercaptoethanol). After 2 minutes at 23°C, 220 μL of this mixture was used for HPLC analysis. The assay is linear for at least 2 hours.

Enzyme activity was assayed in the high speed supernate obtained from rat livers homogenized in 50 mM potassium phosphate buffer (pH 7.4) containing 0.1 M NaCl. It was also shown to be applicable to extracts from human promyelocytic leukemia HL-60 cells, and mouse embryo fibroblasts.

Other assays reported by Lee et al. (1992) and Goeger and Ganther (1993) deal with the same activity but use different methyl donors.

9.4.23 L-Alanine: Glyoxylate Aminotransferase (Petrarulo et al., 1992)

Located in the peroxisomes of liver, L-alanine-glyoxylate aminotransferase catalyzes the transamination of alanine and glyoxylate to form pyruvate and glycine. A rare inborn error of metabolism manifested as hyperoxaluria is due to a deficiency of this enzyme.

The phenylhydrazone derivative of pyruvate is quantitated by chromatography on a LiChrospher RP-8 column (4 mm × 250 mm, 5 μm). Isocratic elution was performed at room temperature and a flow rate of 1.0 mL/min by using a ternary mixture of 14.7 mM KH_2PO_4, 8.76 mM K_2HPO_4, and HPLC-grade methanol (75:10:15, v/v/v). The eluate was monitored at 314 nm.

The reaction mixture contained 10 mM glyoxylate, 80 mM L-alanine, 100 mM potassium phosphate buffer (pH 8.0), and 100 μM pyridoxal-5'-phosphate in a final volume of 281 μL. The reaction was initiated by adding alanine. After 30 minutes the reaction was stopped by mixing a 50 μL aliquot of the incubated sample with 50 μL of phenylhydrazine solution (1 M in water) and 2.0 mL of water. After 15 minutes of incubation at room temperature, 50 μL of the mixture was used for analysis by HPLC.

The assay was used to determine activity in liver specimens obtained by biopsy. Samples were homogenized by sonication in a solution containing 0.1 M potassium phosphate (pH 8.0) and 0.25 M sucrose.

9.4.24 Tyrosinase (Li et al., 1990)

Li et al. (1990) developed an assay to measure the diphenol oxidase activity of tyrosine by following the conversion of 3,4-dihydroxymandelic acid (DHMA) to 3,4-dihydroxybenzaldehyde (DHBZ). Tyrosinase is involved in the formation of melanotic pigments in a wide variety of plants and animals.

Separation of 3,4-dihydroxymandelic acid and 3,4-dihydroxybenzaldehyde occurred at 40°C on an Altex Ultrasphere ODS column (4.6 mm × 250 mm, 5 μm). The mobile phase at a flow rate of 1.0 mL/min consisted of 0.4 citrate buffer (pH 3.2) containing 0.5 mM Na_2EDTA and 5% (v/v) acetonitrile. An amperometric detector fitted with a glassy carbon electrode was used. The electrode was maintained at +650 mV.

The standard reaction mixture contained 0.3 μg of mushroom tyrosinase in 300 μL of 0.05 M Mops buffer (pH 6.5) and 3.6 μmol of 3,4-dihydroxybenzaldehyde dissolved in 300 μL of the same buffer. The reaction mixture was incubated at 30°C, and aliquots were withdrawn at various times up to 10 minutes, with each added to an equal volume of ice-cold 0.2 M perchloric acid. After centrifugation, 2 to 5 μL was analyzed by HPLC. For single time-point determinations, the reaction volume was reduced to 50 μL. Product formation was linear with time to 10 minutes and with the amount of protein added.

The assay was used to measure the activity of a commercial preparation of mushroom tyrosinase, and the activity in cell-free hemolymph from mosquitoes.

Figure 9.54 shows chromatogram for assay of hemolymph.

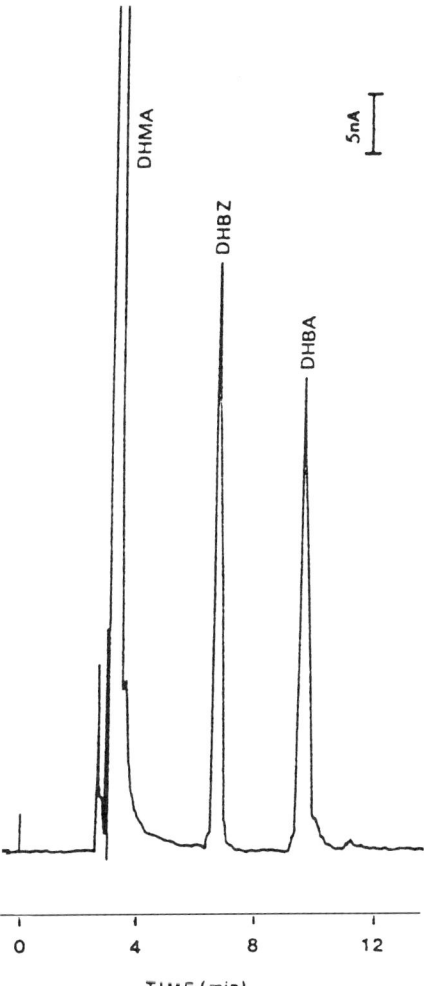

Figure 9.54 Standard mushroom tyrosinase reaction that was stopped by the addition of an equal volume of 0.2 M perchloric acid containing 120 mM DHBZ. After centrifugation, 2 µL of resulting supernatant was injected. (From Li et al., 1990.)

δ-(L-α-Aminoadipyl)-L-Cysteinyl-D-Valine Synthetase (White et al., 1989)

The synthetase that catalyzes the synthesis of the named peptide from L-α-aminoadipic acid, L-cysteine, and L-valine uses ATP as the energy source. The enzyme activity is found in the actinomycete *Streptomyces clavuligerus*.

After the enzymatic reaction, the samples were oxidized with performic acid and then converted to the isoindole derivatives. The resulting sulfonate isoindole derivatives of δ-(L-α-aminoadipyl)-L-cysteinyl-D-valine was separated on a Beckman Ultrasphere ODS column (4.6 mm × 45 mm, 5 μm). The mobile phase was a binary gradient between solvent A [90:9.5:0.5 mixture of 0.1 M sodium acetate buffer (pH 6.25), methanol, and tetrahydrofuran] and solvent B (methanol). Following injection, B was increased from 0 to 15% in 1 minute, and that concentration was held until minute 5, when it was further increased to 100% B by minute 6. The composition was returned to 0% B between minutes 7 and 8. The effluent was monitored by fluorescence using excitation and detection wavelengths in the ranges of 305 to 395 and 420 to 650 nm, respectively.

The enzyme assays were carried out in 3 mL serum vials that could be sparged with nitrogen and sealed with rubber septa. The mixture contained 0.5 mL of crude or fractionated cell extract, and 0.5 mL of a solution containing 150 mM KCl, 45 mM ATP, 45 mM MgCl$_2$, 15 mM EDTA, and 3 mM chloramphenicol in 300 mM Mops buffer (pH 7.2). After sparging with nitrogen, the reaction was started by injecting 0.5 mL of a deaerated solution adjusted to pH 7.2 with KOH and containing 15 mM L-α-aminoadipic acid, 15 mM L-cysteine hydrochloride, and 15 mM L-valine. After incubation at 27°C, the reaction was ended by injecting 0.4 mL of 20% trichloroacetic acid. After centrifugation, a 100 μL portion of the supernate was oxidized by mixing with an equal volume of performic acid. After 2.5 hours at 0°C, 2 mL of water was added and the solution was freeze-dried. The residue was taken up in 100 μL of water. Fluorescent isoindole derivatives were formed by mixing 20 μL of the aqueous sample with 20 μL of homoserine solution (internal standard). Then 40 μL of Fluo-R reagent (mixture of o-phthaldialdehyde and mercaptoethanol from Beckman Instruments) was added. After 1 minute, the reaction was quenched by addition of 120 μL of 0.1 M sodium acetate buffer (pH 6.25) and 20 μL was used for HPLC analysis.

The source of enzyme activity was mycelium from *Streptomyces clavulingerus*. The mycelium were disrupted by sonic oscillation. The supernate was clarified by centrifugation and for best results desalted and equilibrated on a Sephadex G-10 column with 0.1 M Mops buffer (pH 7.0) in 25% (v/v) aqueous glycerol containing 20 mM MgCl$_2$, 30 mM 2-mercaptoethanol, 2 mM dithiothreitol, and 5 mM EDTA.

9.5 POLYAMINE METABOLISM

9.5.1 Ornithine Decarboxylase (Haraguchi et al., 1980; Beeman and Rossomando, 1989)

The enzyme that catalyzes the decarboxylation of ornithine to putrescine is a key factor in the biosynthesis of polyamines; ornithine decarboxylase is involved in the control of cell regulation, differentiation, and growth.

The assay described by Haraguchi et al (1980) involves a preparation on CellexP, a conversion to the fluorescent derivative with fluorescamine, and separation on HPLC.

The column was LiChrosorb RP-18, and the separation was carried out by elution with a gradient of 45 to 80% methanol and 0.1 M ammonium chloride in an acetate buffer (pH. 4.0).

The incubation mixture contained ornithine, pyridoxyl phosphate, and the enzyme. After incubation for 1 hour, the reaction was terminated with perchloric acid. The precipitate was removed by centrifugation, the supernatant extracted with chloroform–methanol (2:1), and the aqueous layer applied to a CellexP column. The putrescine was eluted, reacted with fluorescamine, and quantitated by HPLC.

The enzyme was from rat intestinal mucosa and was partially purified by a 20 to 80% precipitation with ammonium sulfate.

In the assay described by Beeman and Rossomando (1989), a μBondapak C_{18} column (3.9 mm $\times$ 300 mm) was used to separate L-[2,3-^{3}H]ornithine from [1,2-^{3}H]putrescine. The mobile phase contained 0.05 M sodium phosphate (pH adjusted to 3.9 with phosphoric acid) containing 0.01 M SDS and 36% acetonitrile. Fractions were collected and the radioactivity determined by a liquid scintillation counter. The flow rate was 1.0 mL/min, and 0.5 mL fractions were collected.

The reaction mixture contained in a final volume of 1 mL 20 mM sodium phosphate buffer (pH 5.0), 2.5 mM dithiothreitol, 0.1 mM pyridoxal 5'-phosphate, 0.4 μCi L-[2,3-^{3}H]ornithine, 100 mM ornithine hydrochloride, and 0.3 mL of either *E. coli* ornithine decarboxylase or tissue homogenate. The *E. coli* enzyme was assayed at pH 5.0 and 37°C; the pH values of reaction mixtures were adjusted to pH 7.3 before assaying homogenates of mammalian tissues. The reaction was stopped by injection onto the column.

The *E. coli* ornithine decarboxylase was purchased from Sigma. Supernates obtained by centrifugation of homogenates of submandibular glands of mice (homogenized in 20 mM sodium phosphate buffer, pH 7.3), 1.25 mM dithiothreitol, and 10 μM disodium EDTA.

9.5.2 Spermidine Synthetase (Porta et al., 1981)

The enzyme spermidine synthetase catalyzes the conversion of putrescine to spermidine. The separation of the substrate from the product was accomplished on a reversed phase column (C_{18}) using a mobile phase of 60% methanol (v/v). The compounds in the eluent were detected by scintillation counting or by UV at 254 nm as shown in Figure 9.101 (labeled Blank).

The assay was carried out in phosphate buffer with radioactive putrescine, decarboxylated *S*-adenosylmethionine, and enzyme. Reactions were incubated at 37°C for 90 minutes and terminated by addition of perchloric acid. The solutions were clarified by centrifugation, and the polyamines were benzoylated and extracted and then analyzed. Figure 9.55 shows the analysis of samples removed at zero time (blank) and in after 60 minutes incubation (sample) at 37°C. The appearance of radioactive spermidine is shown. The rate of product formation is shown in Figure 9.56.

274 SURVEY OF ENZYMATIC ACTIVITIES ASSAYED BY THE HPLC METHOD

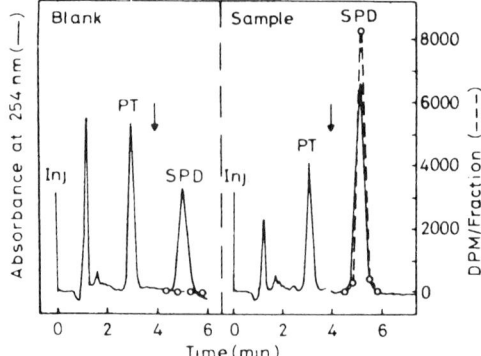

Figure 9.55 HPLC separation of the benzoyl derivatives of putrescine (PT) and spermidine (SPD) contained in a standard reaction sample and in the corresponding blank after 60 minutes of incubation at 37°C. Fractions (0.8 mL) were collected for determination of radioactivity. The arrows indicate a change in detector sensitivity from 0.2 to 0.05 absorbance unit full scale. (From Porta et al., 1981.)

The enzyme was prepared from mouse brain homogenized in phosphate buffer. An S-100 solution was prepared and used as the source of the synthetase.

It is of interest to note the use of the "sensitivity shift" procedures, illustrated in Figure 9.55. Arrows on the figure indicate where detector sensitivity was increased to allow for the detection of low levels of the product.

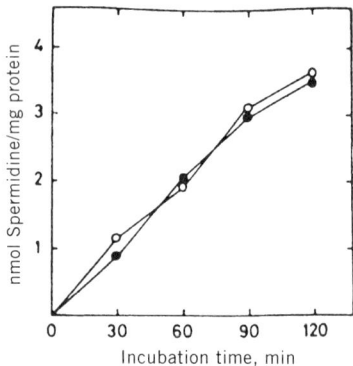

Figure 9.56 Mouse brain spermidine synthetase activity as a function of time. Both the absorbance (○) and the radiometric (●) determination of product formation represent the mean value of duplicate assays. The reproducibility was within 10%. (From Porta et al., 1981.)

9.5.3 Polyamine Oxidase (Halline and Brasitus, 1990)

Polyamine oxidase is a flavin-dependent peroxisomal enzyme involved in the degradation of polyamines to putrescine. This assay measures the conversion of N^1-acetylspermidine to putrescine.

Following derivatization with o-phthalaldehyde, N^1-acetylspermidine, and putrescine are separated on a Beckman Ultrasphere ion-pair column (4.6 mm × 250 mm, 5 μm). Solvent A was composed of 0.1 M sodium acetate and 10 mM octanesulfonic acid adjusted to pH 4.5 with acetic acid. Solvent B contained 0.2 M sodium acetate (pH 4.5)–acetonitrile (10:3, v/v) with 10 mM octanesulfonic acid. A linear gradient from 35% A and 65% B to 100% B was achieved in 10 minutes, followed by continued flow of solvent B for 15 minutes. The flow rate was 1 mL/min. Postminutes, column derivatization of polyamines with o-phthalaldehyde was accomplished with a pump. Fluorescence detection was used, with excitation and emission wavelengths of 340 and 455 nm.

The standard reaction mixture contained 0.1 M glycine (pH 9.5), 5 mM dithiothreitol, 250 μM N^1-acetylspermidine, and 250 to 400 μg of protein in a final volume of 250 μL. The reaction mixtures also contained 0.56 mM aminoguanidine and 0.04 mM pargyline to inhibit diamine oxidase and monoamine oxidase, respectively. The reaction was initiated by adding enzyme and stopped by the addition of 50 μL of 50% trichloroacetic acid. After filtration, 25 to 100 μL aliquots were injected. The reaction was linear with time and protein up to 60 to 500 μg of protein.

Enzyme activities were measured in cytosolic fractions of small intestinal or colonic mucosa from rats.

9.5.4 Diamine Oxidase (Biondi et al., 1989)

Diamine oxidase catalyzes the oxidation of various diamines (e.g., putrescine, cadaverine) to their corresponding aminoaldehydes, which are in equilibrium with their cyclic condensation products (Δ^1 pyrroline and Δ^1-piperideine, respectively). Diamine oxidase activity has been proposed to be a marker of intestinal mucosa integrity.

HPLC analysis was based on the separation of o-aminobenzaldehyde, 2,3-tetramethylene-1,2-dihydroquinazolinium ion (obtained from Δ^1-piperideine) and 2,3-trimethylene-1,2-dihydroquinazolinium ion (obtained from Δ^1-pyrroline) on a LiChrosorb RP-8 column (4 mm × 250 mm, 7 μm). The mobile phase was composed of 0.1 M phosphoric acid adjusted to pH 3 with triethylamine and acetonitrile (85:15, v/v) at a flow rate of 1.0 mL/min. The absorbance was monitored at 465 nm.

Mucosal extracts (0.5 mL) were preincubated at 37°C with 50 μL of 1 mM Δ^1-pyrroline solution and 100 μL of o-aminobenzaldehyde-saturated aqueous solution. The enzymatic reaction was started by adding 50 μL of 0.1 M cadaverine solution in 0.1 phosphate buffer (pH 7.0). After 20 minutes, the reaction was stopped by heating the tube for 2 minutes in a sand bath maintained at 150°C. After centrifugation, 10 μL aliquots were analyzed by HPLC. Quantitation was based on oxidizing known amounts of cadaverine to Δ^1-piperideine by pea seedling diamine oxidase, followed by treatment with o-aminobenzaldehyde to form the quinazolinium ion. The assay was demonstrated to be linear with time to 20 minutes and with volume of extract added to 1.5 mL.

Tissue homogenates were prepared from mucosa scraped from small intestine of dog. The scrapings were homogenized in 10 volumes of ice-cold 0.1 M phosphate buffer (pH 7.0). Supernates were prepared by centrifuging for 60 minutes at 25,000g.

9.6 HEME METABOLISM

9.6.1 δ-Aminolevulinic Acid Synthetase (Tikerpae et al., 1981; Tomokuni et al., 1991)

This mitochondrial enzyme (ALA synthetase) catalyzes the formation of δ-aminolevulinic acid (ALA) from glycine and succinyl–CoA. This is the initial step in heme biosynthesis.

The assay of Tikerpae et al. (1981) incorporates a novel feature wherein radioactive succinyl–CoA is formed from succinate, which in turn is formed from radioactive δ-ketoglutarate. The succinyl–CoA then reacts with glycine to form ALA. For the assay, the ALA is converted to the pyrrole derivative 2-methyl-3-carbethoxy-4-(3-propionic acid) pyrrole.

The assay involves the isolation of the pyrrole by ion-paired, reversed-phase HPLC (Hypersil-SAS) with a mobile phase of methanol and water (45:155, v/v) and 0.005 mol/L 1-heptanesulfonic acid (PIC B-7). The radioactive product was detected by scintillation counting. The separation obtained when the compounds were in distilled water is shown in Figure 9.57a. However, when the analysis was carried out on freeze-dried samples, the salts present led to the results shown in Figure 9.57b.

The reaction mixture contained glycine, coenzyme A, buffer, α-ketoglutarate, and bone marrow lysate. After incubation for 1 hour, the reaction was terminated by addition of 10% TCA. The samples were chilled and clarified by centrifugation, and the pyrrole formed. After processing, the pyrrole was isolated by HPLC and its radioactivity was quantitated.

The enzyme was obtained from bone marrow cells. The cells were harvested and pelleted by centrifugation at 2500g for 5 minutes. The pellets were washed, resuspended, counted, and disrupted by sonication to release mitochondria. This lysate was used directly as the source of the enzyme.

In the assay described by Tomokuni et al. (1991), the product, δ-aminolevulinic acid, is formed from glycine and endogenously generated succinyl–CoA.

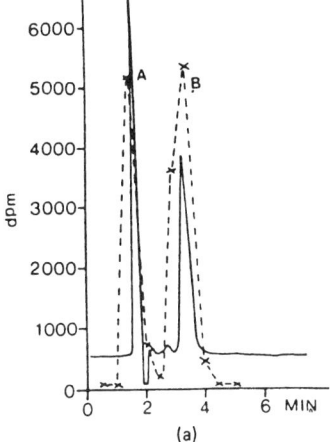 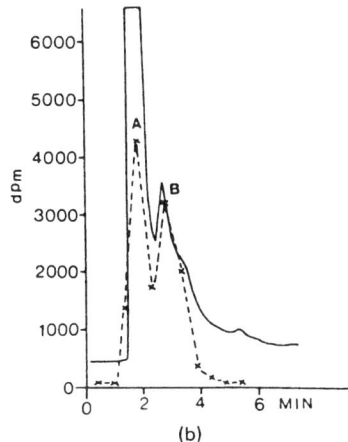

Figure 9.57 Separation of [5-^{14}C]α-ketoglutarate (A) and [4-^{14}C]aminolevulinic acid (B) by HPLC on a Hypersil column with methanol–aqueous PIC B-7 (50:150, v/v) as the mobile phase at a flow rate of 1 mL/min. (*a*) Mixture in distilled water. (*b*) Mixture in presence of salts recovered from a freeze-dried incubation mixture. Solid lines, spectrophotometric monitoring at 278 nm and 0.05 AUFS; dashed lines, radioactive distribution (0.5 mL fractions counted). (From Tikerpae et al., 1981.)

The fluorimetric determination of δ-aminolevulinic acid is based on derivatization with acetylacetone and formaldehyde. The fluorescent derivative was separated at ambient temperature from other reaction components on a Unisil NQ C_{18} column (4.6 mm × 250 mm). The mobile phase was composed of methanol–water–glacial acetic acid (600:400:10, v/v). Detection is by fluorescence (excitation, 370 nm; emission, 460 nm).

To a 20 mL Erlenmeyer flask, 1 mL of 0.15 *M* Tris-HCl buffer (pH 7.2), 0.1 mL of 0.2 *M* disodium EDTA, and 0.5 mL of liver homogenate were added. After preincubation for 5 minutes at 37°C, 0.4 mL of 0.5 *M* glycine was added. Shaking continued for 30 minutes before the reaction was terminated by adding 0.5 mL of 25% trichloroacetic acid. The reaction mixture was centrifuged and the supernate diluted fivefold with distilled water. To 0.1 mL of the supernate, 2.5 mL of distilled water, 0.4 mL of acetylacetone, and 1 mL of 10% formaldehyde solution were added. This mixture was heated for 10 minutes at 100°C and then kept in an ice bath until analysis of 20 μL by HPLC. Product formed was linear with liver homogenate up to 0.5 mL, and with time up to 60 minutes.

The source of enzyme was livers from mice homogenized in 3 volumes of ice-cold 0.9% NaCl.

9.6.2 5-Aminolevulinate Dehydrase (Crowne et al., 1981)

5-Aminolevulinic acid dehydrase (ALA dehydrase) is the second enzyme of the heme biosynthetic pathway. It catalyzes the condensations of two molecules of ALA to form porphyrobilinogen (PBG).

The assay involves the separation of ALA from PBG by ion-paired, reversed, phase HPLC (Hypersil SAS) with a mobile phase of methanol–water (22:78, v/v) and PIC B-7 (0.005 M 1-heptanesulfonic acid) adjusted to pH 3.5. An internal standard of 2-methyl-3-carbomethoxy-4-(3-propionic acid)pyrrole was used. All three compounds were readily separated in 6 minutes (Fig. 9.58). Detection was at 240 nm.

Reaction mixtures contained substrate (ALA) and buffer; the reaction was started by the addition of lysate. TCA was added to terminate the reaction. The incubation solution was clarified by centrifugation, and a constant volume removed and mixed with the internal standard. A known volume was injected for analysis. Figure 9.59 compares the profiles of two samples after incubation with enzyme and with blank. The appearance of the reaction product PBG is observed in the test profile (Fig. 9.59A) only.

Whole blood was hemolyzed in water, and the lysate was used as the enzyme solution.

9.6.3 Uroporphyrinogen Decarboxylase (James and Marks, 1989)

An enzyme involved in heme biosynthesis, uroporphyrinogen decarboxylase has been found to be inhibited by a variety of xenobiotics. This assay measures the conversion of pentacarboxylporphyrinogen I to coproporphyrinogen I.

Coproporphyrinogen I was separated from other assay components by chromatography on a Partisil 5 column (4.6 mm × 250 mm) column. The mobile phase was composed of heptane–ethyl acetate–dichloromethane–methanol (60:25:14:1). Quantitation was by a UV fluorescence detector (excitation, 405 nm; detection, 605 nm). At a flow rate of 1.1 mL/min, the

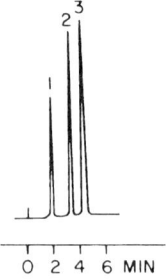

Figure 9.58 Separation of ALA (1), PBG (2), and internal standard (3). Stationary phase, Hypersil-SAS; mobile phase, methanol–PIC B-7 in water (22:78, v/v); pressure, 80 bar; flow rate, 1.2 mL/min; detection, 240 nm. (From Crowne et al., 1981.)

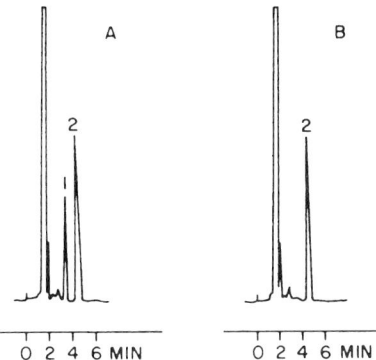

Figure 9.59 Separation of PBG (1) and internal standard (2) in incubation mixture: (A) test and (B) blank. (From Crowne et al., 1981.)

retention times for mesoporphyrin (internal standard) and coproporphyrinogen I were 4.2 and 6.15 minutes, respectively.

Pentacarboxylporphyrinogen I was prepared from pentacarboxylporphyrin I by reduction with 3% sodium amalgam in dilute KOH. The preparation was diluted to a concentration of 6.67 nmol/mL in the assay buffer (sodium phosphate, pH 6.0, containing 5 mM dithiothreitol and 0.1 mM EDTA) and the pH readjusted to pH 6.0 with 1 M HCl. For the assay, 1.5 mL of this solution was placed in the sidearm of a Thunberg tube. One milliliter of the enzyme preparation (1.2 mg protein/mL) and 1.5 mL of the assay buffer were placed in the bottom of the tube. The reaction tube and its contents were made oxygen free by repeatedly evacuating and flushing with O_2-free nitrogen. The reaction was initiated by tipping the contents of the sidearm into the bottom of the Thunberg tube. After 20 minutes, 200 μL of 1% iodine in ethanol was added to terminate the reaction. Mesoporphyrin (0.25 nmol) was added as an internal standard before the assay mixture was freeze-dried. Next 5 mL of 5% H_2SO_4 in methanol was added and the mixture maintained at $-20°C$ for 24 hours. The methylated porphyrins were extracted into dichloromethane, dried under nitrogen, and stored at $-20°C$. The residue was dissolved in a 50:50 mixture of dichloromethane and HPLC mobile phase prior to injection onto the HPLC. The enzyme activity was linear, with protein concentrations ranging from 0.3 to 2.0 mg/mL. At a protein concentration of 1.2 mg/mL, the activity was linear for a reaction time of 50 minutes.

The enzyme preparation was a low speed homogenate prepared from livers of 17-day-old chicken embryos.

9.6.4 Heme Oxygenase (Lincoln et al., 1988)

Heme oxygenase is a microsomal enzyme that catalyzes the first and rate-determining step in the degradation of heme. This assay measures bilirubin, which is produced by the consecutive actions of heme oxygenase and biliverdin reductase.

This assay is based on the separation and quantitation of bilirubin. The column was a C_{18} Novapak radial compression column. Solvent A contained 49% 0.1 M monobasic ammonium phosphate, 40% methanol, 10% acetonitrile, and 1% DMSO, adjusted with H_3PO_4 to pH 3.6. Solvent B contained 79% methanol, 20% acetonitrile, and 1% DMSO. The 16-minute run included a linear gradient from 50 to 100% solvent B in 3 minutes, maintaining 100% B for 13 minutes, followed by a return to starting conditions. The eluate was monitored at 450 nm, and detector response was linear with 0 to 500 pmol of bilirubin.

The 1 mL reaction mixture contained heme oxygenase, 5 mM deferroxamine, 25 μM hemin, 15 μM bovine serum albumin, 1 mM NADPH, 0.1 M potassium phosphate (pH 7.4), and 0.05 mL of 105,000g supernate from 20% (w/v) perfused rat liver homogenate as a source of biliverdin reductase. To stop the reaction, 0.1 mL of the reaction mixture was added to 0.1 mL of ethanol–DMSO (95:5, v/v). After 2 minutes incubation on ice, the mixture was centrifuged and 0.1 mL of the supernate was injected on HPLC. The rate of bilirubin formation was linear up to 0.63 mg protein/0.237 mL assay, and up to 9 minutes incubation at 37°C.

The enzyme preparations assayed were the 10,000g (10 min) supernate fractions from livers of chick embryos and rats, and liver biopsy samples from humans.

9.6.5 Ferrocheletase (Guo et al., 1991)

Ferrocheletase catalyzes the final step in heme synthesis by inserting Fe^{2+} protoporphyrin IX to form heme. The assay uses Zn^{2+} and is sufficiently sensitive to measure ferrocheletase activity in leukocytes.

The product, Zn-mesoporphyrin, mesoporphyrin and Zn-deuteroporphyrin (internal standard) are separated on an ODS-Hypersil column (5 mm × 250 mm). The mobile phase was 88% (v/v) methanol in 1 M ammonium acetate, pH 5.16. Fluorescence detection was used, with excitation at 405 nm and emission at 574 nm. Protoporphyrin was also used as a substrate, in which case excitation and emission wavelengths were 410 and 590 nm, respectively. The amount of product formed was determined from calibration curves.

The standard incubation mixture contained 100 μL of 0.25 M Tris-HCl buffer (pH 7.4) containing 1.75 mM palmitic acid and 1% (w/v) Tween-20; 50 μL of enzyme preparation (about 0.5 mg protein), and 50 μL of 80 or 100 μM zinc acetate solution. After incubating 5 minutes at 37°C, the reaction was started by adding 50 μL of 100 μM mesoporphyrin or protoporphyrin. The incubation was continued for 30 minutes at 37°C in the dark. The reaction was stopped by adding 1 mL of ice-cold methanol–DMSO (8:2, v/v) containing 13 nM Zn-deuteroporphyrin as internal standard. The mixture was cooled on ice and centrifuged before analysis of the supernate by HPLC. The

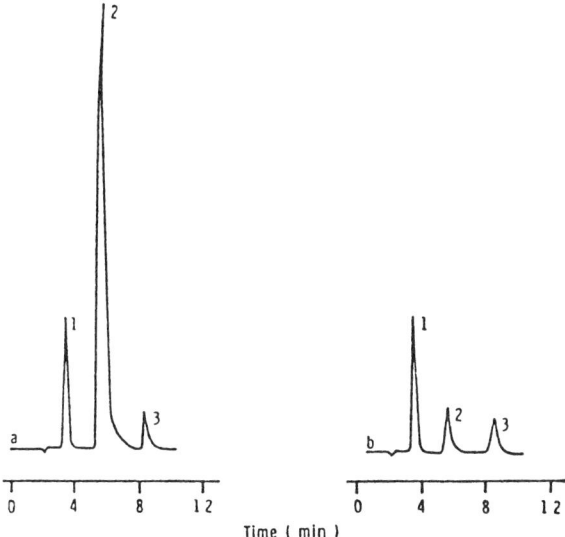

Figure 9.60 HPLC for the measurement of ferrochelatase activity in human leukocytes with mesoporphyrin and Zn^{2+} as substrates. *(a)* Enzyme incubation mixture. *(b)* Blank incubation with boiled leukocytes. Column ODS-Hypersil (250 mm × 5.0 mm i.d.); eluent, 88% (v/v) methanol in 1 M ammonium acetate, pH 5.16. Flow rate, 1.5 mL/min; fluorescence detection, excitation at 403 nm and emission at 574 nm. *Peaks* 1, Zn-deuteroporphyrin (internal standard); 2, Zn-mesoporphyrin; 3, mesoporphyrin. (From Guo et al., 1991.)

rate of nonenzymatic formation of metal chelate was determined in parallel by omitting the enzyme or by using boiled, inactivated enzyme. With Zn-mesoporphyrin formation, the assay was linear for up to 70 minutes and proportional to leukocyte protein concentration up to 0.7 mg per assay.

Enzyme activity was measured in leukocytes isolated from heparinized blood.

Figure 9.60 shows chromatogram for mesoporphyrin.

9.6.6 Protoporphyrinogen Oxidase (Guo et al., 1991)

Protoporphyrinogen oxidase catalyzes the six-electron oxidation of protoporphyrinogen IX to protoporphyrin IX. Application of the assay to leukocytes provides a more convenient enzyme source than liver tissues in the diagnosis of porphyrias.

The product, protoporphyrin IX, is separated from mesoporphyrin, the internal standard, on an ODS-Hypersil column (5 mm × 250 mm). The mobile phase was 88% (v/v) methanol in 1 M ammonium acetate, pH 5.16. Fluorescence detection was used with excitation at 400 nm and emission at 618 nm.

Protoporphyrinogen IX was prepared fresh by reduction of protoporphyrin IX with sodium and amalgam. The incubation mixture contained 100 μL of leukocyte suspension (about 0.5 mg of protein) and 50 μL of 0.25 M Tris-HCl buffer (pH 8.6) containing 5 mM EDTA, 5 mM glutathione, and 1% Tween-20 (w/v). After preincubation at 37°C in the dark for 5 minutes, the reaction was started by adding 100 μL of approximately 35 μM protoporphyrinogen IX substrate. After 10 minutes, the reaction was stopped by adding 1 mL of ice-cold methanol–DMSO (8:2, v/v) containing 42 nM mesoporphyrin as internal standard. The mixture was centrifuged, and the resulting supernate was flushed with nitrogen before HPLC analysis. A boiled leukocyte suspension was used in a parallel assay to correct for formation of protoporphyrin IX by autoxidation. Protoporphyrin IX formation was linear up to 10 minutes and was proportional with leukocyte protein up to 1 mg.

Enzyme activity was measured in leukocytes isolated from heparinized blood.

Figure 9.61 shows chromatogram for this assay.

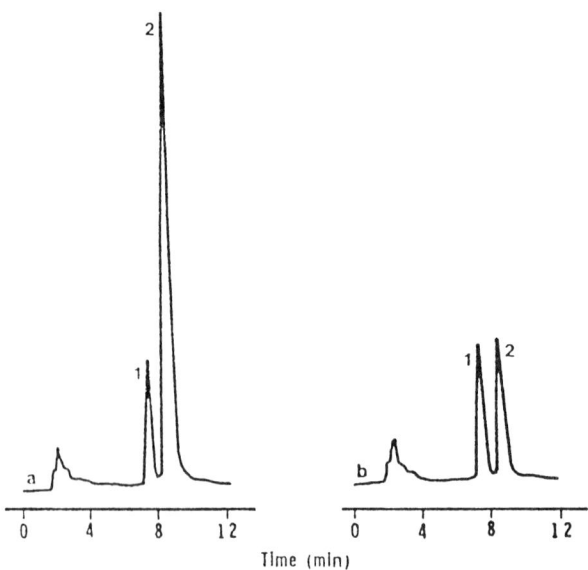

Figure 9.61 HPLC for the determination of protoporphyrinogen oxidase activity in human leukocytes. (a) Enzyme incubation mixture. (b) Blank incubation with boiled enzyme. Column ODS-Hypersil (250 mm × 5.0 mm i.d.); eluent, 88% (v/v) methanol in 1 M ammonium acetate, pH 5.16. Flow rate, 1.5 mL/min; fluorescence detection, excitation at 400 nm and emission at 618 nm. *Peaks:* 1, mesoporphyrin (internal standard); 2, protoporphyrin IX. (From Guo et al., 1991.)

9.7 CARBOHYDRATE METABOLISM

9.7.1 β-Galactosidase (Naoi and Yagi, 1981)

β-Galactosidase (β-D-galactoside galactohydrolase, EC 3.2.1.23) catalyzes the release of galactose from a variety of substrates including glycosphingolipids. The galactosidase specific for the release of terminal galactose from glycosphingolipids was studied using a derivative of a galactocerebroside, 1-O-galactosyl-2-N-DANS-sphingosine, as the substrate. The product of the β-galactosidase reaction, N-DANS-sphingosine, was measured by means of HPLC.

The product was separated from the substrate on a normal phase silica gel column (Zorbax Sil) and eluted with methanol at 35°C as shown in Figure 9.62*I,II*. The concentration of the reactants was determined by the fluorescence intensity at 535 nm, with an excitation wavelength at 340 nm.

The reaction mixture included the substrate, galactosyl-N-DANS-sphinogosine, suspended in citrate buffer and dispersed by sonication. The reaction was started by the addition of the substrate to the enzyme solution. The reaction mixture was incubated at 37°C for 30 minutes, and the reaction was terminated by the addition of a mixture of chloroform–methanol (2:1, v/v). After centrifugation, the lower chloroform phase was recovered and evaporated to dryness, and the residue was dissolved in methanol and analyzed by HPLC. Figure 9.62*III* shows a chromatogram of a sample. The smaller of

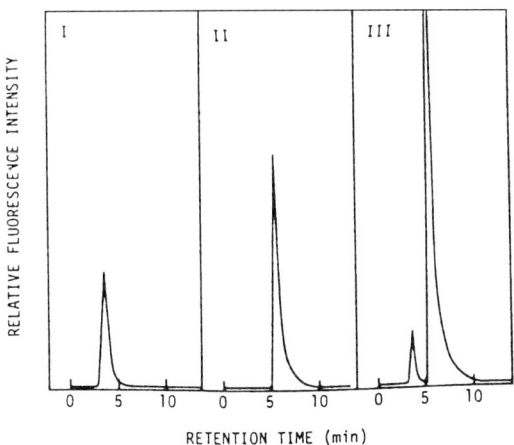

Figure 9.62 HPLC profiles of N-DANS-sphingosine and galacytosyl-N-DANS-sphingosine. Purified N-DANS-sphingosine. *(I)* or galacytosyl-N-DANS-sphingosine *(II)*. After reaction of crude β-galactosidase with galacytosyl-N-DANS-sphingosine, the sample was applied on HPLC *(III)*. (From Naoi and Yagi, 1981.)

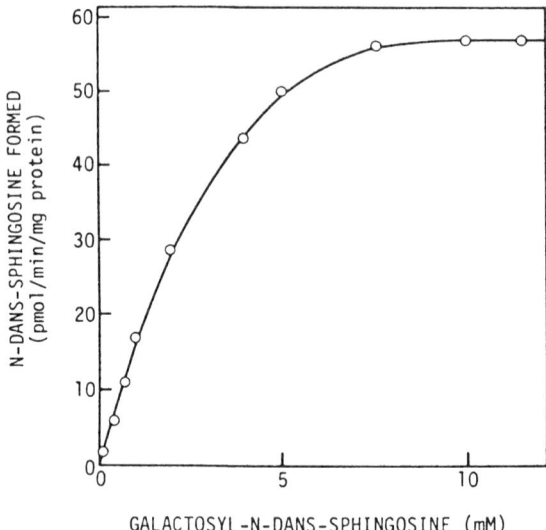

Figure 9.63 Relation between substrate concentration and rate of hydrolysis of galactosyl-N-DANS-sphingosine by crude β-galactosidase. Galactosyl-N-DANS-sphingosine was hydrolyzed by crude β-galactosidase (0.7 mg), and the formation of N-DANS-sphingosine was measured by HPLC. (From Naoi and Yagi, 1981.)

the two peaks, the product, is observed. Figure 9.63 shows the relationship between substrate concentration and rate of hydrolysis as determined by the HPLC method.

The β-galactosidase was prepared from rat brain. The rat brain was homogenized in detergent and the homogenate centrifuged at 20,000g for 20 minutes. The supernatant solution was made 60% saturated in ammonium sulfate. The precipitated protein was recovered, dissolved in detergent, and dialyzed against the same detergent. The dialysate was centrifuged, and the supernatant solution was used as the enzyme fraction.

9.7.2 Lactose-Lysine β-Galactosidase (Schreuder and Welling, 1983)

One particular β-galactosidase cleaves the compound lactose-lysine into β-galactose and fructose-lysine. This activity is of interest because it has been suggested as a possible marker for the presence of bacteria, since it does not appear to be present in germ-free animals.

The reaction is followed by separation of the substrate, lactose-lysine, from the product, fructose-lysine, on a cation-exchange resin (Durrum DC6A) using an isocratic mobile phase of pyridine–acetic acid–water (6:60:176, v/v). o-Phthalaldehyde derivatives were formed and detected by fluorescence.

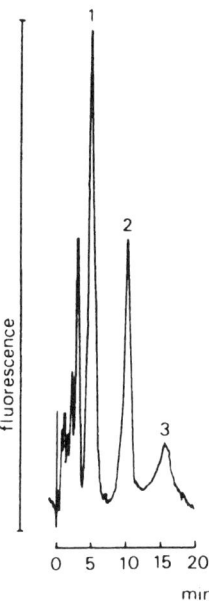

Figure 9.64 Chromatogram obtained after injection of a β-galactosidase incubation mixture onto a 45 mm $\times$ 3.6 mm cation-exchange column. Mobile phase: pyridine–acetic acid–water (6:60:176, v/v). Flow rate, 0.4 mL/min; temperature, 50°C. In front of lactose-lysine some free amino acids elute which are the result of proteolytic activity in the intestinal enzyme preparation. *Peaks:* 1, lactose-lysine; 2, β-alanine (internal standard); 3, fructose-lysine. (From Schreuder and Welling, 1983.)

The reaction was carried out with ε-lactose-lysine as the substrate using a sodium phosphate buffer (pH 7.5). The reaction was started by the addition of the enzyme; after incubation for 1 hour, it was terminated with methanol containing β-alanine as an internal standard. Precipitated protein was removed by centrifugation, and samples of the supernatant solution were injected and analyzed by HPLC. A chromatogram showing the analysis of a reaction mixture is given in Figure 9.64.

The enzymes were obtained from mouse intestine. The intestinal segments were cut into pieces, homogenized with a glass rod, sonicated, and centrifuged. The supernatant solution was dialyzed and stored frozen for later use.

9.7.3 Arylsulfatase B (*N*-Acetylgalactosamine 4-Sulfatase) (Fluharty et al., 1982)

Arylsulfatase B catalyzes the hydrolysis of the sulfate from UDP-Gal-NAc-4-sulfate to form UDP-GalNAc and sulfate. This activity is found in normal fibroblasts but not in Maroteaux–Lamy fibroblasts.

The substrate was separated from the product using ion-paired reversed-phase HPLC (Ultrasphere-ODS). The samples were eluted isocratically at room temperature using a mobile phase composed of methanol and 20 mM potassium phosphate (pH 2.5), (40:60, v/v) containing 15 mM tetrabutylammonium phosphate. The eluant was monitored at 262 nm.

The reaction was carried out in a solution containing UDP-GalNAc-4-S, acetate buffer at pH 3.5, and the arylsulfatase B. Reactions were incubated at 37°C for 30 minutes and terminated by heating in a boiling water bath for 1 minute. Longer heating resulted in the formation of a new component, which overlapped the product peak on the HPLC. The heated material was centrifuged at 8000g for 2 minutes, and a sample was used directly for analysis. Figure 9.65 shows chromatograms of samples taken at zero time and after 30 minutes of incubation. The appearance of the new peak at about 4.5 minutes indicates the formation of the reaction product. The time course of sulfatase activity is shown in Figure 9.66.

The enzyme was prepared from human liver. This substrate is not hydrolyzed by arylsulfatase A.

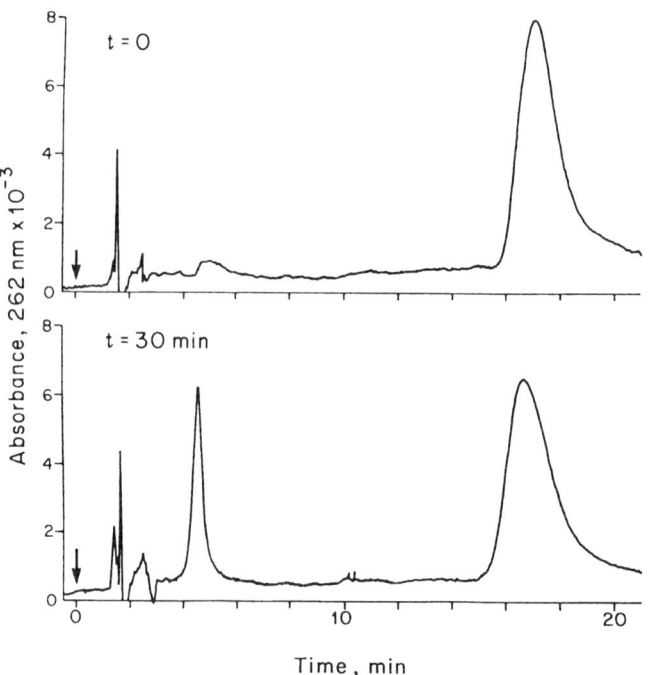

Figure 9.65 HPLC analysis of arylsulfatase B reaction with UDP-GalNAc-4-S The reaction mixture contained 5 mU of enzyme. (From Fluharty et al., 1982).

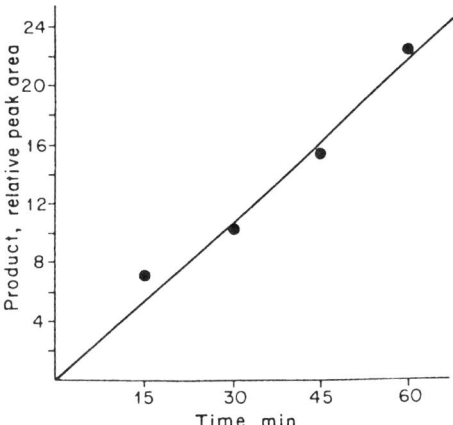

Figure 9.66 Time course of arylsulfatase B reactions with UDP-GalNAc-4-S. The reaction mixture contained 3 mU of enzyme. (From Fluharty et al., 1982.)

9.7.4 Galactosyltransferase (Hymes and Mullinax, 1984)

Galactosyltransferase catalyzes the transfer of galactose from UDP-galactose to either glucose or N-acetylgalactosamine or to N-acetylglucosamine when this is a terminal residue of complex oligosaccharides. The activity occurs in tissues and body fluids of many different types.

Most assays for this activity either determine the amount of radioactive product formed from radioactive UDP-Gal or measure the amount of UDP present using a coupled assay in which the UDP formed is coupled to the formation of NAD or NADH.

The HPLC assay developed for this activity is based on the isocratic separation of UDP, UDP-Gal, and UMP on an amino-bonded column (μBondapak NH$_2$ column) eluted with 0.167 M KH$_2$PO$_4$ (pH 4.0). The separation is shown in Figure 9.67. Detection was at 260 nm.

The reaction mixture contained N-acetylglucosamine, UDP-Gal, MnCl$_2$, and buffer at pH 8.0. The reaction was started by the addition of the enzyme. Samples were transferred at intervals to cacodylate buffer on ice (pH 6.5) to terminate the reaction. Samples (10 μL) were analyzed by HPLC. The conversion of UDP-Gal to UDP is shown in Figure 9.68. Each panel represents a different time point, from 0 to 60 minutes. During the incubation, the disappearance of the substrate and the formation of the two products is seen.

The enzyme was obtained from commercial sources and human serum.

9.7.5 Uridine Diphosphate Glucuronosyltransferase (Matsui and Nagai, 1980; To and Wells, 1984)

Uridine diphosphate glucuronosyltransferase catalyzes the transfer of glucuronic acid from uridine diphosphate glucuronic acid (UDPGA) to various

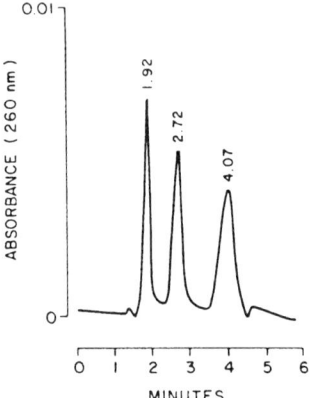

Figure 9.67 Isocratic HPLC separation and UV absorption detection of UMP, UDP-Gal, and UDP as sequentially eluted. A mixture containing 500 pmol of each compound was injected. (From Hymes and Mullinax, 1984.)

substrates, such as 4-nitrophenol (4-NP), phenolphthalein (P), and testosterone.

Matsui and Nagair (1980) separated the substrates and the product by reversed-phase HPLC (styrene–divinylbenzene copolymers) at 40°C with a methanol–water (65%, v/v) mobile phase containing 0.01 N HCl as shown in Figure 9.69. Absorbance at 300 nm was measured, and quantitation was carried out from peaks height measurements.

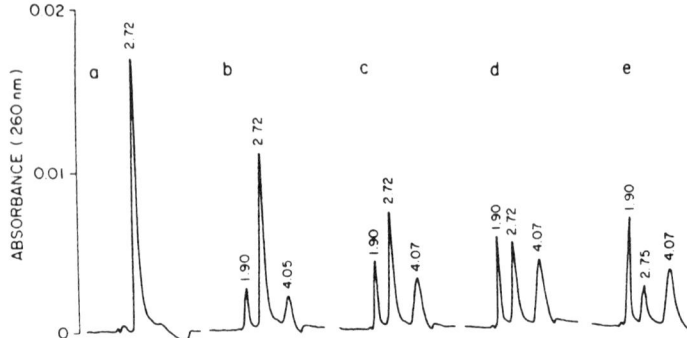

Figure 9.68 Conversion of UDP-galactose to UMP and UDP. Aliquots of galactosyltransferase reaction (glucose substrate, bovine milk enzyme) were removed at 0, 15, 30, 45, and 60 minutes and assayed by HPLC with UV detection (a–e, respectively). Peaks representing UMP, UDP-Gal, and UDP are detected in later samples. (From Hymes and Mullinax, 1984.)

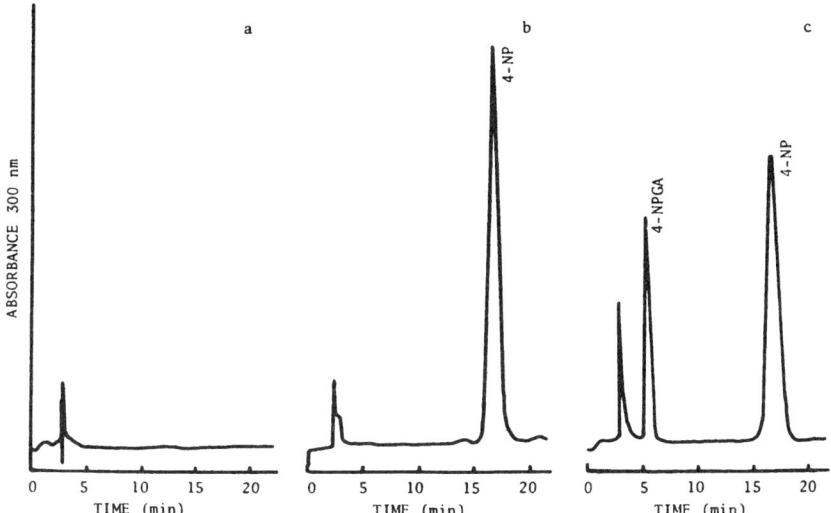

Figure 9.69 Chromatograms recorded in the assay of UDPGT toward 4-NP. (*a*) Assay without 4-NP. (*b*) Assay without UDPGA. (*c*) Assay with all ingredients. Chromatographic conditions: column, Hitachi gel No. 3010, 3 mm × 500 mm, mobile phase, 0.01 N HCl in 65% (v/v) methanol; detection, 300 nm; flow rate, 0.8 mL/min; pressure, 60 kg/cm^2; column temperature, 40°C. (From Matsui and Nagai, 1980.)

The assay mixture contained, in a final volume of 1 mL, 0.4 to 1.2 mg of the microsomal fraction protein, UDPGA as donor, and either 4-NP or P as the acceptor. The reactions were terminated in a boiling water bath (1–5 min). Methanol was added, and the solution was clarified by centrifugation. Samples of the supernatant solution were analyzed directly. Figure 9.69 shows chromatograms obtained from samples of reaction mixtures without substrates (*a* and *b*) and the complete reaction mixture (*c*). The formation of product is clearly seen.

The enzyme was prepared from rat liver. A homogenate was prepared in sucrose with a Teflon–glass homogenizer. A series of differential centrifugations (2000g for 10 min, 16,000g for 45 min, and 105,000g for 60 min) produced a microsomal pellet that was used as the enzyme source.

In a study by To and Wells (1984), the glucuronic acid was transferred from UDPGA to the acceptor, α-naphthol. In this assay the substrate α-naphthol was separated from the reaction product α-naphthol glucuronide by reversed-phase HPLC (C$_{18}$ column) with a solvent of 0.1 M acetic acid–methanol (55:45, v/v). Absorbance was monitored at 240 nm.

The reaction mixture contained the substrate, α-naphthol (5.0 mM), in Tris-HCl buffer (pH 7.4) with 10 mM magnesium chloride and DMSD. The second substrate, UDPGA, was added, and the reaction mixture was preincubated for 5 minutes. The microsomal protein was then added as the source

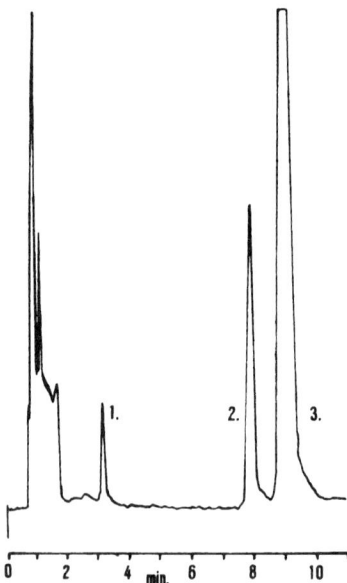

Figure 9.70 Chromatogram of the HPLC resolution of α-naphthol glucuronide. *Peaks:* 1, α-naphthol glucuronide; 2, β-naphthol, the internal standard; 3, α-naphthol, the substrate. Solvent: 0.1 M acetic acid–methanol (55:45, v/v); flow rate 1.5 mL/min; wavelength, 240 nm. (From To and Wells, 1984.)

of enzyme activity. Incubations were at 37°C for up to 20 minutes. The reactions were terminated by the addition of ice-cold methanol, and any insoluble material was removed by centrifugation. The supernatant solution was dried under nitrogen, redissolved in methanol, and injected onto the HPLC column for analysis. Figure 9.70 shows a chromatogram illustrating the separation of the substrate α-naphthol and the reaction product, glucuronide. Figure 9.71 shows the rate of glucuronide formation during the incubation.

The enzyme was obtained from mice. Hepatic microsomal fractions were used for these studies.

9.7.6 α-Amylase (1,4-α-D-Glucaglucanohydrolase, EC 3.2.2.1) (Haegele et al., 1981)

α-Amylase is a hydrolase that cleaves 1,4-α glycosidic bonds and is important in the diagnosis of pancreatitis.

This assay involves the use of maltoheptose (seven glucose residues in a linear 1,4-α linkage) as the substrate. The activity degrades the substrate to smaller oligosaccharides, which are then subjected to α-glucosidase treatment to generate glucose. The α-glucosidase and the hexokinase are used as "indicator reactions."

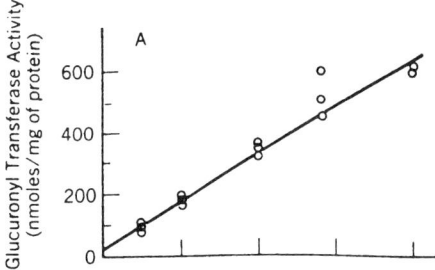

Figure 9.71 Effect of incubation time on enzyme activity. Glucuronyltransferase activity was measured by the amount of α-naphthol glucuronide produced after varying incubation time periods. (From To and Wells, 1984.)

The assay relies on the separation of maltose, maltotriose, maltotetrose, maltopentose, maltohexose, and maltoheptose. These compounds were separated by partition chromatography on a cation-exchange column (Nucleosil 10 SA) with a mobile phase of acetonitrile–water (72.5:27.5) eluted isocratically. The detector was a differential refractometer.

The enzyme assays were carried out in a phosphate buffer (pH 7.0), at 30°C with maltoheptose as the substrate. At intervals after the addition of the substrate to a solution of the amylase and indicator enzymes, samples were removed and diluted into a slurry of a mixed-bed ion exchanger to stop the reaction. The oligosaccharides were recovered and concentrated by lyophilization. The samples were dissolved in acetonitrile–water (1:1, v/v) and injected onto the HPLC column for analysis. The results of the assay are shown in Figure 9.72. The amount of glucose (peak 1) is seen to increase when these data are plotted as a function of time, as shown in Figure 9.73.

The α-amylase was purified from human pancreas.

9.7.7 Lysosomal Activities (Sandman, 1983)

The HPLC method has been used to assay a number of activities usually found associated with lysosomal vesicles. All these assays utilize the fluorometric compound 4-methylumbelliferone (4-MU). When carbohydrates, lipids, phosphates, or sulfates are conjugated with 4-MU, these compounds can be used as substrates for glycosidase, lipases, acid phosphatases, and arylsulfatases. The activity is determined by the release of 4-MU.

The separation of free 4-MU (the reaction product) from the conjugate (the substrate) was carried out by reversed phase HPLC on a PRP-1 column (styrene–divinylbenzene copolymer). A guard column was also used. The column was eluted isocratically for all the assays except the sulfatases, with 0.04 M glycine–sodium hydroxide buffer (pH 10.32), in aqueous methanol. For the sulfatase activity separations, the glycine buffer with 20% methanol

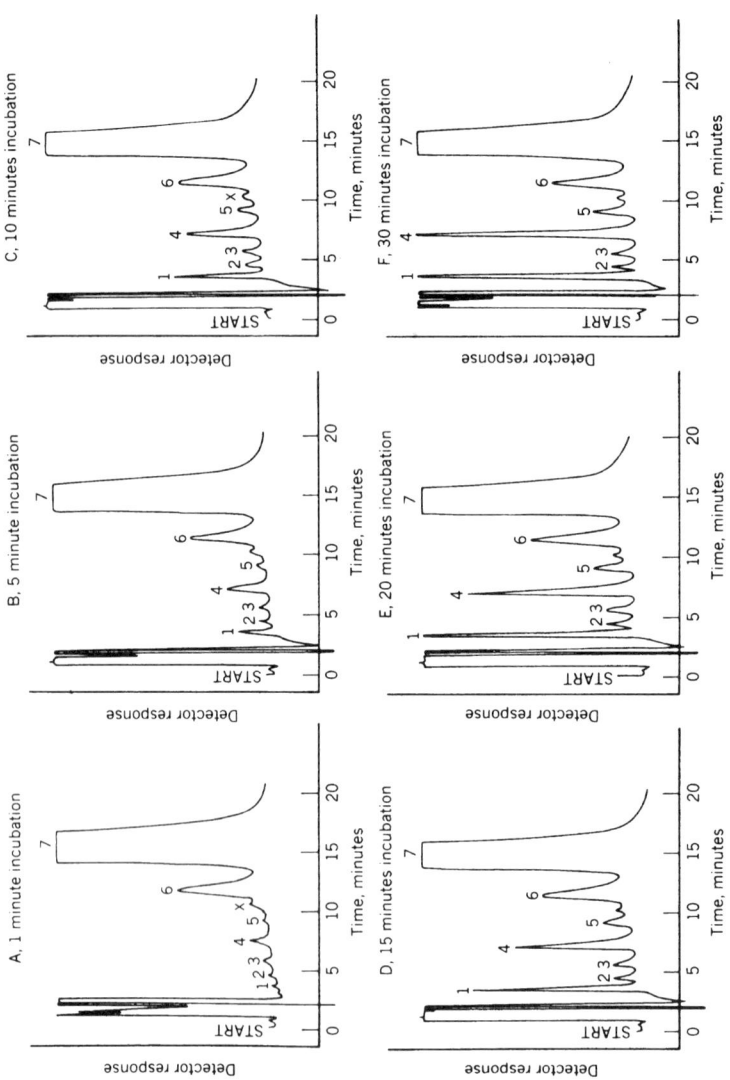

Figure 9.72 Chromatograms of the action patterns of maltoheptaose after the indicated periods of incubation with α-amylase and α-glucosidase. *Peaks:* 1, glucose; 2, maltose; 3, maltotriose; 4, maltotetraose; 5, maltopentaose; (x) compound A; 6, maltohexaose; 7, maltoheptaose. (*A*) Pure maltoheptaose used for the assay. (*B*) Blank sample before the addition of substrate. (*C–H*) Chromatograms after 1, 5, 10, 15, 20, and 30 minutes, respectively, of incubation. Chromatographic conditions: column, 10 μm Nucleosil SA (250 mm × 4 mm); solvent, acetonitrile–water (72.5:27.5); flow rate, 0.7 mL/min; temperature, 27°C; detection, differential refractometer, full scale = 2×10^{-6} refractive index units. (From Haegel et al., 1981.)

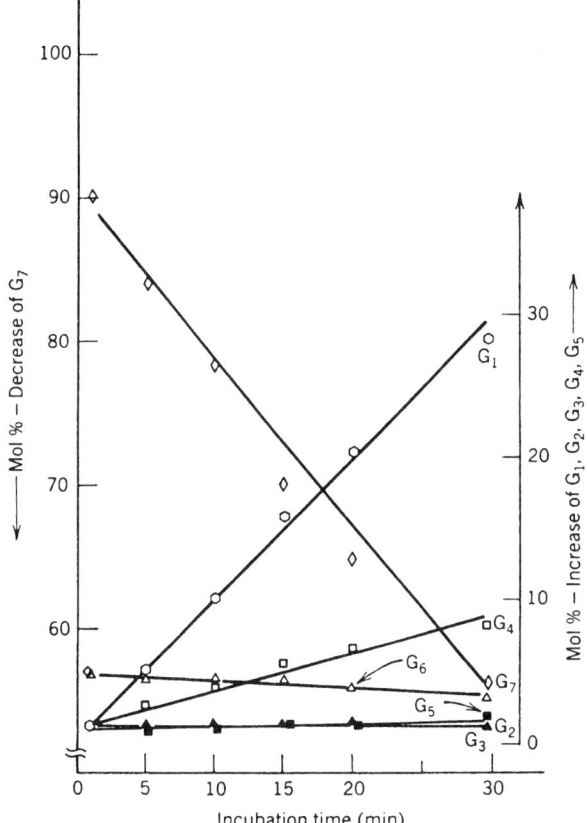

Figure 9.73 Kinetics of the coupled enzyme system α-amylase–α-glucosidase, with maltoheptaose in mole percent. (From Haegele et al., 1981.)

was used as mobile phase. The fluorescence of the eluent was monitored with excitation and emission wavelengths of 360 and 455 nm, respectively.

The conditions for enzymatic analysis varied with the enzyme under study. In general, the reactions were started by the addition of the enzyme activity, and incubations were at 37°C. The reaction was stopped by the addition of 100% methanol, and precipitated protein was removed by centrifugation. Samples of the supernatant solution were analyzed by HPLC. A chromatogram of a glycosidase activity is shown in Figure 9.74. The peak of free 4-MU is noted.

The enzyme was obtained from urine samples.

9.7.8 Sialidase (Omichi and Ikenaka, 1982)

Sialidases, which are found in microorganisms and animal tissue, cleave sialic acid residues from terminal saccharide residues of carbohydrates. An as-

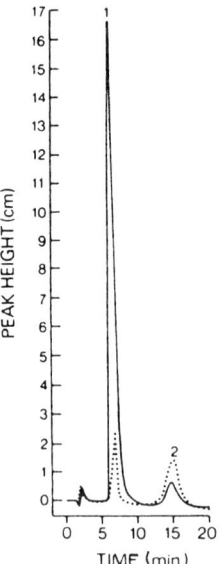

Figure 9.74 Chromatogram of urinary β-N-acetyl-D-glucosaminidase (NAG) assay reaction mixture (solid curve) and assay blank (dotted curve). *Peaks:* 1, 4 MU; 2, 4 MU-NAG. Enzyme source: tenfold diluted urine. Detector sensitivity for the assay blank was twice that of the assay reaction mixture. (From Sandman, 1983.)

say was developed for this activity using the fluorogenic compound α-D-N-acetylneuraminyl-β-D-(2→3)galactopyranosyl-(1→4)-1-deoxy-1-[(2-pyridyl)-aminol]-D-glucitol (PA-sialyllactose).

The assay method involves the separation of the substrate, PA-sialyllactose, from the product, PA-lactose. This is accomplished by gel filtration HPLC on a TSK-Gel LS 410 column. Elution was carried out with a mobile phase of 0.1 M acetic acid. The compounds were detected with a fluorescence detector using excitation and emission wavelengths of 320 and 400 nm, respectively. The separation of the substrate and product is shown in Figure 9.75.

The reaction mixture contained, in 15 μL, the PA-sialyllactose in a sodium acetate buffer at pH 5.0. The reaction was started with the sialidase preparation and incubated at 37°C. At appropriate intervals, 20 μL samples were removed and injected directly for HPLC analysis. The results of a reaction are shown in Figure 9.76. The appearance of the PA-lactose peak is an indication of sialidase activity.

The activity was obtained from urine.

9.7.9 Cytidine Monophosphate–Sialic Acid Synthetase (Petrie and Korytnyk, 1983)

Cytidine monophosphate–sialic acid synthetase (EC 2.7.7.43) catalyzes the activation of N-acetylneuraminic acid (NANA) by CTP to form CMP-NANA.

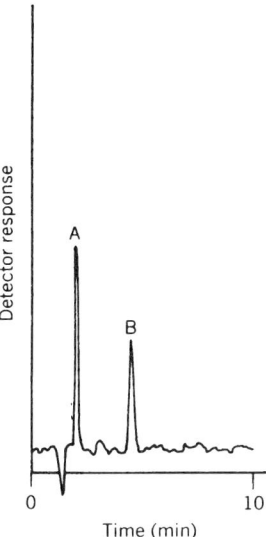

Figure 9.75 Separation of PA-lactose and PA-sialyllactose by HPLC. *Peaks:* A, PA-lactose (0.6 pmol); B, PA-sialyllactose (0.6 pmol). (From Omichi and Ikenaka, 1982.)

The HPLC assay method developed for this activity involves the separation of the three compounds NANA, CTP, and CMP-NANA. However, at the wavelength used for detection, only the last two are detected.

In this assay, separation is carried out by ion-paired reversed-phase HPLC using an Ultrasphere-ODS-IP column with gradient elution from 5% acetonitrile, in a buffer solution (pH 7.5) of 5 mM tetrabutylammonium phosphate and 5 mM sodium phosphate, to 10% acetonitrile in the same buffer. Detection of CMP and CMP-NANA was at 270 nm. The separation obtained is shown in Figure 9.77. The reaction mixture, into which the lyophilized enzyme was homogenized immediately before the assay, contained CTP, glutathione, $MgCl_2$, and Tris-HCl at pH 9.0. The reaction was started by the addition of NANA to the enzyme solution. Incubations were for 30 minutes and were terminated by the addition of cold acetonitrile. The solutions were clarified by centrifugation and by passage through small columns (ODS). The eluant was analyzed by HPLC.

The synthetase was prepared from calf brains by homogenization, centrifugation (1500g for 20 min), and lyophilization of the supernatant solution.

9.7.10 Succinyl–CoA Synthetase (Lambeth and Muhonen, 1993)

Succinyl–CoA synthetase participates in the Krebs cycle by catalyzing the reaction between succinyl–CoA, phosphate, and a nucleoside diphosphate to

296 SURVEY OF ENZYMATIC ACTIVITIES ASSAYED BY THE HPLC METHOD

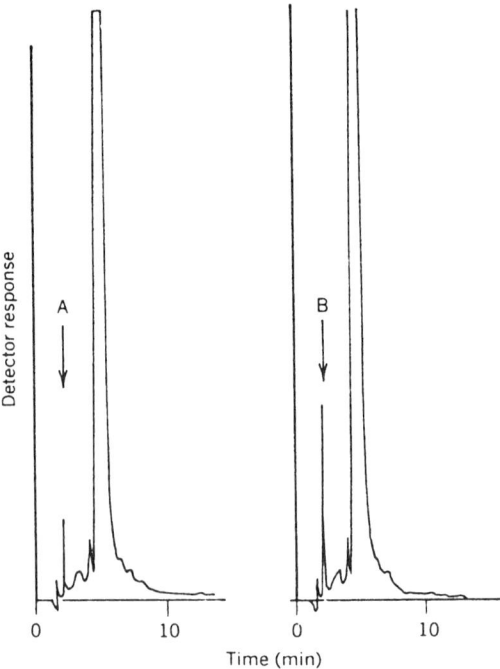

Figure 9.76 Chromatograms of the digest of PA-sialyllactose by urine. PA-sialyllactose (0.167 mM) was incubated with dialyzed urine in 0.1 M sodium acetate buffer (pH 5.0) at 37°C for 2 hours. (A) Substrate + urine heated at 100°C for 10 minutes (control). (B) Substrate + urine. The arrows show the elution position of PA-lactose. (From Omichi and Ikenaka, 1982.)

form the nucleoside triphosphate, succinate, and coenzyme A. In later evolving species, the enzyme is highly specific for GTP and GDP. The enzyme in more primitive species uses ATP and ADP.

Succinyl–CoA, CoA, GTP, GDP, and GMP are separated on a Beckman C_{18} Ultrasphere IP column (4.6 × 250 mm, 5 μm). The mobile phase adjusted to pH 4 contained 20% acetonitrile, 100 mM potassium phosphate, 20 mM potassium acetate, and 5 mM tetrabutylammonium hydroxide. The absorbance of the monitor was followed at 254 nm. ATP-specific enzyme was assayed by substituting ATP for GTP.

The reaction mixture contained in a final volume of 0.3 mL; 50 mM succinate, 33 mM Na-Hepes, (pH 7.4), 5 mM $MgCl_2$, 1 mM CoA, 1 mM GTP, and 5 μg of oligomycin (to inhibit ATPase). The reaction was initiated by adding mitochondria or partially purified enzyme. After 5 minutes of incubation at 30°C, the reaction was stopped by lowering the pH to below 3.5 by adding 0.2 mL of 0.2 M formic acid or 0.2 M HCl. Aliquots of the resulting mixtures were analyzed directly by HPLC. When 10% or less of the CoA was converted

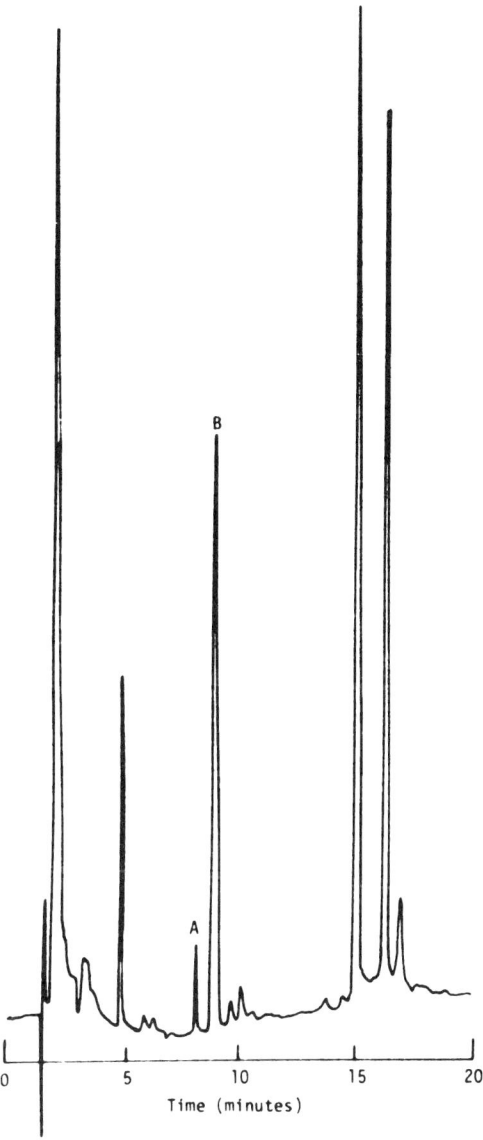

Figure 9.77 HPLC chromatogram of a typical enzyme assay sample with NANA and CTP as substrates. *Peaks:* A, CMP; B, CMP-NANA. (From Petrie and Korytnyk, 1983.)

to succinyl–CoA, the reaction was linear for 10 minutes and with up to 15 μg of mitochondrial protein.

The assay has been used with mitochondrial extracts prepared from freeze-thawed mitochondria from rabbit and pigeon, and to monitor enzyme activity during purification.

Figure 9.78 illustrates the chromatographic separations and the linearity of the assay.

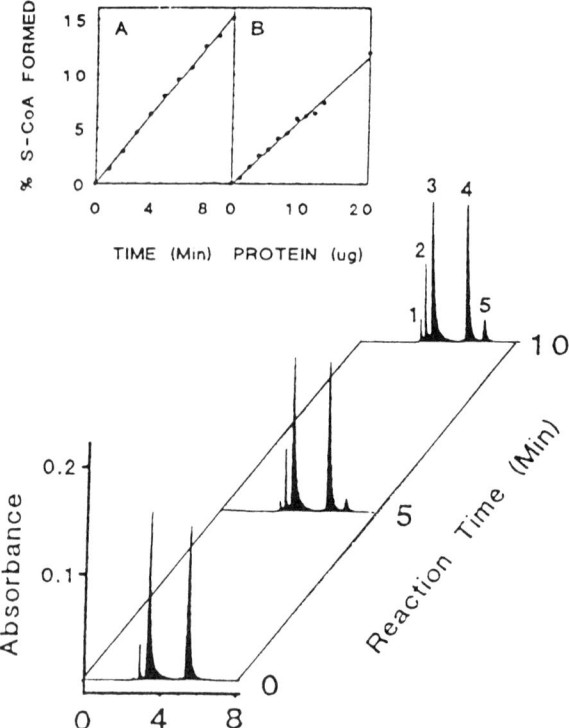

Figure 9.78 HPLC-based assay for succinyl–CoA synthetase from rabbit liver mitochondria. One hundred microliters of diluted supernate was added to 3.0 mL containing 50 mM succinate, 33 mM Na-Hepes, pH 7.4, 5 mM MgCl$_2$, 1 mM CoA, 1 mM GTP, and 5 μg of oligomycin. The reaction was run at 30°C and aliquots of 0.30 mL were removed at 1-minute intervals and transferred to autosampler vials containing 0.2 mL of 0.2 M formic acid. Nucleotides were separated by HPLC. The UV detector was set to 254 nm. The chromatographic profiles for the reaction after 0, 5, and 10 minutes are shown. *Peaks* (top profile) 1, GMP; 2, GDP; 3, GTP; 4, CoA; and 5, succinyl–CoA. (*A*) Linearity of succinyl–CoA (S-CoA) formation, expressed as the percentage conversion of CoA to succinyl–CoA, with time. (*B*) Linearity of S-CoA formation with micrograms of protein added for 5 minute assays carried out in a volume of 0.30 mL. The latter assays were carried out in duplicate. The values shown are averages of the duplicate assays. (From Lambeth and Muhonen, 1993.)

9.7.11 α-Ketoglutarate Dehydrogenase (Shylaja et al., 1990)

α-Ketoglutarate dehydrogenase is a large, multienzyme complex that participates in the Krebs cycle by catalyzing the oxidation of α-ketoglutarate to succinyl–CoA. This assay is based on quantitation of succinyl–CoA.

Succinyl–CoA is separated on a YMC-Pack C-8 column (6 mm × 300 mm, 5 μm). The mobile phase was composed of 0.02 M sodium acetate–0.02 M citrate buffer (pH 3.5), 0.1 mM disodium EDTA, and 15% methanol. The column effluent was monitored at 254 nm.

The reaction mixture contained 80 μL of 130 mM Hepes–67 mM Tris buffer (pH 7.4); 10 μL each (to give final concentration of 1 mM) of NAD, thiamine pyrophosphate, coenzyme A, $MgCl_2$, and dithiothreitol; 20 μL of tissue extract or enzyme source, and 30 μL of bovine serum albumin (1 mg). The reaction was started by adding 20 μL of α-ketoglutarate to give a final concentration of 10 mM. After incubation at 30°C for 1, 5, or 20 minutes for purified enzyme from bovine heart, brain, or liver mitochondria, or platelet homogenates, the reaction was stopped by addition of 20 μL of 60% perchloric acid and the denatured protein was removed by centrifugation. A 10 μL aliquot was used for HPLC analysis.

Sources of enzyme activity were purified bovine heart α-ketoglutarate dehydrogenase from Sigma, mitochondria isolated from brains and livers of adult Mongolian gerbils (2–4 mg protein/mL, stored at −80°C until use), and platelets isolated from heparinized blood that had been homogenized by sonication before use.

Figure 9.79 shows representative chromatograms.

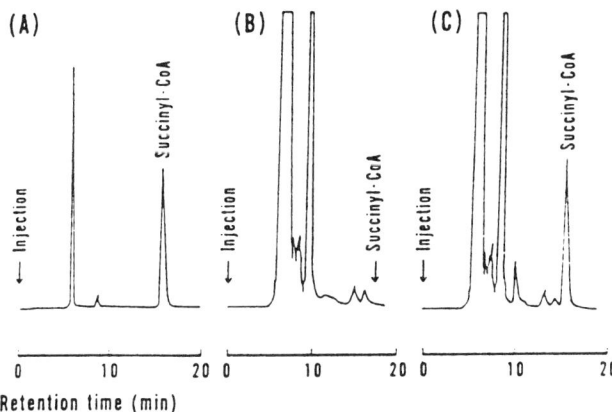

Figure 9.79 Chromatograms of (A) standard succinyl–CoA, obtained by injecting 0.6 nmol of standard solution (retention time for succinyl–CoA peak, 16.3 min), (B) blank, and (C) sample (20 μL of brain mitochondria containing 0.068 mg of protein). Arrow in the blank points to the retention time of succinyl–CoA. (From Shylaja et al., 1990.)

9.7.12 Sucrose Phosphate Synthetase (Salvucci and Crafts-Brandner, 1991)

Sucrose phosphate synthetase catalyzes the reaction of UDP-glucose with fructose-6-P to form sucrose-6-P and UDP. This step is the penultimate step in the synthesis of sucrose in leaves. The chromatographic method can be applied to many UDP-glucose-requiring enzymes. The method eliminates the need for treatment of reaction mixtures with alkaline phosphatase.

Radiolabeled UDP-glucose and sucrose-P were separated on a Selectispher-10 boronate column (5 mm × 250 mm). The mobile phase was 0.12 M sodium phosphate (pH 7.6) delivered at a rate of 1 mL/min. The column eluent was monitored for absorbance at 262 nm, and for radioactivity by a radioactive flow-through detector.

Sucrose phosphate synthetase was assayed at 30°C in a total of 50 μL containing 50 mM Hepes-KOH (pH 7.5), 15 mM MgCl$_2$, 1 mM EDTA, 6 mM UDP-[^{14}C]glucose (1 mCi/mmol), 3 mM fructose-6-P, 15 mM glucose-6-P, and enzyme. Immediately prior to termination by boiling, 90 μL of 50 mM EDTA was added to prevent metal-dependent hydrolysis of UDP-glucose. After 3 minutes at 100°C, the quenched assay was cooled and treated with 0.1 mL of Dowex AG-50W to remove components that might decrease the life of the HPLC column. The mixture was centrifuged, and an aliquot of the supernate was taken for analysis by HPLC. Salvucci and Crafts-Brandner also described assays for sucrose synthetase and UDP-glucose pyrophosphorylase.

Sources of enzyme were tobacco leaf tissue or red beet tubers homogenized at 4°C in 50 mM Hepes-KOH (pH 7.2), 5 mM MgCl$_2$, 1 mM EDTA, 25 mM mercaptoethanol, 1% (w/v)polyvinylpyrollidone-40, 1 mM phenylmethylsulfonyl fluoride, and 10 μM leupeptin. Supernates obtained by centrifugation were desalted and equilibrated with a buffer containing all but the last three components in the homogenization medium.

9.7.13 6-Phosphogluconate Dehydratase (Taha and Deits, 1994)

6-Phosphogluconate dehydratase participates in the Entner–Doudoroff pathway, which plays a primary role in glucose metabolism in many microorganisms. The enzyme catalyzes the dehydration of 6-phosphogluconate to form 2-keto-3-deoxy-6-phosphogluconate.

The product, 2-keto-3-deoxy-6-phosphogluconate, was separated by chromatography at room temperature and a flow rate of 1 mL/min on a Dionex CarboPac PA-1 column (4 mm × 250 mm). The mobile phase was composed of 24 mM NaOH and 300 mM sodium acetate for 5 minutes, followed by a linear gradient to 700 mM sodium acetate in 10 minutes. A linear gradient back to the initial conditions was run in 5 minutes. Pulsed amperometric detection was used with a pulse train consisting of a 480 ms detection pulse at +80 mV, followed by pulses of 120 ms at +600 mV and 60 ms at −600 mV.

The reaction mixture was composed of 10 mM Bicine (pH 8.0) containing 50 mM NaCl, 10 mM β-mercaptoethanol, 2 mM MnCl$_2$, and 1.2 mM 6-

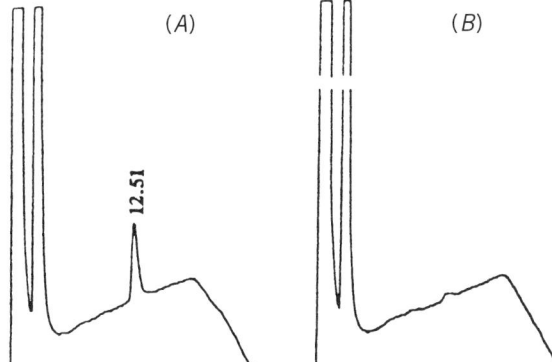

Figure 9.80 Representative chromatographs of Entner–Doudoroff metabolites. (a) 6-Phosphogluconate dehydratase-catalyzed formation of 2-keto-3-deoxy-6-phosphogluconate (12.51 min). (b) Blank run for 6-phosphogluconate dehydratase-catalyzed formation of 2-keto-3-deoxy-6-phosphogluconate, (KDPG); 6-phosphogluconate omitted from assay. (From Taha and Deits, 1994.)

phosphogluconate. Samples were quenched by the addition of trichloroacetic acid to give a final concentration of 5%. After centrifugation, a volume equivalent to 9% of the sample volume, containing 4 M NaOH and 2 M sodium acetate, was added to neutralize the trichloroacetic acid and lower the pH to about 5. A 20 μL sample was used for HPLC analysis. In contrast to a spectrophotometric coupled-enzyme assay, which showed an initial lag phase, the HPLC-based assay was linear for up to 2 minutes. Product formation was also linear with the amount of enzyme added.

Enzyme used in the assays was purified from *Azotobacter vinelandii*. A modification of the assay was used to follow the approach to equilibrium of the 2-keto-3-deoxy-6-phosphogluconate aldolase reaction.

Figure 9.80 shows representative chromatograms.

9.8 STEROID METABOLISM

9.8.1 Δ^5-3β-Hydroxysteroid Dehydrogenase (Suzuki et al., 1980)

The activity Δ^5-3β-hydroxysteroid dehydrogenase catalyzes the conversion of pregnenolone to progesterone, the progestational hormone of the placenta and corpus luteum. The product has an absorption maximum at 240 nm, and therefore detection can be readily carried out by UV absorbances near this wavelength.

In this assay the amount of product formed during the reaction was determined on a reversed-phase HPLC (μBondapak) column containing cyanopro-

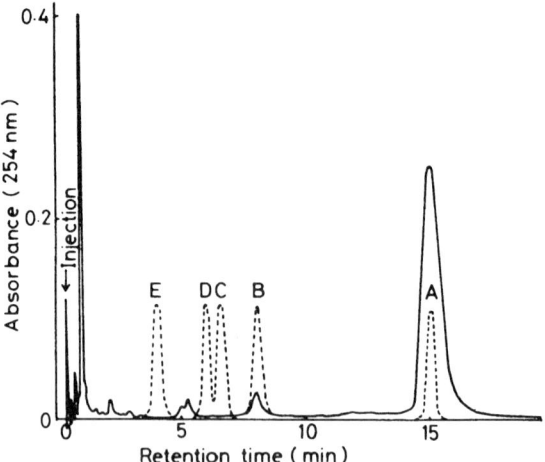

Figure 9.81 HPLC chromatograms of steroids. The elution profiles of five standard Δ^4-3-oxosteroids are illustrated as the broken line. Peaks = A, progesterone; B, 17 α-hydroxyprogesterone; C, androstenedione; D, testosterone; E 11-deoxycortisol. The solid line is a chromatogram of a defatted extract obtained by incubation of pregnenolone with ovarian homogenates. (From Suzuki et al., 1980.)

pylsilane as the functional group. The column was eluted isocratically with a mixture of acetonitrile and water (1 : 4, v/v). Detection was at 254 nm.

The substrate, pregnenolone (158 nmol) dissolved in propylene glycol, was added to the incubation flask containing the enzyme preparation and NAD in a final volume of 5 mL. After an incubation period of 60 minutes, the reaction was terminated by addition of 15 μL of dichloromethane, and radioactive progesterone was added as a recovery standard. The organic phase was recovered, dried, and redissolved in 70% ethanol, and a sample was analyzed by HPLC. Figure 9.81 shows an analysis of an incubation mixture with and without incubation with the ovarian preparation. The substrate (not seen) is converted exclusively to progesterone.

The ovaries of rats were used in the preparation of the active fraction.

9.8.2 11-β-Hydroxylase and 18-Hydroxylase (Gallant et al., 1978)

The 11-β-hydroxylase found in the adrenal cortex catalyzes the hydroxylation of 11-deoxycorticosterone to corticosterone. The enzyme requires NAD as a cofactor and contains heme as the prosthetic group.

The substrate, 11-deoxycorticosterone, was separated from the two reaction products, corticosterone (11-β-hydroxylation) and 18-hydroxyl-11-deoxycorticosterone (18-hydroxylation), on reversed-phase HPLC (MicroPak silica) with a mobile phase of 16% tetrahydrofuran in water. Figure 9.82A shows a chromatogram of the separation of the authentic steroids.

Hydroxylase activity was determined in a reaction mixture containing mitochondrial protein and 11-deoxycorticosterone (60 μM). The reaction was started by the addition of 10 mM isocitrate as the source of reducing equivalents. At intervals during the incubation, samples were removed

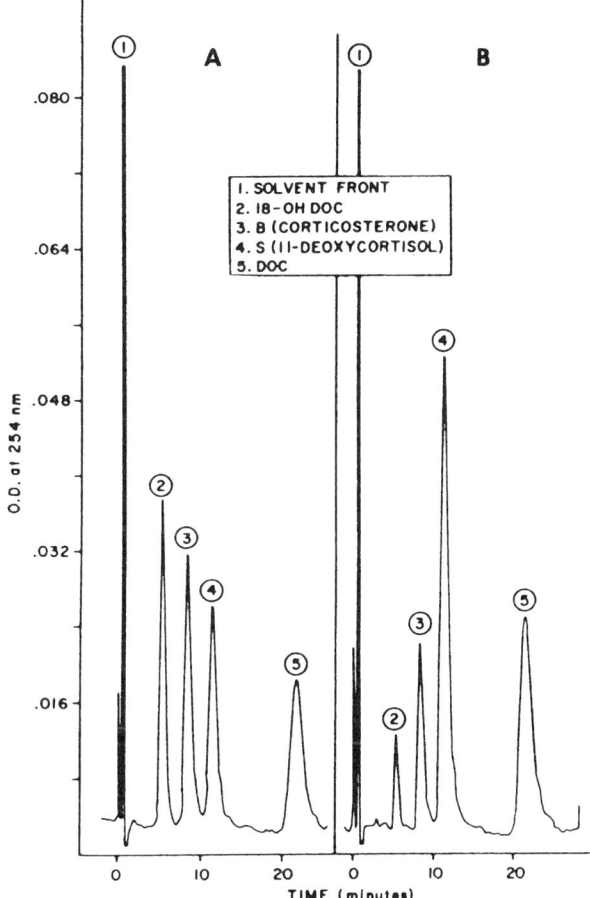

Figure 9.82 (A) HPLC chromatogram obtained with 10 μL of a mixture of authentic steroid standards (1.0 μg of each) on a Varian 0.22 mm × 25 cm MicroPak silica column with 16% tetrahydrofuran in water. The flow rate was 60 mL/h at 300 atm. (B) Typical chromatographic profile obtained with extracts of rat adrenal mitochondria incubated with 60 μM 11-deoxycorticosterone. An aliquot (0.8 mL) was removed at 5 minutes of incubation and extracted with 10 mL of methylene chloride containing 10 μg of internal standard (11-deoxycortisol). The extract was evaporated to dryness and solubilized with 50 μL of ethanol. Then 10 μL was injected onto the column. (From Gallant et al., 1978.)

and placed into methylene chloride containing 11-deoxycortisol as an internal standard and extracted. After further processing of the extracts, the samples were injected, and the amount of 11-β- and 18-hydroxylation reaction products were determined from peak areas. Figure 9.82B shows a chromatogram of a sample taken after 5 minutes of incubation. The two products and the unreacted substrate as well as the internal standard are shown.

The hydroxylase activity was assayed using intact mitochondria obtained from the adrenal cortex of the rat.

9.8.3 25-Hydroxyvitamin D_3-1α-Hydroxylase (Tanaka and DeLuca, 1981)

This hydroxylase carries out the 1-hydroxylation of the compound 25-hydroxyvitamin D_3 (25-OH-D_3) to form the product 1,25-dihydroxyvitamin D_3 [1,25-$(OH)_2$-D_3]. The HPLC assay developed replaces those using radiolabeled substrates.

Separation was carried out by normal phase HPLC on a Zorbax Sil column with a solvent of 10% 2-propanol in hexane. The column was eluted isocratically and monitored at 254 nm.

The reaction mixture contained 1 mL of tissue homogenate, sucrose, Tris–acetate (pH 7.4), magnesium acetate, and sodium succinate. The reaction was initiated by the addition of the substrate 25-OH-D_3 and at intervals was stopped by the addition of a methanol–chloroform mixture (2 : 1). The sample was prepurified and analyzed by HPLC.

The hydroxylase was obtained from both chicken kidney and liver by homogenization.

9.8.4 Cholesterol 7α-Hydroxylase (Hylemon et al., 1989)

Cholesterol 7α-hydroxylase, a cytochrome P450–dependent enzyme, catalyzes the first and rate-limiting step in the biosynthesis of bile acids from cholesterol. Potential mechanisms for regulation of the enzyme have been extensively studied.

In work published in 1989, Hylemon et al. used cholesterol oxidase to convert 7α-hydroxycholesterol to 7α-hydroxy-4-cholesten-3-one. Cholesterol that remained was converted to 4-cholesten-3-one. 7β-Cholesterol, which was added as an internal steroid recovery standard, was oxidized to 7β-hydroxy-4-cholesten-3-one. These steroid products were analyzed by C_{18} reversed-phase chromatography on an Altex Ultrasil-ODS column (4.6 mm × 25 cm) using 70 : 30 (v/v) mixture of acetonitrile and methanol (Fig. 9.83). The eluate was monitored at 240 nm, and the amount of product determined from a calibration curve.

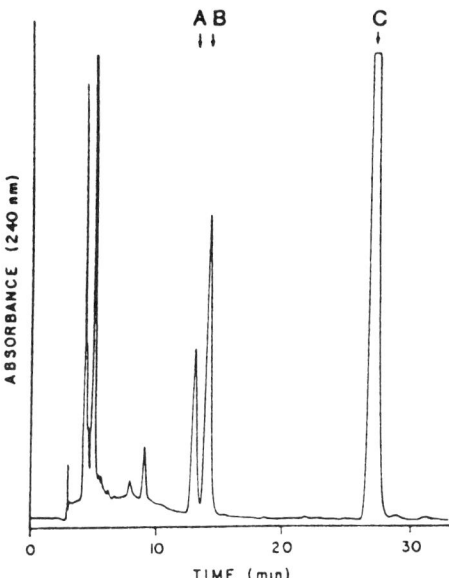

Figure 9.83 Separation of 7α-hydroxy-4-cholesten-3-one (A), 7β-hydroxy-4-cholesten-3-one (B), and 4-cholesten-3-one (C) by C-18 reversed-phase HPLC.

One milligram of microsomal protein was added to 0.1 M potassium phosphate buffer (pH 7.4), 50 mM NaF, 5 mM dithiothreitol, 1 mM EDTA, 20% glycerol, and 0.015% CHAPS. After a 5-minute incubation at 37°C, the reaction was initiated by adding a NADPH regenerating system consisting of (final concentrations) 5 mM isocitrate, 5 mM MgCl$_2$, 0.5 mM NADPH, and 0.075 unit of isocitrate dehydrogenase. The final volume was 1.0 mL. The reaction was allowed to proceed for 20 minutes before being terminated by adding 30 µL of 20% sodium cholate and 1 µg 7β-hydroxycholesterol as an internal recovery standard. Then, 44 µL of 0.1% cholesterol oxidase in 10 mM potassium phosphate buffer containing 1 mM dithiothreitol and 20% glycerol was added. The mixture was incubated 10 minutes at 37°C and the reaction terminated by adding 2 mL of 95% ethanol. Cholesterol metabolites were extracted by adding 6 mL of petroleum ether, vortexing and centrifuging. The ether layer was collected and dried under nitrogen at 40°C. The extraction was repeated three times and the residues resuspended in 0.10 mL of the mobile phase; 20 µL was injected. Cholesterol 7 α-hydroxylase activity was linear for 0.5 to 3 mg of microsomal protein and for up to 40 minutes using 1 mg of microsomal protein.

Microsomes from rat liver were prepared in 100 mM potassium phosphate buffer (pH 7.2) containing 100 mM sucrose, 50 mM KCl, 50 mM NaF,

5 mM EGTA, 3 mM dithiothreitol, 1 mM EDTA, 1 mM PMSF, and 100 μM leupeptin.

9.8.5 3β-Hydroxy Δ^5-C_{27}-steroid Oxidoreductase (Hylemon et al., 1991)

The second step in the synthesis of bile acids, according to Hylemon et al. (1991), is the conversion of 7α-hydroxycholesterol to 7α-hydroxy-4-cholesten-3-one by NAD$^+$-dependent 3β-hydroxy-Δ^5-C_{27}-steroid oxidoreductase. This enzyme is located in the endoplasmic reticulum of liver, and its catalysis of the 3β-hydroxy group also results in isomerization of the double bond from Δ^5 to Δ^4.

The steroid products are separated by reversed-phase chromatography on a Beckman Ultrasphere ODS (4.6 mm × 25 cm, 5 μM) column equlibrated with 70:30 (v/v) acetonitrile–methanol. The absorbance is monitored at 240 nm and the amount of product is determined from a calibration curve.

One milligram of microsomal protein is added to 0.1 M potassium phosphate buffer (pH 7.4) containing 50 mM NaF, 10 mM dithiothreitol, 1 mM EDTA, 20% glycerol (v/v), 150 μM 5-cholestene-3β, 7α-diol, and 0.915% CHAPS. The reaction is initiated by 1 mM NAD$^+$ to give a final reaction volume of 1.0 mL. After incubation at 37°C for 5 minutes, the reaction is terminated by adding 2 mL of 95% ethanol. An internal recovery standard, 4-cholesten-3-one (3 μg in methanol) is also added. The steroid products are extracted into 5 mL of petroleum ether (repeated twice). After the ether has been removed at 40°C under a stream of nitrogen, the products are dissolved in 100 μL of mobile phase and 20 μL is injected into the column. The amount of product formed is linear with protein (to 1.5 mg) and with time (up to 10 min, 1 mg protein). The assay is much more sensitive than the direct spectrophotometric assay, and it avoids the use of thin-layer chromatography and radioisotopes described in other methods.

The source of enzyme is rat liver microsomes prepared by standard techniques.

Figure 9.84 shows a representative chromatogram.

9.8.6 Cytochrome P450$_{scc}$ (Sugano et al., 1989)

The side chain cleavage of cholesterol, producing pregnenolone, is catalyzed by cytochrome P450$_{scc}$. This is the initial step in the biosynthesis of several steroid hormones. In this assay, the initial product, pregnenolone, is quantitatively converted to progesterone by treatment with cholesterol oxidase, which increases by about 10-fold the sensitivity of the assay.

Cholestenone (derived from unreacted cholesterol) and progesterone were separated from deoxycorticosterone acetate (internal standard) by normal phase chromatography on a TSK-gel silica 150 column (4 mm × 250 mm)

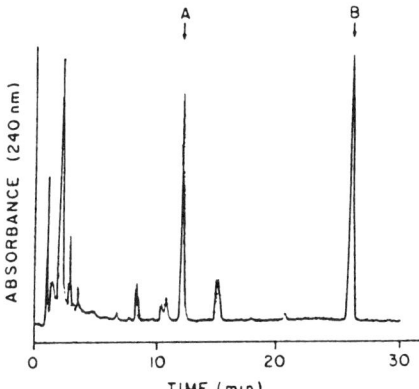

Figure 9.84 Separation of 7α-hydroxy-4-cholesten-3-one (A) from 4-cholesten-3-one (B) by C_{18} reversed-phase HPLC. Microsomes (1 mg) prepared from cholestyramine-fed rats. (From Hylemon et al., 1991.)

from Tosoh. The mobile phase of *n*-hexane–isopropanol (100 : 2, v/v) was run at a flow rate of 1.5 mL/min. The effluent was monitored at 240 nm.

The cholesterol side chain cleavage reaction was carried out in 0.9 mL containing 20 mM potassium phosphate buffer (pH 7.4), 0.3% (w/v) Tween 20, 100 nmol of cholesterol dispersed in 10 μL of ethanol, 70 pmol of cytochrome $P450_{scc}$, 10 nmol of adrenodoxin, 1 nmol of adrenodoxin reductase, 5 μmol G6P, 0.5 unit of G6P dehydrogenase, and 4 μmol of $MgCl_2$. The reaction was initiated by adding 100 nmol of NADPH. After incubating at 37°C for 5 minutes, the reaction was terminated by heat treatment. Pregnenolone was converted to progesterone by adding 100 μL of cholesterol oxidase (0.4 U) dissolved in 20 mM potassium phosphate buffer (pH 7.4) containing 1% sodium cholate. After incubation at 37°C for 10 minutes, the steroids were extracted into dichloromethane. Deoxycorticosterone acetate (5 nmol) was added as an internal standard. The extracts were dried under nitrogen before reconstitution and analysis by normal phase HPLC. The amount of product was determined from a standard curve. Production of progesterone was linear with amounts of P450 up to 150 pmol.

Cytochrome $P450_{scc}$ was purified from mitochondria isolated from bovine adrenal cortex by a published protocol.

Figure 9.85 shows chromatograms.

9.8.7 Steroid 17α-Hydroxylase/C_{17-20} Lyase (Cytochrome $P450_{21scc}$) (Schatzman et al., 1988)

Cytochrome $P450_{21scc}$ is involved in the syntheses of both glucocorticoids and androgens.

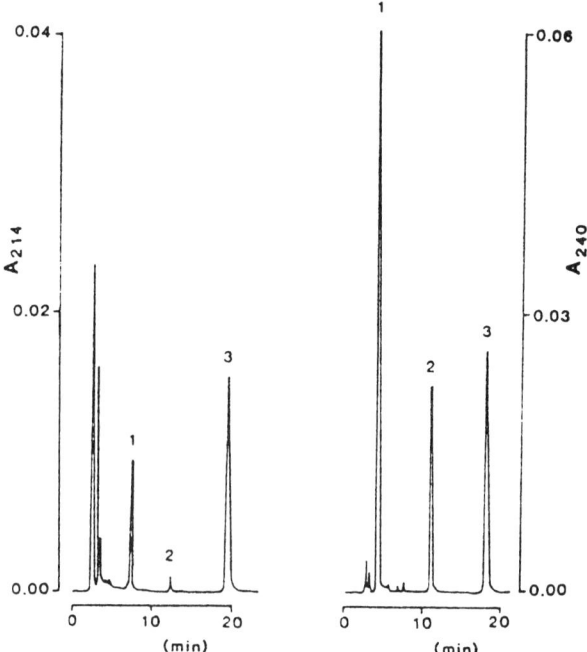

Figure 9.85 Normal phase HPLC profiles of the reaction product of the cholesterol side chain cleavage system. Peaks were identified on the basis of their retention times. (A) Without cholesterol oxidase treatment. Cholesterol (100 nmol) was incubated with cytochrome P450$_{scc}$ (70 pmol) in the presence of adrenodoxin, adrenodoxin reductase, and an NADPH-generating system. Monitoring was at 214 nm. *Peaks:* 1, cholesterol; 2, pregnenolone; 3, deoxycorticosterone acetate (internal standard) (B) The reaction mixture of (A) was further incubated with cholesterol oxidase at 37°C for 10 minutes. Monitoring was at 240 nm. *Peaks:* 1, cholestenone; 2, progesterone; 3, deoxycorticosterone acetate (internal standard). (From Sugano et al., 1989.)

Three variants of one protocol are used to separate 17α-hydroxypregnenolone and its metabolites, 17 α-hydroxyprogesterone and its metabolites, and pregnenolone and its metabolites. The compounds have been separated on a silica gel column using a THF–hexane mobile phase in conjunction with a Flo-One Model HS radiometric detector from Radiomatic Instruments.

The C_{17-20} lyase activity of the enzyme is determined by using 0.8 μM [7-^{3}H] 17α-hydroxypregnenolone (human) or [1,2-^{3}H] 17α-hydroxyprogesterone (rat), 1 mM NADPH, 5 mM glucose-6-phosphate, glucose-6-phosphate dehydrogenase (1 I.U/mL), and 0.02 mg of microsomal protein, in 100 μL of phosphate buffer, pH 7.4. Incubations were carried out at 34°C for 6 minutes. Reactions were stopped by adding 5 mL of 2 : 1 CHCl$_3$–methanol, 0.9 mL water, and 5 μg each of cold substrate and reaction products as carriers in 5 μL of CHCl$_3$.

The assay for 17 α-hydroxylase was carried out under the same conditions except pregnenolone (human) or progesterone (rat) was used as the substrate. For either assay, the quenched reactions are shaken for 5 minutes and then centrifuged to separate the layers. The upper phase was discarded and the interface was washed with $CHCl_3/MeOH/H_2O$ (3 : 48 : 47). Following removal of the wash, the lower phase was evaporated to dryness at 40°C under nitrogen. The extracted metabolites were dissolved in the mobile phase.

The enzyme preparations used were microsomes prepared from human or rat testes.

9.9 PURINE METABOLISM

9.9.1 Nicotinate Phosphoribosyltransferase (Hanna and Sloan, 1980)

Nicotinate phosphoribosyltransferase catalyzes the formation of nicotinate mononucleotide (N_aMN) and pyrophosphate from 5-phosphoribosyl-α-D-pyrophosphate (PRibPP) and free nicotinic acid. The reaction requires the hydrolysis of ATP to ADP.

In this study, the reactants ATP, N_aMN, and ADP were separated by reversed-phase HPLC (μBondapak C_{18}) eluted isocratically with a mobile phase of 25 mM $(NH_4)_3PO_4$ (pH 8.0). Figure 9.86 illustrates the separation obtained under these conditions.

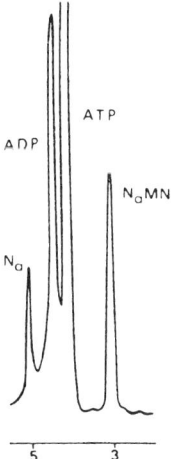

Figure 9.86 Elutions of a stock solution of ATP, ADP, nicotinate, and N_aMN through a μBondapak C_{18} column using 25 mM $(NH_4)PO_4$ at pH of 8.0. Stock solutions of each reactant were used to assign the peaks. Elution conditions: 5 μL sample injection volumes, 0.7 mL/min flow rate, 25°C. (From Hanna and Sloan, 1980.)

310 SURVEY OF ENZYMATIC ACTIVITIES ASSAYED BY THE HPLC METHOD

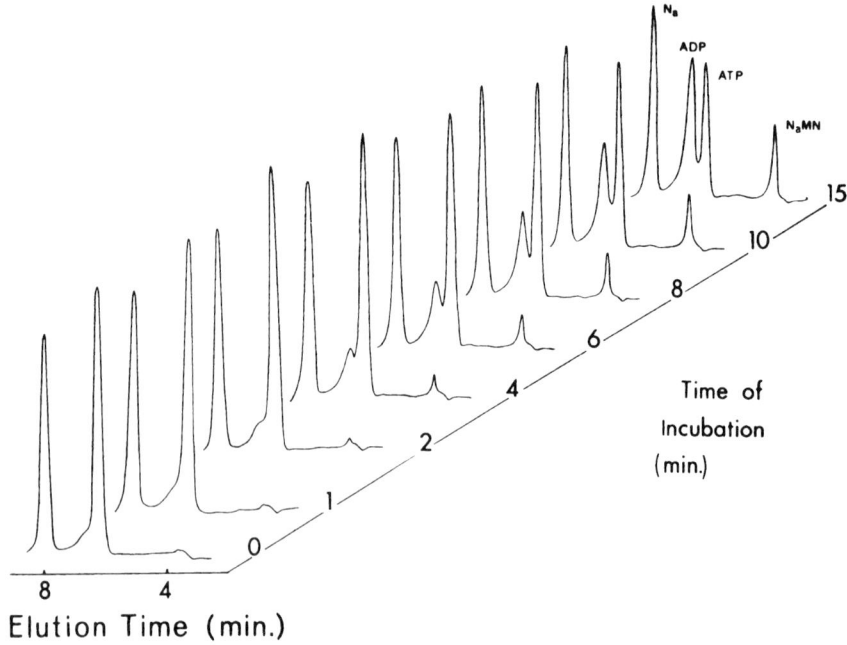

Figure 9.87 Elutions of the nicotinate phosphoribosyltransferase (N-PRTase) assay solution through a μBondapak C_{18} column after various enzyme incubation times. The incubation mixture contained 5 mM $MgCl_2$, 100 μM nicotinate, 75 μM ATP, 30 μM PRibPP, and 25 μL of 4 mg/mL N-PRTase in 50 mM Tris-HCl (pH 8). Elution conditions: 5 μL sample injection volumes, 0.7 mL/min flow rate, 25 mM $(NH_4)PO_4$ (pH 8) elution buffer, 25°C. (From Hanna and Sloan, 1980.)

The assay mixture contained in a final volume of 5 mL of $MgCl_2$, nicotinate, PRibPP, ATP, and Tris-HCl (pH 8.0). The reaction was started by the addition of enzyme (100 μg). At suitable intervals, 0.5 mL samples were removed and placed in a boiling water bath to terminate the reaction. After centrifugation, the supernatant solutions were filtered. Volumes of 5 μL were removed and analyzed. Figure 9.87 shows a chromatogram of samples removed at various times during the incubation. The formation of the two products, ADP and N_aMN, is clearly observed.

The enzyme was purified from yeast and was free of N_aMN adenyltransferase, ATPase, and adenylate kinase activities.

9.9.2 5′-Nucleotidase (Sakai et al., 1982; Amici et al., 1994)

We discuss two assays for the measurement of pyrimidine 5′-nucleotidase activity. In the first, as described by Sakai et al. (1982), a pyrimidine nucleoside 5′-phosphate is hydrolyzed to form the corresponding pyrimidine nucleoside.

The activity is found in erythrocytes, platelets, and lymphocytes, and determination of its value aids in diagnosis of some blood disorders. In this assay, which can readily be used for purine and pyrimidine 5′- and 3′-nucleotidase activities, the nucleoside monophosphate (the substrate) was separated from the nucleoside (the product) using ion-pair reversed-phase HPLC with a mobile phase of 5% methanol–5 mM potassium dihydrogen phosphate; 0.25 mM 1-decanesulfonic acid was also added to the mobile phase. The elution was carried out at room temperature and the eluent monitored at 254 nm.

The reaction mixture contained Tris-HCl-buffered UMP (purified free of uridine by ion-exchange chromatography) as the substrate and $MgCl_2$. The reaction was started by the addition of the enzyme, and the incubation was carried out at 37°C for 60 minutes. The reaction was terminated by placing the reaction tubes in a boiling water bath for 3 minutes. After dilution and centrifugation, the supernatant solution was analyzed by HPLC.

Figure 9.88 shows the separation of UMP and uridine with and without the enzyme. Plasma and erythrocytes from both normal subjects and those with a variety of pathologies were used as enzyme sources. The formation of uridine during the reaction was observed.

Erythrocytes from normal subjects were collected, washed, and lysed by dilution in distilled water. This lysate solution was used directly as the source of the enzyme.

In the assay described by Amici et al. (1994), a wide variety of pyrimidine and purine nucleoside 5′-monophosphates were separated from their nucleosides by chromatography on a Supelco LC_{18} guard column (4.6 mm × 20 mm, 5 μm). The short column allows separations in less than a minute. The mobile phase was 0.1 M potassium phosphate buffer (pH 6.0) except in the case of adenosine and deoxyadenosine, when 5% methanol was also included. The flow rate was 2 mL/min. Compounds were detected by monitoring the effluent at 254 nm, although sensitivity could be improved in some cases by using a different wavelength.

The standard assay mixture contained 50 mM Tris-HCl (pH 7.5), 1 mM $MgCl_2$, 1 mM dithiothreitol, 0.1 to 10 mM substrate, and the appropriate amount of enzyme in a final volume of 500 μL. Cytidine monophosphate (CMP) at an initial concentration of 0.2 mM was used to assay red blood cell lysates. Incubations were carried out at 37°C for either 30 or 60 minutes before the reaction was stopped by adding 100 μL of assay mixture to 50 μL of ice-cold 1.2 M perchloric acid. After the tubes had cooled for 10 minutes in an ice bath, proteins were removed by centrifugation and 130 μL of the supernate was neutralized by the addition of 35 μL of 1 M K_2CO_3. Potassium perchlorate was removed by centrifugation before analysis by HPLC.

Enzyme activity was measured in lysed erythrocytes isolated from heparinized blood.

Figure 9.89 shows HPLC chromatographic analysis of an enzymatic reaction.

312 SURVEY OF ENZYMATIC ACTIVITIES ASSAYED BY THE HPLC METHOD

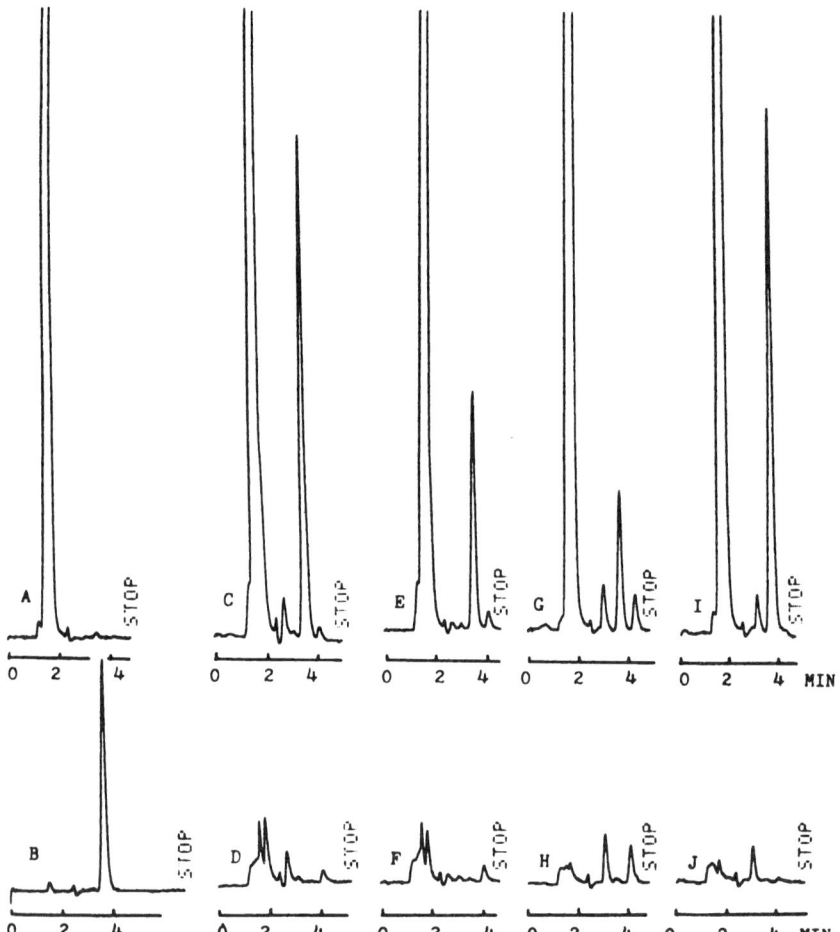

Figure 9.88 Separation of uridine from UMP and blood components. (*A, B*) Chromatograms of UMP in the reaction mixture without enzyme and uridine standard (50 μM), respectively. Samples were incubated with (*C, E, G, I*) or without (*D, F, H, J*) UMP. (*C, D*) Erythrocytes from a normal subject. (*E, F*) Erythrocytes from a lead-poisoned subject. (*G, H*) Plasma from a person suffering from hepatobiliary disease. (From Sakai et al., 1982.)

9.9.3 Alkaline and Acid Phosphatase (Rossomando et al., 1981a, 1983; Togari et al., 1987)

Alkaline and acid phosphatase are organ-specific enzymes that are assayed in the diagnosis of many diseases. These activities are phosphomonoesterases that dephosphorylate a number of compounds, including nucleoside monophosphates, to their respective nucleosides and free phosphates. However, such dephosphorylations have traditionally been assayed with 4-nitrophenyl

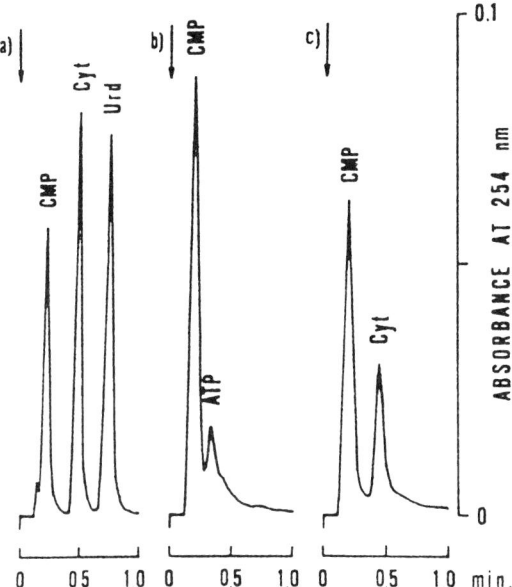

Figure 9.89 Pyrimidine 5'-nucleotidase assay in undialyzed human erythrocyte lysate obtained by 1:5 dilution of packed cells with deionized water. (*a*) Separation of 1 nmol of CMP, cytidine, and uridine as the standards. Separation of the assay mixture, containing 50 mM Tris-HCl, pH 7.5, 0.2 mM CMP, 1 mM MgCl$_2$, 1 mM DTT, and 20 μL of lysate in 0.5 mL of total volume (*b*) at time zero and (*c*) after 30 minutes of incubation (*c*). (From Amici et al., 1994.)

phosphate as the substrate, using a continuous spectrophotometric method as described in Chapter 1 (Section 1.3.1). More recently, the HPLC method has been used together with a nucleoside monophosphate such as AMP as the substrate. In an assay of Rossomando et al. (1983), the formation of adenosine during the course of the reaction was monitored at 254 nm.

Substrate and product were separated by reversed-phase HPLC (μBondapak) using a mobile phase of a phosphate buffer at pH 5.5 with 1% methanol. The column was eluted isocratically, and the detection was at 254 nm.

The reaction mixture contained the substrate and buffer, and the reaction was started by the addition of the enzyme. In one study, the substrate was formycin 5'-monophosphate, a fluorescent analog of AMP. The formation of formycin A, the analog of adenosine, is shown in Figure 9.90 as a function of incubation time.

In an interesting application of this assay to the question of reaction mechanisms, the substrate AMP was replaced by the thioanalog 5'-deoxy-5'-thioadenosine monophosphate [A(S)MP]. The structure of A(S)MP is shown in Figure 9.91. (This analog is available from Calbiochem-Behring.) With this analog it was possible to explore the question of whether the enzyme cleaved between the C-5' and the oxygen atoms or between the oxygen atom and the

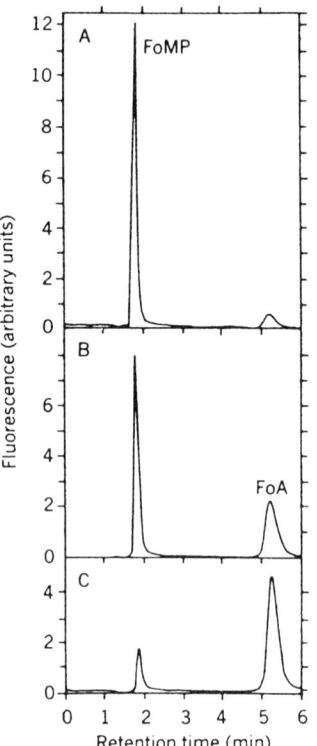

Figure 9.90 HPLC analysis of a reaction mixture containing AMP and alkaline phosphatase. Tracings obtained of reaction mixture (A) immediately after the addition of enzyme, (B) after 10 minutes, and (C) after 15 minutes. (From Rossomando et al., 1981a.)

phosphorus atom. These alternatives are illustrated in Figure 9.92. It is clear that the site of cleavage can be distinguished, since the alternatives will produce different reaction products: in one case thioadenosine and phosphate, in the other, thiophosphate and adenosine.

Since thioadenosine is readily separated from adenosine by HPLC, the use of this analog together with the HPLC assay method allowed the site of cleavage to be established. As shown in Figure 9.93A, an analysis of the incubation mixture during the course of the reaction revealed the formation of thioadenosine. No adenosine was detected. These findings supported the conclusion that the site of cleavage was the bond between the sulfur and the phosphate.

This analog proved to be useful in another way as well when it was found that the analog was not a substrate for a 5'-nucleotidase, since, as shown in Figure 9.93B, incubation of the thio analog with this activity produced no

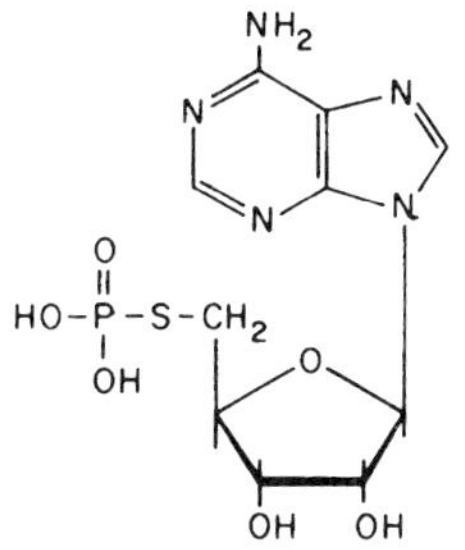

5'-DEOXY-5'-THIOADENOSINE 5'-MONOPHOSPHATE
[A(S)MP]

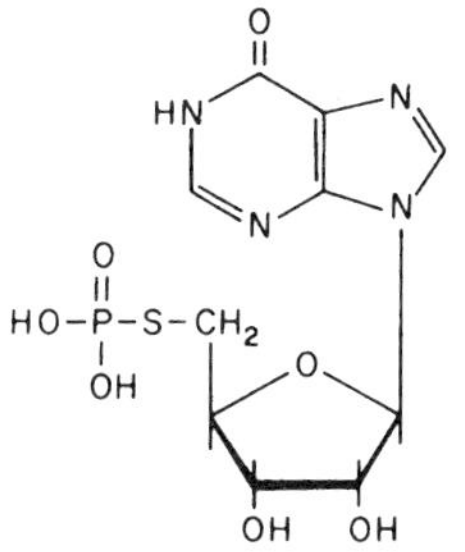

5'-DEOXY-5'-THIOINOSINE 5'-MONOPHOSPHATE
[I(S)MP]

Figure 9.91 Structure of the thio analogs of AMP and IMP in which sulfur replaces the bridge oxygen between the 5' carbon and the phosphorus. (From Rossomando et al., 1983.)

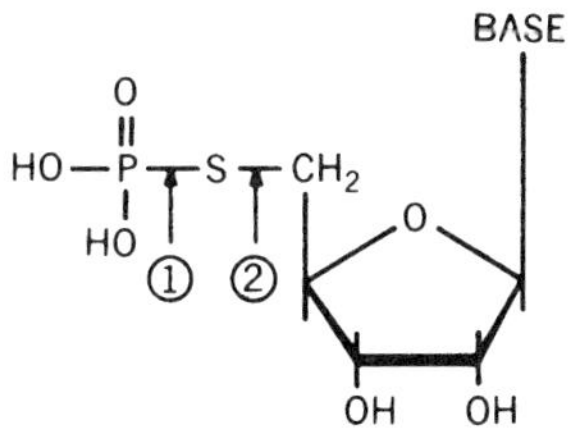

1. A(S)MP $\xrightarrow{PME}$ (S)Ado + P_i
2. A(S)MP $\xrightarrow{PME}$ Ado + (S)P_i

Figure 9.92 Sites of bond cleavage by phosphomonoesterases. Arrow 1 indicates cleavage of the bond between the phosphorus and sulfur atoms. The products are shown in reaction (1). Arrow 2 indicates cleavage between the carbon and the sulfur, and the reaction products are shown in reaction (2).

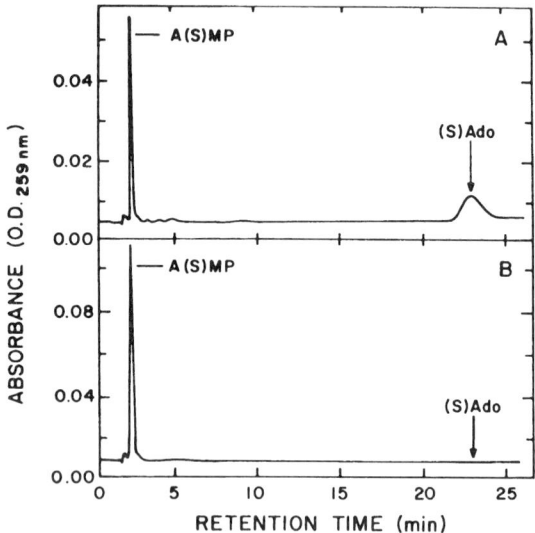

Figure 9.93 HPLC chromatograms of phosphomonoesterase hydrolysis of A(S)MP. (A) Chromatogram obtained from calf intestinal mucosa alkaline phosphatase hydrolysis of A(S)MP. In a reaction volume of 100 μL containing 100 mM Tris-HCl (pH 8.1), 300 μM A(S)MP, and 20 mM MgCl$_2$, the reaction was initiated by addition of 2 μg of enzyme and incubated at 30°C for 6 hours. A 20 μL sample was then injected onto the HPLC column and analyzed. (B) Chromatogram obtained from snake venom 5'-nucleotidase incubated with A(S)MP. In a reaction volume of 100 μL containing 100 mM Tris-Cl (pH 8.1), 300 μM A(S)MP, and 20 mM MgCl$_2$, the reaction was initiated by addition of 6 μg of enzyme and the reaction mixture incubated at 30°C for 60 minutes, and a 20 μL sample was injected onto the HPLC column and analyzed. (From Rossomando et al., 1983.)

reaction product. These results suggest the possibility that this thio analog and the HPLC assay method may be useful for discriminating between the 5'-nucleotidase and alkaline phosphatase activities.

As mentioned, the foregoing activities can also be assayed using phenylphosphate as substrate. In the assay described by Togari et al. (1987), electrochemical detection provided high sensitivity.

In this assay, the product, phenol, was separated by chromatography on a Develosil ODS-7 analytical column (4.6 mm × 150 mm). The mobile phase was a 70:30 mixture of 10 mM acetate buffer (pH 4.0) and methanol at a flow rate of 1.5 mL/min. For detection of phenol, the electrode potential was set at 1.2 V against an Ag/AgCl reference electrode.

The standard incubation mixture for assay of alkaline phosphatase, contained in a total 200 μL: 5 mM disodium phenylphosphate, 50 mM carbonate buffer (pH 10.2), saliva, and distilled water. The reaction was started by addition of saliva and was carried out at 37°C for 30 minutes. The reaction was terminated by adding 50 μL of 25% trichloroacetic acid. After centrifugation, a 20 μL aliquot of the supernate was analyzed by HPLC. The assay of acid phosphatase was carried out in the same manner except that 50 mM citrate (pH 4.8) was used as the buffer. The formation of phenol was linear for up to 60 minutes and 40 μL of saliva as enzyme source.

Whole saliva and duct saliva from parotid and sublingual glands were used as the sources of the acid and alkaline phosphatases studied.

9.9.4 Adenosine Deaminase (Uberti et al., 1977; Hartwick et al., 1978)

Adenosine deaminase catalyzes the deamination of adenosine to form inosine and ammonia. The inosine (Ino) can be degraded further to hypoxanthine (Hyp) by nucleoside phosphorylase, an activity often present in extracts. Therefore, in many cases, the assay involves a determination of either the loss of adenosine (Ado) or the formation of both inosine and hypoxanthine. An early study by Uberti et al., 1977, was followed by another by Hartwick et al., 1978.

In the study by (Hartwick et al., 1978), the compounds were separated by reversed-phase HPLC using columns prepacked with C_{18} μBondapak or Partisil ODS. Compounds were eluted isocratically using a mobile phase of methanol and 10 mM KH_2PO_4 (14:86, v/v) with no further pH adjustment. The separations obtained using these conditions are shown in Figure 9.94A.

The activity was obtained from a lysate of red blood cells. The reactions were terminated with a boiling water bath (45 s), and the samples clarified by centrifugation. Samples of 5 μL were analyzed. Figure 9.94B, C, D shows the chromatographic profiles obtained after incubation times of 3, 30, and 50 minutes, respectively, with the enzyme. The loss of adenosine is noted, but the effect of the nucleoside phosphorylase is seen, since hypoxanthine, not inosine, is the final product.

9.9.5 AMP Deaminase (Jahngen and Rossomando, 1984; Raffin and Thebault, 1991)

The enzyme adenylic acid deaminase catalyzes the deamination of AMP to IMP and ammonia. For the HPLC method, the assay involves the separation of the substrate, AMP, from the reaction product IMP. The enzyme is found in muscle.

In the assay of Jahngen and Rossomando (1984), AMP and IMP were separated by means of ion-paired reverse-phased HPLC on a C_{18} (μBondapak) column with a mobile phase of 65 mM potassium phosphate (pH 3.6), 1 mM tetra-n-butylammonium phosphate, and 4% methanol. The column was eluted isocratically, and the eluent was monitored at 254 nm. When the formycin analogs were used, detection was at 295 nm; at this wavelength there was no

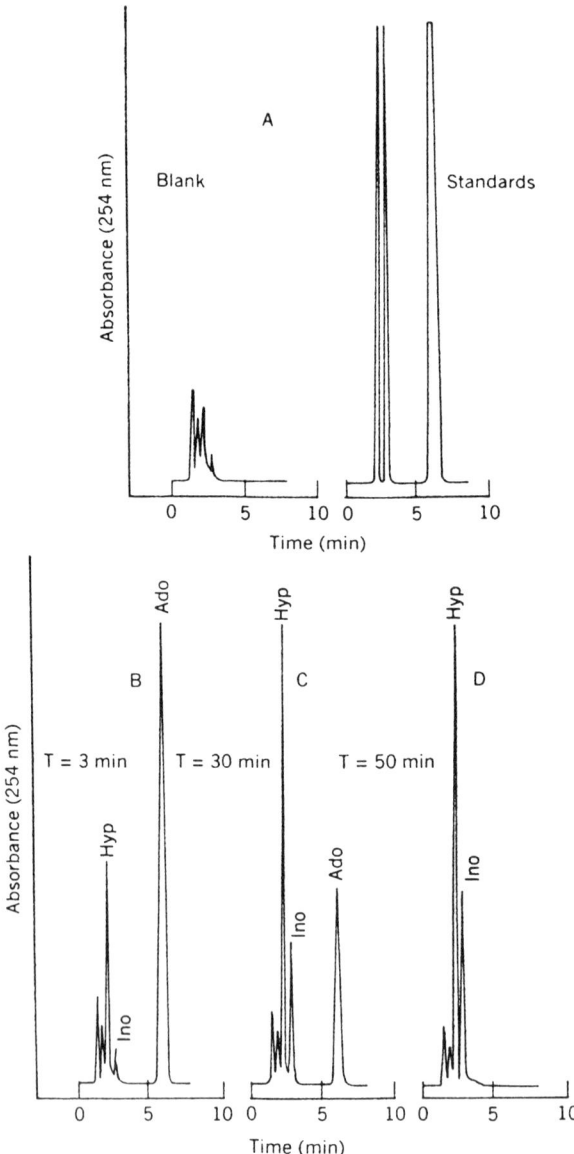

Figure 9.94 The separation of the components of human erythrocytes by reversed-phase HPLC. (*A*) Blank erythrocyte lysate with three standards: Hyp, Ino, and Ado. (*B*, *C*, and *D*) Decrease in the substrate peak area (Ado) as a function of time. Chromatographic conditions: isocratic elution; flow rate, 2.0 mL/min, mobile phase: 0.01 F KH_2PO_4 (pH unadjusted) and methanol (86:14, v/v). In each chromatogram, the injection volume was 5 µL, at an attenuation of 64 on a Hewlett-Packard integrator. (From Hartwick et al., 1978.)

interference from the ATP present in the reaction mixture, since the formycin has a maximum at 298 nm and ATP at 265 nm.

The reaction mixture contained, in a final volume of 250 μL, imidazole-HCl (pH 6.8) as the buffer, either 250 nmol AMP or the formycin analog formycin 5'-AMP (FoMP) as substrate, and the activators ATP and KCl. The reaction was started by the addition of the enzyme, and at intervals samples were withdrawn from the reaction tube and injected directly onto the HPLC column for analysis.

Figure 9.95 shows a series of chromatograms representing the analysis of samples taken from an incubation mixture. As expected, the chromatogram of the sample of the reaction mixture taken before the start of the reaction shows only the substrate, in this case FoMP. Chromatograms of the incubation mixture sampled after the start of the reaction show the formation of the reaction product formycin 5'-IMP. Comparison of the peak heights of both substrate and product show clearly that the loss of substrate can be completely accounted for by the appearance of product. When these area values were converted to units of concentration, the rate curves shown in the inset were obtained.

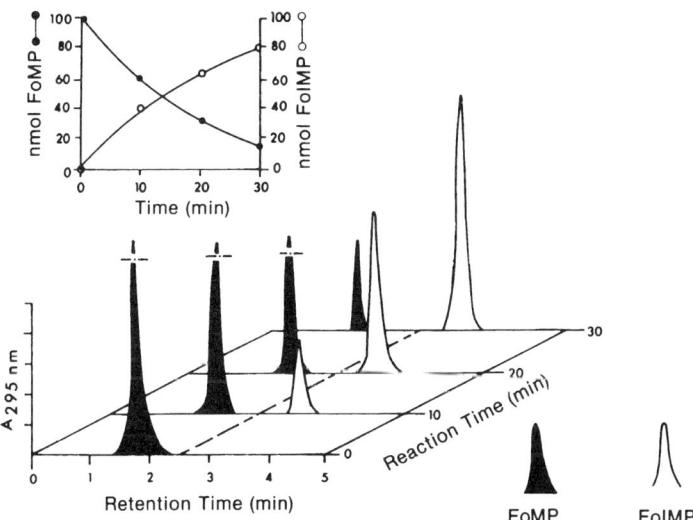

Figure 9.95 HPLC elution profiles of an adenylate deaminase incubation mixture. The reaction mixture contained 15 μmol of imidazole-HCl (pH 6.8), 250 nmol of FoMP, 250 nmol of ATP, and 5 μmol of KCl in a final volume of 250 μL. The reaction was initiated by the addition of activity obtained from the S-100 fraction and incubated at 37°C. At 10-minute intervals, 25 μL samples were injected onto the HPLC column. There is a decrease in the FoMP peak (retention time, 1.7 min) and a significant rise in the peak corresponding to FoIMP (retention time, 3.1 min) up to 30 minutes. *Inset:* Graphical representation of the first 30 minutes of the reaction. (From Jahngen and Rossomando, 1984.)

The activity was prepared from the microorganism *Dictyostelium discoideum*. Cells were lysed, and an S-100 fraction was prepared and used as the source of the deaminase activity.

In the study by Raffin and Thebault (1991), AMP and IMP were separated on a Partisphere SAX column (4.6 mm × 125 mm, 5μm). The mobile phase was composed of 40 mM potassium phosphate buffer (pH 3.5) and the flow rate was 1 mL/min. The effluent was monitored at 254 nm and peaks were quantified using peak heights and standard solutions of AMP and IMP.

The incubation mixture contained in a final volume of 2 mL: 50 mM cacodylic acid (pH titrated to 6.5 with 10 M KOH), 150 mM KCl, and 10 mM ATP. The reaction was started by adding 100 μL of diluted enzyme. The mixture was incubated at 25°C, and samples 500 μL of the enzyme incubation mixture were drawn after 10, 20, and 30 and added to 250 μL of 6% (w/v) trichloroacetic acid. After centrifugation, 500 μL of the supernate was added to the same volume of 0.5 M tri-*n*-octylamine in Freon. The mixture was recentrifuged, and the upper layer was used for HPLC analysis.

The assay was used with elasmobranch fish and shrimp muscle homogenates prepared by homogenizing the tissue in 0.089 M potassium phosphate (pH 6.5), 0.18 M KCl, and 0.1 mM dithiothreitol.

9.9.6 Cyclic Nucleotide Phosphodiesterase (Tsukada et al., 1980)

The enzyme 2',3'-cyclic nucleotide 3'-phosphodiesterase has been suggested as a marker for myelin. The activity catalyzes the degradation of 2',3'-cyclic AMP to 2'-AMP or 2',3'-cyclic CMP to 2'-CMP.

These compounds were separated on a Teflon column packed with silica gel–NH$_2$ (ODS C$_{18}$ LiChrosorb NH$_2$). The mobile phase was a solution of 25 mM KH$_2$PO$_4$–25 mM K$_2$HPO$_4$ (1:1, v/v, pH 6.8). Only 1 μL samples were required for analysis. Detection was at 254 nm. The separations obtained with standards on this column are illustrated in Figure 9.96A.

The reaction mixture contained the diesterase activity from brain homogenate (about 10 μg of protein), Tris-HCl buffer (pH 7.4), and MgSO$_4$. The reaction was started by addition of substrate. The reactions were terminated at suitable intervals with ethanol and extracted with chloroform. After centrifugation for 10 minutes, 1 μL of the supernatant solution was analyzed. Figure 9.96B shows the results of the assay, illustrating an increase in the height of the peak of the reaction product, 2'-AMP, with incubation.

Cerebral tissues were homogenized with a glass homogenizer and used as the source of the diesterase activity.

9.9.7 ATP Pyrophosphohydrolase (Rossomando et al., 1981a; Rossomando and Jahngen, 1983)

ATP pyrophosphohydrolase catalyzes the hydrolysis of ATP to AMP and pyrophosphate (PP$_i$). It is an activity that may be involved in several functions,

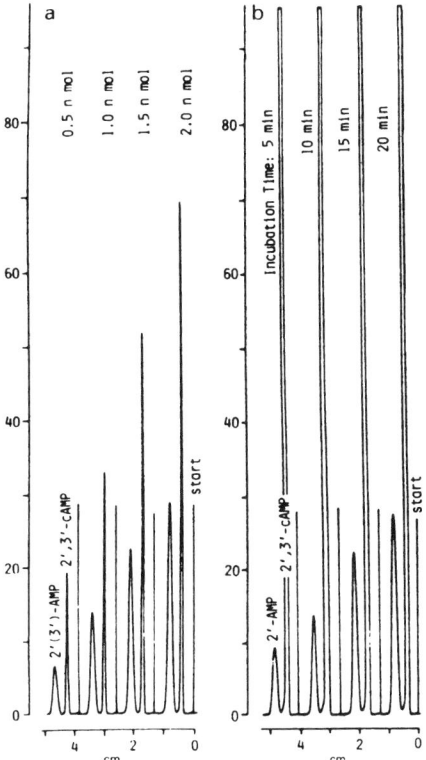

Figure 9.96 Separation and determination of 2′(3′)-AMP. Assay conditions: sample, 1 μL; column, 100 mm × 0.5 mm, silica gel–NH_2 (ODS-C_{18}; Merck LiChrosorb NH_2); mobile phase, 8 μL/min; range, 0.16 (absorbance) = 100; chart speed, 1 mm/min. (A) Assay of standard solutions. (B) Assay of enzyme activities. Enzyme: Rat cerebral homogenate (9.5 μg protein). Substrate: 20 mM 2′,3′-cAMP. Preincubation: 5 minutes at 37°C. Incubation: 37°C for 5, 10, 15, and 20 minutes. (From Tsukada et al., 1980.)

including calcification, the polymerzation of actin, and the regulation of ATP levels.

The separation of product from substrate was accomplished using ion-paired, reversed-phase HPLC on a C_{18} (μBondapak) column with a mobile phase of 65 mM potassium phosphate and 1 mM tetrabutylammonium phosphate adjusted to pH 3.6 with phosphoric acid and 1.5% acetonitrile. The column was eluted isocratically and monitored at 254 nm. The separations obtained are shown in Figure 9.97.

The reaction mixture contained the substrate ATP, $MnCl_2$, and a sodium acetate buffer at pH 6.0. Reactions were started by the addition of enzyme, and incubations were at 30°C. At intervals samples were withdrawn and injected onto the HPLC column for analysis. Chromatograms obtained showed

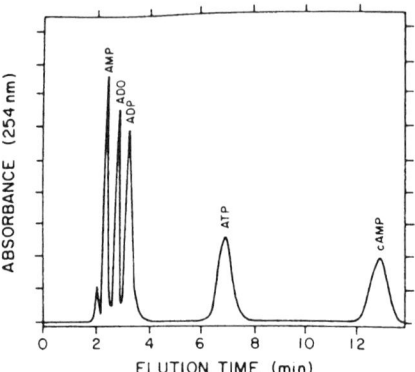

Figure 9.97 Separation of adenine nucleotides and adenosine by ion-paired, reversed-phase HPLC. Standards of AMP, adenosine, ADP, ATP, and cAMP (approximately 2 nmol of each) in Tris-HCl (pH 7.4) were injected onto a C_{18} μBondapak reversed-phase column (300 mm × 7.8 mm) and eluted with 65 mM KH_2PO_4 (pH 3.6), 1 mM tetrabutylammonium phosphate, and 2% acetonitrile. The flow rate was 2 mL/min, and detection was at 254 nm. (From Rossomando, 1987.)

a time-dependent increase in the amount of AMP and a corresponding decline in the level of the substrate ATP.

The activity was prepared from the microorganism *Dictyostelium discoideum*. Cells were lysed and an S-100 supernatant solution prepared.

9.9.8 Hypoxanthine Guanine Phosphoribosyltransferase (Ali and Sloan, 1982; Jahngen and Rossomando, 1984)

Hypoxanthine guanosine phosphoribosyltransferase (HGPRT) catalyzes the formation of IMP and pyrophosphate (PP_i) from hypoxanthine (Hyp) and phosphoribosylpyrophosphate (PRibPP) as shown in reaction (1):

$$Hyp + PRibPP \rightarrow IMP + PP_i \qquad (1)$$

This enzyme represents a principal route for the return, or salvage, of purines such as hypoxanthine, adenine, and guanine to the monophosphate level. The activity requires a metal, preferably magnesium.

Measurement of the activity of this enzyme by the HPLC assay method requires separation of the hypoxanthine and the IMP, which can be easily accomplished by several methods, including reversed-phase or ion-exchange HPLC. Jahngen and Rossomando (1984) used a reversed-phase C_{18} column with a mobile phase of 10 mM potassium phosphate (pH 5.6), and 10% methanol. Detection was at 254 nm, and the separation obtained is shown in Figure 9.98.

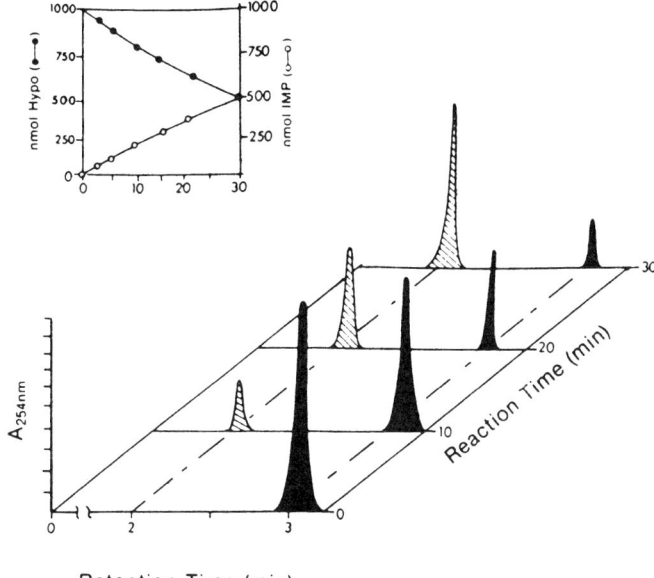

Figure 9.98 HPLC elution profiles of an incubation mixture to study hypoxanthine/guanine phosphoribosyltransferase. The reaction was initiated by the addition of the enzyme mixture, and aliquots were injected onto the HPLC column at 10-minute intervals as indicated on the z axis. The solid peaks represent hypoxanthine, which decreases with time, while the hatched peaks describe the formation of IMP. (From Jahngen and Rossomando, 1984.)

The reaction mixture contained Tris-HCl as buffer (pH 8.4), $MgCl_2$, and the two substrates, hypoxanthine and PRibPP. The reaction was started by the addition of enzyme, and samples were removed at intervals and injected onto the HPLC for analysis. The results of this assay are shown in Figure 9.98, where the appearance of the IMP is observed as well as the disappearance of the hypoxanthine.

The activity for this study was obtained from the microorganism *Dictyostelium discoideum*. An S-100 supernatant solution was prepared and used throughout as the source of HGPRT activity.

9.9.9 Nucleoside Phosphorylase (Halfpenny and Brown, 1980)

Nucleoside phosphorylase catalyzes the reversible conversion of a purine riboside such as inosine to a purine base such as hypoxanthine and ribose-1-phosphate. Free phosphate is also required as a substrate.

The assay involves the separation of reactants by reversed-phase HPLC on a C_{18} (Partisil 5 ODS) column with a mobile phase of 0.02 F KH_2PO_4 (pH

4.2) and 3% methanol applied isocratically. Absorbance measurements were at 254 nm. The separations obtained are shown in Figure 9.99.

The reaction was carried out in the following manner. Stock blood was transferred into test tubes, water was added, and the solution was frozen and thawed to lyse the cells. At zero time, excess xanthine oxidase was added as a coupling enzyme to convert all the hypoxanthine that was formed during the reaction to uric acid. The reaction was started by the addition of inosine in a phosphate buffer.

The complete reaction mixture was incubated for 10 minutes at 25°C, and the reaction was terminated by immersing the reaction tubes in a boiling water bath for 1 minute. The solutions were clarified by centrifugation, and samples of the supernatant solution were analyzed by HPLC.

The reverse reaction was assayed as well, using hypoxanthine and glucose-1-phosphate as substrates in a Tris-HCl buffer at pH 7.4. The results of this assay are shown in Figure 9.100. The chromatograms, taken at various times,

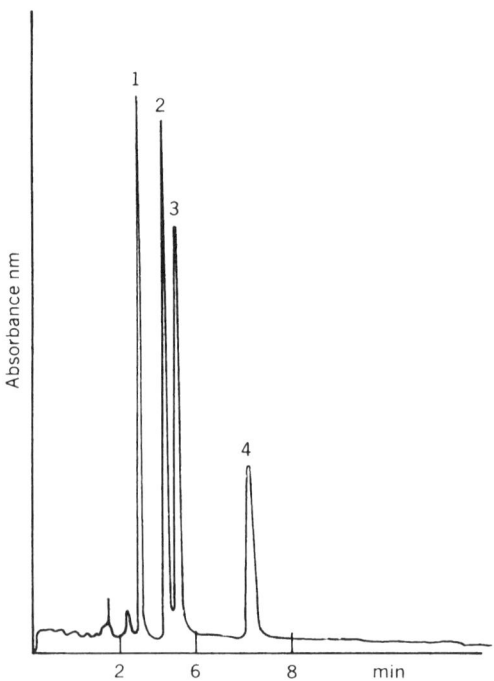

Figure 9.99 Separation of the components of the reaction studied (catalysis by nucleoside phosphorylase of a purine riboside–base conversion) by HPLC. Chromatographic conditions: isocratic elution; flow rate, 2 mL/min; 0.02 F KH_2PO_4 (pH 4.2) and 3% methanol. *Peaks:* 1, uric acid; 2, hypoxanthine; 3, xanthine; 4, inosine. (From Halfpenny and Brown, 1980.)

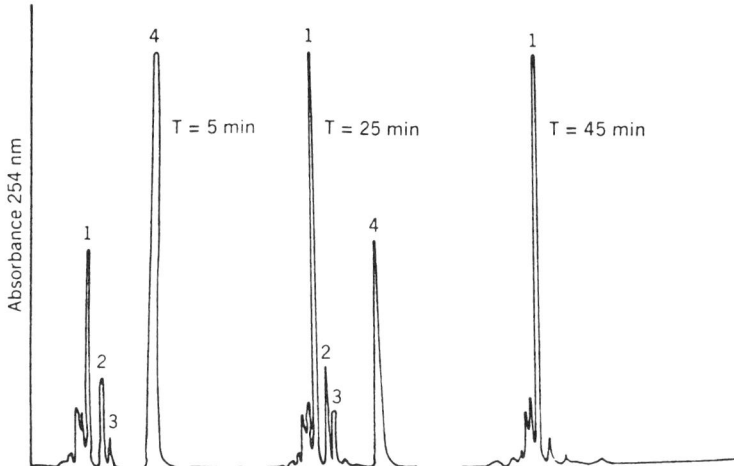

Figure 9.100 Reaction of PNPase as a function of time. Chromatograms at various time intervals show the decrease of the substrate inosine (4) and the increase of the products uric acid (1), hypoxanthine (2), and xanthine (3). (From Halfpenny and Brown, 1980.)

show the decrease in the substrate inosine and the increase in the product uric acid.

Red blood cells were lysed by dilution and one cycle of freezing and thawing and used a source of the enzyme.

9.9.10 Creatine Kinase (Danielson and Huth, 1980)

Creatine kinase catalyzes the reversible reaction whereby ADP + phosphocreatine form ATP + creatine. One HPLC method developed for this activity involved the direct determination of the ATP formed.

The reactants were separated from products by ion-paired reversed-phase HPLC (RP-18 LiChrosorb). The mobile phase consisted of an 88% mixture of 0.1 M KH_2PO_4–0.025 M butylammonium hydrogen sulfate and 12% methanol. To this was added enough 0.75 N NaOH to adjust the pH to 6.8. The separation of ATP, ADP, and AMP is shown in Figure 9.101.

The reaction mixture contained ADP, AMP, and KF (the last to inhibit any adenylate kinase activity), phosphocreatine, and magnesium, at concentrations 10-fold excess of ADP. The reaction was started by the addition of enzyme and terminated with a boiling water bath.

After the mixture had been allowed to cool, a 10 μL sample was analyzed by HPLC. The results of one assay are shown in Figure 9.101. The appearance of the ATP is clearly noted and is evidence of creatine kinase activity.

Prepared creatine kinase was obtained from a commercial source.

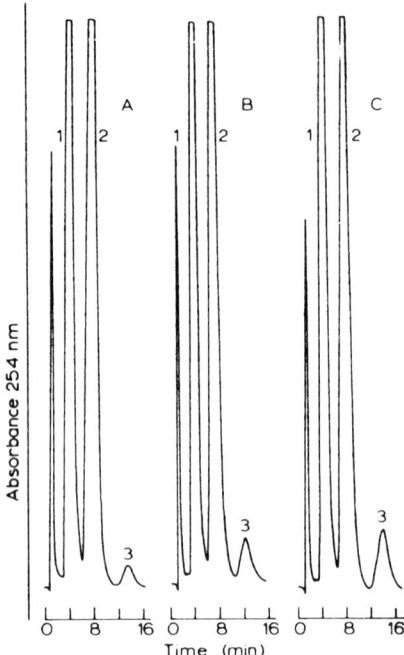

Figure 9.101 Separation of AMP (1), ADP (2), and ATP (3) for reaction times of (*A*) 10 minutes, (*B*) 20 minutes, and (*C*) 30 minutes using 0.04 μg/mL creatine kinase. Chromatographic conditions: column, LiChrosorb C_{18}; flow rate, 2.0 mL/min; temperature, ambient; detection, UV 254 nm, 0.04 absorbance unit full scale; injection volume, 10 μL. (From Danielson and Huth, 1980.)

9.9.11 Adenosine Kinase (Dye and Rossomando, 1982; Jahngen and Rossomando, 1984)

Adenosine kinase catalyzes the transfer of phosphate from ATP to adenosine (Ado) to form AMP and ADP. The separation of the reactants, Ado and ATP, from the products, AMP and ADP, can be accomplished by reversed-phase HPLC (C_{18}) with isocratic elution with a mobile phase of 0.1 *M* potassium phosphate (pH 5.5) and 10% methanol. Detection depends on the substrate. In this assay, it is useful to replace the substrate adenosine with the fluorescent analog formycin A (FoA) and to monitor the column eluent with a fluorescence detector. Thus, ATP and any of its hydrolytic products will not be detected.

The reaction mixture contained Tris-HCl (pH 7.4) as the buffer, plus ATP, FoA, MgCl, and KCl. The reaction mixture also contained EHNA (*erythro*-9,2-hydroxy-3-nonyladenine), an inhibitor of the secondary reaction catalyzed by adenosine deaminase. The reaction was started by the addition of the

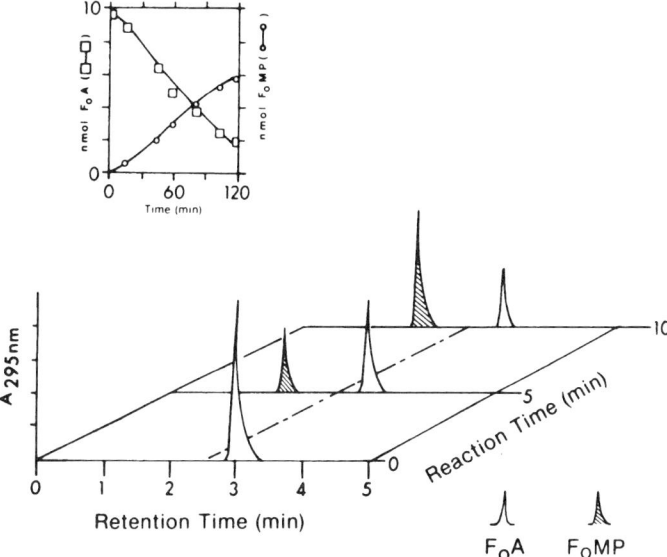

Figure 9.102 HPLC elution profiles of an incubation mixture to study adenosine kinase. The reaction was initiated by the addition of the enzyme mixture (S-100), and samples were removed and analyzed by HPLC at 5-minute intervals. *Inset:* Time-dependent utilization of FoA and the formation of FoMP as determined by integration of the respective peaks from the chromatograms. (From Jahngen and Rossomando, 1984.)

enzyme preparation and terminated by injecting samples directly onto the HPLC column. The results of the reaction are shown in Figure 9.102.

The enzyme was prepared from mouse liver. The liver was disrupted by grinding, and any insoluble material was removed by centrifugation at 30,000g for 30 minutes to form an S-30 fraction, which was used throughout the study.

9.9.12 Adenylate Cyclase (Rossomando et al., 1981b; Reysz et al., 1987)

Adenylate cyclase catalyzes the formation of 3′,5′-cyclic AMP (cAMP) from ATP.

In the assay described by Rossomando et al. (1981b), cAMP is separated from the ATP substrate by reversed-phase HPLC on a C_{18} (μBondapak) column with a mobile phase of 0.01 M potassium phosphate (pH 5.5) with 10% methanol. The column was eluted isocratically. The detectors for the eluent depended on the substrate (see below).

The separations shown in Figure 9.103 was obtained for the fluorescent analog of ATP, formycin ATP (FoTP), which was used instead of ATP because

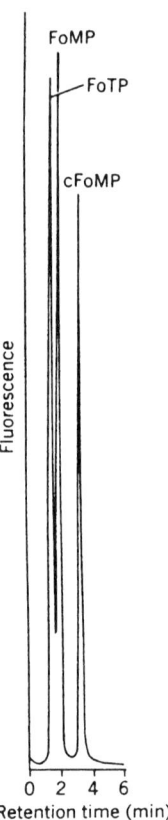

Figure 9.103 Separation of FoTP, FoMP, and cFoMP by HPLC. Operating conditions: 0.01 M KH_2PO_4 (pH 5.5) buffer with 10% methanol; flow rate, 2 mL/min; room temperature; μBondapak C_{18} column packing; fluorescence, excitation at 300 nm and emission above 320 nm. Approximately 10 μg of each component was injected. (From Rossomando *et al.,* 1981a.)

of the greater sensitivity of fluorescence detectors. Therefore, by substituting a fluorometer for the UV spectrophotometer, an increased sensitivity of five- to tenfold was achieved, and lower levels of the reaction product could be detected. A calibration curve obtained with formycin is shown in Figure 9.104.

The reaction mixture contained FoTP and $MgCl_2$, with Tris-HCl (pH 7.5) as the buffer. The reaction was started by the addition of the enzyme, and samples were removed at intervals and injected onto the HPLC column for analysis. Chromatograms were obtained, and the peak for cyclic FoMP (cFoMP) was integrated to determine the amount of cFoMP formed as a function of reaction time. These data are shown in Figure 9.105; insets show the chromatograms obtained on two samples.

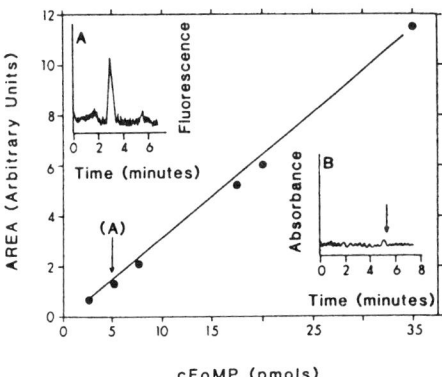

Figure 9.104 Calibration of fluorometric detector with cFoMP. Solutions were prepared ranging from 0.4 to 7 µM, and 5 µL of each was injected onto the column. Areas of the cFoMP peaks were determined from tracings obtained at each concentration and are plotted (in arbitrary units) as a function of amount of cFoMP injected. (A) Tracing obtained after injection of 5 pmol of cFoMP. (B) Tracing obtained with 15 pmol of cAMP (arrow indicates retention time of authentic cAMP). (From Rossomando et al., 1981b.)

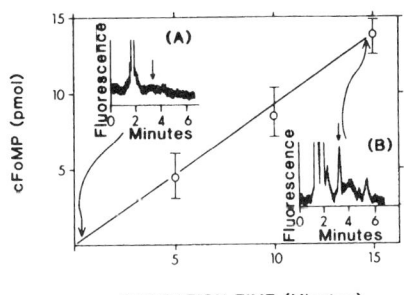

Figure 9.105 Kinetics of cFoMP formation as determined by fluorometry. Adenylate cyclase activity was determined at 0.3 mM FoTP with 100 µg of membrane protein in a final reaction volume of 100 µL. Reactions were terminated, and cFoMP was purified and then analyzed by HPLC. *Insets:* Representative HPLC profiles obtained at (A) 30 seconds and (B) 15 minutes after start of the reaction. Arrows indicate the retention time observed after injection of authentic cFoMP. The area under the curves was determined by integration, and amount of cFoMP present was determined from a calibration curve. Data obtained from HPLC assay gave values within the error bars. (From Rossomando et al., 1981b.)

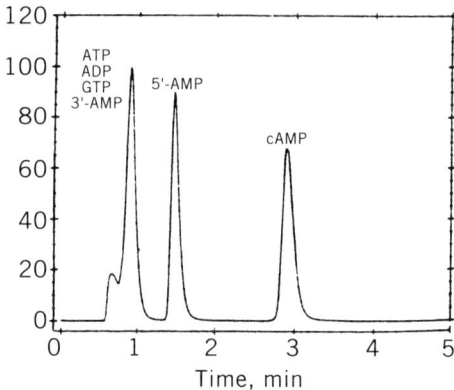

Figure 9.106 HPLC separation of cAMP from other nucleotides. Fifty microliters of a sample containing 1 mM ATP, 5'-AMP, and cAMP was injected onto the column and eluted. The elution times of several other nucleotides are also shown; each nucleotide was also chromatographed separately and in this way identified. The data were scaled such that the peak eluting at 1 minute would be 100. The small peak at 0.7 minutes represents the elution time of unretained substances. (From Reysz et al., 1987.)

The membrane-bound activity was obtained from crude homogenate of rat osteosarcoma cells.

In the assay by Reysz et al. (1987), cyclic AMP is separated from other nucleotides including 5'-AMP by chromatography on an Adsorbosphere C_8 column (4.6 mm × 100 mm). The elution protocol consisted of running solvent A (0.1 M KH_2PO_4, pH 3.0) at a flow rate of 2 mL/min for 5 minutes, followed by a rapid change (within 0.1 min) to 50% solvent B (methanol) and holding at that composition until 9 minutes. The mobile phase was returned to starting conditions in 0.1 minutes and reequilibrating for 1 minute before the next sample. Detection was at 260 nm.

The enzyme assay in a final volume of 0.6 mL contained 40 MM Tris-HCl (pH 7.5), 3.3 mM $MgSO_4$, 10 mM NaF, 10 mM theophylline, 1 mM $CaCl_2$, 1 mM EGTA, 10 μM GTP, 3 mM ATP, 0.3 mg/mL of bovine serum albumin, and enzyme. The reaction was started by adding enzyme and terminated after 30 minutes at 30°C by the addition of 0.6 mL of 4% (w/v) trichloroacetic acid. The resulting mixture was centrifuged followed by analysis of 50 μL of the supernate by HPLC.

The assay was used for measuring the activity of adenylate cyclase in microsomes prepared from sea urchin eggs and frozen bovine brain.

Figure 9.106 shows a representative HPLC separation.

9.9.13 cAMP Phosphodiesterase (Rossomando et al., 1981a; Spoto et al., 1991)

The enzyme cAMP phosphodiesterase catalyzes the formation of AMP from cyclic AMP as shown in reaction (1):

$$\text{cAMP} \to \text{AMP} \qquad (1)$$

$$\text{AMP} \to \text{Ado} + P_i \qquad (2)$$

Reaction (2), the formation of adenosine (Ado) from AMP, is catalyzed by 5′-nucleotidase, an activity often present together with the diesterase. It is useful to be able to measure the activity of this enzyme as well.

In the HPLC assay of Rossomando et al. (1981a), the compounds AMP, cAMP, and adenosine are separated by reversed-phase HPLC on C_{18} (μBondapak) with a mobile phase of 10 mM KH_2PO_4 (pH 5.5) containing 1% methanol. The separation of cAMP and AMP obtained with a mobile phase of a phosphate buffer and 10% methanol is shown in Figure 9.107.

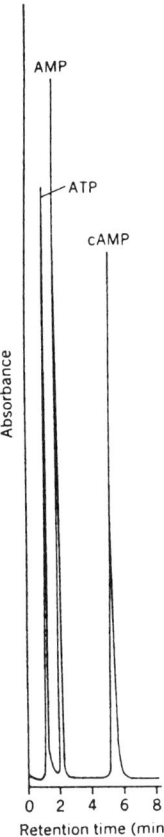

Figure 9.107 Separation of ATP, AMP, and cAMP by HPLC. Operating conditions: 0.01 M KH_2PO_4 (pH 5.5) buffer with 10% methanol; flow rate, 2 mL/min; room temperature, μBondapak C_{18} column packing; absorbance at 260 nm. Approximately 10 μg of each component was injected. Sensitivity of detection, 0.5 AUFS. (From Rossomando et al., 1981a.)

Adenosine elutes at about 10 minutes (not shown), which clearly allows the measurement of the level of each of these components in an incubation mixture during the course of reaction.

The reaction mixture contained cyclic formycin monophosphate, an analog of cAMP, as the substrate, Tris-HCl (pH 7.5) as buffer, and MgCl$_2$. The reaction was started by the addition of the enzyme. Samples were removed at intervals and injected directly onto the reversed-phase column for analysis. Figure 9.108 shows chromatograms after 10 and 30 minutes of incubation. While the amount of cFoMP substrate in the incubation mixture has declined and the amount of product FoMP has increased, the amount of formycin A (FoA), the analog of adenosine, has remained unchanged. When the area of each peak is plotted as a function of reaction time, the data shown in the central inset are obtained. Although these data clearly illustrate the activity of the cyclic phosphodiesterase, they also show the absence of any 5'-nucleotidase.

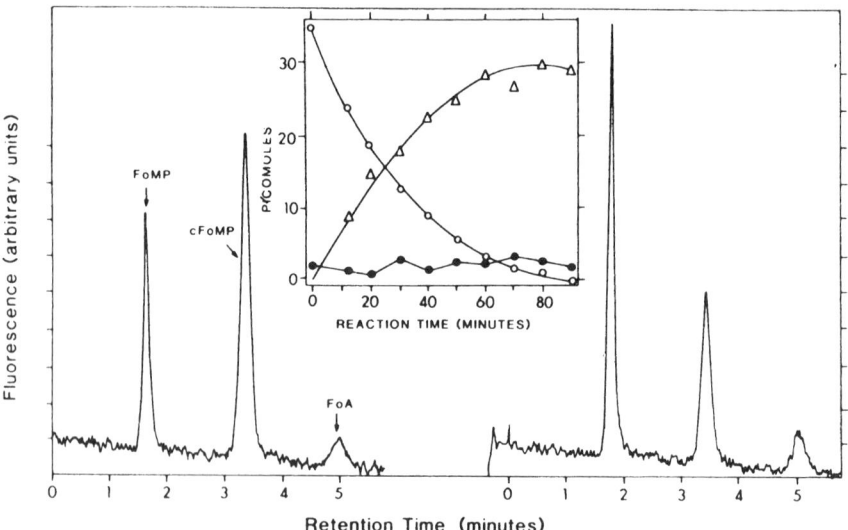

Figure 9.108 Degradation of cFoMP by 3'-5'-cyclic AMP phosphodiesterase. The reaction mixture contained, in a final volume of 50 μL, 10 mM Tris-HCl (pH 7.5), 1 μM MgCl$_2$, 1 mM cyclic nucleotide, and 50 μg of enzyme. Incubations were at 37°C. At intervals, samples were removed from the reaction mixture (20 μL) and analyzed by HPLC using a μBondapak C$_{18}$ column preceded by a guard column. Flow rate was 2 mL/min. Fluorescence of column effluent was monitored. Profiles obtained after incubation for (A) 10 minutes and (B) 30 minutes. Areas under cFoMp peak were determined at several time points from profiles such as those shown in (A) and (B). *Inset:* Amounts of cFoMP (○), FoMP (△), and FoA (●) as determined from tracings obtained at times shown. (From Rossomando et al., 1981a.)

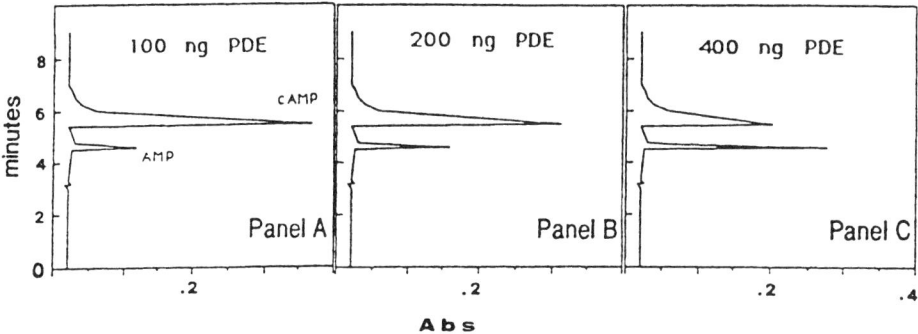

Figure 9.109 HPLC separation of AMP and cAMP in the reaction mixtures containing different concentrations of cAMP phosphodiesterase (PDE): (A) 100 ng, (B) 200 ng, and (C) 400 ng. Incubation time was 45 minutes; 1.5×10^{-4} M cAMP. (From Spoto et al., 1991.)

The activity was purified from *Dictyostelium discoideum* by differential centrifugation and gel filtration.

In the assay described by Spoto et al. (1991), cAMP and AMP were separated on a LiChrosorb RP-18 column (4 mm × 250 mm). The mobile phase at pH 6.0 contained 0.2 M phosphate buffer, 25 mM tetrabutylammonium hydroxide, and 15% (v/v) acetonitrile. Detection was performed at 254 nm.

The enzymatic reaction mixture contained 0.1 M Tris-HCl buffer (pH 8.0), 10 mM MgCl$_2$, and 0.1 M KCl at 37°C. The final volume ranged between 500 and 100 µL. The reaction was initiated by cAMP to give a final concentration of 0.15 to 0.2 mM. The reaction was terminated by heating the tubes in a boiling water bath for 3 minutes. Samples were clarified by centrifugation or filtration before injecting into the HPLC system.

Cyclic AMP phosphodiesterase was partially purified from rat brain and from rat heart. The assay was applicable at different levels of enzyme purification.

Figure 9.109 shows a representative series of chromatograms.

9.9.14 Adenylate Kinase (Rossomando and Jahngen, 1983)

Adenylate kinase (myokinase) catalyzes the following reversible reaction.

$$\text{ATP} + \text{AMP} \rightarrow 2\text{ADP} \tag{1}$$

In one assay developed to measure this activity, ion-paired reversed-phase HPLC (C$_{18}$) was used to separate the reactants from the products. The mobile phase was 0.065 M potassium phosphate with 1 mM tetrabutylammonium phosphate adjusted to pH 3.6 with phosphoric acid and 3% acetonitrile. The column was eluted isocratically and monitored at 254 nm and with a radio-

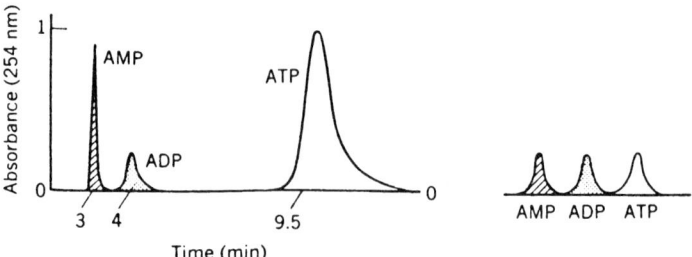

Figure 9.110 Separation of adenine nucleotides and adenosine by paired-ion reversed-phase HPLC. Standards of AMP, adenosine, ADP, ATP, and cAMP (approximately 2 nmol each) in Tris-HCl (pH 7.4) were injected onto a C_{18} μBondapak reversed-phase column (300 mm × 7.8 mm) and eluted with 65 mM KH_2PO_4 (pH 3.6), 1 mM tetrabutylammonium phosphate, and 3% acetonitrile. The flow rate was 2 mL/min, and detection was at 254 nm. (From Rossomando and Jahngen, 1983.)

chemical detector. The reaction mixture contained Tris-HCl buffer, ATP, and AMP, and, when assayed in the reverse direction, ADP was the substrate. Under these conditions, the separation shown in Figure 9.110 was obtained.

The reaction mixture contained the substrates unlabeled ATP and radioactive [³H]AMP, buffered with Tris-HCl (pH 7.4). The reaction was started by addition of enzyme and terminated by injection of samples directly onto the HPLC column.

The results are summarized in Figure 9.111, where Figure 9.111A shows the chromatographic profile of the reaction mixture before the addition of the enzyme: The two substrates in the reaction mix, [³H]AMP and ATP, are detected. After initiation of the reaction, ADP is formed, and after the reaction has proceeded in the reverse direction, the amount of radioactive ATP increases steadily until almost all the radioactivity originally present in the AMP is recovered in the ATP.

The activity was obtained from *Dictyostelium discoideum.*

9.9.15 Adenylosuccinate Synthetase (Jahngen and Rossomando, 1984)

The enzyme adenylosuccinate synthetase condenses IMP with aspartic acid to form adenylosuccinate (sAMP). GTP participates directly in the reaction process, and during the course of the reaction GDP is formed.

For the HPLC assay, IMP, sAMP, GTP, GDP, and AMP are separated by ion-paired, reversed-phase HPLC (C_{18}) with a mobile phase of 65 mM potassium phosphate (pH 4.4), 1 mM tetrabutylammonium sulfate, and 10% methanol. Detection was at 254 nm. A representative chromatogram is shown in Figure 9.112.

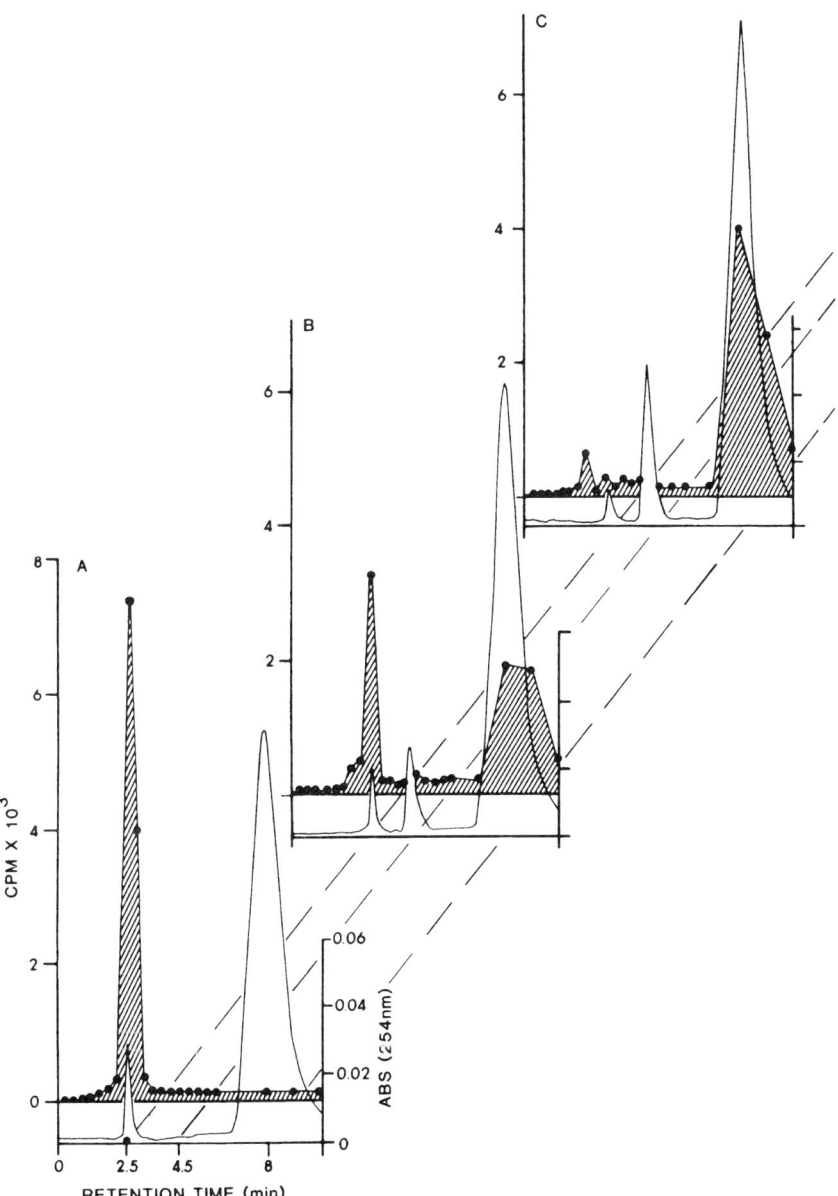

Figure 9.111 Adenylate kinase activity measured by the HPLC assay method. The assay mixture contained 200 μM ATP, 20 μM [^{3}H]AMP (approximately 120,000 cpm), 50 mM Tris-HCl (pH 7.4), and 32 μg of protein of the enzyme from *Dictyostelium discoideum*. Samples (20 μL) were injected and the radioactivity monitored continuously with a Berthold LB 503 detector using PICO-Fluor 30 scintillant. Absorbance was measured at 254 nm. Three representative time points were shown: (A) 1 minute (B) 30 minutes, and (C) 60 minutes after initiation of the reaction. (From Rossomando, 1987.)

336 SURVEY OF ENZYMATIC ACTIVITIES ASSAYED BY THE HPLC METHOD

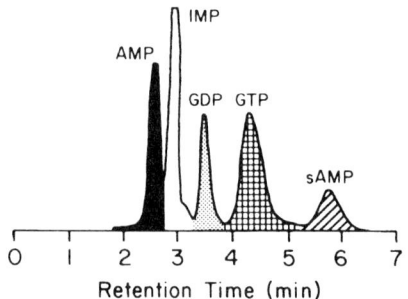

Figure 9.112 Separation of substrates and products of reaction catalyzed by adenylosuccinate synthetase. Column: Prepacked C_{18} μBondapak, 10 μm particle size. Mobile phase: 65 mM potassium phosphate, 1 mM tetrabutylammonium phosphate, 10% methanol at pH 4.4. Absorbance was measured at 254 nm. (From Rossomando, 1987.)

In addition to the enzyme the reaction mixture contained Hepes buffer, IMP, GTP, MgCl, and creatine phosphate and phosphocreatine kinase (a regeneration system for GTP). The reaction was initiated by adding aspartic acid. Samples were removed at intervals, and the reactions were terminated by direct injection onto the HPLC column. Figure 9.113 shows chromatograms of samples removed at 0, 5, and 10 minutes of incubation. The disappearance of IMP and GTP and the appearance of GTP and sAMP can be noted.

The enzyme was prepared from *Dictyostelium discoideum*. The cells were lysed, and S-100 solutions were prepared from the lysate by differential centrifugation. Samples of this solution were used in the assay.

9.9.16 Dinucleoside Polyphosphate Pyrophosphohydrolase (Garrison et al., 1982)

This activity (and the specific diadenosine tetraphosphate pyrophosphohydrolase activity) catalyzes the symmetrical hydrolysis of Ap_4A to release 2 molecules of ADP. In the HPLC method developed for this assay, the substrate was separated from the product on an anion-exchange column (Partisil PXS 10/25 SAX), and the amounts of both compounds determined. A precolumn (Solvecon) was placed in line between the mixing chamber and the injection valve for mobile phase conditioning. Nucleotides were eluted with a pH and ionic gradient as follows: The column was initially equilibrated with 50 mM ammonium phosphate (pH 5.2). After sample injection, the concentration of the second buffer, composed of 1 M ammonium phosphate (pH 5.7), was brought to 5%. After 10 minutes, the concentration of the second buffer was increased linearly from 5% to 45% over a 30-minute period. The separation obtained is shown in Figure 9.114.

The reaction mixture contained HEPES-NaOH (pH 7.5), the substrate (Ap_4A), and the enzyme. After a 10-minute incubation, the reactions were

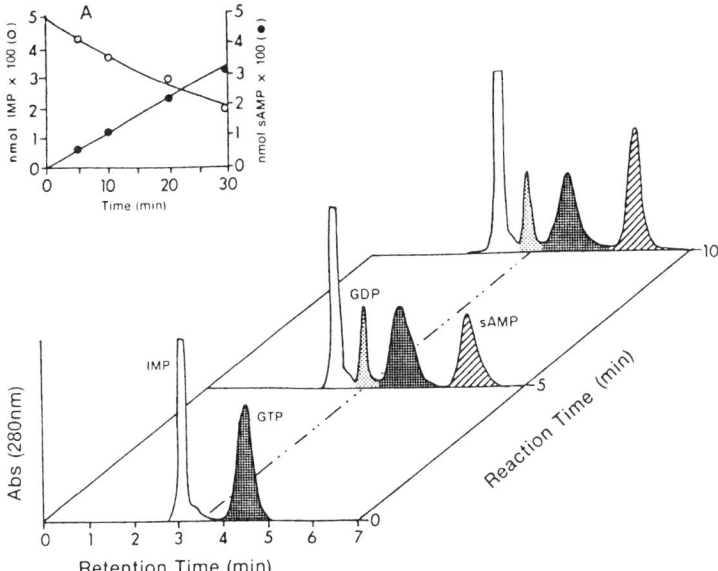

Figure 9.113 HPLC elution profiles of adenylosuccinate synthetase incubation mixtures. This reaction was initiated by the addition of 1.25 μmol of aspartate (pH 7.4). At 5-minute intervals, 20 μL samples were injected onto the HPLC reversed-phase column and eluted. *Inset:* Time-dependent utilization of IMP and the formation of sAMP, as determined by integration of the respective peaks from the HPLC chromatograms. (From Jahngen and Rossomando, 1984.)

stopped by quick-freezing on dry ice or by injecting samples directly onto the HPLC column.

The enzyme was purified to homogeneity from *Physarum polycephalum*.

9.9.17 NAD Glycohydrolase (Pietta et al., 1983)

The NAD glycohydrolase (EC 3.2.2.5) in this study catalyzes the hydrolysis of NAD^+ to form nicotinamide, adenosine diphosphate ribose (ADPR), and H^+. The assay developed for this activity follows the disappearance of the substrate NAD^+ and the production of nicotinamide.

The separations were accomplished by reversed-phase HPLC on a C_{18} (μBondapak) column with a precolumn packed with C_{18} Corasil, using a mobile phase of 0.01 M diammonium hydrogen phosphate–acetonitrile (100:5) at pH 5.5 adjusted with 20% phosphoric acid. Detection was at 259 nm. The separation obtained is shown in Figure 9.115.

The reaction mixture contained NAD in 0.2 M phosphate buffer (pH 7.0), and was equilibrated at 37°C. The reaction was started by addition of the hydrolase, and at intervals it was terminated by diluting a sample in the mobile

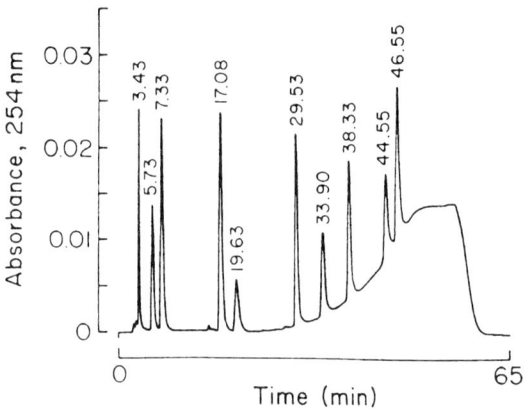

Figure 9.114 Chromatographic resolution of diadenine and monoadenine nucleotides by HPLC: 0.8 to 1.2 nmol of each was separated on a 4.6 mm × 25 cm column of Whatman Partisil PXS 10/25 SAX resin with buffer A, 50 mM ammonium phosphate (pH 5.2) and buffer B, 1 M ammonium phosphate (pH 5.7) by gradient elution. Flow rate, 1.5 mL/min; temperature, 23 to 24°C. Nucleotides and retention times (min): cAMP, 3.43; AMP, 5.73; Ap$_2$A, 7.33; Ap$_3$A, 17.08; ADP, 19.63; Ap$_4$A, 19.53; ATP, 33.9; Ap$_5$A, 38.33; Ap$_4$, 44.55; Ap$_6$A, 46.55. (From Garrison et al., 1982.) [Reprinted with permission from *Biochemistry,* **21:**6129–6133 (1982) American Chemical Society.]

phase. Each sample was then analyzed by HPLC. Intervals as short as 6 minutes (the time required for a complete HPLC run and quantitation) were used. For shorter intervals the reaction was terminated with 1 M HCl (pH 3.0) and diluted with mobile phase prior to injection.

The hydrolase was prepared from *Neurospora crassa.* A crude extract was obtained, solubilized by extraction with KCl, and used throughout the study.

9.9.18 Assay of Enzymes Involved in Cytokinin Metabolism (Chism et al., 1984)

Cytokinins, N^6-substituted derivatives of adenine, are important in the regulation of many processes in plant tissues. Chism et al. (1984) reported an HPLC assay measuring the hydroxylation of 6-(3-methylbut-2-enylamino)purine (IPA) and 6-(3-methylbut-2-enylamino)-8-hydroxypurine by cytokinin nucleosidases and adenosine nucleosidases.

The separations were carried out by reversed-phase HPLC (C$_{18}$ μBondapak) using a mobile phase of 40% methanol–water. The column was eluted isocratically and monitored at 275 nm.

Reaction conditions for hydroxylation of IPA included a phosphate buffer (pH 7.4) and the partially purified xanthine oxidase added to start the reaction. The reaction was stopped at 10-minute intervals with methanol, clarified by centrifugation, filtered, and analyzed by HPLC. The results of an assay are shown in Figure 9.116.

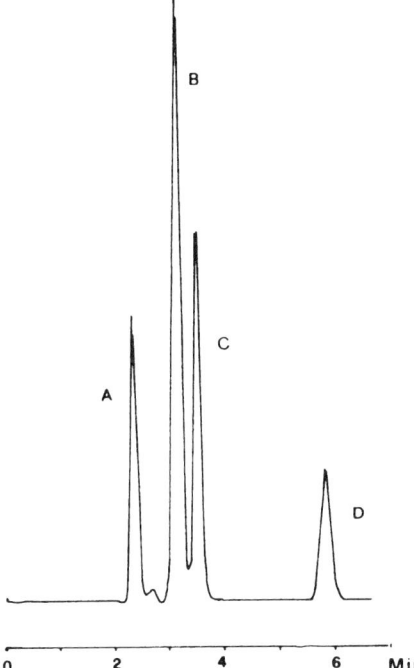

Figure 9.115 HPLC chromatogram of ADPR (A), NAD$^+$ (B), 4-aminobenzoic acid (C), and nicotinamide (D). (From Pietta et al., 1983.)

The xanthine oxidase for the hydroxylation was obtained by procedures published earlier.

9.9.19 Xanthine Oxidase (Sasaoka et al., 1988)

Xanthine oxidase is a flavoenzyme that also contains molybdenum and iron. It shows activity toward a variety of substrates including purines, pyrimidines, aldehydes, and NADH. In this assay, isoxanthopterin is produced from 2-amino-4-hydroxypteridine.

Isoxanthopterin was separated from 2-amino-4-hydroxypteridine by chromatography on a Unisil, ODS column (4.6 mm × 50 mm, 3 μm). The mobile phase was 0.1 M phosphate buffer (pH 2.0). A fluorescence detector was used, with excitation and detection wavelengths 340 and 410 nm, respectively.

The standard reaction mixture contained in a final volume of 250 μL: 50 μL of 0.5 M sodium phosphate buffer (pH 7.0), 25 μL of 200 μM AHP solution, 25 μL of 10 μM 2,6-dichlorophenolindolphenol sodium, 10 μL of enzyme sample, and 10 μL of 10 mg/mL bovine serum albumin. After a 10-minute incubation at 37°C, the reaction was stopped by adding 25 μL of 60% perchloric acid and cooling in an ice bath for 10 minutes. The mixture was centrifuged before injection of 10 μL of the supernate into the HPLC system.

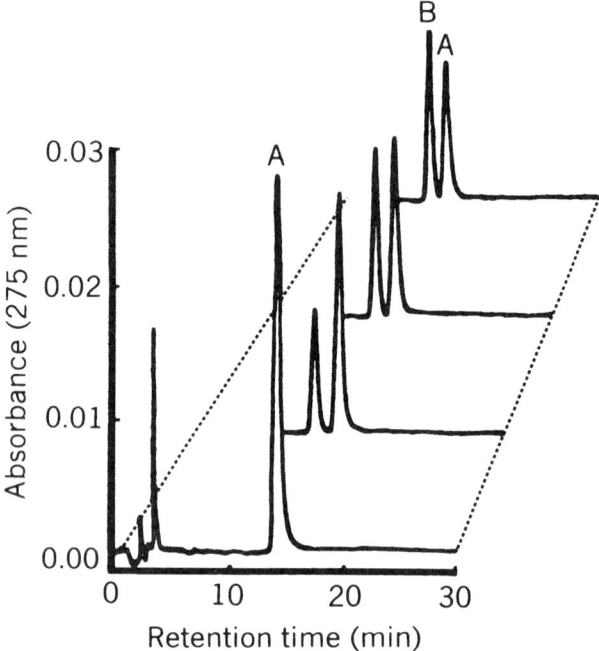

Figure 9.116 Reversed-phase HPLC for monitoring the formation of 8-OH IPA from IPA. Aliquots (20 μL) of the stopped reaction mixture were injected onto a μBondapak C_{18} column and eluted with 40% aqueous methanol at a flow rate of 1 mL/min. IPA (peak A) is eluted after 8-OH IPA (peak B), which is absent at zero time. The chromatograms are from aliquots taken at 0, 30, 60, and 90 minutes. (From Chism et al., 1984.)

The assay was linear for up to 30 minutes and with up to 40 μg of the soluble fraction of rat liver.

Sources of enzyme were the supernates from rat liver, kidney, and brain obtained by homogenizing these tissues in 4 volumes of 100 mM sodium phosphate buffer (pH 7.0) followed by centrifugation at 15,000g for 30 minutes.

Figure 9.117 shows HPLC separations.

9.9.20 Phosphoribosylpyrophosphate Synthetase (Sakuma et al., 1991)

The synthesis of phosphoribosylpyrophosphate from ATP and ribose-5-phosphate is catalyzed by phosphoribosylpyrophosphate synthetase. Superactivity of this enzyme may be a cause of overproduction of uric acid, which may result in gout.

Assay of the enzyme is based on the separation of ATP, ADP, and AMP on a Tosoh ODS-120A column (4.6 mm × 250 mm, 5 μm). The mobile phase

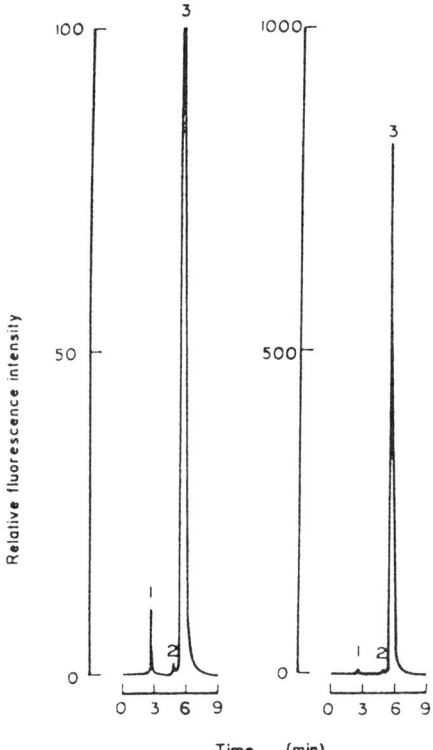

Figure 9.117 HPLC elution pattern of the standard compounds (10 pmol each). *Peaks:* 1, 2-amino-4-hydroxypteridine (AHP); 2, lumazine (LZ); 3, isoxanthopterin (IXP). (From Sasaoka et al., 1988.)

was 40 mM sodium phosphate buffer (pH 5.6) containing 1% (v/v) methanol. Quantitation is based on the ADP produced, which requires the subsequent activity of adenylate kinase acting on the AMP produced by phosphoribosylpyrophosphate. The absorbance of the effluent is followed at 260 nm.

The reaction mixture contained 40 mM sodium phosphate buffer (pH 7.4), 1 mM ribose-5-phosphate, 1.4 mM ATP, 6 mM MgCl$_2$, 1 mM reduced glutathione, and 1.8 IU/mL adenylate kinase in a final volume of 2.0 mL. The reaction was started by adding ATP. After further incubation at 37°C, 0.5 mL of the reaction mixtures was withdrawn at 10 and 30 minutes and added to tubes containing 0.5 mL of 0.6 M perchloric acid. After centrifugation, the supernate was neutralized by adding an equal volume of 0.4 M Na$_2$HPO$_4$, and aliquots were used for HPLC analysis. Product ADP increased linearly with time for 40 minutes.

The enzyme source was thrice-washed red blood cells that were then lysed by freeze-thawing, treated with charcoal and dextran, and centrifuged. The supernates were used as enzyme samples.

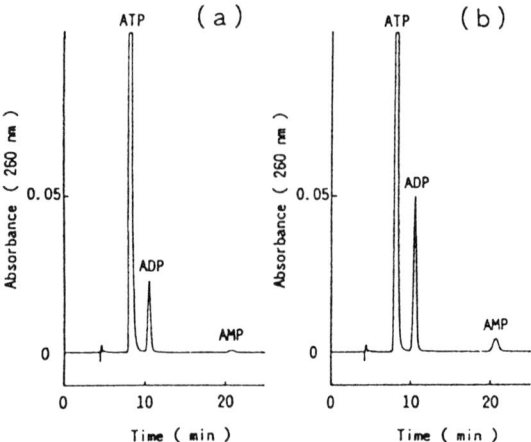

Figure 9.118 HPLC profiles of the reaction mixtures after the enzyme reaction of erythrocyte phosphoribosylpyrophosphate synthetase from a healthy subject. Incubation times were 10 minutes (*a*) and 30 minutes (*b*). (From Sakuura, et al., 1991.)

Figure 9.118 shows a chromatogram. Note that because a reversed phase column is used, ATP elutes first and AMP last.

9.9.21 Guanase (Canepari et al., 1993)

Guanase catalyzes the hydrolytic deamination of guanine to xanthine and ammonia. Its activity in serum is strongly elevated in patients with liver disease.

Xanthine is separated from guanine on a Supelcosil LC_{18} column (4.6 mm × 250 mm, 5 μm). The mobile phase was 0.05 M acetate buffer (pH 6.0). The effluent was monitored at 270 nm.

The reaction mixture contained 200 μL of substrate solution (25 mg dissolved in 10 mL of 1 M NaOH and diluted to 100 mL), 3 mL of Tris-HCl buffer, and 20 μL of enzyme solution or serum sample. Water was added to give a final volume of 5.0 mL. After incubation at 37°C for 20 minutes, the reaction was stopped with 30 μL of 70% perchloric acid in an ice bath. Aliquots of 20 μL were injected into the HPLC system. The reaction was linear for up to an hour.

The enzyme source, a guanase preparation from Sigma, was diluted to a concentration of 0.2 mg enzyme/mL in 0.025 M Tris-HCl (pH 8.0). Enzyme was also added to bovine serum to simulate the levels present in pathological serum.

Figure 9.119 shows chromatograms for three reaction times.

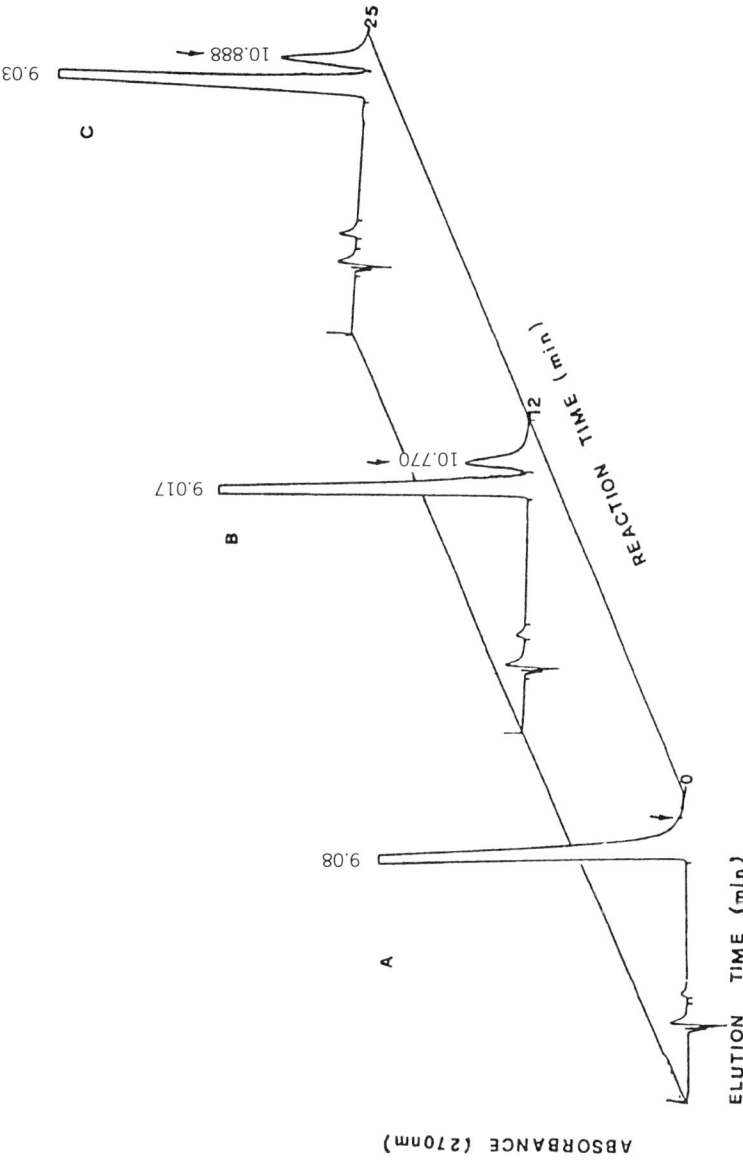

Figure 9.119 Time course of formation of xanthine from guanine as measured by guanase activity. The reaction mixture contained, in a final volume of 5 mL, 0.45 mmol of Tris-HCl (pH 8.0), 0.33 µmol of guanine, and 20 µL of guanase solution (4 µg). A 20 µL aliquot of sample was injected. Chromatograms were obtained after incubation for (*A*) 0 minutes, (*B*) 12 minutes, and (*C*) 25 minutes. The arrows indicate the elution time for the reaction product (xanthine). (From Canepari et al., 1993.)

9.9.22 Urate Oxidase (Greenberg and Hershfield, 1989)

Some mammals have urate oxidase, which metabolizes uric acid to a much more soluble product, allantoin. Administration of the enzyme to humans could be useful in the treatment of gout. This assay was designed to measure the activity of urate oxidase bound to polyethylene glycol.

Urate and allantoin were separated on a μBondapack C_{18} column (3.9 mm $\times$ 300 mm). The mobile phase was 50 mM ammonium phosphate (pH 3.0). The effluent was monitored at 290 nm for quantification of uric acid or at 218 nm for detection of allantoin, or mixed in-line with scintillation fluid for quantitation by a flow-through detector.

The reaction mixture contained in a final volume of 50 μL: 5 to 10 μL of enzyme sample, 35 μL of 0.1 M Na borate (pH 8.5), and 0.25 μCi (4.55 nmol) [2-^{14}C]uric acid, plus unlabeled uric acid to give a final concentration of 300 μM urate. The reaction was carried out at 37°C and was started by the addition of enzyme or substrate. Reactions were terminated at varying times by adding 20 μL of 1 M perchloric acid. After the tubes had been on ice for 5 minutes, the supernate was analyzed by HPLC. The reaction was linear in 10-minute assays when up to 20% of the substrate was converted to product.

The enzyme preparation used was uricase modified with polyethylene glycol, from Enzon, or plasma from patients is treated with PEC-uricase.

9.9.23 Glutamine: 5-Phosphoribosyl-1-pyrophosphate Amidotransferase (Taha and Deits, 1993)

In the enzyme catalysis of the first committed step in the *de novo* synthesis of purines, an amino group from L-glutamine is transferred to 5-phosphoribosyl-1-pyrophosphate to form glutamate and 5-phosphoribosyl-1-amine. The assay includes glycinamide ribonucleotide synthetase, which converts 5-phosphoribosyl-1-amine to glycinamide ribonucleotide, which is the reaction product quantitated.

Glycinamide ribonucleotide is separated on a Dionex CarboPac PA-1 column (4 mm $\times$ 250 mm). A three-solvent system was employed consisting of 0.4 M sodium hydroxide (eluent A), 2.0 M sodium acetate (eluent B), and water (eluent C). Resolution of the glycinamide ribonucleotide peak was obtained by using 33% A, 10% B, and 57% C. To remove tighter eluting compounds at the end of the run, each 6-minute isocratic run (1 mL/min) was followed by a 4-minute gradient to 33% A, 35% B, 34% C, followed by a 5-minute reequilibration period. Detection was by pulsed amperometric detection, with the pulse train consisting of a 480 ms detection pulse at +80 mV, followed by pulses of 120 ms at +600 mV and 60 ms at −600 mV.

Activity was determined in a 100 μL reaction mixture containing 10 mM Bicine (pH 8), 50 mM KCl, 3 mM magnesium acetate, 2 mM ATP, 3 mM glycine, 10 mM L-glutamine, 1 mM 5-phosphoribosyl-1-pyrophosphate, 3 to

5 units of glycinamide ribonucleotide synthetase, and 5 to 30 µL of extract. The assay proceeded for 30 minutes at ambient temperature before being stopped by the addition of 100 µL of acetonitrile–1 M HCl, 9:1. The samples were centrifuged for 5 minutes to remove precipitated proteins before analysis by injection of 25 µL into the HPLC system.

The assay was demonstrated to be suitable for measurement of enzyme activity in extracts prepared from *Azotobacter vinelandii* and soybean nodules.

9.9.24 Thiopurine Methyltransferase (Lennard and Singleton, 1994)

Thiopurine methyltransferase S-methylates 6-mercaptopurine and 6-thioguanine, both of which are used as antileukemic drugs. The source of the methyl group is S-adenosyl-L-methionine.

Separation of 6-mercaptopurine and 6-methylmercaptopurine was accomplished on a Waters Resolve C_{18} cartridge (8 mm × 100 mm, 5 µm). The mobile phase contained 20% methanol (v/v) and 100 mM triethylamine with the pH adjusted to 3.2 with orthophosphoric acid. The mobile phase also contained 0.5 mM dithiothreitol, which was added just before use. A photodiode-array detector was used with quantitation based on absorption at 303 nm.

Enzyme assays were conducted in a 10 mL screw-neck glass test tube containing 100 µL of lysate, 90 µL of a 250 µg/mL solution of 6-mercaptopurine in 0.01 M HCl, and 15 µL of 250 mM sodium phosphate buffer (pH 9.2). Reactions were initiated by the addition of 32 µL of a 3:1 mixture of 250 µM S-adenosyl-L-methionine and 30 mM dithiothreitol. The final pH was 7.5. After a 1-hour incubation at 37°C, the reaction was stopped by the addition of 850 µL of ice-cold 3.5 mM dithiothreitol and 50 µL of 1.5 M H_2SO_4. The tubes were then heated at 100°C for 2 hours. To each tube, 500 µL of 3.4 M NaOH was added, immediately followed by 8 mL of toluene–amyl alcohol–phenyl mercuric acetate. The tubes were shaken for 10 minutes and centrifuged. Then 6 mL of the toluene layer was transferred to a glass-stoppered conical test tube and 0.2 mL of 0.1 M HCl added. After vortex-mixing and centrifuging, the toluene layer was discarded. Samples (50 µL in 0.1 M HCl) were used for HPLC analysis. Product formation was linear for up to 120 minutes and 150 µL of lysate.

The source of enzyme assayed was red cell lysates prepared from whole blood.

9.9.25 NAD Pyrophosphorylase (Paulik et al., 1991)

NAD pyrophosphorylase catalyzes a reversible reaction in which the nucleotidyl moiety of ATP is transferred to nicotinamide mononucleotide to form NAD and pyrophosphate. The enzyme activity has been suggested as a target for cancer chemotherapy and is thought to be critical for cell survival.

NAD was separated from other nucleotides by chromatography on a Waters Partisil 10-SAX column (8 mm × 100 mm). The column was equilibrated with 5 mM ammonium phosphate buffer (pH 2.9). After injection, the same solvent was used for 10 minutes at a flow rate of 1 mL/min, and then for 2 minutes at 1.5 mL/min. A mobile phase was then switched to 0.65 M ammonium phosphate (pH 3.7) for 8 minutes at a flow rate of 2 mL/min, after which the column was reequilibrated with the initial solvent. Fractions from the NAD area were collected and counted by scintillation spectrometry.

The assay mixture in a total volume of 50 μL contained 50 mM glycylglycine buffer (pH 7.6), 6 mM ATP, 4 mM [4-^{3}H]nicotinamide mononucleotide (2 mCi/mmol, 8 μCi/mL), 60 mM MgCl$_2$, 20 mM nicotinamide, and enzyme extract. The mixture was incubated at 37°C for 10 or 20 minutes before the reaction was stopped by the addition of 50 μL of 20% trichloroacetic acid. The samples were cooled on ice before 10 minutes and centrifuged, and the supernates were neutralized with 0.5 M tri-n-octylamine in Freon before analysis by HPLC.

The assay was used to determine NAD pyrophosphorylase activity in extracts prepared from human chronic myelogenous leukemia cells, and in extracts prepared from rat tissues.

9.9.26 Nucleoside Diphosphate Kinase (Lambeth and Muhonen, 1993)

Nucleoside diphosphate kinase is relatively nonspecific enzyme that transfers a phosphoryl group from a variety of nucleoside triphosphates to nucleoside diphosphates. Its biological function is presumably to use ATP to phosphorylate the various ribo- and deoxyribonucleoside diphosphates to form the triphosphate derivatives need by the cell.

Chromatographic conditions were optimized to follow the reaction in which ATP phosphorylates UDP. These nucleotides, and the products, ADP and UTP, were separated on a Beckman C$_{18}$ Ultrasphere IP column (4.6 mm × 250 mm, 5 μm). The mobile phase was adjusted to pH 5.0 with H$_3$PO$_4$ and contained 100 mM potassium phosphate, 20 mM potassium acetate, 5 mM tetrabutylammonium hydroxide, and 12% acetonitrile. The effluent was monitored at 254 nm, and quantitation was based on the percentage conversion of UDP to UTP.

The reaction mixture contained in a volume of 0.30 mL: 33 mM N-Hepes (pH 7.4) 5 mM MgCl$_2$, 1 mM ATP, 1 mM UDP, and 1.6 μg of oligomycin. Reactions were run at 30°C for 5 minutes and were stopped by adding 0.2 mL of 0.2 M formic acid. Aliquots of these mixtures were injected directly into the HPLC system.

The assay was suitable for measuring the activity in intact and detergent-disrupted mitochondria isolated from rat, rabbit, and pigeon tissues.

Figure 9.120 shows an HPLC analysis of this reaction.

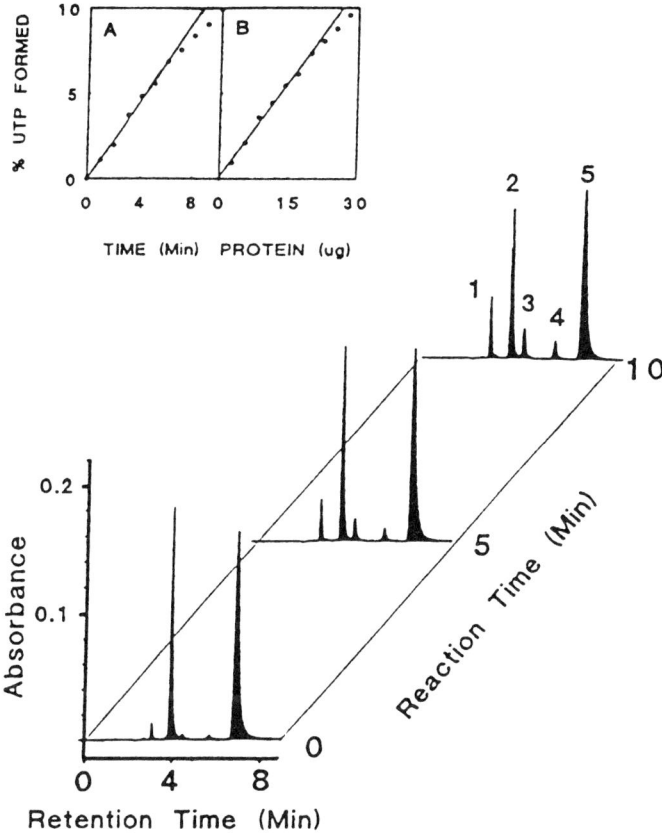

Figure 9.120 HPLC-based assay for nucleoside diphosphate kinase from rabbit liver mitochondria. One hundred microliters of 10-fold diluted supernate (140 μg of protein), obtained by centrifugation after freeze, thawing a suspension of rabbit liver mitochondria (100 mg of protein/mL), was added to 3.0 mL containing 1 mM UDP, 1 mM ATP, 5 mM MgCl$_2$, 33 mM Hepes (pH 7.4), and 5 μg of oligomycin. The reaction was run at 30°C and aliquots of 0.30 mL were removed at 1-minute intervals and transferred to autosampler vials containing 0.2 mL of 0.2 M formic acid. The chromatographic profiles for the reaction after 0, 5, and 10 minutes are shown. *Peaks:* 1, UMP; 2, UDP; 3, ADP; 4, UTP; 5, ATP. (*A*) Linearity of UTP formation, expressed as percentage conversion of UDP to UTP, with time. (*B*) Linearity of UTP formation with micrograms of protein added for 5-minute assays carried out in a volume of 0.30 mL. The latter assays were carried out in duplicate, using 1.4 to 42 μg of protein [14 μg corresponds to the amount (in 0.3 mL) used in the data for *A*]. The values shown are the averages for the duplicate assays. (From Lambeth and Muhonen, 1993.)

9.9.27 ATPase (Barret et al., 1993)

A variety of ATPase activities are present in biological systems. This assay is novel in that it uses hydrophilic interaction chromatography to measure release of labeled ^{32}P from [γ-^{32}P] ATP.

Reaction mixtures are chromatographed on a PolyHydroxyethyl Aspartamide column (4.6 mm × 100 mm) from Poly LC (Columbia, MD). Inorganic phosphate is not adsorbed to the column under the initial conditions (40 mM triethylamine–phosphoric acid, pH 2.8), while elution of ATP requires 800 mM triethylamine–phosphoric acid (pH 2.8). Initial conditions are maintained for 16 minutes and total run time was 30 minutes including reequilibration. Column effluent was monitored by a flow-through radioactivity monitor.

Samples containing enzyme were mixed with a solution of sonicated asolectin (0.5 mg/mL) and preincubated for 5 minutes at 30°C in a final volume of 100 μL containing 25 mM Mes-HCl (pH 6.0) and 50 mM NaCl. The reaction was started by the addition of 60 μL of [γ-^{32}P]ATP (400 μCi/mmol, 100 μM, final). After 10 minutes of incubation at 30°C, the reaction was stopped by addition of 500 μL of 20 mM triethylamine–phosphoric acid (pH 2.8) in acetonitrile.

The assay was used in inhibition studies of F_1-ATPase partially purified from rat liver mitochondria, and the ATPase activity of plasma membranes.

9.10 OXYGENATIONS

9.10.1 Acetanilide 4-Hydroxylase (Guenthner et al., 1979)

Acetanilide 4-hydroxylase (A4H) is a microsomal monooxygenase activity that can be followed by means of the model substrate acetanilide. In the HPLC assay developed for this activity, the conversion of acetanilide to 4-hydroxyacetanilide was followed.

The separation of the substrate and the product was accomplished by reversed-phase HPLC on an ODS column. The column was eluted isocratically with a mobile phase of methanol–water (33:67 v/v). The compounds were detected at 254 nm. The separations obtained are shown in Figure 9.121. Radiolabeled substrates were also used, and the eluent was assayed for radioactivity on fractions collected during the elution.

The reaction mixture contained Tris-HCl buffer (pH 7.6), $MgCl_2$, water, NADPH, and NADH. The microsomal suspension was added, followed by radiolabeled acetanilide to start the reaction. After 10 minutes, the reaction was terminated with ice-cold ethyl acetate, extracted, dried, and analyzed by HPLC. Formation of product is shown chromatographically and as a function of time of incubation in Figures 9.122 and 9.123, respectively.

The enzyme activity was obtained using a microsomal suspension prepared from mouse liver.

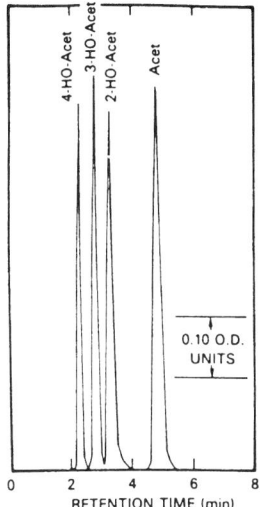

Figure 9.121 HPLC chromatogram of 4-hydroxy-, 3-hydroxy-, and 2-hydroxyacetanilide and acetanilide. About 12.5 μg of each of the four compounds in a total volume of 25 μL of methanol was injected. Approximately 0.33 mL fractions were collected. (From Guenthner et al., 1979.)

9.10.2 Ceruloplasmin (Richards, 1983)

Ceruloplasmin, a protein from the α-globulin fraction of human plasma, is usually considered to be the major copper transport protein. However, it also catalyzes the oxidation of biogenic amines, including catecholamines, adrenaline, noradrenaline, dopamine, and the indoleamine 5-hydroxytryptamine (5HT).

The assay for oxidation of these amines involves the use of any of the compounds listed above as the substrate and their separation from their respective products, aminochromes, by reversed-phase, ion-paired HPLC (ODS-Hypersil). The column was eluted isocratically using a mobile phase containing 50 mM potassium phosphate buffer (pH 5.5), 2 mM sodium heptanesulfonate, and methanol, at concentrations of 7.5 to 17.5% depending on the substrate. The detection was at 300 nm.

Activity was assayed in a reaction mixture (1.2 mL final volume) containing the substrate and buffer. The reaction was started by the addition of the enzyme and, after incubation at 37°C for 45 minutes was terminated by the addition of a 15 mM sodium azide solution. Samples were removed and injected directly onto the HPLC column for analysis. To generate products to be used as standards, oxidation of each of the amines to the corresponding aminochrome was carried out by incubation of the substrate with 30 mM potassium hexacyanoferrate(III) solution for 1 hour at room temperature.

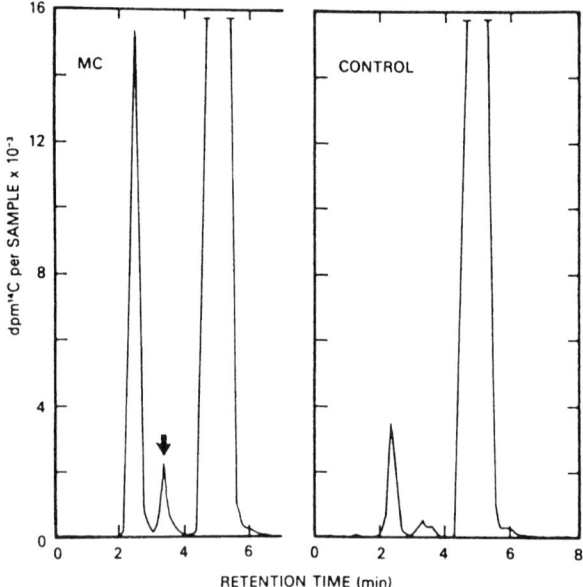

Figure 9.122 HPLC chromatogram of acetanilide and its metabolites generated by hepatic microsomes from mice induced with 3-methylcholanthrene (left) and control (right) mice. Approximately 0.33 mL fractions were collected. Arrow indicates formation of 2-OH acetanilide. (From Guenthner et al., 1979.)

Figure 9.124 shows a chromatogram of a sample following incubation of adrenaline with ceruloplasmin. The appearance of the peak of the reaction product, adrenochrome, indicates the activity of the enzyme. Figure 9.125 shows product formation following the oxidation of adrenaline and 5HT.

The enzyme was obtained from commercial sources or from human serum.

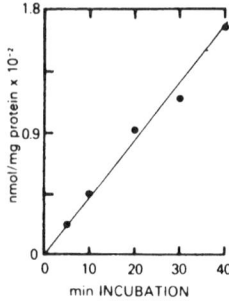

Figure 9.123 Formation of 4-hydroxyacetanilide in nanomoles per milligram protein as a function of incubation time. For these studies, the entire 4-hydroxyacetanilide peak (approximately 1.5 mL) was collected. (From Guenthner et al., 1979.)

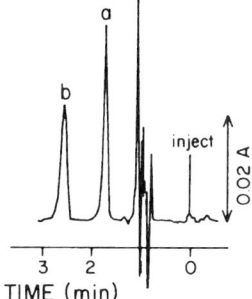

Figure 9.124 Chromatogram obtained following incubation of adrenaline with ceruloplasmin. *Peaks:* a, adrenochrome; b, adrenaline. (From Richards, 1983.)

9.10.3 Aryl Hydrocarbon Hydroxylase (EC 1.14.14.2) (Tulliez and Durand, 1981)

Aryl hydrocarbon hydroxylase (AHH) is part of the microsomal mixed-function oxidase system involved in the detoxification of polycyclic aromatic hydrocarbons. In the HPLC assay developed for the AHH activity, benzo[a]pyrene (BaP) is used as the substrate, and the activity is determined by measuring the unreacted BaP during the reaction.

The quantitation of BaP was accomplished by reversed-phase HPLC (LiChrosorb RP 18) with a mobile phase of 10% water in acetonitrile. The column was eluted isocratically, and the BaP was determined with a fluorometer using a 366 nm excitation and monitoring emission at 385 nm.

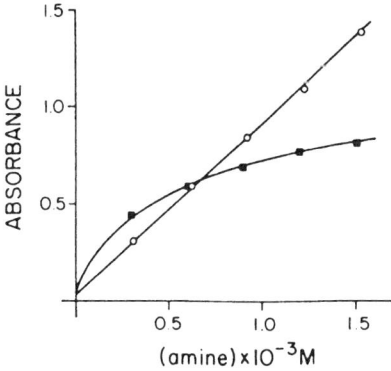

Figure 9.125 Relationship between amine concentration and oxidation product formation at a fixed level of ceruloplasmin. Oxidation of adrenaline (■) was monitored at 300 nm; oxidation of 5-hydroxytryptamine (○) was monitored at 315 nm. (From Richards, 1983.)

352 SURVEY OF ENZYMATIC ACTIVITIES ASSAYED BY THE HPLC METHOD

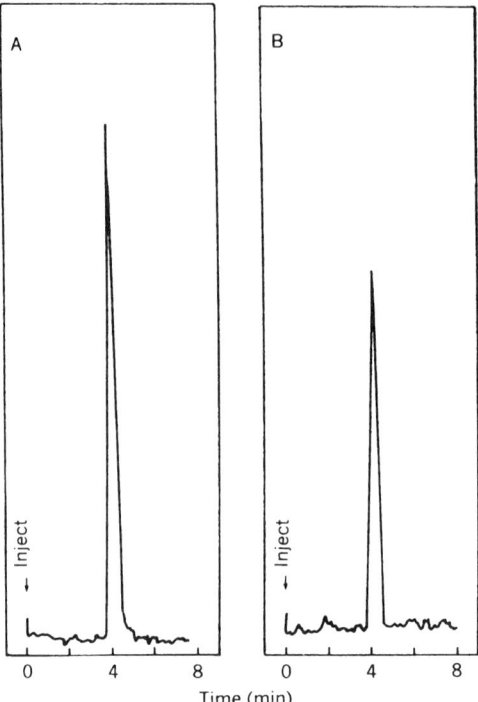

Figure 9.126 Liquid chromatograms of 10 μL samples of diluted incubation mixtures: (A) blank incubation and (B) assay incubation. After the 30-minute incubation period, dilution with acetonitrile was such that the BaP concentration in the control flask (A) was 1 μM; the same dilution was applied to the assay flask (B). The difference in the response gives the hydroxylation rate. (From Tulliez and Durand, 1981.)

The reaction mixture contained Tris-HCl buffer (pH 7.4), glucose-6-phosphate, NADP, glucose-6-phosphate dehydrogenase, and the microsomes. The mixture was preincubated, and the reaction was started by the addition of BaP dissolved in acetone. The mixture was incubated for 30 minutes at 37°C and terminated by the addition of cold acetone. Samples were diluted and injected for analysis. An example of the assay results is shown in Figure 9.126.

The activity was from rat microsomes.

9.10.4 Hepatic Microsomal Testosterone Hydroxylase (van der Hoeven, 1984)

Testosterone has been used as a model substrate for different cytochrome P450 monoxygenase activities. As a result of these oxidation reactions, multiple chemically related products are formed. In the assay developed for this activity, seven products are distinguished.

The separation of the substrate testosterone from the numerous reaction products was accomplished by reversed-phase HPLC on a C_{18} column preceded by a guard column. The compounds were separated by a combined isocratic and gradient elution procedure. The eluent was monitored at 240 nm. The separations obtained with these HPLC conditions are shown in Figure 9.127.

The reaction mixture contained 0.05 M Hepes-Na buffer (pH 7.4), $MgCl_2$, microsomes, and 1 mM testosterone. The reaction was initiated by the addition of the NADPH-generating solution, incubated for 10 minutes at 37°C, and terminated (after addition of the internal standard corticosterone in methanol) with dichlormethane. The dichlormethane extract was dried and analyzed by HPLC. A representative chromatogram of the metabolites of testosterone is shown in Figure 9.128 and the rates of formation of the different metabolites are shown in Figure 9.129.

The microsomal hydroxylase was prepared from male rates. Hepatic microsomes were isolated by treatment of a postmitochondrial fraction with polyethylene glycol 6000.

9.11 PTERIN METABOLISM

9.11.1 Folic Acid Cleaving Enzyme (DeWit et al., 1983)

This activity (FAS) cleaves folic acid into pterin-6-aldehyde and *p*-aminobenzoylglutamic acid, and the HPLC assay developed for it involves the separation of the two.

The separation was carried out on a reversed phase HPLC column (LiChrosorb 10 RP-18) or on an anion exchanger (Partisil PXS 10/25 SAX). For pterin-6-aldehyde, the mobile phase was a 15 mM phosphate buffer (pH 6.0) with 10% methanol. For *p*-aminobenzoylglutamic acid, a 0.1M NH_3 solution containing 0.2M NaCl, 20% (v/v) 1-propanol, and 10% (v/v) acetonitrile adjusted to pH 5.32 with acetic acid was used. Both columns were eluted isocratically, and the detection was by UV at 254 nm and by liquid scintillation counting.

The mixture contained FAS, radioactive folic acid, a potassium phosphate buffer (pH 6.0), EDTA, and DTT to prevent oxidation of reaction products. The reaction was terminated by the addition of an ice-cold anion-exchange resin suspension in ammonium carbonate and ethanol. Representative results are shown in Figure 9.130 for the two chromatographic systems.

The enzyme activity was from the cellular slime mold *Dictyostelium minutum*.

9.11.2 Dihydrofolate Reductase (Reinhard et al., 1984)

Dihydrofolate reductase (EC 1.5.1.3) catalyzes the reduction of the 5,6 double bond in the H_2-folate to form H_4-folate. The activity also converts H_2-biopterin

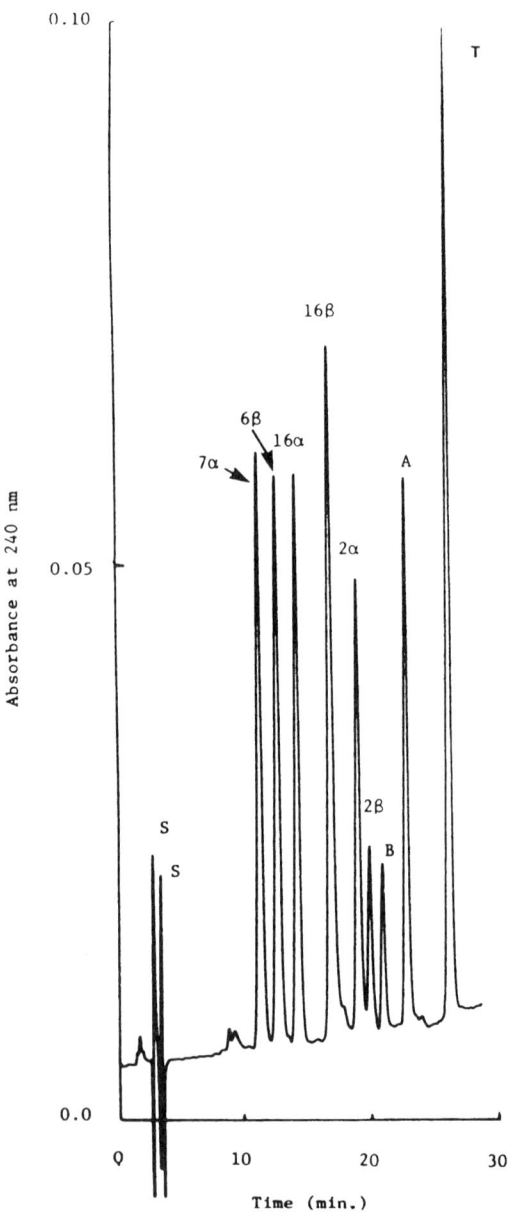

Figure 9.127 Separation of authentic standards of testosterone, its metabolites, and the internal standard corticosterone by HPLC: S, solvent; 2α, 2β, 6β, 7α, 16α, and 16β are the hydroxylated metabolites of testosterone; A is androstenedione, B is corticosterone, and T is testosterone. (From van der Hoeven, 1984.)

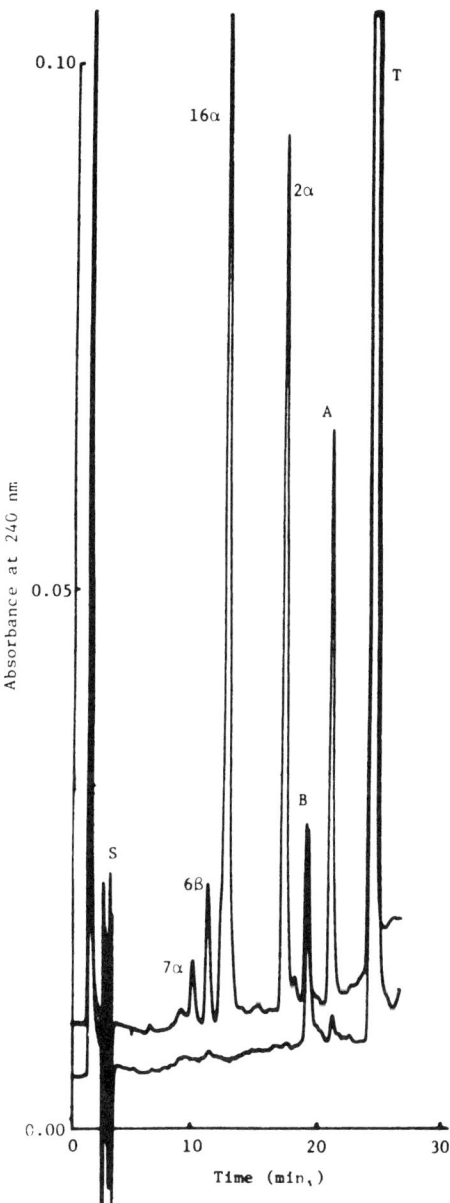

Figure 9.128 Typical chromatogram of the metabolites of testosterone formed by 1.0 mg of liver microsomal protein in 1.0 mL of reaction mixture incubated for 10 minutes at 37°C with (upper trace) and without (lower trace) a source of NADPH. (From van der Hoeven, 1984.)

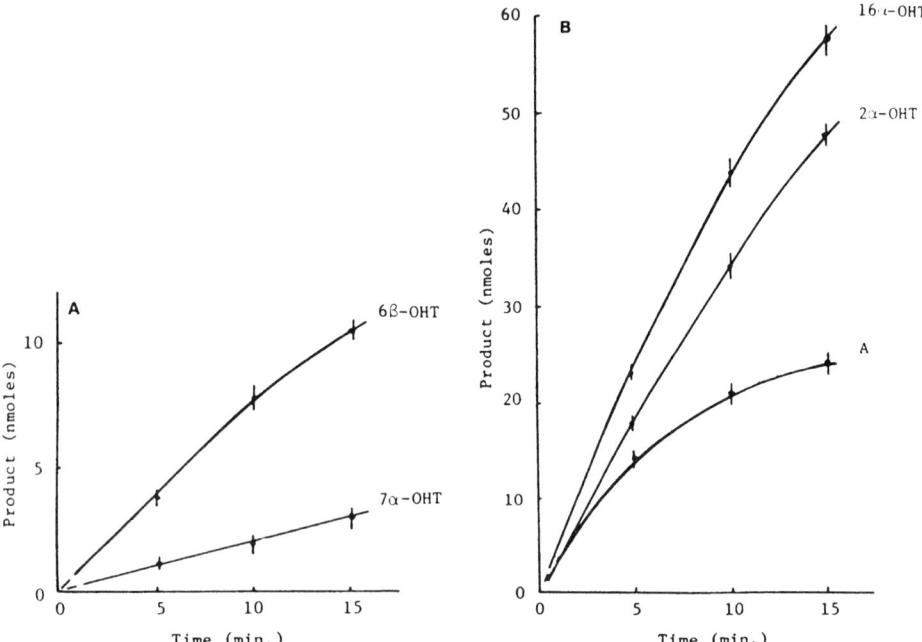

Figure 9.129 Amount of hydroxylated reaction product formed by 0.75 mg of hepatic microsomal protein in 1.0 mL of reaction mixture incubated at 37°C for different lengths of time. 2α, 6β, 7α, and 16α-OHT are the hydroxylated metabolites of testosterone; A is androstenedione. (From van der Hoeven, 1984.)

to H_4-biopterin, and the H_2-biopterin reductase may be the same enzyme. An HPLC assay for dihydrofolate reductase has been developed using H_2-biopterin as the substrate.

The separation of the substrate from the product was achieved by reversed-phase HPLC using an ODS column. The mobile phase consisted of an aqueous solution containing 0.5% acetonitrile and 0.1% tetrahydrofuran (v/v). The eluent was monitored fluorometrically with 350 and 450 nm excitation and emission wavelengths, respectively.

The assay mixture contained imidazole (pH 7.2), KCl, NADPH, glucose-6-phosphate, DTT, glucose-6-phosphate dehydrogenase, and H_2-biopterin. Samples were incubated for 60 minutes with tissue extract and terminated with 2 N TCA. After centrifugation (15,000g for 5 min), the H_4-biopterin was oxidized to pterin with iodine, the reaction was terminated with ascorbic acid, and samples were injected onto the HPLC column for analysis. An example of an assay is shown in Figure 9.131, and a composite of the rate of product formation is shown in Figure 9.132.

The activity was prepared from adult rat brain after homogenization, centrifugation, and desalting on G-25 Sephadex.

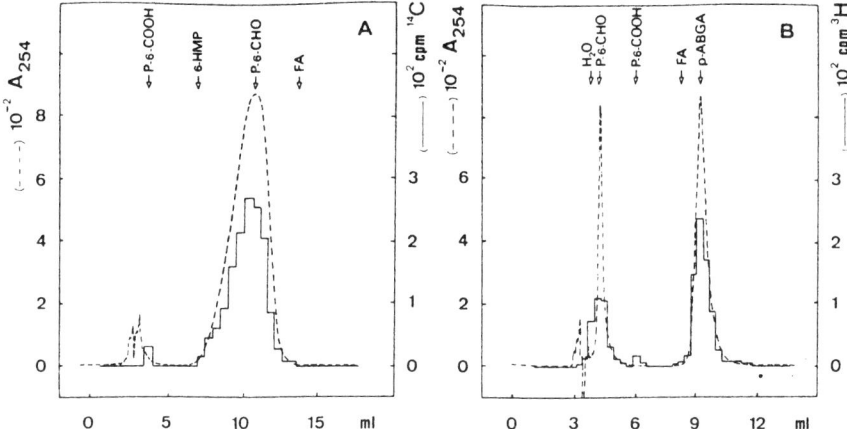

Figure 9.130 Identification of the products of folic acid C_9–N_{10} cleavage. (*A*) [2-^{14}C]folic acid was incubated with FAS and coinjected with pterin-6-aldehyde. Radioactivity was measured in 0.5 mL fractions. (*B*) [7,9,3′,5′-^{3}H]folic acid was incubated with FAS and coinjected with pterin-6-aldehyde and *p*-aminobenzoylglutamic acid. Radioactivity was determined in 0.5 mL fractions. Arrows indicate retention volumes of related compounds. Abbreviations: P-6-COOH, pterin-6-carboxylic acid; 6-HMP, 6-hydroxymethylpterin; P-6-CHO, pterin-6-aldehyde; FA, folic acid; p-ABGA, *p*-aminobenzoylglutamic acid. (From De Witt et al., 1983.)

9.11.3 Guanosine Triphosphate Cyclohydrolase I (Blau and Niederwieser, 1983)

D-*erythro*-7,8-Dihydroneopterin triphosphate synthetase, or GTP cyclohydrolase I (EC 3.5.4.16), catalyzes the formation of D-*erythro*-dihydroneopterin triphosphate (NH$_2$TP) from GTP. This activity is required for the synthesis of tetrahydrobiopterin. The HPLC assay developed for this activity involves the direct measurement of neopterin phosphates after separation from GTP and its other hydrolytic products.

Separation was carried out by ion-paired reversed-phase HPLC on a Li-Chrosorb RP-8 column with a mobile phase of isopropanol–triethylamine–85% phosphoric acid–water (3:10:3:984 v/v) at a final pH of 7.0. The column was eluted isocratically, and the nucleotides were detected at 254 and 287 nm and the pterins at 365 and 446 nm excitation and emission wavelengths, respectively. The separations obtained are shown in Figure 9.133.

The assay mixture contained GTP, EDTA, KCl, 10% glycerol, Tris-HCl (pH 7.8), and hydrolase. The reaction was incubated at 37°C for 90 minutes in the dark and was terminated by the addition of 1 *M* HCl containing 1% iodine and 2% KI. After 30 minutes, proteins were removed by centrifugation and excess iodine was removed with ascorbic acid. Samples were injected for analysis. Figure 9.134 shows the results of an assay, while Figure 9.135 shows the activity in a homogenate as a function of incubation time.

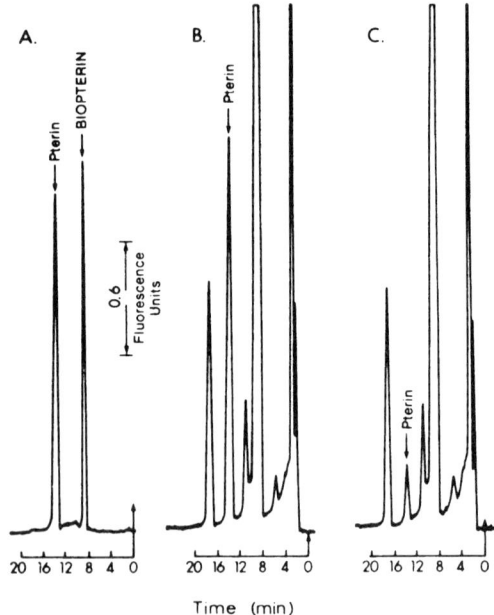

Figure 9.131 Chromatograms of enzyme activity. (A) Standard of 4 pmol biopterin and 10 pmol pterin; (B) brain extract incubated under assay conditions for 60 min; (C) boiled extract. The peak eluting at 18 min is imidazole. (From Reinhard et al., 1984.)

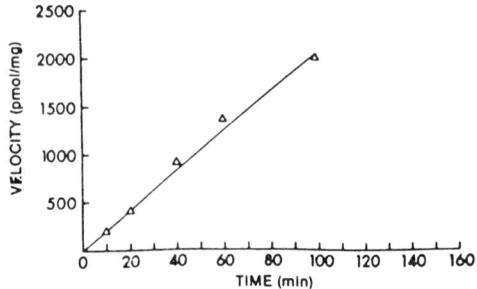

Figure 9.132 Effect of time on product formation. Desalted rat brain supernatants (510 µg of protein) were incubated for the times given on the ordinate. Boiled tissue blanks were subtracted. The concentration of H_2-biopterin was 250 µM. (From Reinhard et al., 1984.)

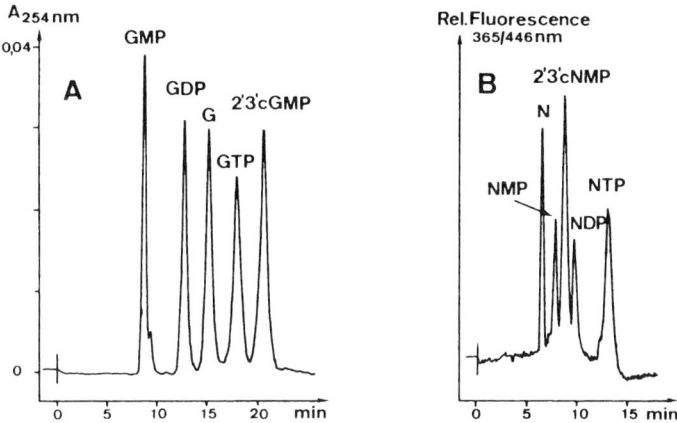

Figure 9.133 HPLC separation of (A) guanosine nucleotides and guanosine, injected amount 5 to 10 nmol each in 5 µL, and (B) neopterin phosphates. The mixture of neopterin phosphates injected was produced by partial hydrolysis of 16 pmol of NTP with alkaline phosphatase and addition of 2′,3′-cNMP. (From Blau and Niederwieser, 1983.)

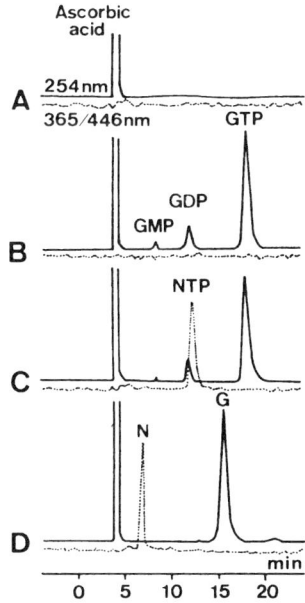

Figure 9.134 (A) HPLC of enzyme blank, mixture incubated without substrate; (B) substrate blank, mixture incubated without enzyme; (C) assay mixture; and (D) assay mixture after treatment with alkaline phosphatase. Solid lines represent UV absorption at 254 nm. Dotted lines represent fluorescence at excitation (365 nm) and emission (446 nm). (From Blau and Niederwieser, 1983.)

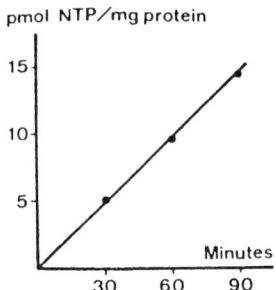

Figure 9.135 GTP cyclohydrolase I activity in rat liver homogenate as a function of incubation time. (From Blau and Niederwieser, 1983.)

The hydrolase activity was obtained from homogenized rat liver, centrifuged, and applied to a Sephadex G-25 column. The eluate was used as the hydrolase.

9.12 LIPID METABOLISM

9.12.1 Retinal Oxidase (Hupert et al., 1991)

Retinoic acid is a metabolic product of vitamin A that supports the growth and differentiation of epithelial tissues. Retinoic acid is formed in the cytosol by the reversible oxidation of retinol to retinal, and the irreversible oxidation of retinal to retinoic acid. There is controversy as to whether retinal is oxidized by "retinal dehydrogenase," which is linked to NAD^+, or by retinal oxidase.

Retinoic acid, retinol, and retinal are separated on a reversed-phase C_{18} column (4.6 mm × 150 mm, 5 μm). The mobile phase consisted of acetonitrile–1% ammonium acetate (60:40) for 6.0 minutes, a linear gradient to 100% acetonitrile in 0.5 minute, then a hold at 100% acetonitrile for 3.0 minutes, followed by a linear gradient back to the initial conditions in 0.5 minute. The system was allowed to equilibrate for 4.0 minutes before the next injection was made. The flow rate was 2.0 mL/min throughout. The column effluent was monitored at 340 nm.

Retinal oxidase was assayed in a medium containing 0.1 M phosphate buffer (pH 7.7) and 10 μL of 25 mM retinal dispersed in acetone containing 5% Triton X-100. The reaction was initiated by adding 20 to 40 μL of reconstituted ammonium sulfate precipitate or cytosol. The final volume was 500 μL. The reaction was continued at 37°C for 30 minutes and was stopped by freezing at −70°C, whereupon 50 μL was injected directly onto the column. Enzyme activity was linear with protein concentration up to 2.4 mg/mL, and with time up to 30 minutes.

Figure 9.136 HPLC elution pattern for retinoic acid, retinol, and retinal. Assay conditions: retinal, 500 μM; NAD, 1.0 mM; 0.65 mg protein; 0.1 M sodium phosphate buffer, pH 7.7; incubated at 37°C for 40 minutes. Retinoic acid (1.0 nmol), retinol (3.6 nmol), and retinal (23 nmol) eluted at 6.1, 10.4, and 11.2 minutes, respectively. The arrow indicates a change in attenuation (1/32) after 9.0 minutes. (From Hupert et al., 1991.)

The enzyme can be assayed in cytosol prepared by conventional means, or in the dissolved pellet obtained from treatment of cytosol with ammonium sulfate added to 45% of saturation.

Figure 9.136 shows representative chromatograms.

9.12.2 Serum Cholinesterase (Miller and Blank, 1991)

Serum cholinesterase (acetylcholine acylhydrolase, EC 3.1.1.8) refers to a family of at least 15 isozymes found in blood and many tissues of animals. In this assay, acetylcholine is used as the substrate, and a postcolumn reactor containing choline oxidase produces electrochemically active H_2O_2 upon oxidation of choline.

Choline is separated from ethylhomocholine (internal standard) and acetylcholine by chromatography on an ODS cartridge (3.2 mm $\times$ 100 mm) from Bioanalytical Systems. The mobile phase consisted of 20 mM Tris buffer (pH 7.50), containing 4.6 mL glacial acetic acid, 1.0 mM tetramethylammonium chloride, 200 μM octyl sodium sulfate, 6.0 mM sodium azide, 67 μM EDTA, and 2.0% acetonitrile. The 3.5 cm postcolumn reactor contained 60 units of acetylcholine esterase and 60 units of choline oxidase. Hydrogen peroxide was monitored electrochemically at a platinum electrode with $E_{app} = 0.50$ V versus Ag/AgCl. The assay is linear with up to 10 μL of serum and for the first 20 minutes of incubation.

The enzyme assay was prepared by mixing 500 μL of 0.100 M phosphate buffer (pH 7.20), 50 μL of ethylhomocholine, 25 μL of 500 mM acetylcholine, and 25 μL of serum diluted 1:4 with isotonic saline. Stock solutions of the substrate and internal standard were prepared in 10 mM acetate buffer (pH 4.5). The assay proceeded for 10 minutes ambient temperature (22°C) before being stopped by the addition of 25 μL of 2.50 M perchloric acid. This amount of perchloric acid did not stop the reaction instantaneously, but a correction can be made by running a zero-time control. The samples were filtered through 0.45 μm filters before injection of 20 μL aliquots into the HPLC.

Serum was obtained from blood acquired by cardiac puncture of mouse.

9.12.3 Carnitine Palmitoyltransferase I (Takeyama et al., 1989)

Carnitine palmitoyltransferases I and II catalyze the transfer of long-chain acyl coenzyme A into mitochondria. The I isozyme is located on the cytosol side of the inner membrane and catalyzes the formation acylcarnitine from acyl–CoA and carnitine. After acylcarnitine crosses the inner membrane, it is converted back to acyl–CoA by the action of the II isozyme. This assay measures the activity of carnitine palmitoyltransferase I in intact mitochondria.

The enzyme was assayed by quantitating the coenzyme A released. The assays are run in the presence and absence of L-carnitine to correct for enzymatic and nonenzymatic hydrolysis of palmitoyl–CoA. The separation was made at 40°C on a LiChrosorb RP-18 column (4.0 mm × 250 mm, 5 μm). The mobile phase was prepared from solvent A [220 mM NaH$_2$PO$_4$ and 0.05% (v/v) β-thiodiglycol] and solvent B [125 mM NaH$_2$PO$_4$, 43% (v/v) methanol, 0.9% (v/v) chloroform, and 0.05% (v/v) β-thiodiglycol]. The composition of the mobile phase was 12% solvent B at time 0, 15% B at 8 minutes, 65% B at 28.5 minutes, and 100% B at 29 to 31 minutes. The mobile phase was returned to starting conditions within 1 minute and maintained there for 28 minutes before injecting the next sample. The column effluent was monitored at 254 nm.

The enzyme assay contained in a final volume of 1.0 mL, 75 mM KCl, 50 mM mannitol, 25 mM Hepes (pH 7.0), 0.2 mM EGTA, 2 mM KCN, 5 mM dithiothreitol, 1.75 mg fatty-acid-free bovine serum albumin, 30 to 120 μM palmitoyl–CoA, and 0.5 mM L-carnitine. After 2 minutes of preincubation at 25°C, the reaction was initiated by adding 100 μL (0.1–0.3 mg protein) of mitochondrial suspension. The reactions were terminated after 5 minutes by the addition of 50 μL of 60% (v/v) perchloric acid. The supernates were adjusted to pH 2 to 3 by adding 2 M potassium phosphate, cooled on ice, and then centrifuged to remove potassium perchlorate. After filtering, 50 μL of the filtrate was injected directly into the chromatograph.

The source of enzyme was rat liver mitochondria suspended in 0.25 M sucrose containing 2 mM Hepes (pH 7.4) and 1 mM EGTA.

9.12.4 Fatty Acid ω-Hydroxylase (Romano et al., 1988; Yamada et al., 1991)

Fatty acid ω-hydroxylation is involved in the metabolism of prostaglandins and leukotrienes, and is the first step of dicarboxylic acid formation in the cell. In this assay, lauric acid is hydroxylated to form 12-hydroxylauric acid, which is then fluorescence-labeled on the carboxyl group with 3-bromomethyl-7-methoxy-1,4-benzoxazin-2-one.

The hydroxylated products of lauric acid, 12-hydroxylauric acid and 11-hydroxylauric acid, were separated from each other and 10-hydroxycapric acid (the internal standard) by chromatography on a Waters-type ODS column (6 × 150 mm, 5 μm). The mobile phase was aqueous 58% acetonitrile containing 1% acetic acid. Detection was by measuring fluorescence at 430 nm with excitation at 355 nm.

The reaction contained 0.1 mM sodium laurate, 0.7 mM NADPH, 50 mM Tris-HCl (pH 7.5) and microsomes (0.5 mg from liver, or 0.3 mg from kidney) in a final volume of 1.0 mL. The reaction was initiated by adding NADPH and terminated after 10 minutes at 37°C by adding 0.25 mL of 4 M HCl. The internal standard, 0.25 mL of 2.5 μg/mL 10-hydroxycapric acid, was then added. The mixture was extracted with 4.5 mL of diethyl ether. The organic phase was washed once with 4 mL of water and evaporated to dryness under reduced pressure. The residue was dissolved in 0.1 mL of acetonitrile containing 10 mM 3-bromomethyl-7-methoxy-1,4-benzoxazin-2-one and 0.1 mL of 15 mg/mL 18-crown-6 in acetonitrile saturated with K_2CO_3. The resulting solution was allowed to stand at 40°C for 30 minutes before 0.05 mL of 2% acetic acid was added. Two-microliter aliquots were used for HPLC analysis.

The source of enzyme was microsomes prepared from liver and kidney cortex of rats. Some animals were treated with dehydroepiandrosterone to increase ω-hydroxylation activity.

9.12.5 Acyl-CoA: Alcohol Transacylase (Garver et al., 1992)

Garver et al. (1992) developed an assay that measures the activity of the acyl-CoA: alcohol transacylase involved in the biosynthesis of storage liquid wax esters in jojoba and some microbes and algae.

Components of the reaction mixture labeled with ^{14}C (octadecenoyl-CoA, octadecenoic acid, and dodecanoyl octadecenoate) are separated on a Waters μBondapak reversed-phase C_{18} column (2 mm × 300 mm). The mobile phase at the beginning was 80:20 (v/v) methanol-water. The percentage of methanol was increased to 100% over a 30-minute period and held there for 10 minutes

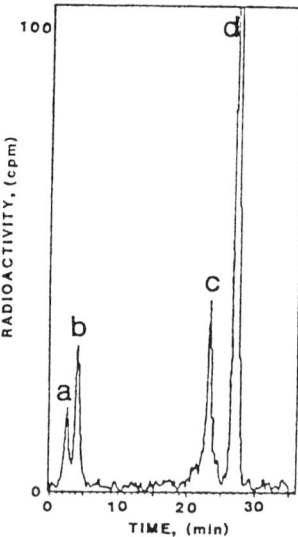

Figure 9.137 Separation of four ^{14}C-labeled components produced after incubating [^{14}C]octadecenoyl–CoA and dodecanol with a crude homogenate of jojoba cotelydons at 35°C for 20 minutes. Radiolabeled compounds were detected with a flow-through radiochemical detector. *Peaks:* a, octadecenoyl–CoA; b, octadecenoic acid; c, unidentified product; d, dodecanoyl octadecenoate ($C_{30:1}$). (From Garver et al., 1992.)

to elute longer chain wax esters present in the homogenate. A radioactive flow-through detector was used to monitor the effluent.

The enzyme was assayed at 35°C for 20 minutes in 259 μL containing 50 μmol Tris-HCl (pH 8.0), 1.25 μmol $MgCl_2$, 125 nmol dodecanol, 25 nmol ATP, 25.5 nmol [^{14}C]octadecenoyl–CoA, and 0 to 165 μg of enzyme protein extract. Reactions were terminated with 100 μL of glacial acetic acid. The ^{14}C-labeled wax ester was extracted with 1 mL of 6.7% Na_2So_4 and 1 mL of hexane. The upper organic phase was evaporated with a stream of nitrogen. The dried product was resuspended in 200 μL of acetone and filtered before injecting into the HPLC. The separation obtained for labeled compounds is shown in Figure 9.137.

The source of enzyme was jojoba cotyledons that were frozen with liquid nitrogen and then powdered by an Omni-Mixer homogenizer. The powdered cotyledons were homogenized and extracted by using a mortar and an isotonic homogenization buffer. The homogenate was filtered through four layers of cheesecloth.

9.12.6 Lipase (Maurich et al., 1991)

Lipases are found widely in nature and catalyze the hydrolysis of esters of glycerol and long-chain fatty acids. This assay shows much improved sensitivity by measuring the *p*-nitrophenol liberated from its ester with lauric acid.

Separation of p-nitrophenol from the internal standard, 2,4-dinitroaniline, is achieved on LiChrosorb RP-18 column (4.6 mm × 250 mm, 10 μm) from Perkin-Elmer. The mobile phase was a 55:45 (v/v) mixture of water and acetonitrile. The effluent was monitored at 300 nm.

Lipase was assayed in a final volume of 1.4 mL containing 1.2 mL of substrate suspension [8 mg of p-nitrophenyllaurate dissolved in 10 mL of acetone, which was then added with agitation to 50 mL of 0.2 M phosphate buffer (pH 7.4), 10 mL of 25 mM sodium taurocholate, 10 mL of 70 mM NaCl, and 20 mL of water] and enzyme in 0.1 M phosphate buffer (pH 7.4). This mixture was incubated at 37°C for 30 minutes before the reaction was stopped by adding 0.1 mL of 5 M HCl. Then 100 μL of internal standard (0.2 mg/mL of 2,4-dinitroaniline in ethanol) was added. After filtration, samples are injected into the HPLC. The assay is linear with up to 2.5 μg protein and 30 minutes incubation.

The enzyme source was a commerical pharmaceutical preparation from pancreas of pig.

9.13 MODIFICATION OF PROTEINS AND PEPTIDES

9.13.1 Tyrosine Protein Kinase (Ferry et al., 1990; Boutin et al., 1992)

Phosphorylation of tyrosyl residues in regulatory proteins may play a role in cancer. The assay of Ferry et al. (1990), measures the phosphorylation of tyrosine in angiotensin II, which serves as a convenient substrate.

[^{32}P]Phosphotyrosyl–angiotensin II is separated from labeled ATP and other compounds by chromatography on either μBondapak phenyl or μBondapak C_{18} (both 4 mm × 300 mm columns from Waters). A gradient program was used as follows: 0 to 5 minutes, 90% A (0.04 M sodium phosphate, pH 7.4, 5 mM tetrabutylammonium phosphate) and 10% B (5 mM tetrabutylammonium phosphate in acetonitrile); 5 to 20 minutes, linear gradient up to 30% B; 20 to 22 minutes, up to 70% B; 22 to 27 minutes, hold at 70% B; return to 100% A in 3 minutes and hold for 2 minutes. The flow rate was 1.5 mL/min. An on-line radioactive monitor detected labeled compounds by using Cerenkov radiation.

The reaction mixture contained in a final volume of 100 μL: 40 μL of buffer [20 mM Hepes-NaOH (pH 7.4), 5 mM $MnCl_2$, 5 mM $MgCl_2$; 10 μL of [γ-^{32}ATP] (10 μM, final)]; 10 μL angiotensin II (3 mg/mL final); 20 μL of saturated NaCl solution and 20 μL of enzyme source. The reaction was started by adding ATP. Incubations were carried out for 30 minutes at 30°C and were stopped by adding 500 μL of 50% trichloroacetic acid. After centrifugation, 30 μL of the supernate was injected.

The enzyme source was HL-60, a human promyelocytic cell line. Cells were washed in phosphate-buffered saline and suspended and sonicated in a low ionic strength extraction buffer. After centrifugation, the supernate was used for the assays.

Several isoforms of tyrosine protein kinase play roles in cellular growth and differentiation. The assay of Boutin et al. (1992), exploits the use of hydrophilic interaction chromatography to separate hydrophilic peptides.

Separation of [γ-^{32}ATP] from labeled peptides has been accomplished for over 100 peptides of various sequences by using HPLC columns and solid-phase extraction cartridges packed with PolyHydroxyethyl Aspartamide (PolyLC, Columbia, MD). ATP is strongly retained on such columns. In most cases, labeled ATP was removed by passing the incubation media through a disposable solid phase extraction column (10 mm × 4.6 mm) filled with PolyHydroxyethyl Aspartamide. The extraction column was preequilibrated with 3 mL of 20 mM triethylamine phosphate (pH 2.8) in 50% acetonitrile. The assay medium (600 μL) was passed through the cartridge, followed by 1 mL of 20 mM aqueous triethylamine phosphate (pH 2.8). ATP was eluted with 2 mL of 800 mM aqueous triethylamine phosphate (pH 2.8). An aliquot of 30 μL was analyzed on a PolyHydroxyethyl Aspartamide HPLC column (4.6 mm × 200 mm, 5 μm) using the following protocol: 0 to 5 minutes, 100% A (4 mM triethylamine phosphate, pH 2.8, containing 90% acetonitrile), 5 to 20 minutes, linear gradient to 100% B (10 mM triethylamine phosphate (pH 2.8). Phosphorylated peptides were detected by an on-line radioactivity monitor.

The standard assay used angiotensin II and contained in a final volume of 100 μL: 40 μL of buffer (20 mM Hepes-NaOH, pH 7.4, 5 mM MnCl$_2$, 5 mM MgCl$_2$), 10 μL of ATP (100 μM, containing 30 μC/ mL [γ-^{32}ATP]), 10 μL of angiotensin II (10 mg/mL), 20 μL of saturated NaCl solution, and 20 μL of enzyme source. The reaction was started by adding ATP. Incubations were carried out for 30 minutes at 30°C and were stopped by adding 500 μL of 20 mM triethylamine phosphate (pH 2.8)–acetonitrile (50 : 50).

The enzyme source was tyrosine protein kinase purified 2300-fold from HL-60, a human promyelocytic cell line.

Fig. 9.138 shows representative chromatographic separations.

9.13.2 Adenosine Diphosphate–Ribosylarginine Hydrolase (Kim and Graves, 1990)

ADP ribosylation of proteins is an important posttranslational modification reaction. Cholera, pertussis, and diphtheria toxins all act by attaching the ADP–ribose moiety to a target protein. There is increasing evidence that an ADP–ribosylation cycle may be present in a number of animal cells.

An artificial substrate labeled with a strong UV-absorbing chromophore was synthesized. This substrate and its product, obtained by hydrolysis of the ADP–ribosyl moiety, were separated on a Zorbax-ODS reversed-phase C$_{18}$ column (4.6 mm × 250 mm). The initial mobile phase was 70% 0.0435 M sodium acetate (pH 4.0) and 30% acetonitrile. The percentage of acetonitrile

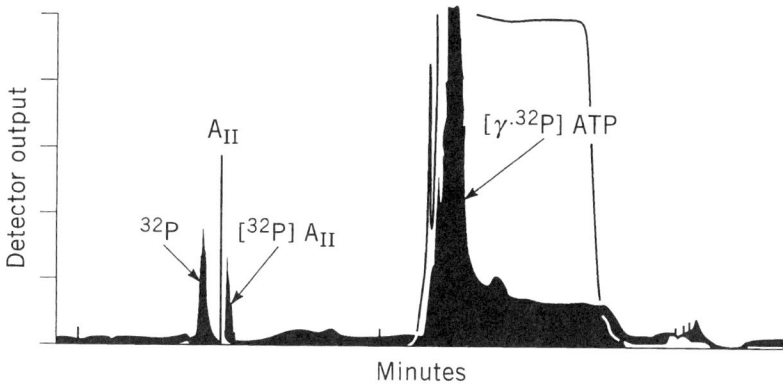

Figure 9.138 Analysis of angiotensin II phosphorylation by tyrosine protein kinase from HL-60. A 30-μL aliquot of incubation medium was analyzed. The black profile shows the radioactivity monitor (2500 mV full-scale output). The open profile shows the absorbance at 280 nm (sensitivity of detection 0.1 AUFS). (From Boutin et al., 1992.)

was increased linearly to 70% in 20 minutes and then increased to 100% in 5 minutes. The column eluate was monitored at 475 nm.

ADP–ribosylarginine hydrolase was assayed in 100 μL containing 20 mM potassium phosphate (pH 7.2), 15 mM magnesium chloride, 5 mM dithiothreitol, and 15 μM ADP-R-DABS-AME. The reaction was initiated with 10 μL of enzyme solution. After incubation at 30°C for 30 minutes, the reaction was stopped by addition of 100 μL of ice-cold 10% (w/v) trichloroacetic acid. After centrifugation, the supernate was injected onto the column.

The sources of enzyme used in this study were a high speed supernate from homogenized chicken muscle cells, and enzyme purified 80-fold from skeletal muscle of rat.

Figure 9.139 shows that the enantiomers of the substrate are separated, indicating the one that is reactive.

9.13.3 Peptidylglycine α-Amidating Monoxygenase (Chikuma et al., 1991)

Several biologically active peptides have a C-terminal α-amide structure that is formed by the action of a copper- and ascorbate-dependent monoxygenase on a C-terminal glycine.

Substrate, Dabsyl-Gly-Phe-Gly, and product, Dabsyl-Gly-Phe-NH$_2$, are separated at 35°C by reversed-phase chromatography on a TSK gel ODS-80TM column (74.6 mm × 150 mm). The mobile phase was composed of 41:59 (v/v) 0.01 M sodium acetate (pH 4.0)–acetonitrile. The effluent was monitored at 460 nm.

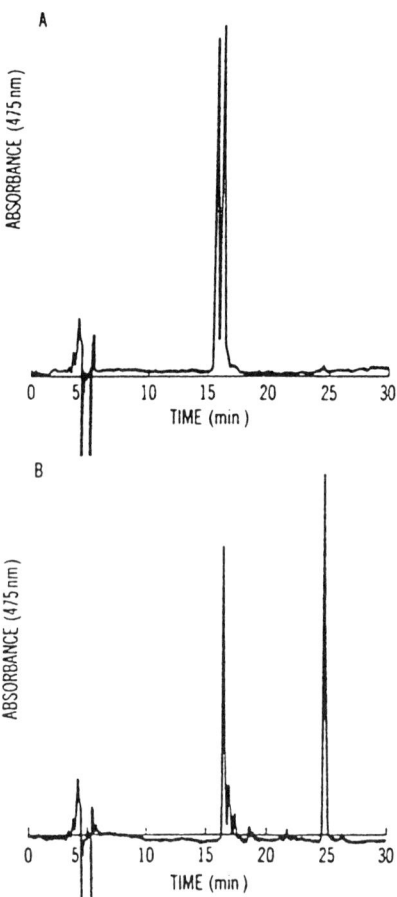

Figure 9.139 HPLC elution profile of ADP–ribosylarginine hydrolase assay: 200 μL of reaction mixture containing 20 mM phosphate (pH 7.5), 15 mM magnesium chloride, and 20 μM ADP-R-DABS-AME was initiated by adding 10 μL (50 μg of protein) of high-speed supernatant of homogenized chicken muscle cells. A 100 μL aliquot was taken from the reaction mixture at the indicated time points and processed. Reaction times were (A) 0 hours and (B) 2 hours. Sensitivity of detection was 0.2 AUFS. (From Kim and Graves, 1990.)

The reaction mixture contained 50 mM sodium acetate (pH 5.4), 200 mM NaCl, 0.5 mM NEM, 100 μg/mL beef liver catalase, 50 μM Dabsyl-Gly-Phe-Gly, 2 mM ascorbate, and $CuSO_4$ concentrations varying in the micromolar range depending on enzyme source being assayed, and enzyme.

The final volume was 250 μL. After incubation at 34°C, the reaction was terminated by heating in boiling water for 5 minutes. After centrifugation, Dabsyl-N-Leu was added to the supernate as an internal standard, and an aliquot of the mixture was used for HPLC analysis.

The assay was used to measure of the activities of the enzyme in extracts of bovine pituitaries and the saliva of rats.

9.13.4 Myosin Light Chain Kinase (Nakanishi et al., 1991)

Myosin light chain kinase is a Ser/Thr-type protein kinase involved in regulation of smooth muscle. The enzyme is also found in smooth muscle and platelets. This assay uses a synthetic substrate that is not radioactively labeled.

The substrate, KKRPQRATSNVFS-NH$_2$, and its phosphorylated derivative, are separated by chromatography at 40° C on a C$_{18}$ silica column (4.6 mm × 150 mm). The mobile phase contained 18% acetonitrile–0.1% trifluoroacetic–H$_2$O. The column eluate was monitored at 220 nm.

The peptide substrate was phosphorylated in a reaction mixture (0.25 mL) containing 25 mM Tris-HCl (pH 7.5), 0.5 mg/mL bovine serum albumin, 4.0 mM MgCl$_2$, 0.2 mM CaCl$_2$, 2.6 nM calmodulin, 24 μM peptide, 1.5 nM myosin light chain kinase, and 400 μM ATP. The reaction was started by adding ATP. After 30 minutes at 28°C, the reaction was terminated by adding 0.1 mL of 10% acetic acid. Aliquots were directly applied to the HPLC column. The reaction was linear for up to 60 minutes.

Myosin light chain kinase was purified from chicken gizzard smooth muscle.

9.13.5 Transglutaminase (Fink et al., 1992)

Transglutaminase catalyzes cross-linking between peptide chains in fibrin clots, the cornified envelope of epidermal cells, and the vaginal plug from seminal plasma in rodents. Glutaminyl residues serve as acyl donors to peptide-bound lysine. This assay makes use of a small synthetic peptide, benzyloxycarbonyl-L-Gln-Gly, and fluorescent amine monodansylcadaverine as substrates.

The substrates named and their product were resolved a Beckman Ultrasphere-ODS column (4.6 mm × 150 mm, 5 μm), using a mobile phase composed of 50:50 methanol–water. Two detectors were used, a variable wavelength detector and a fluorometer (excitation, 352–360 nm; emission, 480–520 nm). Monodansylcadaverine is retained on the column and is removed with 100% methanol at the end of a series of chromatographic runs. A number of peptide substrates can be used, but changes in the chromatographic conditions are then required.

The enzymatic reaction was carried out at 37°C in 50 μL containing 0.1 M Tris-HCl (pH 7.45), 1 mM EDTA, 10 mM CaCl$_2$, 1 or 2 mM monodansylcadaverine, varying amounts of substrate, and enzyme (0.4–5 μg). Reactions were terminated by the addition of excess EGTA or organic solvent (methanol or acetonitrile). Didansylcadaverine was used as an internal or external standard to calibrate the fluorescence response for quantitation.

The sources of glutaminase studied included guinea pig liver and chicken liver. In the latter case, tissue was homogenized in 0.25 M sucrose and the supernate obtained after centrifugation was used.

9.13.6 Phosphotyrosyl Protein Phosphatase (Nash et al., 1993)

Phosphorylation of tyrosyl residues in proteins play an important role in cell regulation. Dephosphorylation of these residues by protein phosphatases must be equally important. This assay makes use of synthetic phosphorylated peptides that can be monitored fluorometrically to provide a sensitivity comparable to the use of ^{32}P-labeled substrates.

Substrate, N^α-fluorenylmethoxycarbonyl-EEY(P)AA, and the dephosphorylated product were separated on a C_{18} Novapak HPLC column (3 mm × 100 mm) using a mobile phase containing 36% acetonitrile–water–0.1% TFA. The eluate was monitored spectrophotometrically at 265 nm, or fluorimetrically (excitation, 268 nm; emission, 307 nm).

The assay was conducted by mixing 75 μL of substrate (0.21 mM in 0.1 M acetate buffer, pH 4.9) with 10 to 20 μL of enzyme. Aliquots (15–20 μL) were withdrawn at intervals and added to 9 volumes of ice-cold methanol to stop the reaction. After centrifugation, 25 μL aliquots were injected.

The source of enzyme was an extract from the spleen of mouse.

Figure 9.140 shows a representative chromatogram.

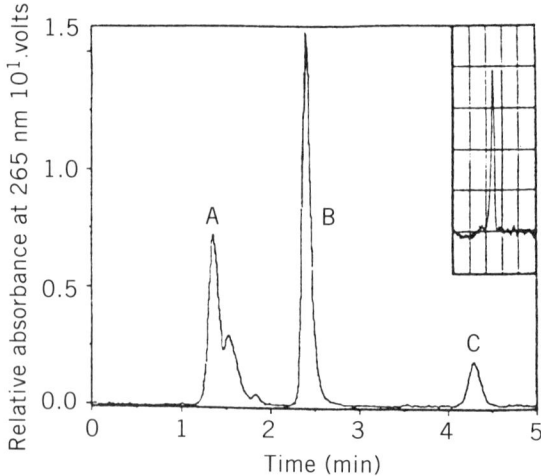

Figure 9.140 HPLC elution profile of a sample taken from a discontinuous assay of crude mouse spleen extract, using N^α-fluorenylmethoxycarbonyl-EEY(P)AA [Fmoc-EEY(P)AA] as substrate. Chromatography was performed on a C_{18} Novapak column (10 cm × 8 mm) eluted isocratically with 36% acetonitrile–water (0.1% TFA) (v/v) at a flow rate of 2 mL/min. Peaks are due to (*A*) methanol and methanol-soluble compounds derived from the sample of crude homogenate, (*B*) Fmoc-EEY(P)AA (1238 pmol), and (*C*) Fmoc-EEYAA (195 pmol). Appropriate controls showed that no interfering compounds eluted in the position of the peptides. *Inset:* Fluorescence monitoring of HPLC of Fmoc-EEYAA (75 fmol) eluted isocratically with 36% acetonitrile–water (0.1% TFA). Excitation and emission wavelengths, 268 and 307 nm, respectively, with gain × 100 and 10 mV chart scale. (From Nash et al., 1993.)

9.13.7 Phosphotyrosine Phosphatases (Madden et al., 1991)

Madden et al. (1991) developed an assay that measures the activity of a phosphotyrosine phosphatase that is active with a chemically synthesized, dansylated phosphopeptide corresponding to the 1143–1153 region of the human placental insulin receptor β subunit.

The substrate, dansyl-Gly-Arg-Asp-Ile-Tyr-Glu-Thr-Asp-Tyr(P)-Tyr-Arg-Lys, and its dephosphorylated product were separated at 30°C on a Spherisorb 3 μm ODS-2 column (4.6 mm × 50 mm). The mobile phase consisted of 20% (w/v) acetonitrile-10 mM sodium phosphate (pH 7.2). The dansyl label was detected by fluorimetry. The limit of detection was 1 pmol/20 μL.

Phosphotyrosine phosphatase activity was measured at 30°C in a total volume of 100 μL containing 24 mM imidazole (pH 7.2), 1 mM EDTA, 1 mM dithiothreitol, 100 μg of bovine serum albumin, 50 μM dansyl phosphopeptide, and extracts containing enzyme activity. After 15 minutes, the reaction was terminated by the addition of 20 μL of 30% (w/v) trichloroacetic acid. The mixture was centrifuged before injection for HPLC analysis. The reaction was linear with respect to both time and enzyme concentration up to at least 30% substrate conversion.

The enzyme used was partially purified from human placental membranes.

9.13.8 Protein Phosphatase 2B (Calcineurin) (Enz et al., 1994)

Calcineurin is a calcium/calmodulin-dependent protein phosphatase that may be involved in neurotransmission and T-cell proliferation. This assay uses nonlabeled peptides and is comparable in sensitivity to a radioactive assay.

The peptide, D-L-D-V-P-I-P-G-R-F-D-R-R-V-S-V-A-A-E, was synthesized by solid phase methods and phosphorylated by protein kinase. Phosphorylated and dephosphorylated peptides were both determined after separation on a C_{18} column, LiChroCART from Merck (4 mm × 125 mm, 5 μm). The mobile phase was 10 mM phosphate in 18% (v/v) acetonitrile adjusted to pH 5.9 with NaOH. The effluent was monitored at 205 nm.

Calcineurin activity was measured at 30°C in a total volume of 150 μL containing 50 mM Tris-HCl (pH 7.0), 0.1 mM EGTA, 0.5 mM dithiothreitol, 0.01% Brij, 0.3 mg/mL bovine serum albumin 0.1 μM calmodulin, 1 mM MnCl$_2$, 1 mM CaCl$_2$, and variable amounts of calcineurin (0.11–0.44 μg/150 μL). Aliquots (20 μL) were taken at intervals and the reaction was stopped with 70 μL of 0.5% perchloric acid containing 0.1% Triton X-100 and the internal standard, probenecid (10 nmol/mL). After centrifugation, a 40 μL aliquot of the supernate was directly applied on the HPLC column. The rate of enzyme activity was calculated from the linear appearance of product over the first 6 minutes.

The enzyme assayed was a commercial preparation of calcineurin purified from bovine brain.

9.14 VITAMIN METABOLISM

9.14.1 Thiamine Triphosphatase (Bettendorff, 1991)

Thiamine triphosphate is present in low concentration in most tissues, but its role remains unknown. Organs and tissues that generate electrical impulses are particularly rich in this compound.

Thiamine triphosphate, diphosphate, and monophosphate were separated as the thiochrome derivatives on Hamilton PRP-1 column. The mobile phase was 15 mM phosphate buffer (pH 8.5) containing 1% tetrahydrofuran. The solvent flow rate was 0.5 mL/min, and a 20 μL sample loop was used. Detection was by fluorescence using excitation and emission wavelengths of 365 and 433 nm, respectively.

The reaction mixture was composed of 50 μL of membrane preparation, 10 mM Hepes-Tris buffer (pH 6.8), 1.5 mM MgCl$_2$, 1.5 mM EGTA, and 0.1 mM thiamine triphosphate in a total volume of 100 μL. After 15 minutes of incubation at 25°C, the reaction was stopped by addition of 500 μL of 6% trichloroacetic acid. The supernate obtained by centrifugation was extracted with 4 volumes of diethyl ether. The thiamine derivatives were transformed into fluorescent thiochromes by the addition of 50 μL of oxidant [4.3 mM K$_3$Fe(CN)$_6$ in 15% NaOH] to 80 μL of sample. Thiamine diphosphatase activity was minimized by using magnesium as the metal and EGTA to chelate calcium.

The source of enzyme was crude or partially purified membranes from the main electrical organ of *Electrophorus electricus*.

9.14.2 Lipoamidase (Hayakawa and Oizumi, 1987)

Lipoamidase is an activity in serum that hydrolyses lipoyllysine. This assay uses lipoyl-4-*p*-aminobenzoate as a substrate that allows for fluorimetric detection.

Lipoyl-4-*p*-aminobenzoate and the product, *p*-aminobenzoate, were separated on a Nucleosil 3C$_{18}$ column (4 mm × 50 mm). Solvent A was 0.1% trifluoroacetic acid and solvent B was methanol. The column was equilibrated with solvent A. In the first minute the composition was changed to 30% B, and then to 100% B by minute 11. This composition was held for 1 minute before returning to the starting conditions. The flow rate was 1 mL/min. Detection was by fluorescence using excitation and emission wavelengths of 276 and 340 nm, respectively.

The reaction mixture was composed of 20 μL of serum or partially purified enzyme and 80 μL of substrate solution containing 50 to 100 μM substrate in 0.1 M sodium phosphate buffer (pH 7.0) containing 10 mM 2-mercaptoethanol. After incubation at 37°C for an appropriate time (up to 100 min), the reaction was stopped by boiling for 1 minute. Then 0.2 mL of methanol was added, followed by centrifugation, and a 10 μL aliquot of the resulting supernate was used for HPLC analysis.

Lipoamidase was purified 15-fold from serum by dialysis against 0.1 M sodium phosphate buffer (pH 6.0) and chromatography on a DEAE-Sephacel column.

9.14.3 Pyridoxal Kinase, Pyridoxamine Oxidase, and Pyridoxal-5'-phosphate Phosphatase (Ubbink and Schnell, 1988)

Three enzymes play an active role in the metabolism of vitamin B_6 in human erythrocytes. Pyridoxal kinase uses ATP to phosphorylate pyridoxine, pyridoxamine, and pyridoxal. Pyridoxamine oxidase oxidizes pyridoxamine-5'-phosphate and pyridoxine-5'-phosphate to pyridoxal-5'-phosphate. The phosphatase activity produces pyridoxal from pyridoxal-5'-phosphate. The assay of the three enzymes required separation of the semicarbazone derivatives of pyridoxal-5'-phosphate and pyridoxal. The mobile phase used by Ubbink and Schnell (1988) contained 2.5% acetonitrile. Detection was by fluorescence.

The reaction mixture for the pyridoxal kinase assay contained 60 μL of hemolysate diluted to 0.4 mL with a solution of 10 mM triethanolamine, 90 mM potassium phosphate, 2 mM magnesium chloride, and 2 mM ATP at pH 7.4. The pyridoxal-5'-phosphate phosphatase assay contained 10 μL of hemolysate diluted to 0.4 mL with a solution containing 50 mM triethanolamine–HCl buffer (pH 7.4) and 5 mM magnesium chloride. The reaction mixture for pyridoxamine oxidase contained 200 μL of hemolysate diluted to 0.4 mL with a solution of 10 mM triethanolamine, 160 mM potassium phosphate buffer (pH 7.4), and 2 mM magnesium chloride. In each assay, the reaction was started by adding 50 μL of the appropriate substrate at a stock concentration of 1.8 mM. Incubation times were 20 minutes for the phosphatase assay, and 45 and 120 minutes for the kinase and oxidase assays, respectively. The reactions were terminated by adding 0.2 mL of 10% trichloroacetic acid and 0.1 mL of 0.5 M semicarbazine. The resulting mixture was heated at 40°C for 30 minutes. The clear supernate obtained after centrifugation as used for HPLC analysis. Product formation was linear for up to 100, 180, and 40 minutes for the phosphatase, kinase, and oxidase assays, respectively.

Enzymes were assayed in hemolysates of erythrocytes isolated from venous blood collected with EDTA as anticoagulant.

9.14.4 Pyridoxine Kinase (Argoudelis, 1990)

All three forms of vitamin B_6 [pyridoxal, pyridoxine, and pyridoxamine] are phosphorylated by a single kinase that uses ATP as the phosphate donor. This assay describes the use of pyridoxamine as the substrate.

Pyridoxine, pyridoxine-5'-phosphate, isopyridoxal (internal standard), ATP, and ADP were separated on a Whatman Partisil-10SCX column (4.6 mm $\times$ 250 mm). The method also resolved pyridoxamine and pyridoxamine-5'-phosphate. The mobile phase was 0.1 M ammonium dihydrogen phosphate (pH 4) at a flow rate of 1.0 mL/min. The excitation wavelength was 290 nm.

The assay contained in a volume of 1 mL: 20 mM potassium phosphate buffer (pH 5.75), 0.08 mM ZnCl$_2$, 0.06 mM KCl, 0.02 mM isopyridoxal (internal standard), 1.2 mM ATP, 0.1 mM pyridoxine, and liver extract as the source of enzyme. To assay the yeast enzyme, ZnCl$_2$ was replaced by 0.1 mM MgCl$_2$ and KCl was omitted. The reaction was started by adding enzyme, and incubations were continued in the dark at 37°C for 90 minutes. The reaction was stopped by heating the test tubes in a boiling water bath for 3 minutes. After centrifugation, an aliquot of the supernate was injected into the HPLC system. The reaction was linear for at least 90 minutes when the rate of pyridoxine phosphate formation was not more than 13 nmol/h.

The sources of enzyme assayed were an extract of rabbit liver acetone powder and an extract of baker's yeast.

9.14.5 Biotinidase (Hayakawa et al., 1993)

Biotinidase hydrolytically cleaves biocytin to produce biotin and lysine. This assay uses a fluorimetric substrate, biotinyl-6-aminoquinoline (BAQ), and is suitable for use with samples such as milk and serum without pretreatment or dialysis.

The product, 6-aminoquinoline, was separated from other assay components by chromatography on a Devolosil ODS column (4.6 mm × 50 mm, 5 μm). Detection was carried out by means of fluorimeter with excitation and emission wavelengths of 350 and 550 nm, respectively.

The reaction mixture contained 45 μL of substrate solution (0.044 mM biotinyl-6-aminoquinoline dissolved in 0.1 M phosphate buffer (pH 7.0) containing 1 mM Na$_4$EDTA and 10 mM 2-mercaptoethanol, and 5 μL of enzyme solution. The reaction was allowed to proceed for an appropriate time at 37°C before being stopped by the addition of 0.10 mL of methanol. After centrifugation, 10 μL of the supernate was injected into the HPLC system. After a lag period during the first 5 minutes, product information was linear with time for the next 20 minutes. The reaction was linear with amounts of human serum up to 100 μL (total reaction volume of 0.5 mL) when incubated for 60 minutes.

The assay was used to measure biotinidase activity in human and porcine milk, and porcine serum.

9.15 XENOBIOTIC METABOLISM

9.15.1 ATP-Sulfurylase (Mina and Rossomando, 1988)

ATP-sulfurylase (ATP:sulfate adenyltransferase) catalyzes the synthesis of adenosine 5'-phosphosulfate, using sulfate and ATP as substrates. This is the first step in the activation of sulfate, with the next step being phosphorylation at the 3' position to give 3'-phosphoadenosine 5'-phosphosulfate, a donor of activated sulfate in numerous biological reactions. The assay described here was suitable for measuring the activity of ATP-sulfurylase in preparations containing ATPase, and kinase and sulfohydrolase activities for adenosine 5'-phosphosulfate.

Adenosine 5'-phosphosulfate was separated from ATP and other adenine nucleotides by chromatography on a Synchropak AX-100 column (4.1 mm × 250 mm). The mobile phase contained 0.1 M sodium phosphate buffer (pH 7.3) and 0.8 M $NaHCO_3$. The effluent profile was obtained by monitoring at 254 nm.

The reaction mixture contained in a final volume of 350 μL: 30 μmol of Tris-HCl (pH 8), 0.9 μmol of ATP, 3 μmol of magnesium sulfate, 6 μmol of sodium fluoride, and 50 μL (2.5 U) of inorganic pyrophosphatase. The reaction was initiated by addition of 50 μL of enzyme solution. Reactions were terminated by removing aliquots, transferring them to glass test tubes, and capping and heating to 155°C in a sand bath for 5 minutes. Precipitated protein was removed by filtration before injection of 50 μL of filtrate onto the HPLC column.

The enzyme preparations assayed were obtained by homogenizing liver and tongue from rat in 0.05 M Tris-HCl (pH 8) at 4°C, to give a final concentration of 1 mg wet weight tissue per milliliter. The homogenate was centrifuged at 30,000g for 10 minutes, and the supernate was centrifuged again for 10 minutes at 30,000g.

Figure 9.141 shows a chromatographic profile.

9.15.2 Sulfotransferase (To and Wells, 1984)

Sulfotransferase catalyzes the transfer of sulfate from the donor molecule adenosine-3'-phosphate-5'-phosphosulfate (PAPS) to an acceptor, β-naphthol, to form the reaction product β-naphthol sulfate.

The assay used for this activity involves the separation of the β-naphthol from the product β-naphthol sulfate by reversed-phase HPLC (C_{18} column) with a mobile phase of 0.1 M acetic acid–acetonitrile (85:15 v/v). Absorbance was monitored at 235 nm.

The reaction mixture contained the substrate β-naphthol (0.125 M) in a phosphate buffer (pH 6.5), 5 mM 2-mercaptoethanol, and 0.2 mM PAPS in 5% acetone (v/v). The mixture was preincubated and the reaction started by the addition of a cytosolic fraction. The suspension was incubated for 10 minutes at 37°C, and the reaction was terminated by the addition of ice-cold methanol. After centrifugation to remove insoluble material, the supernatant solution was dried, redissolved in methanol, and analyzed by HPLC.

Figure 9.142 shows the separation of the substrate from the product, and Figure 9.143 shows the rate of formation of product during the incubation. A hepatic cytosolic fraction obtained from mice was used as the source of activity.

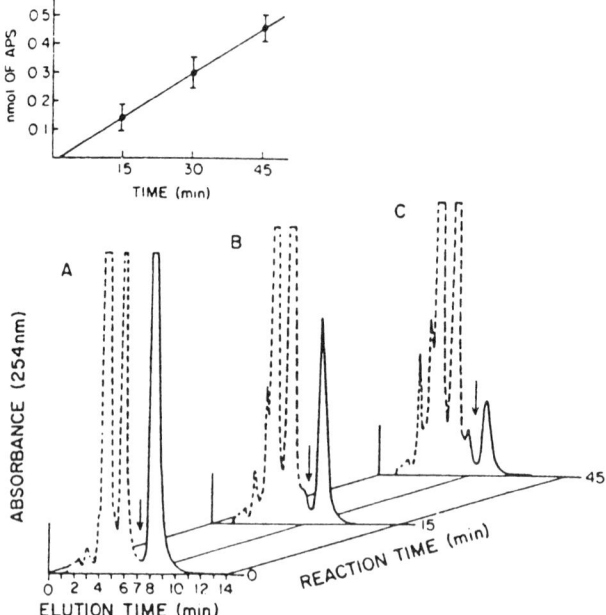

Figure 9.141 Time course of formation of adenosine 5′-phosphosulfate (APS) by activity in rat liver. The reaction mixture contained, in a final volume of 350 μL, 30 μmol of Tris-HCl (pH 8.0), 0.9 μmol of ATP, 3 μmol of magnesium sulfate, 6 μmol of sodium fluoride, and 50 μL of inorganic pyrophosphatase (2.5 U). A 50 μL supernatant sample (19 mg of protein) from rat liver was added to start the reaction. Then samples were removed at intervals, and the reaction was terminated and analyzed. Chromatograms were obtained after incubation for (A) 0 minutes, (B) 15 minutes, and (C) 45 minutes. Arrow indicates elution time for the reaction product APS. (From Mina and Rossomando, 1988.)

9.15.3 Glutathione S-Transferase (Brown et al., 1982; Eaton and Stapleton, 1989; Tracy and O'Leary, 1991)

Glutathione S-transferases catalyze the reaction of electrophiles with glutathione to form thioether conjugates. Substrates include alkyl halides and organophosphorus insecticides. In the HPLC method of Brown et al. (1982), styrene oxide was used as the substrate and the activity was measured by the formation of conjugates between the styrene oxide and the reduced glutathione conjugates.

The identification of the reaction conjugates was carried out by reversed-phase HPLC on a C_{18} (μBondapak) column eluted isocratically with a mobile phase of methanol–glacial acetic acid–water (20:1:79 v/v). The wavelength for detection was 254 nm.

The reaction mixture contained in 2 mL of reduced glutathione (1 mM): sodium phosphate buffer (pH 7.8) and the transferase preparation. The reac-

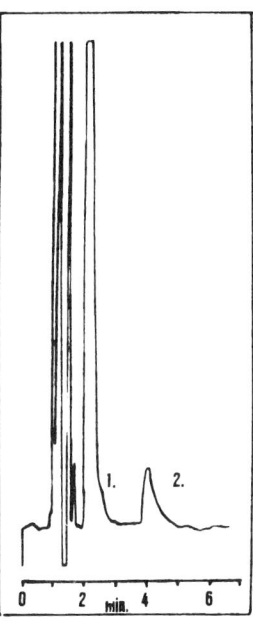

Figure 9.142 Chromatograms of the HPLC separations of β-naphthol sulfate. *Peaks:* 1, acetaminophen, the internal standard; 2, β-naphthol sulfate. Solvent, 0.1 M acetic acid–acetonitrile (85:15, v/v); flow rate, 1.5 mL/min; wavelength, 235 nm. (From To and Wells, 1984.)

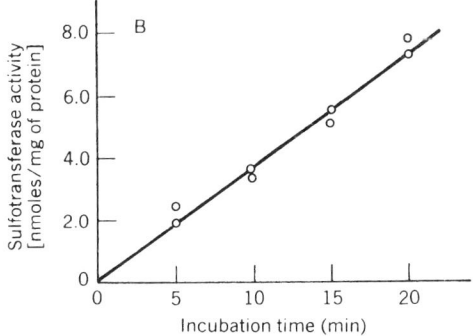

Figure 9.143 Effect of incubation time on enzyme activity. Sulfotransferase activity was measured by the amount of β-naphthol sulfate produced after varying incubation time periods. (From To and Wells, 1984.)

tion mix was preincubated at 37°C, and the styrene oxide was added to start the reaction. Reactions were terminated by adding ice-cold ethyl acetate to extract the unreacted styrene oxide. The aqueous phase was dried and used for the HPLC analysis.

Figure 9.144 shows a chromatogram of the reaction product. Two peaks were observed, and after analysis they were shown to be the conjugates expected.

The transferase was obtained from rat lung and liver cytosol fractions.

The assay method described by Eaton and Stapleton (1989), measures the activities of both cytosolic glutathione S-transferase and microsomal epoxide hydrolase toward benzo[a]pyrene-4,5-oxide as a substrate. These enzymes are important in the biotransformation of many epoxide xenobiotics, including potentially carcinogenic arene oxides.

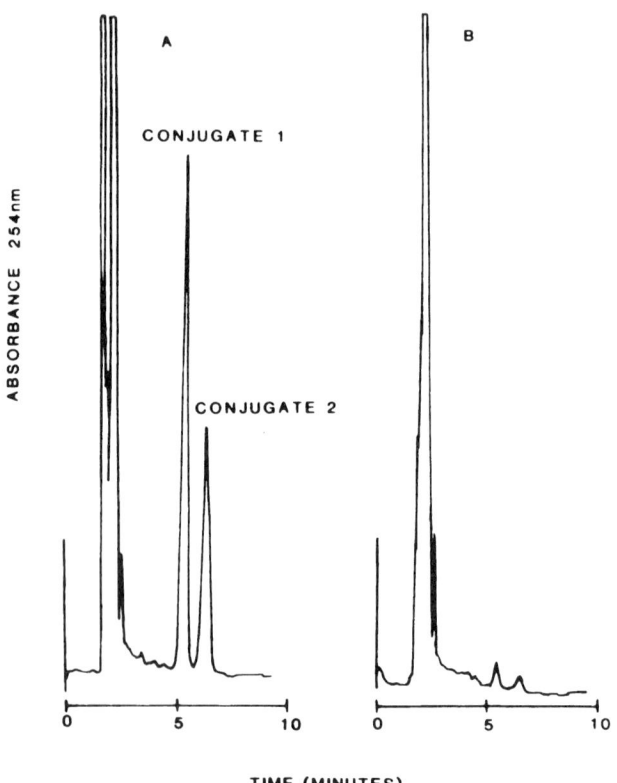

Figure 9.144 HPLC profile of 50 μL of the aqueous phase prepared from an incubation containing styrene oxide (1.0 mM), reduced glutathione (1.0 mM), and (A) 1.0 mg of native liver cytosol or (B) 1.0 mg of boiled liver cytosol. The UV absorbance was monitored at 254 nm. Detector sensitivities were 0.01 and 0.005 AUFS for (A) and (B), respectively. (From Brown et al., 1982.)

The substrate, benzo[a]pyrene-4,5-oxide, and products, the diol and glutathione derivatives of the substrate, were separated on a Alltech Econosphere C_{18} column (4.6 mm $\times$ 150 mm). Three mobile phases are used in the ternary separation: Solvent A contained 0.1% H_3PO_4 adjusted to pH 3.5 with NH_4OH, solvent B was HPLC-grade acetonitrile, and solvent C was water. The column was equilibrated with a mixture of 80% A and 20% B. After injection, solvent A was replaced with 80% water using a 1-minute linear gradient. A linear gradient from 20% B to 100% B was run between 1 and 5 minutes. The mobile phase was then held at 100% B until 7 minutes. The solvents were recycled to 80% water and then to 80% solvent A by 9 minutes. Detection and peak quantitation were performed at 260 nm, or in the case of fluorescence detection, at excitation and emission wavelengths of 241 and 389 nm, respectively. Quantitation was performed from a standard curve constructed from a known quantity of benzo[a]pyrene-4,5-oxide. The glutathione adduct and the diol had essentially the same UV absorbance as the parent compound.

Enzyme assays were performed in 1.5 mL microcentrifuge tubes containing 165 μL of incubation buffer (0.25 M Tris-HCl, 0.1 mM EDTA, pH 7.4), 50 μL of 10 mM glutathione in incubation buffer, and 25 μL of appropriately diluted enzyme. The reaction was initiated by addition of 10 μL of benzo[a]pyrene-4,5-oxide in acetonitrile (0.875 mg/mL). The assay was stopped by addition of 0.25 mL of acetonitrile containing 300 μM 2-methoxynaphthalene as an internal standard. The solution was kept overnight in the dark at 4°C, and then 500 μL of distilled water was added. The mixture was centrifuged before analysis by HPLC. Production of both the glutathione and diol derivatives was linear with time and protein up to 15 minutes and 500 μg/mL, respectively.

Cytosolic and microsomal fractions were prepared by conventional techniques from the livers of adult rats.

The assay of Tracy and O'Leary (1991) measures the formation of S-methylglutathione and S-ethylglutathione, using iodomethane or iodoethane as substrates.

Products and reactants, and N-dinitrophenyl-glutamyl-glycine (internal standard) were separated on a Zorbax C_{18} column (6.2 mm $\times$ 80 mm, 3 μm). Solvent A was 0.3 M acetic acid. Solvent B was 50% (v/v) acetonitrile in 0.3 M acetic acid. The gradient, started at the time of sample injection, was increased linearly from 50% solvent B to 100% solvent B over a 10-minute period and held at 100% solvent B for 2 minutes. Return to initial conditions was made in 5 minutes. The column effluent was monitored at 365 nm.

The enzymatic reaction was carried out in 1.5 mL polypropylene centrifuge tubes containing in a total volume of 250 μL: 125 μmol 3-N-morpholinopropanesulfonate buffer (pH 7.5), 1.25 of μmol glutathione, xenobiotic substrate (1.25 μmol of iodomethane, 1.25 μmol of iodoethane, 0.12 μmol of methyl parathion, or 1.25 of μmol dichlorvos), and enzyme. Reactions were initiated by adding the xenobiotic substrate dissolved in 5 μL of ethanol and terminated after incubation at 30°C by adding 25 μL of ice-cold 60% perchloric acid.

Three time points were used to determine linear reaction velocities, and control reactions contained enzyme that had been boiled for 10 minutes. Five milliliters of γ-Glu-Gly was added as internal standard. Next, 125 μL of ice-cold 2.0 M KOH and 2.4 M KHCO$_3$ was added. When gas evolution slowed, 200 μL of freshly prepared 1.5% (v/v) solution of 1-fluoro-2,4-dinitrobenzene in ethanol was added. The derivatization reaction was allowed to proceed in the dark for at least 8 hours at 22°C. The mixtures were filtered before 5 to 50 μL was analyzed by HPLC.

The enzyme preparation used was obtained by chromatography of the high speed supernate of a phosphate extract of rat liver on a glutathione–agarose affinity column.

9.15.4 Adenosine 3′-Phosphate 5′-Sulfophosphate Sulfotransferase (Schwenn and Jender, 1980)

This sulfotransferase catalyzes the transfer of sulfate from adenosine 3′-phosphate 5′-sulfophosphate (PAPS) to an acceptor thiol.

All compounds, including substrates and both primary and secondary reaction products, were separated by ion-paired, reversed-phase HPLC (LiChrosorb RP-18) with a mobile phase of 9.4% 2-propanol containing tetrabutylammonium hydroxide at several pH values adjusted with phosphoric acid. The separations obtained are shown in Figure 9.145. The column was eluted isocratically and monitored by UV at 254 nm and by liquid scintillation counting.

The reaction mixture contained Tris-Cl (pH 8.0), MgCl$_2$, glutathione as the acceptor, [^{35}S]PAPS, and protein. Samples were withdrawn, and the reaction was terminated by forcing the sample through a microfilter at 6 bar under nitrogen. The filter retained the enzyme protein, and the nucleotides were recovered.

The activity was obtained from plants.

9.15.5 Phenolsulfotransferase (Honkasalo and Nissinen, 1988; Khoo et al., 1990)

Phenolsulfotransferase catalyzes the transfer of active sulfate from 3′-phosphoadensine 5′-phosphosulfate to various phenols and catechols. Honkasalo and Nissinen (1988) developed an assay that is suitable for measuring both the thermolabile (TL) and thermostable (TS) isoforms of phenol sulfotransferase. Both are active toward phenols, while the TL form also conjugates catechols including dopamine.

The product, p-nitrophenylsulfate, is separated from p-nitrophenyl by chromatography on a Viosfer-ODS column (4 mm × 150 mm, 5 μm). The mobile phase contained 25% methanol in 50 mM sodium phosphate (pH 3.0). The flow rate was 1.5 mL/min, and the column effluent was monitored at 300 nm.

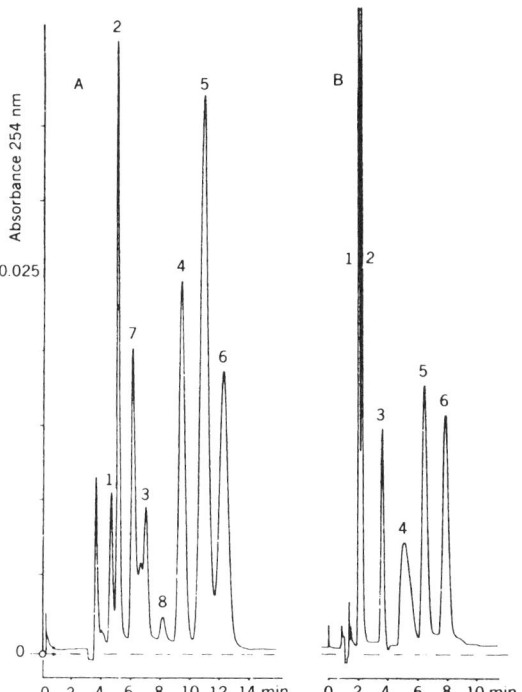

Figure 9.145 Separation of adenine nucleotides (mixture of authentic compounds) at pH 9.4 (A) and pH 8.0 (B). Column: Knauer RP-18/10, operated with a precolumn. Eluent: 9.4% 2-propanol, 3 mM tetrabutylammonium hydroxide, pH adjusted with 1 N phosphoric acid. *Peaks:* 1, 5'-AMP; 2, APS; 3, ADP; 4, ATP; 5, 3'5'-PAP; 6, PAPS; 7, NADP; 8, FAD. (From Schwenn and Jender, 1980.)

The reaction mixture in a total volume of 250 μL contained 50 μL of 10 mM sodium phosphate buffer (pH 7.2), 50 μL of enzyme preparation, and 100 μL of 50 μM 3'-phosphoadenosine 5'-phosphosulfate. The reaction was started by adding 50 μL of 12.5 mM p-nitrophenol (when assaying the thermolabile form) or 50 μL of 50 μM p-NP (when determining the thermostable form). After incubation for 30 minutes at 37°C, the reaction was stopped by the addition of 25 μL of 4 M perchloric acid. After centrifugation, a 20 μL aliquot was injected into the HPLC system. Formation of product was linear with time for both forms of enzyme for up to 45 minutes, and with protein amount up to 0.5 mg.

The enzyme source was livers of female rats. Livers that had been kept frozen at −80°C were homogenized in 1:5 (w/v) ice-cold 10 mM sodium phosphate buffer (pH 7.2) containing 10 mM dithiothreitol. The homogenate was centrifuged at 15,000g for 20 minutes at 4°C, and the supernate was centrifuged at 100,000g for 60 minutes. The assay was found to be applicable to the thermolabile form in human platelets and jejunum, rat brain, and guinea pig ileum. The thermostable form was determined in human jejunum and guinea pig ileum.

In the assay described by Khoo et al. (1990), N-acetyldopamine sulfate was separated within 7 minutes from N-acetyldopamine by chromatography on a Hypersil-ODS microbore column (2.1 mm × 100 mm, 5 µm). The mobile phase contained 1.2 acetic acid and 1 mM EDTA at pH 4.4. The flow rate was 0.6 mL/min. An electrochemical detector was set at a sensitivity of 2 nA and a working potential of 0.8 V. The detector response was calibrated by using a standard of N-acetyldopamine-[^{35}S]sulfate that had been prepared in the laboratory. The standard curve was linear in the range of 10 to 140 pmol.

The enzyme was assayed in a final volume of 200 µL containing 73 µM 3'-phosphoadenosine 5'-phosphosulfate, 50 µM N-acetyldopamine, and 50 mM phosphate buffer (pH 6.0). The reaction was started by addition of 10 to 60 µL of the enzyme preparation. The reaction was stopped after 15 minutes of incubation at 37°C by boiling for 1 minute, followed by the addition of 50 µL of 2 M Tris-HCl buffer (pH 8.6). Unreacted N-acetyldopamine was extracted twice by adsorption onto 10 mg of activated alumina. The supernate obtained after centrifugation was filtered, and 5 to 100 µL aliquots were analyzed by HPLC.

The assay was used for measuring phenolsulfotransferase activities in cytosol prepared from human liver (70–200 µg of protein/assay), in extracts of human platelets (130–400 µg of protein), and in human whole blood subjected to freezing and thawing (0.3–1 mg of protein).

9.15.6 Aryl Sulfotransferase (Duffel et al., 1989)

Aryl sulfotransferases catalyze the transfer of the sulfuryl moiety from 3'-phosphoadenosine 5'-phosphosulfate to phenols, catechols, benzylic alcohols, arylhydroxylamines, and arylhydroxamic acids. This assay measures adenosine 3',5'-diphosphate and is thus suited to quantitate enzyme activity when the sulfate esters formed are chemically unstable.

Separation of 3'-phosphoadenosine 5'-phosphosulfate and adenosine 3',5'-diphosphate is carried out on an Econosphere C$_{18}$ column (4.6 mm × 250 mm, 5 µm) at ambient temperature. The mobile phase was water–methanol (88:12) containing 75 mM KH$_2$PO$_4$, 1 mM 1-octylamine, and 100 mM ammonium chloride. The pH was adjusted to 5.45 with KOH before the addition of methanol. Detection was by measuring absorbance at 254 nm.

Assay mixtures contained 200 µM 3'-phosphoadenosine 5'-phosphosulfate, 0.25 M phosphate buffer (pH 7.0), 8.3 mM 2-mercaptoethanol, acceptor substrate (e.g., 1-naphthalenemethanol) in acetone (< 5% (v/v) final concentration in assay), and 0.5 to 3.0 µg of aryl sulfotransferase IV in a final volume of 30 µL. The reaction was initiated by addition of enzyme. After 30 minutes of incubation, the reaction was terminated by adding 30 µL of methanol and

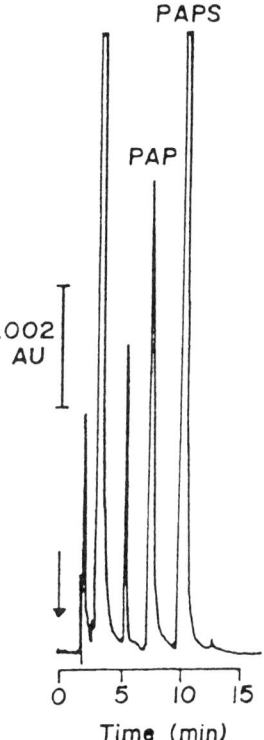

Figure 9.146 HPLC analysis of PAP in an aryl sulfotransferase IV reaction mixture. The reaction mixture and incubation conditions were: 50 μM 1-naphthalenemethanol and 2.9 μg of AST IV. The mobile phase for HPLC analysis contained 12% methanol in 75 mM potassium phosphate (pH 5.45), 100 mM ammonium chloride, and 1.0 mM 1-octylamine. The flow rate was 2.0 mL/min, and detection was at 254 nm, with a full scale sensitivity of 0.02 AU. Sample injection is indicated by an arrow. (From Duffel et al., 1989.)

cooling to 4°C. The mixture was centrifuged, and 20 μL of the supernate was injected.

Aryl sulfotransferase IV (AST IV) was purified to apparent homogeneity from the livers of rats.

Figure 9.146 shows a representative chromatogram.

9.15.7 Cysteine Conjugate β-Lyase (Stijntjes et al., 1992)

Cysteine conjugate β-lyases cleave the carbon–sulfur bond of S-substituted L-cysteine conjugates via a β-elimination reaction to produce pyruvate, ammonia, and the corresponding thiols. These enzymes are found in kidney, liver and intestine of mammals. This assay is applicable to a wide variety of cysteine conjugates as substrates, since a fluorescent derivative of pyruvate is measured.

The quinoxalinol derivative obtained by reaction of pyruvate with o-phenylenediamine is chromatographed on a ChromSphere C_{18} column (3 mm × 100 mm, 5 μm) at a flow rate of 0.4 mL/min using a mobile phase of 45% methanol, 1% acetic acid, and 54% water. The eluate was monitored by a fluorescence spectrometer using excitation and emission wavelengths of 336 and 420 nm, respectively. Calibration curves were constructed by adding known amounts of pyruvate to cytosolic fractions and carrying them through the derivatization and analysis procedures.

Incubation mixtures contained 75 μL of substrate solution (in 50 mM Tris-HCl buffer, pH 8.6) and enzyme solution in the same buffer to give a final volume of 300 μL. Good substrates included S-1,2-dichlorovinyl-L-cysteine used at a concentration of 1 mM. After an incubation at 37°C for 10 minutes, the incubations were terminated by adding 1.0 mL of 12 mM o-phenylenediamine in 3 M HCl. The caps of the incubation vials were closed and pierced with a needle before being placed in a preheated oven at 60°C for 60 minutes. The samples were centrifuged, and 100 μL of the supernate was analyzed by HPLC.

The assay was used to measure enzyme activity in dialyzed cytosolic fractions from rat kidney (protein concentration 2 mg/mL) and in rat kidney mitochondria (protein concentration 3 mg/mL).

9.15.8 UDP–Glucuronyl Transferase (Spahn, 1988)

Transfer of a glucuronic acid moiety from UDP–glucuronic acid is a common reaction in drug metabolism and detoxification pathways. Spahn (1988) described an assay method that is applicable to arylpropionic acids and is capable of separating and quantitating enantiomers.

The conjugates of arylpropionic acids enantiomers with D-glucuronic acid were separated on a Beckman Ultrasphere ODS column (4.6 mm × 250 mm, 5 μm) column. The mobile phase was a 28:72 (v/v) mixture of acetonitrile and 8 mM tetrabutylammonium hydrogen sulfate buffer (pH 2.5). Five minutes after the diastereomeric conjugates had been resolved, the percentage of acetonitrile was increased to 60% to wash out excess substrate. The mode of detection depended on the samples injected. Fluorescence was used for flunoxaprofen (excitation, 305 nm; emission, 355 nm), benoxaprofen (313/365 nm) carprofen (285/350 nm), and idoprofen (275/433 nm). UV absorption was used to detect flurbiprofen (255 nm), naproxen (285 nm), ketoprofen (255 nm), pirprofen (265 nm), and cicloprofen (238 nm).

The reaction mixture contained 1 mg of microsomal protein per milliliter, 50 mM Tris-HCl buffer (pH 7.4), 10 mM MgCl$_2$, 10 mM 1,4-saccharolactone, 1 mM phenylmethylsulfonyl fluoride, 0.04% Triton X-100, 10 mM UDP–glucuronic acid, and 0.1 to 0.4 mM substrate in a final volume of 250 μL. The reaction was initiated by the addition of UDP–glucuronic acid. After 20 minutes at 37°C, the reaction was stopped by cooling the tubes to 0°C. A 50 μL aliquot of the reaction mixture was mixed first with 50 μL of acetonitrile, and then with 400 μL of the mobile phase. After centrifugation, an aliquot of the supernate was injected into the HPLC system. Product formation was shown to be linear for 30 minutes for both enantiomers of the substrates flunoxaprofen and naproxen.

The source of enzyme activity was rat liver microsomes prepared by conventional techniques.

9.15.9 UDP–Glucosyltransferase (Real, et al., 1991)

The enzyme UDP–glucosyltransferase is involved in the detoxication of xenobiotics. Whereas vertebrates carry out glucuronidation by using UDP–glucuronic acid, invertebrates and plants form the glucoside derivatives with UDP–glucose as the donor of the glucosyl moiety.

Separation of 4-nitrophenol and 4-nitrophenol glucoside was obtained on a Tracer Spherisorb ODS-2 column (4.6 mm × 250 mm, 5 μm). The mobile phase was 10 mM potassium phosphate buffer (pH 7.1) in 15% methanol. The absorbance at 295 nm was used for quantitation.

The standard reaction mixture contained in a final volume of 0.25 mL: 16 mM $MgCl_2$, 6.4 mM UDP–glucose, 3.2 mM p-nitrophenol, enzyme extract (100–500 μg of protein) and 0.1 M Tris-HCl buffer (pH 8.0). After a 20-minute incubation at 43°C, the reaction was terminated by the addition of 20 μL of 40% perchloric acid. The supernate obtained by centrifugation was filtered and then subjected to HPLC analysis. Formation of product was linear with time up to 40 minutes and with protein added in the range of 0 to 5.5 mg protein/mL. Alternative substrates for the enzyme include 1-naphthol and 2-naphthol.

The enzyme source was derived from homogenates of *Drosophila melanogaster*.

9.15.10 Ethoxycoumarin *O*-Deethylase (Rosenberg et al., 1990)

An assay for ethoxycoumarin O-deethylase is useful in as a part of a battery of tests for cytochrome P450 activities in a variety of tissues. The deethylase activity results in the conversion of 7-ethoxycoumarin to 7-hydroxycoumarin. A postcolumn pH shift allows the product to be detected by fluorescence.

The product, 7-hydroxycoumarin, was separated on a Nov-Pak C_8 column (3.9 mm × 150 mm). The mobile phase, adjusted to pH 3.5, was a 35:65 mixture of methanol and 1% acetic acid. The flow rate was 1 mL/min. The postcolumn eluate was mixed with 1.0 N NaOH pumped at a flow rate of 0.5 mL/min. Complete mixing occurred via a T-joint followed by 1.5 m of Teflon tubing. The fluorometer was set to use excitation and emission wavelengths of 368 and 456 nm, respectively.

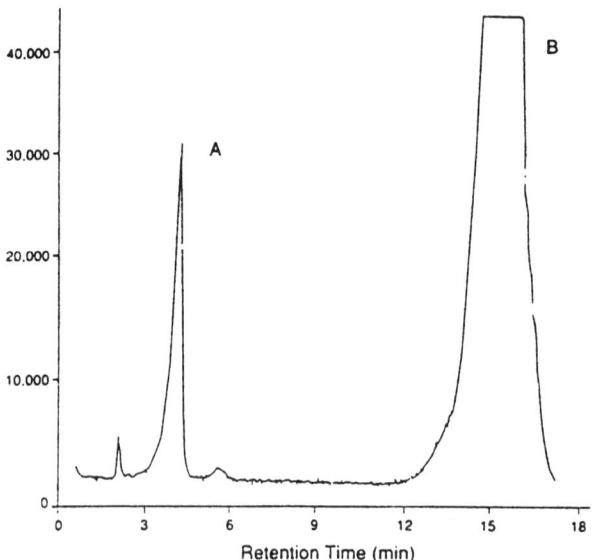

Figure 9.147 HPLC chromatogram of 7-hydroxycoumarin generated from incubation with HepG$_2$ liver cells. A 10,000g supernatant was obtained from approximately 1 × 10^6 cells and incubated (15 min) with 4.2 μM 7-ethoxycoumarin. Following extraction with CHCl$_3$, 7-hydroxycoumarin metabolite formed was quantified by the HPLC-fluorescent method. (From Rosenberg et al., 1990.)

The reaction mixture contained in a total volume of 1.0 mL: 80 μmol of potassium phosphate buffer (pH 7.4), 5 μmol of magnesium chloride, 1 mg of bovine serum albumin, 0.5 μmol of NADPH, 0.5 μmol of NADH, and 0.1 mL of microsomal suspension. The reaction was initiated by the addition of 20 μL of 7-ethoxycoumarin (0.43 μmol) dissolved in 50% methanol. After incubation for 10 to 20 minutes at 37°C, the reaction was stopped by the addition of 125 μL of ice-cold 15% (w/v) trichloroacetic acid. The resulting mixture was extracted with 2 mL of chloroform. An aliquot (0.75 mL) of the organic phase was dried under nitrogen and the residue resuspended in 50 μL of HPLC-grade methanol. Formation of product was linear with up to 5 μg of protein.

The source of the enzyme assayed was microsomes prepared from liver and intestinal epithelial cells.

Figure 9.147 shows a chromatogram.

9.15.11 Cytochrome P450$_{2E1}$ (Duescher and Elfarra, 1993; Tassaneeyakul et al., 1993)

Cytochrome P450$_{2E1}$ is an isoform that is inducible by ethanol. It has been implicated in the activation of numerous low molecular weight toxins and carcinogens. This assay measures the formation of 4-nitrocatechol from 4-nitrophenol.

The substrate and product were separated by chromatography on a Beckman Ultrasphere ODS column (4.8 mm × 250 mm, 5 µm). Solvent A contained 0.1% trifluoroacetic acid (pH 2.5) and 1% acetonitrile. Solvent B consisted of 0.1% trifluoroacetic acid (pH 2.5) and 75% acetonitrile. The percentage of solvent B in the mobile phase was 7.5% for 2 minutes. The percentage was increased to 70% for 2 minutes and held there for 2 more minutes. The gradient was then decreased to 7.5% B over 2 minutes and was maintained there to give a total run time of 10 minutes. The flow rate was 1.5 mL/min, and the injection volume was 100 µL. Quantitation of product formed was performed using the absorbance at 345 nm.

The reaction mixture contained 0.24 mL of 3 mM NADPH prepared in 0.05 M Tris buffer (pH 7.4) containing 5 mM MgCl$_2$ and 20 µL of rat liver microsomes (0.06–0.96 mg/mL). The reaction was started with the addition of 0.24 mL of 200 µM p-nitrophenol. After a 20-minute incubation at 37°C, the reaction was terminated by the addition of 25 µL of trifluoroacetic acid. The resulting mixture was centrifuged and the supernate was filtered before analysis of a 100 µL aliquot by HPLC. Formation of product was linear with time up to 30 minutes with protein in the range of 0.03 to 0.48 mg.

Microsomes were prepared from rat liver and stored at −80°C in 0.1 M potassium phosphate buffer (pH 7.4) containing 0.15 M KCl, 1.5 mM EDTA, and 20% glycerol.

9.15.12 Flavin-Containing Monoxygenase (Cashman and Proudfoot, 1988; Kawaji et al., 1993)

Flavin-containing monoxygenase oxidizes a variety of xenobiotics that contain nucleophilic nitrogen, sulfur, and phosphorus atoms. An early assay for this enzyme was published by Cashman and Proudfoot (1988). A more recent assay, developed by Kawaji et al. (1993) uses benzydamine, a nonsteroidal anti-inflammatory drug, as a substrate. The benzydamine N-oxide formed is fluorescent.

Benzydamine, benzydamine N-oxide, and norbenzydamine were separated by chromatography on a LiChrosorb RP-18 column (4 mm × 150 mm). The mobile phase was methanol-acetonitrile–water-25% NH$_4$OH [50:40:10:0.05 (v/v)]. The flow rate was 1.5 mL/min for 3 minutes, and then 3 mL/min from 3 to 20 minutes. The effluent was monitored for fluorescence using excitation and emission wavelengths of 303 and 377 nm, respectively.

The incubation mixtures contained 0.1 M Tricine–KOH (pH 8.5), 0.5 mM NADPH, enzyme, and benzydamine in concentrations up to 1 mM (K_m = 15 µM). The final volume was 0.3 mL, and the reaction was initiated by adding benzydamine. After a 10-minute incubation at 37°C, the reaction was stopped by adding a twofold volume of methanol. The supernate obtained by centrifugation was analyzed by HPLC. Product formation was linear for 30 minutes and with up to 50 µg/mL of microsomal protein.

Microsomes were prepared from rats. The Michaelis–Menten constant of highly purified enzyme did not differ from the apparent K_m of the enzyme in microsomes. Microsomes also supported formation of norbenzydamine as a result of the presence of cytochrome P450.

9.16 PYRIMIDINE METABOLISM

9.16.1 Dihydropyrimidine Dehydrogenase (Klein and Haas, 1990; Lu et al., 1992)

Dihydropyrimidine dehydrogenase catalyzes the first step in the degradation of uracil and thymine. Inherited deficiencies of the enzyme are known, and the enzyme can play a critical role in cancer chemotherapy, for example, in the metabolism of 5'-fluorouracil.

In the assay by Klein and Haas (1990), the substrate, 5-bromouracil, is separated from the reaction mixture by chromatography on Hypersil ODS 2. The mobile phase was 0.02 M potassium phosphate buffer (pH 5.6)–methanol (94:6, v/v). The column effluent was monitored at 275 nm. The rate of disappearance of substrate was obtained from the slope of the line obtained by plotting the concentrations of 5-bromouracil against the incubation times.

The standard reaction mixture contained 0.20 mL of the enzyme source, 0.25 mL of NADPH (2.5 mg/mL), and 2.30 mL of Sörensen buffer (pH 7.4). The reaction was started by adding 0.05 mL of 5 bromouracil (0.9 mg/mL). At intervals, 50 µL portions of the assay mixture were mixed with 150 µL of 6% trichloroacetic acid. After centrifugation, the supernate was shaken with 200 µL of 0.5 M trioctylamine solution in Freon and centrifuged again. Subsequently, 25 µL of the upper phase was injected. When weak enzymatic activities were being measured, the volume of the reaction mixture was reduced to 1.0 mL.

The source of enzyme was rat liver homogenate.

In the assay by Lu et al. (1992), the catabolites of uracil, thymine, or 5-fluorouracil were separated from the respective parent compounds by chromatography on two 5 µm Hypersil columns (from Jones Chromatography, Littleton, CO) used in tandem. The mobile phase contained 1.5 mM potassium phosphate (pH 8.0 for 5-fluorouracil, pH 8.4 for thymine and uracil) and 5 mM tetrabutylammonium hydrogen sulfate. The column effluent was monitored at an appropriate wavelength in the UV region.

The reaction mixture used to monitor enzyme purification contained 35 mM potassium phosphate (pH 7.4), 2.5 mM magnesium chloride, 10 mM 2-mercaptoethanol, 200 µM NADPH, 20 µM 5-fluorouracil, and enzyme solution. The final volume was 2.0 mL. At various reaction times, 350 µL of reaction sample was taken and added into an equal volume of ethanol. After filtering, an aliquot was analyzed by HPLC.

Enzyme activity was measured in the 100,000g fraction obtained from homogenized frozen human liver.

9.16.2 Dihydroorotic Acid Dehydrogenase (Peters et al., 1987; Ittarat et al., 1992)

The fourth step in the *de novo* synthesis of pyrimidine nucleotides—the conversion of dihydroorotic acid to orotic acid—is catalyzed by dihydroorotic acid dehydrogenase. The enzyme, located on the cytosolic side of the inner membrane of mitochondria, is a target for antitumor agents.

In the assay of Peters et al. (1987), orotic acid was separated from other components of the reaction system by chromatography on a Whatman Partisil-SAX column (4 mm × 250 mm, 10 μm). The mobile phase contained 8 mM potassium phosphate (pH 4.0) and 8 mM KCl. The flow rate was 1.5 mL/min and 20 μL was injected onto the column. The column effluent was monitored at 280 nm.

The reaction mixture consisted of 0.9 mL mitochondrial suspension (0.05–4 mg protein) and 31.5 μL 30 mM L-dihydroorotic acid. Reactions were terminated after 5 to 60 minutes by addition of 180 μL 16% trichloroacetic acid and chilling on ice for 20 minutes. Denatured proteins were removed by centrifugation, and the supernates neutralized with alanine–Freon before analysis by HPLC. The assay was linear with protein in the range of 0.05 to 2 mg protein/mL, and with time up to at least 60 minutes.

Mitochondria from rat liver were isolated by conventional techniques and suspended in 0.1 M Tris-HCl (pH 8.0).

In the assay of Ittarat et al. (1992), dihydroorotic acid and orotic acid were detected and quantitated at 230 nm after separation on a Corasil μBondapak C$_{18}$ column (3.9 mm × 100 mm, 10 μm). The mobile phase contained 3 mM PIC A (from Waters) in 5 mM ammonium dihydrogen phosphate (pH 6.0) containing 5% methanol.

The reaction mixture consisted of 180 μL of 180 μM dihydroorotic acid in 10 mM Hepes-KOH buffer (pH 8.0) and 20 μL of intact parasites. The reaction was incubated at 37°C for 25 minutes and was stopped by boiling in a water bath for 10 minutes. The mixture was centrifuged before analysis by HPLC. Blanks were performed by adding intact parasites to the reaction mixture just before boiling. Assays were linear with time and with protein up to 140 μg.

The malarial parasite, *Plasmodium falciparum,* was grown in red blood cells to the late trophozoite stage. Cells isolated by centrifugation were lysed with saponin, and the intact parasites were resuspended in 10 mM Hepes-KOH buffer (pH 8.0) at a dilution of 1:3.

9.16.3 Cytidine Deaminase (James et al., 1989)

Cytidine deaminase is found in liver and polymorphonuclear leukocytes. It catalyzes the deamination of cytidine and its analogs to the corresponding uridine compounds.

The product, uridine, was separated from cytidine and allopurinol (internal standard) by chromatography on a Hypersil ODS column (4.6 mm × 100 mm, 5 µm). The mobile phase consisted of 100 mM ammonium acetate (adjusted to pH 5.0 with 6 M HCl) containing 1% (v/v) methanol and 1 mM 1-octanesulfonic acid. The effluent was monitored at 262 nm. Uridine production was calculated with reference to a uridine standard, and after correction based on the concentration of the internal standard.

The reaction mixture contained 100 µL of serum sample 400 µL of 0.8 mM cytidine in 50 mM potassium phosphate buffer (pH 7.0). The tubes were capped and incubated for 20 minutes at 56°C. Aliquots of 400 µL were then removed and added to 100 µL of cold 6% (v/v) perchloric acid containing 0.75 mM allopurinol. Mixing was followed by addition of 500 µL of 100 mM ammonium acetate (pH 7.0). The supernate obtained by centrifugation was then analyzed by HPLC.

Serum samples were prepared from whole blood of patients attending rheumatology clinics.

9.16.4 β-Ureidopropionase (Waldmann and Podschun, 1990)

The final step in the degradation of uracil requires the presence of β-ureidopropionase, the enzyme that catalyzes the hydrolysis of N-carbamoyl-β-alanine to β-alanine, CO_2, and ammonia. This assay measures the phenylisothiocyanate (Edman reagent) derivative of β-alanine.

Phenylthiocarbamoyl-β-alanine was purified by chromatography on a LiChrosphere 100 C_{18} column (4.6 mm × 250 mm, 5 µm), using a mobile phase of 50 mM sodium acetate (pH 6.0) containing 15% acetonitrile. The effluent was monitored at 245 nm. Standard samples of β-alanine taken through the derivatization and chromatography procedures were used to prepare a calibration curve.

The incubation mixture contained 10 mM phosphate (pH 7.0), enzyme, and 5 mM N-carbamoyl-β-alanine. The reaction was started by adding the substrate. Incubation was carried out for various times at 37°C and terminated by heating 5 minutes at 95°C. After removal of proteins by centrifugation, aliquots of the supernate were evaporated to dryness. The residue was dissolved in 20 µL of ethanol–water–triethylamine (2 : 2 : 1, v/v) and again evaporated to remove remaining ammonia. After addition of 20 µL of coupling solution (ethanol–water–triethylamine–phenylisothiocyanate, 7 : 1 : 1 : 1, v/v), the mixtures were allowed to stand for 30 minutes at room temperature. The resulting samples were dried and dissolved in the mobile phase and aliquots were applied to the HPLC column. The enzyme reaction was linear for at least 45 minutes, when 0.4% of the substrate was converted to product.

The enzyme used had been purified 1000-fold to apparent homogeneity from cytosol of calf liver.

9.16.5 Dihydroorotase (Mehdi and Wiseman, 1989)

Dihydroorotase catalyzes the intramolecular cyclization of N-carbamyl-L-aspartic acid to L-dihydroorotic acid. In mammals, the activity is present in a trifunctional enzyme that catalyzes the first three steps in the *de novo* synthesis of pyrimidine nucleotides.

Dihydroorotate and carbamyl aspartate are separated on a Nova-Pak C_{18} cartridge (5 mm × 100 mm, 5 μm) with a mobile phase composed of 3.5 mM tetrabutylammonium phosphate (pH 70) and acetonitrile (85:15). A radioactive flow detector was used. The flow rate was 1.5 mL/min, and the flow scintillant was pumped at 5.0 mL/min. Resolution slowly changed (because SDS was used as a stopping agent) but could be restored by washing the column briefly with acetonitrile and water.

Enzyme assays were carried out in a total volume of 100 μL containing 50 mM Hepes (pH 7.4), 5 to 20 μL of enzyme, and L-[6-^{14}C]dihydroorotate. The reaction as continued for 20 to 30 minutes at 37°C, and stopped by addition of 100 μL of 1% SDS. After a brief incubation, 200 μL of the HPLC elution buffer was added and the sample was centrifuged before analysis by HPLC. The rate was linear up to 60 minutes when 2.45 μM dihydroorotate was used as substrate and conversion was less than 20%.

The enzyme used was partially purified from rat liver.

9.16.6 Thymidylate Synthetase (Krungkrai et al., 1989)

Thymidylate synthetase catalyzes the conversion of deoxyuridylate to deoxythymidylate in a folate-dependent reaction. This enzyme is a target for many agents used in chemotherapy of cancer and treatment of infectious diseases.

Deoxythymidylate was separated from other reaction components by chromatography on a LiChrosorb RP-8 column (4.6 mm × 250 mm, 5 μm). The mobile phase contained 2-propanol–85% phosphoric acid–triethylamine–water (3:3:10:984, v/v). The effluent was monitored at 254 nm.

The enzyme assay contained in a total volume of 250 μL: 50 mM Tris-HCl buffer (pH 7.8), 50 mM 2-mercaptoethanol, 5 mM formaldehyde, and 1 mM methylenetetrahydrofolate. The reaction was initiated by adding 25 μL of 1 mM deoxyuridylate. After 30 minutes at 37°C, the reaction was stopped by addition of 250 μL of ice-cold 1 M perchloric acid. After 30 minutes in an ice bath, the reaction mixture was centrifuged and the resulting supernate was neutralized with 0.1 volume of 10 M KOH containing 1 M KH_2PO_4. The supernate obtained by centrifugation was injected onto the HPLC column. Formation of deoxythymidylate was linear with time up to 30 minutes, and with protein in the range of 80 to 720 μg.

The source of enzyme was the human malaria parasite *Plasmodium falciparum*, cultured in red blood cells. Intact parasites were obtained by lysis of the cells with saponin, and the enzyme extract was obtained by freezing and thawing.

9.17 METABOLISM OF COMPLEX SACCHARIDES AND GLYCOPROTEINS

9.17.1 α-L-Fucosidase (Johnson et al., 1990)

The ubiquitous lysosomal glycosidase α-L-fucosidase is involved in the degradation of fucoglycoconjugates including oligosaccharides, glycopeptides, glycoproteins, and glycolipids.

The substrate, fucosyl-G_{M1}, was separated from its product, G_{M1}, by chromatography on a LiChrosorb-NH_2 column (4 mm × 250 mm, 5 μm). The mobile phase was a 25:75 mixture of 5 mM potassium phosphate buffer (pH 5.5) and acetonitrile. Similarly, fucosyl-G_{D1b} was separated from G_{D1b}, except the concentration of phosphate buffer was 20 mM. Gangliosides were detected by their absorbance at 195 nm.

Stable micelles of the gangliosides fucosyl-G_{M1} or fucosyl-G_{D1b} (2.7 and 2.4 nmol, respectively) were obtained by incubation in glass tubes with 4.0 μL of 0.1 M phosphate–citrate buffer (pH 3.2) for several hours. α-L-Fucosidase was dialyzed for several hours to remove L-fucose. The enzyme was added in 10 mM sodium phosphate buffer (pH 5.5) containing 0.02% (w/v) NaN_3 to obtain a final volume of 9.0 μL and a final pH of 3.4. Incubations were carried out at 37°C in a shaking water bath. The reactions were stopped with the addition of 0.1 mL of ice-cold $CHCl_3$–methanol (1:2, v/v). Supernates obtained by centrifugation were dried under vacuum. The residues were redissolved in 50 μL of the mobile phase for analysis by HPLC. Enzyme activity was calculated based on the increase in ganglioside product. The hydrolysis of fucosyl-G_{M1} is linear for 2 to 3 minutes and with amounts of enzyme up to 100 units in 1-minute incubations.

The enzyme preparation used was purified to apparent homogeneity from human liver by affinity chromatography on agarose-ε-aminocaproylfucosamine resin.

Figure 9.148 shows a representative chromatrogram.

9.17.2 α-N-Acetylgalactosaminyltransferase (Iwase et al., 1988)

This assay measures the ability of α-N-acetylgalactosaminyltransferase to transfer the sugar moiety of UDP-N-acetylgalactosamine to an oligosaccharide that bears a fluorescent label. There are two kinds of α-N-acetylgalactosaminyltransferase: One transfers sugar to a protein, the other to an oligosaccharide.

The substrate, pyridylaminolacto-N-fucopentaose I (Fucα1-2Galβ1-3GlcNAcβ1-3Galβ1-4Glc-PA) was separated from its N-acetylgalactosaminylated product by chromatography on Inertsil ODS-5 or ODS-2 column (4.6 mm × 250 mm, 5 μm). The mobile phase contained 0, 0.1, or 0.2% 1-butanol in 10 mM phosphate buffer (pH 3.8). The pyridylaminated derivatives were detected by fluorescence, using excitation and emission wavelength of 320 and 400 nm, respectively.

9.17 METABOLISM OF COMPLEX SACCHARIDES AND GLYCOPROTEINS

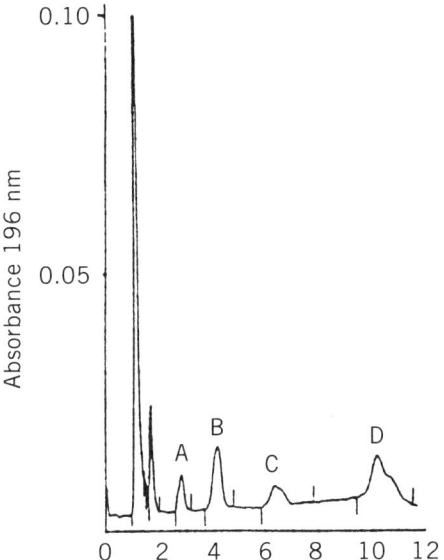

Figure 9.148 HPLC profile of gangliosides (A) G_{M1}, (B) fucosyl-G_{M1}, (C) G_{D1b}, and (D) fucosyl-G_{Db1}. Approximately 2 μg of each ganglioside was eluted with the buffer described in the text from the LiChrosorb-NH_2 column. (From Johnson, et al., 1990.)

The standard reaction mixture contained in a final volume of 100 μL: 50 mM Tris-HCl (pH 7.0), 10 mM $MnCl_2$, 0.5% Triton X-100, 1.0 mM UDP-GalNAc, 20 μL of enzyme, and about 10 pmol of pyridylamino lacto-N-fucopentaose I. The mixture was incubated at 37°C for 2.5 hours, and the reaction was terminated by adding 3 volumes of ethanol. After centrifugation, the supernate was dried and reconstituted in 100 μL of water. A 5 μL aliquot was analyzed by HPLC.

Microsomes were prepared from the antrum and corpus regions of rat stomach. The microsomal fraction from 1 g of wet tissue was suspended in 50 μL of 0.25 M sucrose containing 20 mM Tris-HCl (pH 7.2).

9.17.3 GM_1 Ganglioside β-Galactosidase (Naoi et al., 1988)

The enzyme called GM_1 cleaves the terminal D-β-galactose from glycolipids and glycoproteins. A deficiency of the enzyme in humans results in gangliosidosis. The assay described quantitates the fluorescent compound formed in a postcolumn reaction between galactose and arginine.

Galactose is separated from other components of the reaction mixture by chromatography on a Shimadzu ISA-07/S2504 column (4 mm × 250 mm).

The eluate was mixed with 2% L-arginine and 3% boric acid solution (0.5 mL/min) and the mixture was heated to 150°C in a 10 m reaction coil. The mobile phase was 0.5 M boric acid–NaOH buffer (pH 8.7), flowed at a rate of 0.6 mL/min at 65°C. Cooling was by passage through a 5 m tube, and the fluorescence intensity was determined (excitation and emission wavelengths of 320 and 430 nm, respectively).

The reaction mixture contained 50 μg to 1 mg of protein, 1 mM GM$_1$ in 200 μL of 50 mM citric acid–100 mM sodium phosphate buffer (pH 4.4) containing 100 mM NaCl, and 0.5% sodium taurodeoxycholate. The mixture was incubated at 37°C for 1 hour, and the reaction was terminated by heating at 100°C for 2 minutes. Cooling in an ice bath was followed by addition of 200 μL of the mobile phase solvent. The supernate obtained by centrifugation was filtered before analysis of an aliquot by HPLC. The reaction was linear for up to 1 hour with up to 0.7 mg protein added.

The source of enzyme activity was autopsy samples from human brains that were homogenized with 10 volumes of 10 mM potassium phosphate buffer (pH 7.4) containing 100 mM NaCl and 0.02% sodium azide. The sample was washed with 10 volumes of buffer by centrifugation with a Centricut Type 20 centrifuge tube, which excludes molecules with molecular weights below 20,000.

9.17.4 Aspartylglycosylaminase (Kaartinen and Mononen, 1990)

Aspartylglycosylaminase is a lysosomal enzyme that catalyzes the hydrolysis of the N-glycosidic linkage between N-acetylglucosamine and asparaginyl residues of glycoproteins.

The phenylisothiocyanate derivatives of aspartate, carboxymethylcysteine (internal standard) and 2-acetamido-1-L-β-aspartamido-1,2-dideoxy-β-D-glucose, were separated on a Spherisorb S3 ODS2 column (4.6 mm × 150 mm). The column was equilibrated with solvent A [50 mM sodium acetate (pH 6.4) containing 0.5 mL triethylamine per liter]. After sample injection, a linear gradient was run within 6 minutes to a composition of 80% solvent A and 20% solvent B (500 mL of solvent A, 400 mL of acetonitrile, 100 mL of methanol). After a 4-minute hold at this concentration, a gradient was run to 100% B in 2 minutes followed by another hold for 5 minutes. The flow rate of the mobile phase was 0.8 mL/min. The column eluent was monitored at 254 nm.

The reaction mixture contained in a total volume of 50 μL: 30 nmol of aspartylglucosamine, 30 nmol of internal standard (carboxymethylcysteine), and 5 to 100 μU of aspartylglycosylaminase in 50 mM potassium phosphate buffer (pH 7.5). Aliquots of 15 μL were removed at 0, 3, and 6 hours, and the reaction was terminated by heating in a boiling water bath for 5 minutes. After centrifugation, the supernate was evaporated *in vacuo*. The residue was dissolved in 10 μL of 50% ethanol in water and 10 μL of derivitization mixture (90% ethanol in water–triethanolamine–phenylisothiocyanate (7:2:1, v/v/v).

After incubation for 30 minutes at room temperature, the sample was evaporated to dryness. The residue was dissolved in 100 μL of solvent A and 5 μL was injected into the C_{18} reversed-phase column. Formation of product was linear with time with up to 6 mg of protein (human chorionic villus cell homogenate) added.

Sources of enzyme assayed included human plasma, leukocytes, amniotic fluid, chorionic villus cells, and brain and lung samples obtained at autopsy.

9.17.5 β-Galactosidase and Glycosyltransferase (Willenbrock et al., 1991)

The activity of β-galactosidase is assayed by measuring galactose release from lactose. Glycosyltransferase activity was measured by quantitating the lactose formed from UDP–galactose and glucose. A primary advantage of the HPLC-based method is that all the reaction products can be quantitated, which is important when enzymatic or chemical degradation of substrates and products occurs.

Lactose, glucose, galactose, and fucose (internal standard) were separated by anion-exchange chromatography on a Carbo-Pac PA-1 column (4 mm × 250 mm). The eluent was 200 mM NaOH flowed at a rate of 1 mL/min. Sugars were detected by triple-pulsed amperometry (detector setting 0–10 μA) using the following potentials and durations: $E_1 = 0.01V$ ($t_1 = 480$ ms), $E_2 = 0.6V$ ($t_2 = 180$ ms), and $E_3 = -0.6$ V ($t_3 = 120$ ms). Under the chromatographic conditions, glucose and galactose coeluted.

The reaction mixture for β-galactosidase contained in a final volume of 0.5 mL: 50 mM sodium citrate buffer (pH 3.5), 0.1 U/mL of β-galactosidase, 1 mg/mL of bovine serum albumin, 0.02% sodium azide, 1 mM α-L-fucose (internal standard), and lactose (5–40 mM). The assay was conducted at 30°C. Aliquots of 20 μL were removed at 8-minute intervals up to 48 minutes and added to 1 mL of distilled water at 4°C. The hydrolysis of substrate did not exceed 5%. Samples were desalted by passage through a column containing 0.2 mL each of Dowex AG50W-X12 (H^+ form) and Dowex AG3-X4 (OH^- form). The column was washed with 2 mL of distilled water and the eluate evaporated to dryness and taken up in water to give a lactose concentration of 0.2 mM. A 175 μL aliquot was used for HPLC analysis.

Galactosyltransferase (0.05–0.25 mU) was assayed in 50 μL of 50 mM cacodylate–HCl buffer (pH 7.4) containing 0.4 mM UDP–galactose, 25 mM glucose, 30 mM $MnCl_2$, 0.2 mg/mL of α-lactalbumin, 1.0 mg/mL bovine serum albumin, and 0.1 mM α-L-fucose (internal standard). The reaction was stopped by the addition of 1 mL of cold distilled water. Samples were processed as described above. The mobile phase for HPLC was 100 mM NaOH, and a detector setting of 0 to 3.0 μA was used.

Jack bean β-galactosidase and bovine milk galactosyltransferase were obtained commercially.

9.17.6 Glucose-1-phosphate Thymidylyltransferase (Lindquist et al., 1993)

Glucose-1-phosphate thymidylyltransferase catalyzes the first of four steps in the biosynthesis of dTDP-L-rhamnose. The reaction is between dTTP and α-D-glucose 1-phosphate to form dTDP-D-glucose and PP_i.

Separation of dTTP, dTDP-D-glucose, dTDP, and dTMP (where dTMP, dTDP, and dTTP are the deoxy forms of thymidine mono-, di-, and triphosphate, respectively) occurred on a Supelcosil, LC-SAX column (4.6 mm × 250 mm). The chromatogram was developed at room temperature with a 20 mL linear gradient from 50 to 400 mM potassium phosphate buffer (pH 4.0), at a flow rate of 1.5 mL/min. Quantitation was based on the absorbance at 254 nm, and all thymidine derivatives were monitored.

The reaction mixture contained in a total volume of 300 μL: 15 μmol Tris-HCl buffer (pH 8.0), 3.6 μmol of $MgCl_2$, 7.2 μmol of α-D-glucose-1-phosphate, 1.8 μmol of dTTP, 1.8 units of inorganic pyrophosphatase, and 30 μL of appropriately diluted enzyme. The mixture was incubated at 37°C. Samples (36 μL) were withdrawn at timed intervals up to 20 minutes, and the reaction was terminated by mixing with 1.0 mL of 50 mM potassium phosphate (pH 3.0). Reactions deviated only slightly from linearity for up to 40% conversion of the starting amount of dTTP.

The enzyme preparation was derived from *Salmonella enterica* LT2 cells that were disintegrated by sonication. The cell extract obtained by centrifugation was equilibrated with 20 mM Tris-HCl (pH 8.0), 1 mM $MgCl_2$, and 22% (v/v) glycerol by passage through a Sephadex G-25 column.

9.17.7 CMP-N-Acetylneuraminic Acid: Glycoprotein Sialyltransferase (Spiegel et al., 1992)

Sialyltransferases transfer sialic acid moieties as the terminal residues in oligosaccharide side chains of animal cell coat proteins and gangliosides. In this assay, CMP-N-acetylneuraminic acid was reacted with asialoglycoprotein to form cytidíne 5'-monophosphate (CMP) and the O-acetylneuraminic acid derivative of the protein.

Separation of CMP and CMP-NeuAc was by reversed-phase chromatography on a Beckman C_{18} Ultrasphere-IP column (4.6 mm × 250 mm, 5 μm). The column was eluted at 2.0 mL/min using a mobile phase composed of 9% acetonitrile in 5 mM sodium phosphate buffer (pH 7.5) containing 5 mM tetrabutylammonium phosphate as an ion-pairing agent. The column eluate was monitored at 270 nm.

The reaction mixture contained in a volume of 60 μL: 25 to 125 μg of asialoglycoprotein, 0.6 mM CMP-NeuAc, 50 mM sodium cacodylate hydrochloride (pH 6.0), 0.5% (v/v) Triton CF-54, 0.1% bovine serum albumin, and 0.1 to 1.2 mU of rat liver sialyltransferase. The reactions were terminated by the addition of 40 μL of ice-cold acetonitrile. The protein precipitate was removed by centrifugation, and 20 µL of the supernate was analyzed by HPLC. Formation of CMP increased with time, although not in linear fashion. The reaction was linear with amount of protein added.

Figure 9.149 shows representative chromatogram.

9.17.8 Thyroxine: UDP-glucuronosyltransferase (Ducrotoy et al., 1991)

The degradation of thyroxine in liver occurs primarily by conjugation with glucuronic acid. In this assay, [^{125}I]thyroxine glucuronide was quantitated by on-line radiochemical detection.

Thyroxine and thyroxine glucuronide were separated on a C_{18} Ultrabase column (4.6 mm × 250 mm, 10 µm). Solvent A was a 35:65 (v/v) mixture of methanol and 20 mM potassium phosphate buffer (pH 7.0) containing 1% triethylamine. Solvent B was methanol. A linear gradient from 0 to 100% B was run in 15 minutes. Elution was completed by maintaining 100% B for 5 minutes. Two methods of on-line radiochemical detection was used: liquid scintillation using a 2 mL detection cell, and solid scintillation using lithium scintillator glass in an effective cell volume of 400 µL. The latter method was most convenient.

The assay mixture consisted of 1.3 mg/mL microsomal proteins, 50 mM phosphate buffer (pH 8.0), 10 mM MgCl$_2$, and 10 µM thyroxine (completely solubilized in 1% bovine serum albumin–NaOH) containing 0.5 µCi/nmol [^{125}I]thyroxine. The reaction was initiated by the addition of 2 mM UDP-glucuronic acid. After incubation at 37°C, the reaction was stopped by addition of an equal volume of methanol–orthophosphoric acid (9:1, v/v). Precipitated proteins were removed by centrifugation before injection of an aliquot into the HPLC system. Reactions were linear for at least 60 minutes.

Sources of enzyme assayed were liver microsomes prepared from control and rats treated with Aroclor 1254.

9.17.9 *trans-p*-Coumaroyl Esterase (Borneman et al., 1990)

The degradation of plant cell walls requires *trans-p*-coumaroyl esterase, an important enzyme in the digestion of forages and dietary fiber. The *trans-p*-coumaric acid released by enzymatic hydrolysis was assayed by reversed-phase HPLC.

The *trans-α*- and *trans-β*-anomers of the substrate, *O*-[5-*O*-(*trans-p*-coumaroyl)-α-L-arabinofuranosyl]-(1→3)-*O*-β-D-xylopyranosyl-(1→4)-D-xylopyranose, were separated from each other and from the product, *trans-p*-coumaric acid, by chromatography on a Resolve C_{18} column (3.9 mm × 150 mm). The mobile phase consisted of a 10 mM NaOH solution titrated to pH 3.0 with formic acid. Methanol was added to this solution to give a final concentration of 21% (v/v). The flow rate was 2 mL/min. The absorbance of the elute was monitored at 313 nm.

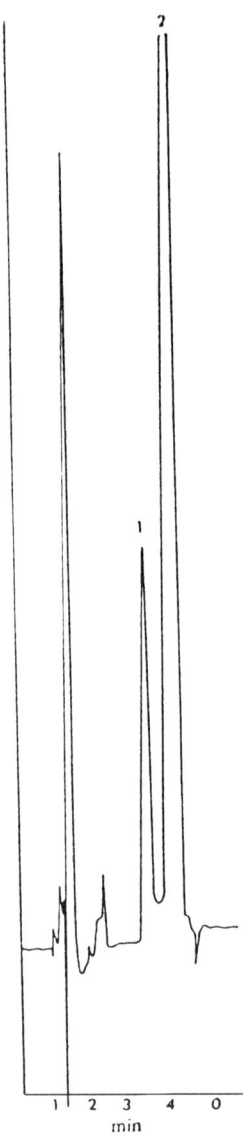

Figure 9.149 Separation of CMP (peak 1) and CMP-NeuAc (peak 2) by reversed-phase, ion-paired HPLC. (From Spiegel et al., 1992.)

The assay mixture was composed of 700 μL of 100 mM bis-trispropane (pH 7.5) containing 40 μg (86.8 μmol) of the substrate and 50 μL of the culture filtrate (40 μg/mL) containing enzyme activity. After a 10-minute incubation at 30°C, the reaction was stopped by the addition of 50 μL of formic acid (20% v/v). Product formation was linear with time for 40 minutes.

The enzyme source was culture filtrate from an anaerobic rumen fungus, *Neocallimastix* MC-2.

9.18 MISCELLANEOUS

9.18.1 Carboxylases (Oizumi and Hayakawa, 1990)

The method described is suitable for the assay of four biotin-containing carboxylases: pyruvate carboxylase, acetyl–coenzyme A carboxylase, propionyl–coenzyme A carboxylase, and 3-methylcrotonyl–coenzyme A carboxylase. The assays do not require radioisotopes and are suitable for use in clinical laboratories.

Substrates and products are separated by reversed-phase chromatography at 45°C on a Nucleosil C_{18} column (4.6 mm × 250 mm). For assay of acetyl–coenzyme A carboxylase, propionyl–coenzyme A carboxylase, and 3-methylcrotonyl–coenzyme A carboxylase, a linear gradient from solvent A (0.1 M sodium phosphate buffer, pH 2.1) to solvent B (methanol–solvent A, 80:20, v/v) was applied in 15 minutes at a flow rate of 1.5 mL/min. Quantitation was based on the absorbance of the product (malonyl–CoA, methylmalonyl–CoA, and 3-methylglutaconyl–CoA, respectively) at 260 nm. For assay of pyruvate carboxylase, pyruvate was separated by isocratic elution using 0.1 M sodium phosphate buffer (pH 2.1) containing 0.1 M sodium sulfate. Quantitation was based on the disappearance of pyruvate as followed at 210 nm.

Acetyl–coenzyme A carboxylase, propionyl–coenzyme A carboxylase, and 3-methylcrotonyl–coenzyme A carboxylase were assayed in a total volume of 40 μL containing 80 mM potassium phosphate (pH 7.0), 3 mM ATP, 5 mM magnesium chloride, 30 mM sodium bicarbonate, 0.5 mM substrate (acetyl–CoA, propionyl–CoA, or methylcrotonyl–CoA), and enzyme solution. The reaction was started by adding enzyme and was allowed to proceed at 37°C for 20 minutes before being stopped by the addition of 80 μL of 6 M HCl. The supernate obtained by centrifugation was used for HPLC analysis. Pyruvate carboxylase was assayed in a reaction volume of 50 μL containing 50 mM Tris-HCl (pH 7.2), 5 mM ATP, 5 mM magnesium sulfate, 1 mM pyruvate, 0.5 mM EDTA, 10 mM potassium bicarbonate, 0.7 mM NADH, 6 μg of malate dehydrogenase, 0.1 mM acetyl–CoA (as activator), and the enzyme. The reaction was started by adding enzyme and was stopped after incubating for 5 minutes at 37°C by adding 0.1 mL of 6 M HCl. After centrifugation, 10 μL of the supernate was used for HPLC analysis. Linear reactions were observed for over 30 minutes for the three acyl–CoA carboxylases, and for over 10 minutes in the case of pyruvate carboxylase.

The assays were developed for guinea pig kidney homogenates prepared in 0.32 M sucrose containing 1 mM sodium phosphate buffer (pH 7.0). The homogenates were stored frozen at $-20°C$ before use. The amounts of homogenate used corresponded to 50 µg of protein for the acetyl and propionyl–CoA carboxylases, and 10 µg for the other two enzymes.

9.18.2 Carbonyl Reductase (Naganuma et al., 1990)

Carbonyl reductase plays a role in reducing a variety of aliphatic, alicyclic, and aromatic aldehydes and ketones, including endogenous ketosteroids and/or prostanoids. This assay allows quantitation of both enantiomers formed from the substrate.

The substrate, 4-(6-methoxy-2-benzoxazolyl)acetophenone, the R and S enantiomers of the product, 4-(6-methoxy-2-benzoxazolyl)phenethyl alcohol, and the internal standard, 3-acetyl-7-(dimethylamino)coumarin, were separated by chromatography on a cellulose-based chiral column (Chiracel OD from Daicel). The mobile phase was a 93:7 mixture (v/v) of n-hexane and 2-propanol, which was used at ambient temperature and a flow rate of 1.2 mL/min. Detection of the enantiomeric alcohols and the internal standard was by fluorescence, with excitation and emission wavelengths of 315 and 375 nm, respectively.

The standard reaction mixture contained 0.1 mL of 10 mM substrate dissolved in acetonitrile, 200 µM NADPH, 0.1 mL of cytosol, and 10 mg/mL bovine serum albumin in 0.1 M potassium phosphate buffer (pH 6.2). The final volume was 2.0 mL. The reaction was continued at 25°C for 10 minutes before being terminated by the addition of 0.2 mL of 2 M HCl. The product was extracted twice with 5 mL of ethyl acetate in the presence of 50 nmol of 3-acetyl-7-dimethylaminocoumarin as the internal standard. After centrifugation, the organic layer was evaporated in vacuo. The residue was taken up in 1.0 mL of the mobile phase and a 5 µL aliquot was used for HPLC analysis. The formation of both enantiomers was linear with time for up to 30 minutes. The rate of formation of the S enantiomer was about ninefold higher than that of the R enantiomer.

Rabbit liver cytosol was prepared from liver homogenized in 50 mM potassium phosphate buffer (pH 7.4) containing 0.5 mM dithiothreitol. The assay was also used to determine activities in kidney, lung, and intestine from rabbit, and the same four tissues from rat.

9.18.3 6-Pyruvoyl Tetrahydropterin Synthetase (Werner et al., 1991)

6-Pyrovoyl tetrahydropterin synthetase catalyzes the second step in the conversion of guanosine triphosphate into tetrahydrobiopterin. The substrate, 4,8-dihydroneopterin triphosphate, is converted to 6-pyruvoyl tetrahydropterin.

Sepiapterin reductase is included in the assay to allow the product to proceed to tetrahydrobiopterin, which is then oxidized to biopterin, which is detected by fluorescence.

The neopterin and biopterin derivatives were separated by chromatography on a LiChrosorb RP-18 column (4 mm × 250 mm, 7 μm). The column was eluted at a flow rate of 0.8 mL/min with 0.015 M potassium phosphate buffer (pH 6.0). To elute biopterin, a pulse of 0.1 mL of 0.6 M potassium phosphate buffer (pH 6.8) was used. Biopterin was measured fluorimetrically by using excitation and emission wavelengths of 353 and 438 nm, respectively.

The standard reaction mixture was composed of 5 μL of Tris-HCl (pH 7.4), 5 μL of 40 mM NADPH, 5 μL of sepiapterin reductase (activity of 400 nmol/min/mL), and 65 μL of cell extract (10–200 μg of protein). The reaction was started by the addition of 20 μL of 0.4 mM 7,8-dihydroneopterin triphosphate. After 30 to 90 minutes of incubation in the dark at 37°C, the reaction was terminated by the addition of 50 μL of a mixture of 0.2 M HCl and 0.02 M KI-I_2 (1:1, v/v). The resulting mixture was incubated for 1 hour in the dark to allow oxidation of tetrahydrobiopterin to biopterin. Excess iodine was destroyed by the addition of 50 μL of 0.02 M ascorbic acid. An aliquot of the mixture was applied to a solid phase cartridge (SCX from Analytichem) that had been preequilibrated with 0.1 M H_3PO_4. The sample was forced through the cartridge with air pressure. The cartridge was then washed with 0.5 mL of 0.1 M H_3PO_4. The eluates were used for HPLC analysis. Assays were linear with up to 150 μg of cellular protein and 90 minutes of incubation.

Enzyme activity was measured in cell-free extracts of normal human fibroblasts and the human transitional cell bladder carcinoma line T-24.

Figure 9.150 shows chromatogram.

9.18.4 Pteroylpolyglutamate Hydrolase (Krungkrai et al., 1987)

Enzymes that hydrolyze the polyglutamate side chain of pteroylpolyglutamates are important in the bioavailability of dietary pteroylpolyglutamates. This assay uses the pentaglutamate derivative as substrate, and all intermediates on the pathway to pteroylmonoglutamate are detected.

The pteroylglutamate derivatives were separated on a μBondapak C_{18} column (2.9 mm × 300 mm, 10 μm). The column was equilibrated with solvent A (0.1 M potassium phosphate buffer, pH 6.0). Following injection of sample, a linear gradient from 75% solvent A and 25% solvent B (solvent A containing 10% acetonitrile) to 15% solvent A and 85% solvent B was applied over a 20-minute time span. The effluent was monitored at 280 nm.

The standard reaction mixture contained in a volume of 250 μL: 200 μL of the enzyme source in 0.1 M acetate buffer (pH 4.5), 25 μL of 0.1 M sodium acetate buffer, and 25 μL of 1.0 M pteroylpentaglutamate. The reaction was started by adding substrate and the reaction was allowed to proceed at 37°C for 1 hour in the dark. The reaction was stopped by the addition of 250 μL

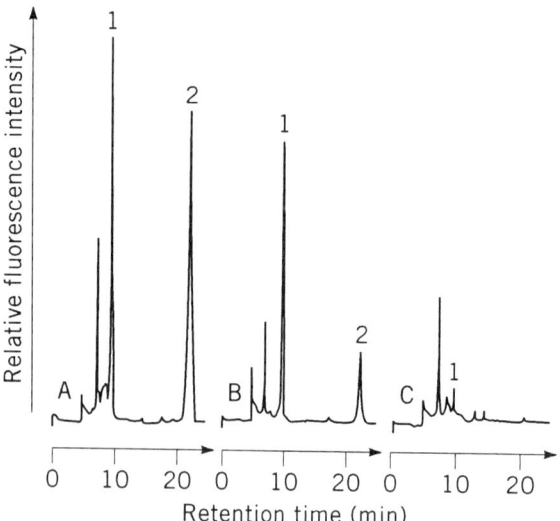

Figure 9.150 HPLC profiles of incubation mixtures for 6-pyruvoyl tetrahydropterin synthetase assays with extracts of (*A*) T 24 cells, (*B*) human dermal fibroblasts, and (*C*) a reagent control. Amounts of 56 μg (*A*) and 60 μg (*B*) of cellular protein or phosphate-buffered saline (*C*) were used in the assay. Incubation time was 70 minutes at 37°C. A 100 μL aliquot of the incubation mixture was used. Fluorescence detection was at an excitation wavelength of 353 nm and an emission wavelength of 438 nm. *Peaks:* 1, neopterin; 2, biopterin. (From Warner et al., 1991.)

of ice-cold 2% ascorbic acid in 0.1 *M* phosphate buffer (pH 6.5) followed by heating in a boiling water bath for 10 minutes. The mixture was cooled in an ice bath and centrifuged before analysis by HPLC. Formation of product was linear for up to 60 minutes and with up to 200 μL of human serum added. Activity was based on disappearance of the substrate and the appearance of all the enzymatic products.

The enzyme preparation was human serum that was equilibrated with 0.1 *M* sodium acetate buffer by chromatography on Sephadex G-25.

9.18.5 Nitrogenase (Bravo et al., 1988)

Nitrogenase reduces N_2 and several nitrogen-containing substrates to ammonia and amines. This assay uses a dabsyl precolumn derivatization method in quantitating the products.

The dabsyl derivatives of ammonia and methylamine are separated from other reaction components by chromatography an Altex Ultrasphere-ODS C_{18} column (4.6 mm × 150 mm, 5 μm). An anion guard column from Bio-Rad (Anion-SA, sulfate form, 4.6 mm × 40 mm) was placed ahead of the main column. For ammonia detection only, an Altex Ultrasphere-ODS column

(4.6 mm × 55 mm) could be used as the main column. The mobile phase used was acetonitrile–methanol–water (3:7:7). The flow rate was 1.5 mL/min. Fluorescence was monitored using 368 and 500 nm as the excitation and emission wavelengths, respectively.

The reaction was stopped by addition of 0.1 mL of 1 N HCl in saturated KIO_3. Following centrifugation, derivatization was performed on a 0.4 mL aliquot by adding 0.4 mL of 0.164 M borate buffer (pH 10.0) followed by 0.4 mL of 1.67 mM dansyl chloride (dissolved in acetone). After 90 minutes of incubation at room temperature, a 20 μL sample was injected directly into the HPLC. To minimize background ammonia and methyl amine, it was important to prepare all reagents in HPLC-grade water.

The source of enzyme was a purified nitrogenase preparation from *Azotobacter vinelandii*.

9.18.6 Strictosidine Synthetase (Pennings et al., 1989)

Strictosidine synthetase catalyzes the stereospecific condensation of tryptamine and the iridoid glucoside secologanin to form strictosidine. The product is the precursor of the monoterpenoid-derived indole and quinoline alkaloids.

Strictosidine, tryptamine, and codeine (internal standard) were separated on a LiChrosorb RP-8 Select B column (4 mm × 250 mm, 7 μm) at a flow rate of 1 mL/min. The mobile phase contained 7 mM sodium dodecyl sulfate and 25 mM sodium phosphate (pH 6.2) in 32% methanol (v/v). The effluent was monitored at 280 nm. Quantitation was based on standard curves for tryptamine and strictosidine.

The reaction mixture contained in a final volume of 0.1 mL, 0.1 M sodium phosphate buffer (pH 6.8), 1.0 mM tryptamine, 5 mM secologanin, and 3 mM dithiothreitol. To inhibit glucosidase activity, 100 mM D(+)-gluconic acid-δ-D-gluconolactone was included. The incubation was started by addition of 10 μL of enzyme. After 30 minutes of incubation at 30°C, the reaction was stopped by addition of 0.1 mL of 5% trichloroacetic acid. Before centrifugation, 25 μL of 8 mM codeine hydrochloride was added as the internal standard. HPLC analysis was performed on 4 μL aliquots. With enzyme purified to a specific activity of 710 pkat per milligram of protein, the reaction was linear with time for 1 hour and with protein up to at least 50 μg of protein during a 20-minute incubation period.

Figure 9.151 shows chromatogram.

9.18.7 Anhydrotetracycline Oxygenase and Tetracycline Dehydrogenase (Neuzil et al., 1989)

Two enzymes catalyze the last two steps of tetracycline biosynthesis: Anhydrotetracycline oxygenase produces dehydrotetracycline, and tetracycline dehydrogenase converts dehydrotetracycline to tetracycline. Usage of a diode-

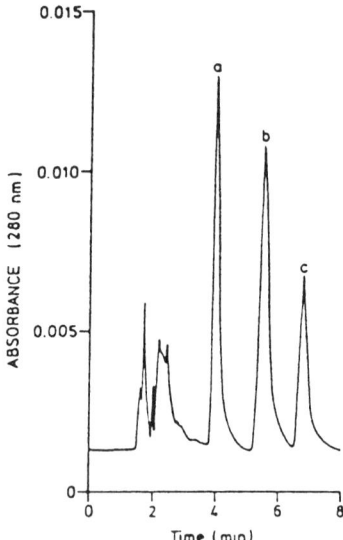

Figure 9.151 Determination of strictosidine synthetase activity by HPLC. Codeine (a), tryptamine (b), and strictosidine (c) were separated on a 4.0 (i.d.) × 250 mm LiChrosorb RP-8 Select B column at a flow rate of 1.0 mL/min. Incubation was for 30 minutes at 30°C with enzyme from *Catharanthus roseus* after ammonium sulfate precipitation (35–50% saturation) and gel filtration on Sephadex G-25, in the presence of 100 mM δ-D-gluconolactone. Injection volume was 8 μL and the UV detector was set at 0.02 AUFS. (From Pennings et al., 1989.)

array detector facilitated the quantification of the reactions by making it possible to monitor the product and substrate at their optimal wavelengths.

Anhydrotetracycline, dehydrotetracycline, and tetracycline were separated at 40°C on a Separon ODS glass-pack column (1 mm × 150 mm, 5 μm). Mobile phase A was a 20:80 mixture of 20 mM EDTA (pH 6.4) and dimethylformamide. Solvent B was methanol. After 0.5 minute on solvent A, a 0.5-minute linear gradient from 0 to 50% B was started, with the higher concentration held for 2 minutes. This was followed by a return to the starting conditions with a 0.5-minute gradient. After a 1.5-minute delay, the next sample was injected. A diode-array detector was used, with peak areas integrated at 440, 400, and 360 nm for anhydrotetracycline, dehydrotetracycline, and tetracycline, respectively.

The anhydrotetracycline oxygenase assay contained 10 μL of anhydrotetracycline oxygenase–stimulating factor [prepared by heat deproteinization of *Streptomyces aureofaciens*, 0.24 mM NADP, 0.6 mM glucose-6-phosphate, 0.2 unit glucose-6-phosphate dehydrogenase, 0.08 M Tris-HCl (pH 7.4), 0.16 mM anhydrotetracycline, and enzyme]. The total volume was 100 μL. The reaction was initiated by adding enzyme. The reactions were carried out in 300 μL HPLC glass microvials placed in the autosampler maintained at 30°C. Reactions were stopped by direct injection onto the column, beginning as early as 0.5 minute after initiation.

The tetracycline dehydrogenase reaction was carried out under identical conditions except anhydrotetracycline was substituted by dehydrotetracycline, and tetracycline dehydrogenase was added. The reaction was often carried out using the anhydrotetracycline reaction mixture after a complete anhydrotetracycline to dehydrotetracycline conversion.

Both reactions were linear over a wide range of amounts of enzyme added. The enzymes assayed were prepared from *S. aureofaciens* and were separated from each other by passing a cell-free extract through a phenyl–Sepharose CL-4B column. Tetracycline dehydrogenase activity was not retained on the column. Anhydrotetracycline oxygenase was liberated using a linear gradient of Lubrol.

9.19 NUCLEIC ACID MODIFICATION AND EXPRESSION

9.19.1 DNA Topoisomerase (Onishi et al., 1993)

Onishi et al. (1993) developed an assay that monitors the relaxation of supercoiled plasmid DNA induced by DNA topoisomerase. Supercoiling of DNA is important in the formation of chromatin and in determining functions of DNA, including replication, transcription, recombination, and repair.

Relaxed and supercoiled DNA samples from plasmid pBR329 were separated by ion-exchange chromatography on a TOSO DEAE-NPR column (4.6 mm × 35 mm) or TOSO DEAE-5PW column (7.5 mm × 75 mm). These columns were equilibrated with 20 mM Tris-HCl buffer (pH 7.5) containing 0.5 M NaCl, and the column was eluted with a linear gradient from 0.5 to 0.65 M NaCl at a flow rate of 1 mL/min. Absorbance was monitored at 260 nm.

The reaction mixture contained 50 mM Tris-HCl buffer (pH 7.5), 120 mM KCl, 10 mM MgCl$_2$, 0.5 mM EDTA, 0.5 mM dithiothreitol, 30 μg/mL bovine serum albumin, and 0.2 μg of pBR329 DNA. After incubation at 37°C for 30 minutes, the reaction was terminated by the addition of Sarkosyl in a final concentration of 1%. The reaction mixture was further incubated at 37°C for 30 minutes after proteinase K had been added to give a concentration of 50 μg/mL. Assays were linear with up to 0.4 μg of topoisomerase I, and with time until about 95% of the substrate had been converted into relaxed DNA.

Topoisomerase I was partially purified from the nuclei of HeLa S3 cells by a procedure that involved lysing the nuclei and precipitating nucleic acid with polyethylene glycol.

Figure 9.152 shows that the separation is obtained in 25 minutes.

9.19.2 Chloramphenicol Acetyltransferase (Davis et al., 1992)

Chloramphenicol acetyltransferase is used as a genetic marker system to study transformation and gene expression in plant protoplasts, cells, and tissues.

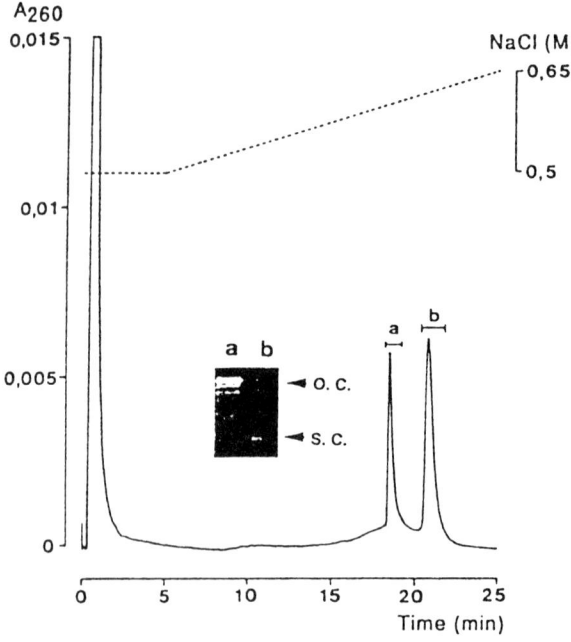

Figure 9.152 Profile of separation of relaxed DNA from supercoiled plasmid pBR329 on a DEAE-NPR column. The reaction was carried out with 1 unit of calf thymus topoisomerase I. Twenty microliters of the reaction mixture was applied on the column and eluted with a linear gradient of 0.5 to 0.65 M NaCl for 30 minutes. DNAs from peaks a and b were collected manually and analyzed by electrophoresis on a 1% agarose gel after ethanol precipitation. *Inset:* O.C., open circular; S.C., supercoiled. (From Onishi et al., 1993.)

Chloramphenicol can be acetylated at the 1 and 3 positions to give monoacetoxy and diacetoxy derivatives.

Chloramphenicol, the 1- and 3-diacetoxy derivatives, and 1,3-diacetoxychloramphenicol were separated by chromatography on a Spherisorb 5 ODS column (5 mm × 250 mm, 5 μm). The mobile phase was composed of ethyl acetate–methanol–distilled water–phosphoric acid (6.0:41:52:5:0.5, v/v/v/v). The effluent was monitored at 278 nm. Enzyme activity was based on the formation of the monoacetoxy derivatives.

The reaction mixture contained enzyme, 0.20 M Tris-HCl (pH 7.8) to give a final volume of 180 μL, 9.5 μL of 1.0 mM chloramphenicol, and 10 μL of 10 mM acetyl–CoA. After incubation for an hour at 37°C, 300 μL of cold ethyl acetate was added to stop the reaction. The aqueous and ethyl acetate phases were separated by centrifugation. The aqueous phase was extracted two more times with ethyl acetate, and the combined ethyl acetate fractions were dried in a fume hood for 16 hours. The residues were redissolved in

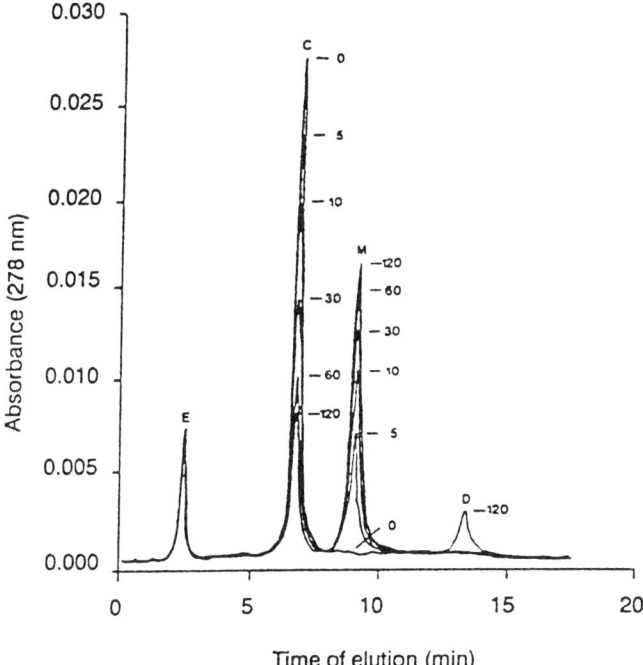

Figure 9.153 Detection of chloramphenicol acyltransferase (CAT) reaction products by HPLC. Triplicate reaction mixtures containing 10 units of CAT were incubated at 37°C. At the times indicated 200 μL aliquots were removed and immediately mixed with 300 μL of ethyl acetate. Samples were extracted twice with ethyl acetate, and the solvent was then allowed to evaporate in a vacuum chamber. The residues were then redissolved in 20 μL of ethyl acetate and injected on top of the HPLC column. *Peaks:* C, chloramphenicol; D, 1,3-diacetoxy chloramphenicol; E, solvent peak, ethyl acetate; M, 1- and 3-monoacetoxy chloramphenicol. Traces from 0-, 5-, 10-, 30-, and 60-minute time points are superimposed on the figure. (From Davis et al., 1992.)

20 μL of ethyl acetate. Formation of the monoacetoxy derivatives was linear for 1 hour when 0.01 to 1.0 unit of enzyme activity was used.

Figures 9.153 and 9.154 show chromatograms.

9.20 SUMMARY AND CONCLUSIONS

This chapter reviewed HPLC methods developed to assay enzyme activity. The enzymes were grouped on the basis of their substrates, but some enzymes could have been put in several groups.

Each report is reviewed according to a standardized format. The reaction is described first, followed by a general statement explaining the basis for the

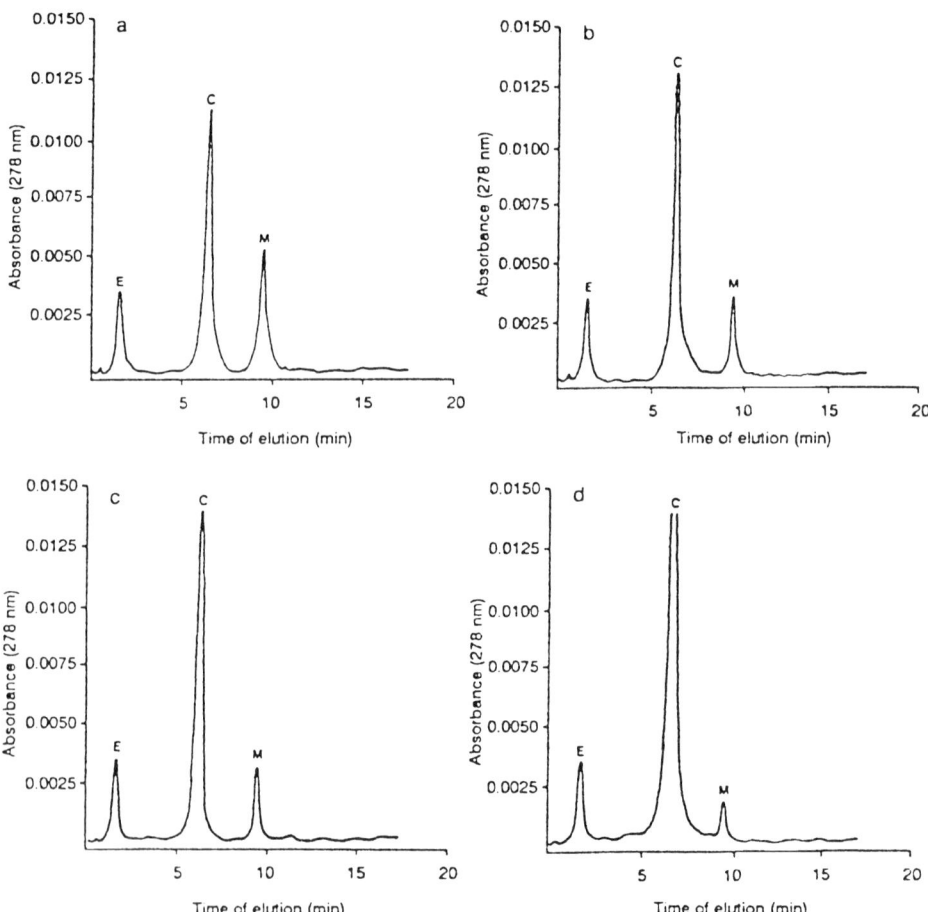

Figure 9.154 Detection of CAT activity in transformed rice and petunia protoplasts by HPLC. (*a*) Petunia leaf, (*b*) Petunia cell suspension, (*c*) rice leaf, and (*d*) rice cell suspension protoplasts electroporated at 500, 625, 825, and 625 V with an 860 μF capacitor, respectively. Protoplasts were transformed by electroporation with 20 μg/mL pDW2. Protein extracts were prepared and used in a 200 μL CAT reaction mixture. Typically 12 μg of crude protein was used per reaction except for rice cell suspension protoplasts, where 20 μg was used. Products were isolated and 1 to 5 μL of purified products was injected on top of the column. *Peaks:* C, chloramphenicol; E, solvent peak, ethyl acetate; M, 1- and 3-monoacetoxy chloramphenicol. (From Davis et al., 1992.)

assay. Next, the method of separation is described, including mention of the stationary phase, the composition of the mobile phase, and the method of elution of the column. The detection method is given, as well.

REFERENCES

Catecholamine metabolism

Beaudouin C, Haurat G, Fraisse L, Souppe J, Renaud B (1993) *J Chromatogr* **613**:51.
Bradley RT, Manowitz P (1988) *Anal Biochem* **173**:33.
D'Erme M, Rosei MA, Fiori A, Di Stazio G (1980) *Anal Biochem* **104**:59.
Feilchenfeld NB, Richter HW, Waddell WH (1982) *Anal Biochem* **122**:124.
Freeman KB, Bulawa MC, Zeng Q, Blank CL (1993) *Anal Biochem* **208**:182.
Haavik J, Flatmark T (1980) *J Chromatogr* **198**:511.
Lee M-K, Nohta H, Ohkura Y (1987) *J Chromatogr* **421**:237.
Mandai M, Iwaki M, Honda Y (1992) *Ophthalmic Res* **24**:228.
Mannens G, Slegers G, Claeys A (1990) *Biochim Biophys Acta* **1037**:1.
Martin RJ, and Downer RGH (1989) *J Chromatogr* **487**:287.
Nagatsu T, Oka K, Kato T (1979) *Anal Biochem* **100**:160.
Naoi M, Takahashi T, Nagatsu T (1988) *J Chromatogr* **427**:229.
Pennings EJ, Van Kempen GM (1979) *Anal Biochem* **98**:452.
Rahman MK, Nagatsu T, Kato T (1980) *J Chromalogr* **221**:265.
Sim MK, Hsu TP (1990) *J Pharmacol Methods* **24**:157.
Smit NM, Pavel S, Kammeyer A, Westerhof W (1990) *Anal Biochem* **190**:286.
Thomas KB, Zawilska J, Iuvone PM (1990) *Anal Biochem* **184**:228.
Trocewicz J, Oka K, Nagatsu T (1982) *J Chromatogr* **227**:407.

Proteinase

Advis JP, Krause JE, McKelvy JF (1982) *Anal Biochem* **125**:41.
Baranowski R, Westenfelder C, Currie BL (1982) *Anal Biochem* **121**:97.
Betageri R, Hopkins JL, Thibeault D, Emmanuel M, Chow GC, Skoog MT, de Dreu P, Cohen KA (1993) *J Biochem Biophys Methods* **27**:191.
Chen CS, Wu SH, Wang KT (1982) *J Chromatogr* **248**:451.
Doig MT, Smiley JW (1993) *J Chromatogr* **613**:145.
Gray RD, Saneii HH (1982) *Anal Biochem* **120**:339.
Grimwood BG, Tarentino AL, Plummer TH, Jr (1988) *Anal Biochem* **170**:264.
Harada M, Hiraoka BY, Mogi M, Fukasawa K, Fukasawa KM (1988) *J Chromatogr* **424**:129.
Harrison R, Teahan J, Stein R (1989) *Anal Biochem* **180**:110.

Hopkins JL, Betageri R, Cohen KA, Emmanuel MJ, Joseph CR, Bax PM, Pallai PV, Skoog MT (1991) *J Biochem Biophys Methods* **23**:107.
Horiuchi M, Fujimura KI, Terashima T, Iso T (1982) *J Chromatogr* **233**:123.
Marceau F, Drumheller A, Gendreau M, Lussier A, St. Pierre S (1983) *J Chromatogr* **266**: 173.
Mousa S, Couri D (1983) *J Chromatogr* **267**:191.
Nakamura-Imago N, Satomura S, Matsuura S, Murakami K (1992) *Clin Chim Acta* **1037**:1.
Ohno M, Kai M, Ohkura Y (1988) *J Chromatogr* **430**:291.
Tamburini PP, Dreyer RN, Hansen J, Letsinger J, Elting J, Gore-Willse A, Daily R, Yoo-Warren H (1990) *Anal Biochem* **186**:363.

Amino acid and peptide metabolism

Aberhart DJ (1988) *Anal Biochem* **169**:350.
Biondi PA, Guidotti L, Negri A, Secchi C (1991) *J Chromatogr* **566**:377.
Chakraborty M, Lahiri P, Anderson GM, Chatterjee D (1991) *J Chromatogr* **571**:235.
Consalvo A, Young SD, Jones BN, Tamburini PP (1988) *Anal Biochem* **175**:131.
Davis AT (1989) *J Chromatogr* **497**:263.
Fasolato C, Galzigna L (1988) *J Chromatogr* **426**:381.
Fukuda H, Yamatodani A, Imamura I, Maeyma K, Watanabe T, Wada H (1991) *J Chromatogr* **567**:459.
Garras A, Djurhuus R, Christensen B, Lillihaug JR, Ueland PM (1991) *Anal Biochem* **199**:112.
Goeger DE, Ganther HE (1993) *Arch Biochem Biophys* **302**:222.
Holmes EW (1988) *Anal Biochem* **172**:518.
Kochhar S, Mehta PK, Christen P (1989) *Anal Biochem* **179**:182.
Krstulovic AM, Matzura C (1992) *J Chromatogr* **176**:217.
Lee K-H, Cava M, Amiri P, Ottoboni T, Lindquist RN (1992) *Arch Biochem Biophys* **292**:77.
Li J, Christensen BM, Tracy JW (1990) *Anal Biochem* **190**:354.
Marques S, Florencio FJ, Candau P (1989) *Anal Biochem* **180**:152.
Martin F, Suzuki A, Hirel B (1982) *Anal Biochem* **125**:24.
Nardi G, Cipollaro M, Loguercio C (1990) *J Chromatogr* **530**:122.
O'Donnell JJ, Sandman RP, Martin SR (1978) *Anal Biochem* **90**:41.
Petrarulo M, Pellegrino S, Marangelia M, Cosseddu D, Linari F (1992) *Clin Chim Acta* **208**:183.
Seifert J (1993) *J Chromatogr* **614**:227.
Singh BK, Szamosi I, Shaner D (1993) *Anal Biochem* **208**:260.
Smyk-Randall EM, Brown OR (1987) *Anal Biochem* **164**:434.
Ubbink JB, Vermaak WJH, Bissbort SH (1991) *J Chromatogr* **566**:369.
Unnithan S, Moraga DA, Schuster SM (1984) *Anal Biochem* **136**:195.

Weir ANC, Bucke C, Holt G, Lilly D, Bull AT (1989) *Anal Biochem* **180**:289.
White RL, DeMarco AC, Shapiro S, Vining LC, Wolfe S (1989) *Anal Biochem* **178**:399.
Wiseman JS, Nichols JS (1990) *Anal Biochem* **184**:55.
Yim JJ, Lim KT, Kim N, Jacobson KB (1987) *J Chromatogr* **419**:296.
Zuo DM, Yu PH (1991) *J Chromatogr* **567**:381.

Polyamine metabolism

Beeman CS, Rossomando EF (1989) *J Chromatogr* **496**:101.
Biondi PA, Secchi C, Negri A, Tedeschi G, Ronchi S (1989) *J Chromatogr* **491**:209.
Halline AG, Brasitus TA (1990) *J Chromatogr* **533**:187.
Haraguchi R, Kai M, Kohashi K, Ohkura Y (1980) *J Chromatogr* **202**:107.
Porta R, Esposito C, Sellinger OZ (1981) *J Chromatogr* **226**:208.

Heme metabolism

Crowne H, Lim CK, Samson D (1981) *J Chromatogr* **223**:421.
Guo R, Lim CK, Peters TJ (1991) *J Chromatogr* **566**:383.
James CA, Marks GS (1989) *Can J Physiol Pharmacol* **67**:246.
Lincoln BC, Mayer A, Bonkovsky HL (1988) *Anal Biochem* **170**:485.
Tikerpae J, Samson D, Lim CK (1981) *Clin Chim Acta* **113**:65.
Tomokuni K, Hirai T, Ichiba M (1991) *J Chromatogr* **567**:65.

Carbohydrate metabolism

Fluharty AL, Glick JA, Samaan GF, Kihara H (1982) *Anal Biochem* **121**:310.
Haegele EO, Schaich E, Rauscher F, Lehmann P, Grassi M (1981) *J Chromatogr* **223**:69.
Hymes AJ, Mullinax F (1984) *Anal Biochem* **139**:68.
Lambeth DO, Muhonen WW (1993) *Anal Biochem* **209**:192.
Matsui M, Nagai F (1980) *Anal Biochem* **105**:141.
Naoi M, Yagi K (1981) *Anal Biochem* **116**:98.
Omichi K, Ikenaka T (1982) *J Chromatogr* **230**:415.
Petrie CR III, Korytnyk W (1983) *Anal Biochem* **131**:153.
Salvucci ME, Crafts-Brandner SJ (1991) *Anal Biochem* **194**:365.
Sandman R (1983) *J Chromatogr* **272**:67.
Schreuder HA, Welling GW (1983) *J Chromatogr* **278**:275.
Shylaja N, Maehara M, Watanabe K (1990) *Anal Biochem* **191**:223.
Taha SM, Deits TL (1994) *Anal Biochem* **219**:115.
To ECA, Wells PG (1984) *J Chromatogr* **301**:282.

Steroid metabolism

Gallant S, Bruckheimei SM, Brownie AC (1978) *Anal Biochem* **89:**196.
Hylemon PB, Raynor L, Bohdan PM, Vlahcevic ZR (1991) *Anal Biochem* **193:**256.
Hylemon PB, Studer EJ, Pandak WM, Heuman DM, Vlahcevic ZR, Chiang JYL (1989) *Anal Biochem* **182:**212.
Schatzman GL, Laughlin ME, Blohm TR (1988) *Anal Biochem* **175:**219–226.
Sugano S, Morishima N, Ikeda H, Horie S (1989) *Anal Biochem* **182:**327.
Suzuki K, Kadowaki A, Tamaoki B (1980) *J Endocrinol Invest* **4:**441.
Tanaka Y, Deluca HF (1981) *Anal Biochem* **110:**102.

Purine metabolism

Ali LZ, Sloan DL (1982) *J Biol Chem* **257:**1149.
Amici A, Emanuelli M, Raffaelli N, Ruggieri S, Magni G (1994) *Anal Biochem* **216:**171.
Barret J-M, Ernould A-P, Rouillon M-H, Ferry G, Genton A, Boutin JA (1993) *Chem Biol Interactions* **86:**17.
Canepari S, Carunchio V, Girelli AM, Messina A (1993) *J Chromatogr* **616:**25.
Chism GW, Long AR, Rolle R (1984) *J Chromatogr* **317:**263.
Danielson ND, Huth JA (1980) *J Chromatogr* **221:**39.
Dye F, Rossomando EF (1982) *Biosci Rep* **2:**229.
Garrison PN, Roberson GM, Culver CA, Bames LD (1982) *Biochemistry* **21:**6129.
Greenberg ML, Hershfield MS (1989) *Anal Biochem* **176:**290.
Halfpenny AP, Brown PR (1980) *J Chromatogr* **199:**275.
Hanna L, Sloan DL (1980) *Anal Biochem* **103:**230.
Hartwick R, Jeffries A, Krstulovic A, Brown PR (1978) *J Chromatogr Sci* **16:**427.
Jahngen EG, Rossomando EF (1984) *Anal Biochem* **137:**493.
Lambeth DO, Muhonen WW (1993) *Anal Biochem* **209:**192.
Lennard L, Singleton HJ (1994) *Biochem* **34:**12445.
Paulik E, Jayaram HN, Weber G (1991) *Anal Biochem* **197:**143.
Pietta P, Pace M, Menegus F (1983) *Anal Biochem* **131:**533.
Raffin JP, Thebault MT (1991) *Comp Biochem Physiol* **99B:**125.
Reysz LJ, Carroll AG, Jarrett HW (1987) *Anal Biochem* **166:**107.
Rossomando EF (1987) *High Performance Liquid Chromatography in Enzymatic Analysis*, 1st ed. John Wiley & Sons, Inc., New York, Ch. 5.
Rossomando EF, Jahngen JH (1983) *J Biol Chem* **258:**7653.
Rossomando EF, Cordis GA, Markham GD (1983) *Arch Biochem Biophys* **220:**71.
Rossomando EF, Jahngen JH, Eccleston JF (1981a) *Anal Biochem* **116:**80.
Rossomando EF, Jahngen JH, Eccleston JF (1981b) *Proc Natl Acad Sci (US)* **78:**2278.
Sakai T, Yanagihara S, Ushio K (1982) *J Chromatogr* **239:**717.
Sakuma R, Nishina T, Yamanaka H, Kamatani N, Nishioka K, Maeda M, Tsuji A (1991) *Clin Chim Acta* **203:**143.

Sasaoka T, Kaneda N, Nagatsu T (1988) *J Chromatogr* **424**:392.

Spoto G, Whitehead E, Ferraro A, Di Terlizzi PM, Turano C, Riva F (1991) *Anal Biochem* **196**:207.

Taha TSM, Deits TL (1993) *Anal Biochem* **213**:323.

Togari A, Sakai J, Matsumoto S (1987) *J Chromatogr* **417**:41.

Tsukada Y, Nagai K, Suda H (1980) *J Neurochem* **34**:1019.

Uberti J, Lightbody JJ, Johnson RM (1977) *Anal Biochem* **80**:1.

Oxygenations

Guenthner TM, Negishi M, Nebert DW (1979) *Anal Biochem* **96**:201.

Richards DA (1983) *J Chromatogr* **256**:71.

Tulliez JE, Durand EF (1981) *J Chromatogr* **219**:411.

van der Hoeven T (1984) *Anal Biochem* **138**:57.

Pterin metabolism

Blau N, Niederwieser A (1983) *Anal Biochem* **128**:446.

De Wit RJW, van der Velden RJ, Konijn TM (1983) *J Bacteriol* **154**:859.

Reinhard JF Jr, Chao JY, Smith GK, Duch DS, Nichol CA (1984) *Anal Biochem* **140**:548.

Lipid metabolism

Garver WS, Kemp JD, Kuehn GD (1992) *Anal Biochem* **207**:335.

Hupert J, Mobarhan S, Layden TJ, Papa VM, Lucchesi D (1991) *Biochem Cell Biol* **69**:509.

Maurich V, Zacchigna M, Pitotti A (1991) *J Chromatogr* **566**:453.

Miller RB, and Blank CL (1991) *Anal Biochem* **196**:377.

Romano MD, Straub KM, Yodis LAP, Eckardt RD, Newton JF (1988) *Anal Biochem* **170**:83.

Takeyama N, Matsuo N, Takagi D, Takaya T (1989) *J Chromatogr* **491**:69.

Yamada J, Sakuma M, Suga T (1991). *Anal Biochem* **199**:132.

Modification of proteins and peptides

Boutin JA, Ernould AP, Ferry G, Genton A, Alpert AJ (1992) *J Chromatogr* **583**:137.

Chikuma T, Hanaoka K, Loh YP, Kato T, Ishii Y (1991) *Anal Biochem* **198**:263.

Enz A, Shapiro G, Chappuis A, Dattler A (1994) *Anal Biochem* **216**:147.

Ferry G, Ernould A-P, Genton A, Boutin JA (1990) *Anal Biochem* **190**:32.
Fink ML, Shao YY, Kersh GJ (1992) *Anal Biochem* **201**:270.
Kim E-S, Graves DJ (1990) *Anal Biochem* **187**:251.
Madden JA, Bird MI, Man Y, Raven T, Myles DD (1991) *Anal Biochem* **199**:210.
Nakanishi S, Kase H, Matsuda Y (1991) *Anal Biochem* **195**:313.
Nash K, Feldmuller M, de Jersey J, Alewood P, Hamilton S (1993) *Anal Biochem* **213**:303.

Vitamin metabolism

Argoudelis CJ (1990) *J Chromatogr* **526**:25.
Bettendorff L (1991) *J Chromatogr* **566**:397.
Hayakawa K, Oizumi J (1987) *J Chromatogr* **423**:304.
Hayakawa K, Yoshikawa K, Oizumi J, Yamauchi K (1993) *J Chromatogr* **617**:29.
Ubbink JB, Schnell AM (1988) *J Chromatogr* **431**:406.

Xenobiotic Metabolism

Brown DL Jr, Boda W, Stone MP, Buckpitt AR (1982) *J Chromatogr* **231**:265.
Cashman JR, Proudfoot J (1988) *Anal Biochem* **175**:274.
Duescher RJ, Elfarra AA (1993) *Anal Biochem* **212**:311.
Duffel MW, Binder TP, Rao SI (1989) *Anal Biochem* **183**:320.
Eaton DL, Stapleton PL (1989) *Anal Biochem* **178**:153.
Honkasalo T, Nissinen E (1988) *J Chromatogr* **424**:136.
Kawaji A, Ohara K, Takabatake E (1993) *Anal Biochem* **214**:409.
Khoo BY, Sit KH, Wong KP (1990) *Clin Chim Acta* **194**:219.
Mina M, Rossomando EF (1988) *J Chromatogr* **433**:63.
Real MD, Ferre J, Chapa FJ (1991) *Anal Biochem* **194**:349.
Rosenberg DW, Roque H, Kappas A (1990) *Anal Biochem* **191**:354.
Schwenn JD, Jender HG (1980) *J Chromatogr* **193**:285.
Spahn H (1988) *J Chromatogr* **430**:368.
Stijntjes GJ, te Koppele JM, Vermeulen NPE (1992) *Anal Biochem* **206**:334.
Tassaneeyakul W, Veronese ME, Birkett DJ, Miners JO (1993) *J Chromatogr* **616**:73.
To ECA, Wells PG (1984) *J Chromatogr* **301**:282.
Tracy JW, O'Leary KA (1991) *Anal Biochem* **193**:1.

Pyrimidine metabolism

Ittarat I, Webster HK, Yuthavong Y (1992) *J Chromatogr* **582**:57.
James IT, Herbert K, Perrett D, Thompson PW (1989) *J Chromatogr* **495**:105.

Klein CM, Haas HJ (1990) *J Chromatogr* **529**:431.
Krungkrai J, Yuthavong Y, Webster HK (1989) *J Chromatogr* **487**:51.
Lu ZH, Zhang R, Diasio RB (1992) *J Biol Chem* **267**:17102.
Mehdi S, Wiseman JS (1989) *Anal Biochem* **176**:105.
Peters GJ, Laurensse E, Leyva A, Pinedo HM (1987) *Anal Biochem* **161**:32.
Waldmann G, Podschun B (1990) *Anal Biochem* **188**:233.

Metabolism of complex saccharides and glycoproteins

Borneman WS, Hartley RD, Himmelsbach DS, Ljungdahl LG (1990) *Anal Biochem* **190**:129.
Ducrotoy G, Richert L, De Sandro V, Lurier D, Pacaud E (1991) *J Chromatogr* **566**:415.
Iwase H, Ishii I, Saito T, Ohara S, Hotta K (1988) *Anal Biochem* **173**:317.
Johnson SW, Masserini M, Alhadeff JA (1990) *Anal Biochem* **189**:209.
Kaartinen V, Mononen I (1990) *Anal Biochem* **190**:98.
Lindquist L, Kaiser R, Reeves PR, Lindberg AA (1993) *Eur J Biochem* **211**:763.
Naoi M, Kondoh M, Mutoh T, Takahashi T, Kojima T, Hirooka T, Nagatsu T (1988) *J Chromatogr* **426**:75.
Spiegel LB, Hadjimichael J, Rossomando EF (1992) *J Chromatogr* **573**:23.
Willenbrock FW, Neville DCA, Jacob FS, Scudder P (1991) *Glycobiology* **1**:223.

Miscellaneous

Bravo M, Eran H, Zhang FX, McKenna CE (1988) *Anal Biochem* **175**:482.
Krungkrai J, Yuthavong Y, Webster HK (1987) *J Chromatogr* **417**:47.
Naganuma H, Kondo J-I, Kawahara Y (1990) *J Chromatogr* **532**:65.
Neuzil J, Novotna J, Vancurova I, Behal V, Hostalek Z (1989) *Anal Biochem* **181**:125.
Oizumi J, Hayakawa K (1990) *J Chromatogr* **592**:55.
Pennings EJM, van den Bosch RA, van der Heijden R, Stevens LH, Duine JA, Verpoorte R (1989) *Anal Biochem* **176**:412.
Werner ER, Werner-Felmayer G, Fuchs D, Hausen A, Reibnegger G, Wels G, Yim JU, Pfleiderer W, Wachter H (1991) *J Chromatogr* **570**:43.

Nucleic acid modification and expression

Davis AS, Davey MR, Clothier RC, Cocking EC (1992) *Anal Biochem* **201**:87.
Onishi Y, Azuma Y, Kizaki H (1993) *Anal Biochem* **210**:63.

GENERAL REFERENCES

Catecholamine metabolism

Allenmark S, Ali Qureshi G (1981) *J Chromatogr* **223**:188.
Boonyarat D, Kojima K, Nagatsu T (1983) *J Chromatogr* **274**:331.
Fujita K, Nagatsu T, Maruta K, Teradaira R, Beppu H, Tsuji Y, Kato T (1977) *Anal Biochem* **82**:130.
Koh S, Arai M, Kawai S, Okamato J (1981) *J Chromatogr* **226**:461.
Matsui H, Kato T, Yamamoto C, Fujita K, Nagatsu T (1981) *J Neurochem* **37**:289.
Nissinen E (1985) *Anal Biochem* **144**:247.
Nohta H, Ohtsubo K, Zaitsu K, Ohkura Y (1982) *J Chromatogr* **227**:415.
Zaitsu K, Okada Y, Nohta H, Kohashi K, Ohkura Y (1981) *J Chromatogr* **211**:129.

Amino acid metabolism

Blume D, Saunders JA (1981) *Anal Biochem* **114**:97.
Hill JA, Kitto GB (1985) *J Chromatogr* **337**:397.
Imamura I, Maeyama K, Watanabe T, Wagde H (1984) *Anal Biochem* **139**:444.
Matsumoto K, Imanari T, Amura Z (1975) *Chem Pharm Bull (Tokyo)* **23**:1110.
Rahman MK, Nagatsu T, Kato T (1980) *J Chromatogr* **221**:265.
Trocewicz J, Oka K, Nagatsu T (1982) *J Chromatogr* **227**:407.
Tsuruta Y, Ishida S, Kchashi R, Ohkura Y (1981) *Chem Pharm Bull (Tokyo)* **29**:3398.

Polyamines

Chabannes BE, Bidard JN, Sarda NN, Cronenberger LA (1979) *J Chromatogr* **170**:430.
Pietta P, Calatroni A, Colombo R (1982) *J Chromatogr* **243**:123.

Heme biosynthesis

Cantoni L, Ruggieri R, DalFiume D, Rizzardini M (1982) *J Chromatogr* **229**:311.
Francis JE, Smith AG (1984) *Anal Biochem* **138**:404.

Carbohydrate metabolism

Iwase H, Morinaga T, Li YT, Li SC (1981) *Anal Biochem* **113**:93.
Kiuchi K, Mutoh T, Naoi M (1984) *Anal Biochem* **140**:146.

Liu Z, Franklin MR (1984) *Anal Biochem* **142**:340.
Stahl RL, Liebes LF, Farber CM, Silber R (1983) *Anal Biochem* **131**:341.

Purine metabolism

Fairbanks LD, Goday A, Morris GS, Brolsma MFJ, Simmonds HA, Gibson T (1983) *J Chromatogr* **276**:427.
Hartwick R, Brown PR (1976) *J Chromatogr* **126**:679.
Krstulovic AM, Hartwick RA, Brown PR (1979) *J Chromatogr* **163**:19.
Ronca-Testoni S, Lucacchini A (1981) *Ital J Biochem* **30**:190.
Rylance HJ, Wallace RC, Nuki G (1982) *Clin Chim Acta* **121**:159.
Vasquez B, Bieber AL (1978) *Anal Biochem* **84**:504.

Oxygenations

Bacliani M, Felici M, Luna M, Artemi F (1983) *Anal Biochem* **133**:275.
Pietta P, Calatroni A, Pace M (1982) *J Chromatogr* **241**:409.
Yamazoe Y, Kamataki T, Kato R (1981) *Anal Biochem* **111**:126.

CHAPTER 10
Multienzyme Systems

OVERVIEW

This chapter describes the use of the HPLC method to assay the activity of several enzymes simultaneously. The examples include several different enzymes that can use the same substrate and form the same product, a single enzyme that can use different substrates to form different products, and two different activities using the same substrate to form different products. In another example the use of the HPLC method to study metabolic pathways is described through a series of reconstitution studies, and finally the HPLC method is applied to the anabolism of adenosine.

10.1 INTRODUCTION

Many investigators use enzymatic activities in a variety of ways—to characterize the stage of growth or development of a cell, to indicate the capacity to perform a specific physiological function, and to serve as a measure of gene function (e.g., when mutants are being screened for the absence of an activity). In such cases, it is often necessary to determine the enzyme activity without purifying the material first. Thus, assays for this purpose are often carried out in intact cells or in cell-free lysates. Attempts to measure an activity under such conditions, however, may founder when it becomes evident that the assay methods usually used were designed to measure only a single component, either the substrate or the product, during the course of a reaction. Clearly, such assay methods are not well suited for quantitating the activity of one enzyme in the presence of others that can compete for the substrate or otherwise consume the product.

Since the HPLC assay method can measure several related components simultaneously, it is possible to monitor the activity of a single enzyme among multiple catalysts. In addition, because many of the enzymes that process related compounds represent a "pathway," the HPLC method makes it possible to study the "flow" of a compound through such a multienzyme complex. This chapter explores the application of the HPLC assay method to the study of metabolic pathways and presents experimental results of some studies.

In one type of study, the HPLC assay method has been applied to the question of assaying the activity of two enzymes that can use the same substrate but form different products. In another, the HPLC method was applied to the case of related two different enzymes that can use the same substrate and form the same product. It was also applied to the case a single enzyme that can use either of two substrates to form two different products. In a separate application, the so-called reconstitution approach, a multienzyme pathway is reconstructed by the addition of enzymes one after the other to form a multienzyme complex and to thereby reconstitute a "naturally occurring" multienzyme complex. And finally, this approach allows us to use the HPLC method to study an intact, naturally occurring multienzyme system.

10.2 ASSAY OF TWO ACTIVITIES FORMING DIFFERENT PRODUCTS FROM THE SAME SUBSTRATE (Pahuja et al., 1981)

Glutamic acid is used by two different activities: glutamine synthetase and glutamic acid decarboxylase.

Glutamine synthetase and glutamic acid decarboxylase catalyze the conversion of glutamic acid to glutamine and γ-aminobutyric acid (GABA), respectively. Since both activities are involved in the metabolism of this amino acid, it would be useful to be able to study both simultaneously. And indeed, an HPLC method was developed to measure both activities in crude extracts.

The separation of glutamic acid and the two reaction products glutamine and GABA was accomplished by reverse-phased HPLC (ODS-5 column) eluted with $100 mM$ potassium dihydrogen phosphate (pH 2.1) and a gradient of 0 to 40% acetonitrile. The compounds were derivatized with dansyl chloride, and the absorbance was monitored at 206 nm. Compounds were also detected by scintillation counting. Figure 10.1 shows the separation obtained using these conditions.

For the glutamine synthetase, the reaction mixture contained imidazole (pH 7.4), NH_4Cl, ATP, $MgCl_2$, and [^3H]glutamic acid. The reaction was started with the enzyme and terminated by cooling in an ice bath. Then 1 N acetic acid was added, the insoluble material was removed by centrifugation, and a sample was taken for dansylation and analysis. The results of a synthetase assay are shown in Figure 10.2. Glutamine forms, but no GABA is seen.

In contrast, when ATP is omitted and pyridoxal phosphate is added for the decarboxylase assay, the formation of GABA can be detected (Fig. 10.3). The activities, were prepared by homogenizing bovine brain. After centrifugation at 1000g to remove debris, the supernatant solution was used directly for experiments. Bovine retinas were also prepared in the same manner.

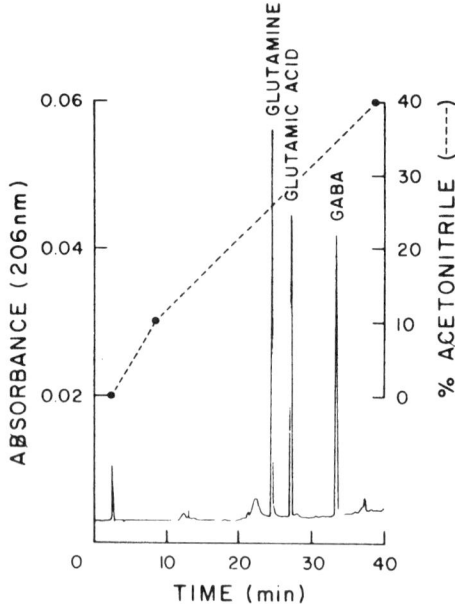

Figure 10.1 Chromatogram showing the separation of 0.5 μg of dansylated glutamine, glutamic acid, and GABA. Dashed line indicates the mobile phase gradient (% acetonitrile) used. (From Pahuja et al., 1981.)

10.3 ASSAY OF TWO ACTIVITIES FORMING THE SAME PRODUCT FROM THE SAME SUBSTRATE (Rossomando and Jahngen, 1983)

Now we turn to a single substrate used by two separate activities to form the same product. This example involves the hydrolysis of ATP according to the following reactions:

$$ATP \rightarrow ADP + P_i \tag{1}$$

$$ATP \rightarrow AMP + PP_i \tag{2}$$

$$ATP + AMP \rightleftharpoons 2ADP \tag{3}$$

In reaction (1), which is catalyzed by a phosphohydrolase, or as it is more commonly called, an ATPase, the β,γ-phosphoanhydride bond of the ATP is cleaved, with the result that ADP and inorganic phosphate (P_i) are formed. Reaction (2), catalyzed by a pyrophosphohydrolase, cleaves the α,β bond of the ATP, with the result that AMP and pyrophosphate (PP_i) are formed. Finally, (3) is the well-known reaction catalyzed by adenylate kinase, or myokinase, where ATP and AMP react to form two ADP molecules. Since reaction

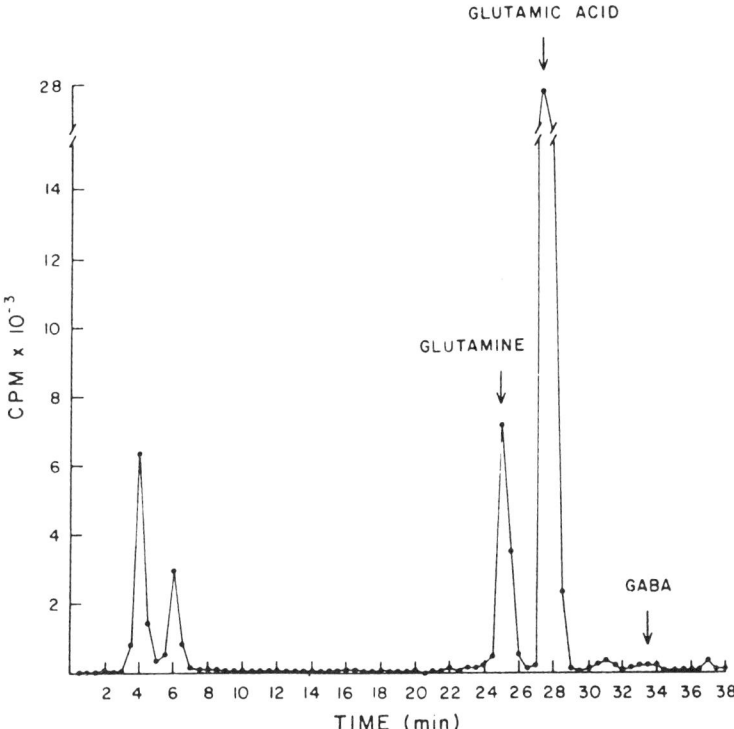

Figure 10.2 Chromatogram of the reaction mixture using 0.32 mg of crude brain extract. (From Pahuja et al., 1981.)

(3) is reversible, with an equilibrium constant of about 1, it could as easily have been written with ADP as the "substrate."

These three reactions are of interest because when ATP is incubated with a preparation containing all three activities and the formation of ADP is observed, it must be determined which of the three reaction pathways was involved. Has ADP been formed directly by reaction (1) or indirectly by a combination of reactions (2) and (3)?

Free phosphate has often been measured to identify the reaction pathway responsible for the formation of ADP in a given case. However, in most preparations that are partially purified, the presence of another enzyme activity, inorganic pyrophosphatase, which catalyzes the hydrolysis of PP_i to $2P_i$, means that measurements of free phosphate levels will not be conclusive.

Other approaches to the problem of establishing the pathway have involved the use of inhibitors presumed to be specific for ATPases. However, since in many cases the effect of such inhibitors on the activity of the pyrophosphohydrolase has not been studied, some uncertainty creeps into the results of these studies as well. And finally, additional purification could always be carried out.

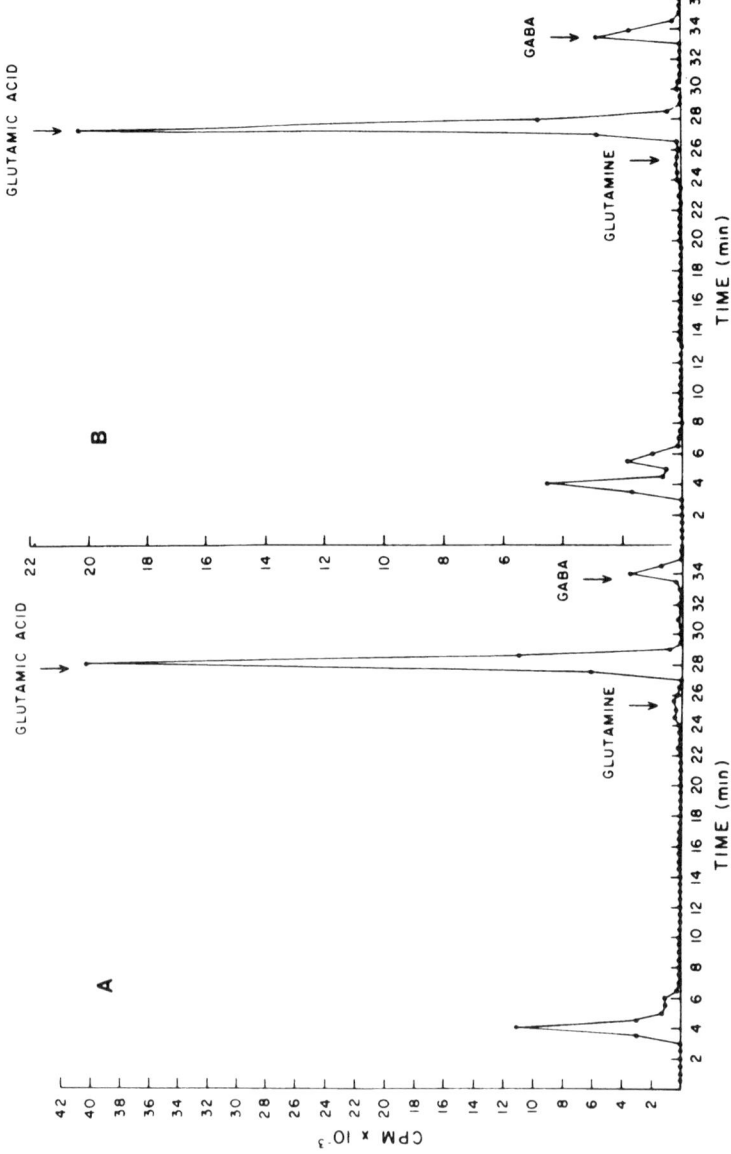

Figure 10.3 Chromatograms of the reaction mixture of (A) crude brain extract under glutamic acid decarboxylase (GAD) assay conditions; 0.17 mg of crude enzyme protein plus assay mixture was incubated for 1 hour (B) Crude retinal extract under GAD assay conditions; 0.30 mg of crude enzyme protein plus assay mixture was incubated for 2 hours. (From Pahuja et al., 1981.)

The HPLC enzyme assay method provides an alternative procedure with which to approach this problem. Clearly, to be able to assay these activities by the HPLC method, it is necessary to separate ATP, ADP, and AMP. This separation can be easily accomplished by ion-exchange HPLC eluted isocratically with a mobile phase containing a phosphate buffer and sufficient concentration of salt to elute the ATP. Under these conditions, the order of elution of the compounds would be AMP first, ADP next, and ATP last.

To apply this method to the problem at hand, the AMP kinase activity should be measured first. Clearly, it would be advantageous to carry out this assay under conditions similar to those that would be present when only ATP was added to the complex. Thus, one might set up a reaction mixture with about 1 mM ATP and with AMP in the nanomolar concentration range: that is, at a concentration that would be expected if the AMP had been derived from reaction (2). In addition, to enable us to follow its fate, in reaction (3) the AMP should be added to the incubation mixture in a radiolabeled form. Reaction (3) is started by the addition of the enzyme complex, and samples should be removed from the incubation mixture at suitable intervals and injected onto the HPLC column for analysis. After separation, the eluent should be monitored by both radiochemical and UV (254 nm) detectors.

If an AMP kinase is active, the following reactions would be expected to occur during the incubation:

$$ATP + {}^*AMP \rightarrow ADP + {}^*ADP \qquad (4)$$

$$ADP + {}^*ADP \rightarrow AMP + {}^*ATP \qquad (5)$$

where the asterisk indicates a radioactive compound. During the reaction, the products, *ATP and unlabeled AMP, would be formed. What about the ADP? Several routes are available for its formation, and one, of course, is reaction (4). However, two others can also occur:

$$ATP + AMP \rightarrow 2ADP \qquad (6)$$

$${}^*ATP + {}^*AMP \rightarrow 2{}^*ADP \qquad (7)$$

Both reactions (6) and (7) would take place after reactions (4) and (5) had occurred and would utilize their products. For example, in reaction (6), the unlabeled AMP formed in reaction (5) would react with unlabeled ATP that remained from the original substrate. As a result of reaction (6), two unlabeled ADP molecules would be formed, and therefore we would expect to see a small peak of unlabeled ADP on the chromatogram.

In contrast, in reaction (7), the radioactive *ATP that was formed during reaction (5) would react with the radioactive *AMP that remained unreacted from reaction (4). However, detection of the reaction products of reaction (7) would be hindered by the excess unlabeled ATP still present in the reaction mixture, which would be expected to reduce the specific activity of any labeled

*ATP formed during reaction (7), thereby making the formation of radioactive *ADP a low probability event.

Finally, it is noted that with a chromatographic system capable of separating ATP, ADP, and AMP and two monitors (to detect both radiolabeled and unlabeled compounds), the presence of an ATPase would be readily picked up, since such an activity would produce an excessive amount of unlabeled ADP with a corresponding reduction in the level of the unlabeled ATP. Of course, the availability of inhibitors of the myokinase such as P^1,P^5-diadenosine pentaphosphate, makes it possible to test more directly the conclusion that the ADP formed is due to this activity, since as a result of its inhibition there would be an increase in the amount of AMP recovered and a proportional decline in the level of ATP.

Experiments have been carried out to demonstrate the method described above by means of a multienzyme complex. This complex was assayed first for the AMP kinase activity. A reaction mixture was prepared containing unlabeled ATP (1 mM) and radioactive [^{3}H]AMP only. The reaction was started by the addition of the complex, and samples were removed and analyzed by HPLC. Chromatographic profiles, each representing the analysis of a sample removed from an incubation mixture at increasing times after the start of the incubation, are shown in Figure 10.4. Both optical density and radioactivity were determined.

The profile of a sample taken from the incubation mixture early in the incubation shows two peaks, one of labeled *AMP with a retention time of 2 minutes and the other unlabeled ATP with a retention time of 9 minutes. No products have been formed. The next sample, obtained after a 10 minute incubation, reveals the presence of small amounts of labeled *ADP, indicating that the AMP kinase is active [see reaction (3)], since an ATPase would, of course, produce unlabeled ADP [see reaction (1)]. Samples taken after additional incubation reveal the continued loss of radiolabeled *AMP and an increase in radiolabeled *ATP. These observations indicate that reaction (4) has been functioning and that the AMP kinase is now operating in the reverse direction, that is, according to reaction (5).

While unlabeled ADP is detected, the amount can be explained in terms of ADP from reaction (6) and not by the activity of an ATPase, which would have been expected to produce greater amounts of ADP and reduce the ATP level.

With the presence of an AMP kinase in the complex established, it is necessary to determine whether an ATP pyrophosphohydrolase activity is present in the complex to catalyze reaction (2). For these experiments, a reaction mixture is prepared with unlabeled ATP only and the formation of AMP and ADP is followed. Since the myokinase is present, as well as any AMP formed from the pyrophosphohydrolase, the remaining ATP will be used by the myokinase, and ADP will be formed. Of course, this ADP might be formed directly by an ATPase according to reaction (1), and the AMP

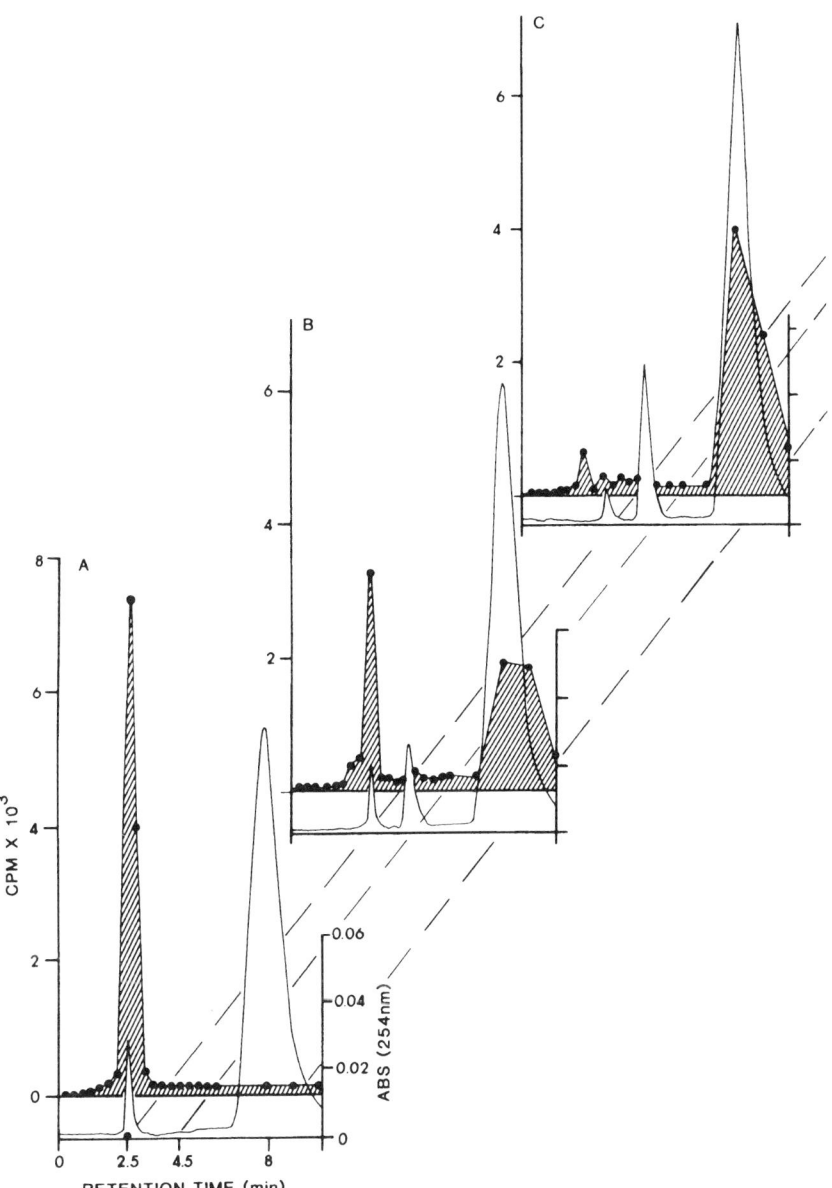

Figure 10.4 Adenylate kinase activity measured by the HPLC assay method. The assay mixture contained 200 μM ATP, 20 μM [^{3}H]AMP (approximately 120,000 cpm), 50 mM Tris-HCl (pH 7.4), and 32 μg of protein of the enzyme from *Dictyostelium discoideum*. Samples (20 μL) were injected, and the radioactivity was monitored with a Berthold LB 503 detector using PICO-Fluor 30 as the scintillant. Absorbance was measured at 254 nm.

formed from it in a reaction catalyzed by an additional activity, an apyrase. These possibilities will be explored.

The foreground panel of Figure 10.5 shows the HPLC analysis of a sample removed from a reaction mixture after a 20-minute incubation. The reaction mixture contained only unlabeled ATP as the substrate, and the enzyme complex. The chromatogram shows that a significant amount of ATP remains and that AMP and trace amounts of ADP are also present.

Additional experiments were undertaken in which an inhibitor of myokinase activity was added to a similar reaction mixture. Incubations were again carried out, and samples were removed and analyzed after 20 minutes. Figure 10.5 shows profiles (three background chromatograms) in which the myokinase activity was progressively inhibited: As inhibition of the myokinase increased, the amount of ADP recovered declined, and the amount of AMP increased proportionately. The area of each of the peaks (ADP and AMP) was determined, and these data (inset) illustrate the proportionality between the decline in ADP and the increase in AMP. Clearly, these results rule out the pathway for the formation of AMP from ADP, but they are consistent with the formation of ADP as a result of the combined actions of reactions (2) and (3). In addition, these data suggest that an ATPase activity is present and would account for the formation of the unlabeled ADP observed in the original experiment.

10.4 FORMATION OF TWO SEPARATE PRODUCTS FROM TWO SEPARATE SUBSTRATES BY THE SAME ACTIVITY
(Ali and Sloan, 1982)

Next we consider the utilization of two different substrates by the same activity to form two different products.

The enzyme hypoxanthine/guanine phosphoribosyltransferase (HG-PRTase) is an example of an enzyme activity that can utilize either of two substrates. In this case, both hypoxanthine and guanine can be acceptors of the phosphate donated by phosphoribosyl pyrophosphate (PRibPP). Since it is possible to separate hypoxanthine from guanine and the IMP from GMP, it is possible to study the utilization of one substrate in the presence of the other, a condition that parallels that expected in a cell.

In this assay both the formation of product and the loss of substrate were followed. The compounds were separated on an ion-exchange HPLC column equilibrated with $0.5M$ phosphate buffer of pH 4.0. The eluent was monitored at 254 nm.

The reaction mixture contained 50 μM guanine, 50 μM hypoxanthine, 100 μM PRibPP, and 1 mM MgCl$_2$ in potassium phosphate (pH 7.4). The reaction was initiated by the addition of the HGPRTase activity. An intervals the reaction was terminated by heating in a boiling water bath for 1 minute. Denatured protein was removed by centrifugation, and the sample was purified

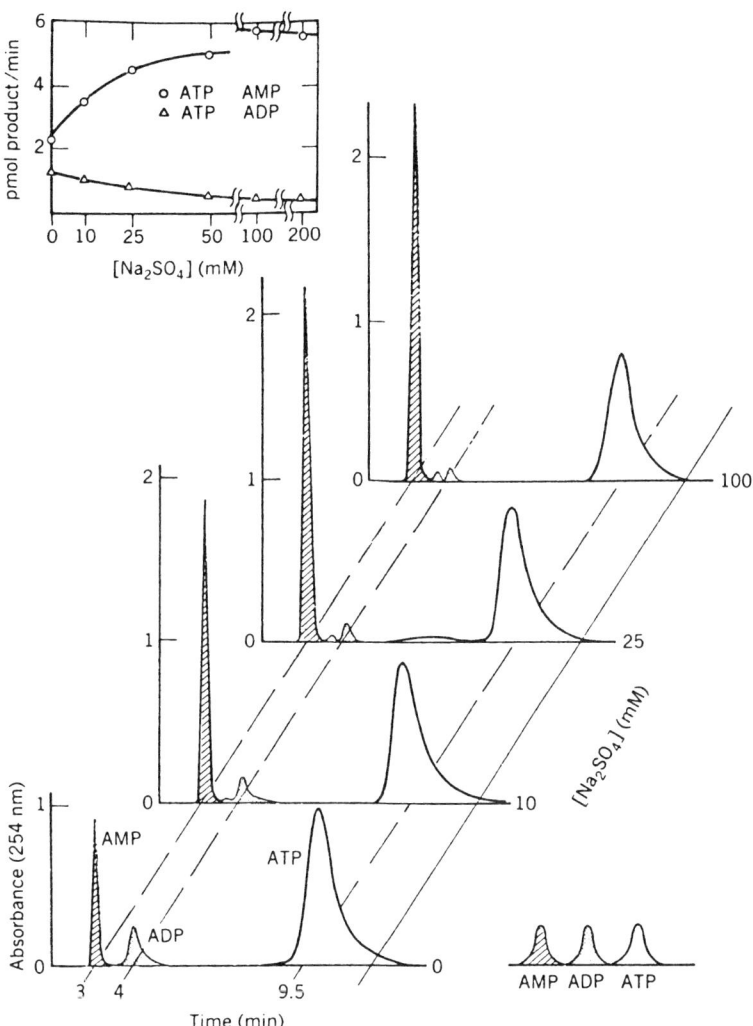

Figure 10.5 Effects of Na$_2$SO$_4$ on fate of ATP. A 30 μg sample of the extract was added to reaction mixtures containing, in a total volume of 200 μL, 100 μM ATP, 0.4 mM Mn^{2+}, 50 mM Tris-HCl (pH 7.4), and 0 to 200 mM sodium sulfate. The reaction was incubated at 31°C for 2 hours and terminated by heating to 155°C. Samples were analyzed by reversed-phase HPLC. Chromatograms represent an analysis of each reaction mixture 2 hours after the start of the reaction at four sulfate concentrations. *Inset:* Initial rate of AMP and ADP formation from ATP. (From Rossomando and Jahngen, 1983.)

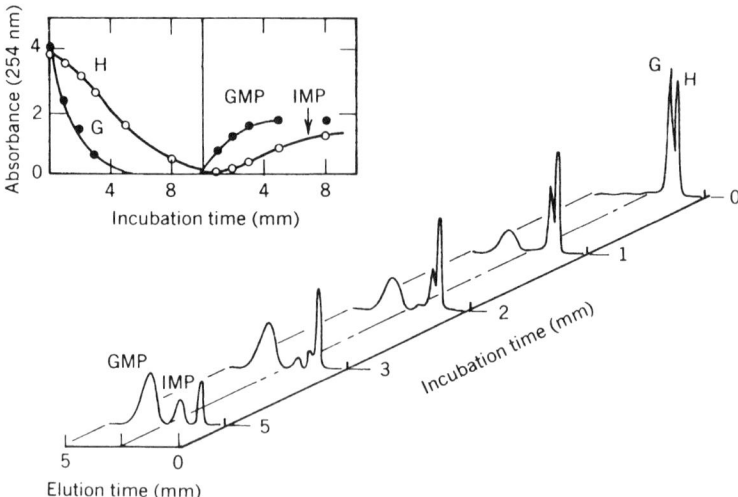

Figure 10.6 HPLC elution profiles of an incubation mixture made up of 2 nmol of hypoxanthine/guanine phosphoribosyltransferase, 50 μM guanine (G), 50 μM hypoxanthine (H), 100 μM PRibPP, and 1 mM MgCl$_2$ in potassium phosphate (pH 7.4). At time intervals of 0 to 5 minutes, aliquots of the mixture were injected onto the HPLC ion-exchange column and eluted. *Inset:* Time-dependent utilization of H and G and formation of GMP and IMP as determined by the absorbance of each peak at 254 nm. (From Ali and Sloan, 1982.)

further by filtration through a 0.45 μm type HA Millipore filter and injected for analysis.

The results of an experiment are shown in Figure 10.6. The formation of GMP and IMP from guanine and hypoxanthine, respectively, can be followed. With this method it was possible to track the initial rates of formation of IMP and GMP separately, the initial rates of both determined simultaneously, and the rate of PRibPP utilization with a fixed ratio of hypoxanthine and guanine.

The HGPRTase activity was purified from yeast.

10.5 ASSAY OF A MULTIENZYME COMPLEX BY THE RECONSTITUTION METHOD (Jahngen and Rossomando, 1984)

10.5.1 The Salvage Pathway: The Formation and Fate of IMP

"Salvage pathway" is a useful term to refer to that collection of biochemical reactions whose transformations result in the phosphorylation of purines. As a consequence of this phosphorylation, purines are not secreted by cells but, in fact, are returned to the cellular metabolic pool. One of these salvage enzymes is hypoxanthine–guanine phosphoribosyltransferase (HGPRTase),

10.5 ASSAY OF A MULTIENZYME COMPLEX BY THE RECONSTITUTION METHOD

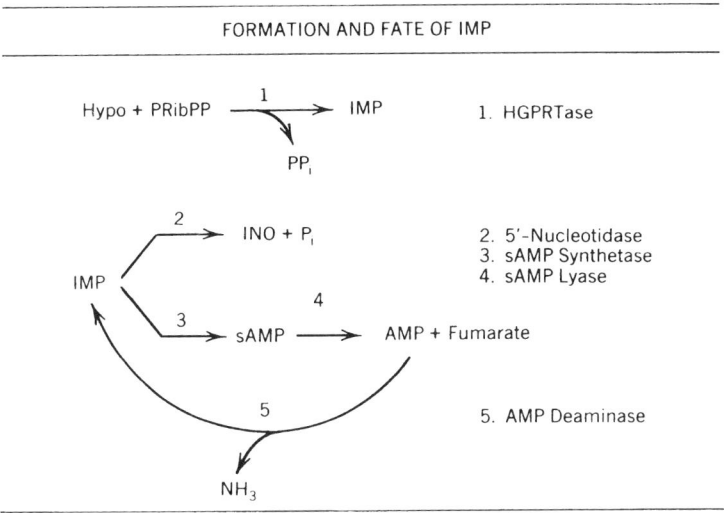

Figure 10.7 Schematic representation of the formation and fate of IMP. Formation of IMP is catalyzed by the enzyme hypoxanthine/guanine phosphoribosyl tranferase (1) from the substrate hypoxanthine (Hypo) and phosphoribosyl pyrophosphate (PRibPP). IMP is shown undergoing several reactions: the first (2) is catalyzed by 5'-nucleotidase to form inosine (INO) and orthophosphate (P_i); the other (3) is a two-step reaction catalyzed by sAMP synthetase to form adenylosuccinate (sAMP) and (4) by the enzyme sAMP lyase to convert sAMP to AMP and fumarate. Finally, (5) the deamination of AMP to IMP and NH_3 is catalyzed by AMP deaminase.

an activity that catalyzes the transfer of phosphoribose from phosphoribosylpyrosphosphate (PRibPP) to hypoxanthine, forming IMP and PP_i. Following its formation, IMP can undergo several other reactions, as illustrated in Figure 10.7. As a prelude to studies on the salvage of hypoxanthine by an intact multienzyme complex, a number of reconstitution studies were undertaken using purified enzymes to assess the suitability of the HPLC enzyme assay system to carry out such studies on an intact multienzyme complex.

10.5.2 The Degradation of IMP to Inosine

The enzyme 5'-nucleotidase dephosphorylates IMP to inosine and P_i. Thus, since this reaction represents a possible fate for the IMP formed by the transferase (Fig. 10.7), reconstitution studies were undertaken with the nucleotidase. These studies were carried out using the HPLC assay method developed for the HGPRTase activity. A reaction mixture was prepared that contained hypoxanthine and PRibPP as substrates. The reaction was started by the addition of purified HGPRTase enzyme. Samples were removed and were analyzed by HPLC. The chromatographic profiles obtained at 0, 10, 20, and

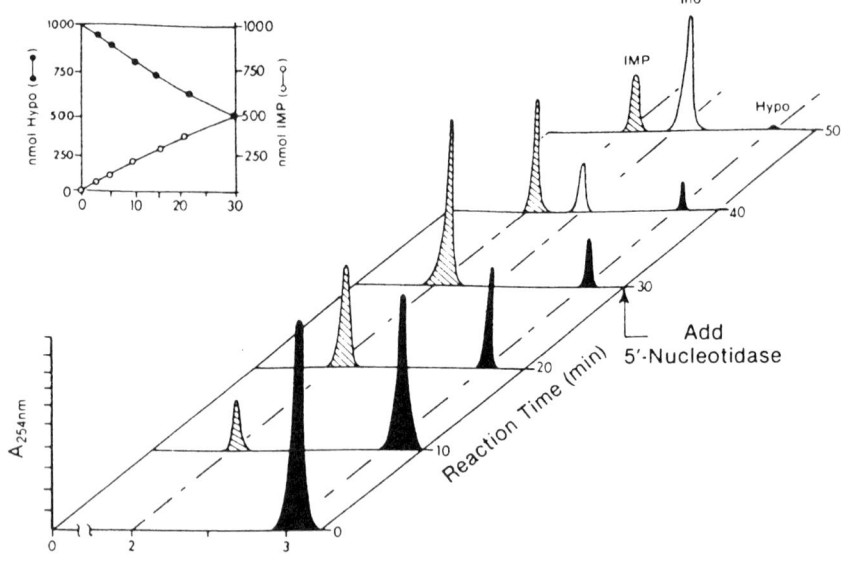

Figure 10.8 HPLC elution profiles of an incubation mixture to study hypoxanthine/guanine phosphoribosyltransferase activity. The reaction was initiated by the addition of the enzyme mixture, and aliquots were injected onto the HPLC column at 10-minute intervals. The solid peaks represent hypoxanthine, and the hatched peaks IMP. At 30 minutes, a 5'-nucleotidase activity was added to the reaction mixture. (From Jahngen and Rossomando, 1984.)

30 minutes after the start of the reaction (Fig. 10.8) indicate the disappearance of hypoxanthine and the appearance of IMP, confirming the HGPRTase reaction.

Immediately after removal of the 30-minute sample, the 5'-nucleotidase enzyme was added to the incubation tube. The incubation was continued, and samples were again removed for analysis. The HPLC profile in Figure 10.8, obtained from a sample removed after an additional 10 minutes of incubation, shows an increase in inosine, the product of the 5'-nucleotidase activity, as well as a decline in IMP. Hypoxanthine continues to decline after the reconstitution of the system.

10.5.3 The Conversion of IMP to AMP

Dephosphorylation is not the only fate of IMP. An alternative fate is the formation of AMP by two reactions also illustrated in Figure 10.7. The first reaction involves the condensation of IMP with aspartic acid to form the compound adenylosuccinate (sAMP). This reaction, catalyzed by the enzyme

10.5 ASSAY OF A MULTIENZYME COMPLEX BY THE RECONSTITUTION METHOD

sAMP synthetase, is followed by another catalyzed by sAMP lyase, in which sAMP is cleaved to form AMP and fumaric acid. Since the HPLC assay method can separate all these components, it would appear to be suitable for the study of such a multienzyme complex in which hypoxanthine could be salvaged to AMP. First, however, reconstitution studies were again carried out to test the capacity of the HPLC assay method to follow just the processing of IMP to AMP.

A reaction mixture containing IMP, aspartic acid, and GTP was prepared, and the reaction was started by the addition of purified sAMP synthetase. As the incubation proceeded, samples were removed and analyzed by HPLC. As shown in Figure 10.9, the chromatograms reveal the presence of the substrates and the formation of sAMP. After incubation for 20 minutes, the multienzyme system was reconstituted by the addition of a sample of sAMP lyase to the reaction. The reconstituted system was incubated for an additional 10 minutes, and a sample removed at that time was analyzed by HPLC. The profile (Fig. 10.9) illustrates a decline in the level of sAMP and the appearance of a new peak, AMP, confirming the successful reconstitution of this two-enzyme system.

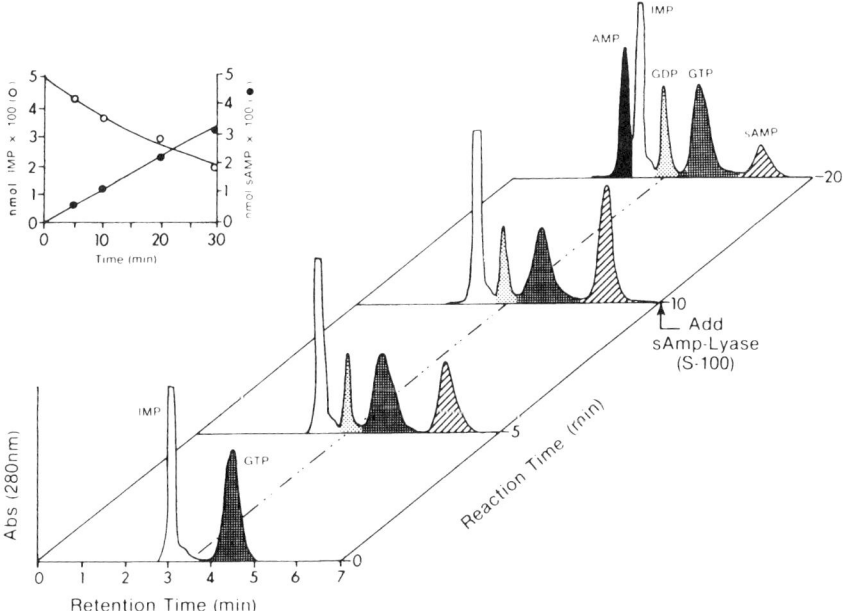

Figure 10.9 HPLC elution profiles of adenylosuccinate synthetase incubation mixtures. The reaction was initiated by the addition of 1.25 μmol of aspartate (pH 7.4). At 5-minute intervals, 20 μL samples were injected onto the HPLC reversed-phase column and eluted. After 10 minutes of incubation (arrow), sAMP lyase was added to the incubation mixture along with 10 nmol of EDTA. *Inset:* Time-dependent utilization of IMP and the formation of sAMP, as determined by integration of the respective peaks from the HPLC chromatograms. (From Jahngen and Rossomando, 1984.)

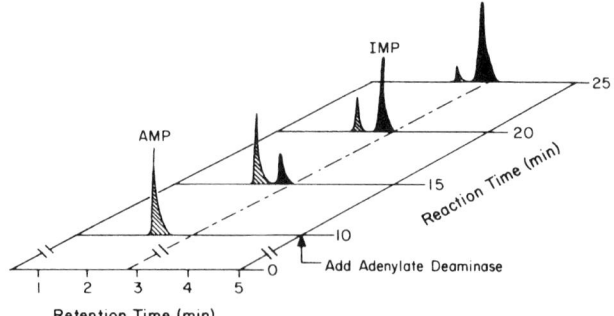

Figure 10.10 AMP was formed from adenosine and ATP in a reaction catalyzed by adenosine kinase. After a 10-minute incubation, adenylate deaminase was added to the reaction mixture, and samples were taken and analyzed by HPLC. Samples were analyzed every 5 minutes.

10.5.4 The Return of AMP to IMP

While the formation of AMP from hypoxanthine is surely a good start, the salvage will be truly successful only if the AMP is converted to ATP. However, prior to continued phosphorylation, the AMP formed by the two-enzyme reaction sequence described above can undergo another fate—deamination to form IMP and ammonia (see Fig. 10.7). Since the HPLC method is able to separate AMP and IMP, reconstitution experiments were again undertaken to determine whether the HPLC could follow this reaction as well. A reaction mixture was prepared, AMP formed, and an AMP deaminase was added to the reaction mixture. Samples were again removed and, as shown in Figure 10.10, the addition of the AMP deaminase resulted in the conversion of AMP to IMP. Thus, this reaction sequence also can be followed.

10.6 ASSAY OF A MULTIENZYME COMPLEX USING THE HPLC METHOD (Dye and Rossomando, 1982)

Adenosine can also be salvaged to AMP by a phosphorylation reaction catalyzed by adenosine kinase. The reaction requires ATP, which is converted to ADP during the reaction.

Following its formation, the AMP can have at least two fates. These are illustrated in Figure 10.11, where the reactions are shown. Reaction 2 of this figure is the deamination of AMP to IMP discussed above, a reaction catalyzed by AMP deaminase. Reaction 4 is the adenylate kinase reaction in which ATP is involved and two ADP molecules are formed. This reaction was also discussed above.

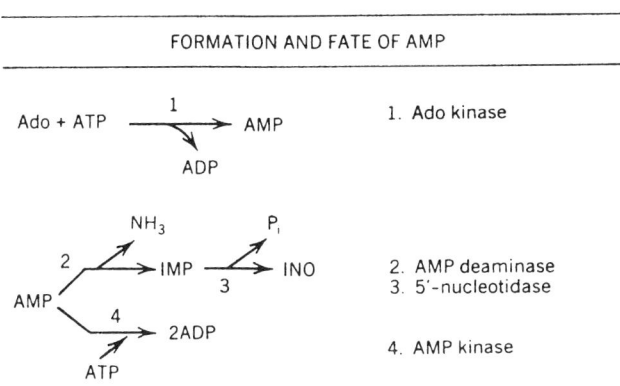

Figure 10.11 AMP can be formed by adenosine kinase (1) in a reaction that uses ATP as the phosphate donor and forms ADP as the second reaction product. Alternatively, AMP can be deaminated to IMP by the enzyme AMP deaminase (2) and converted to inosine (INO) by a 5'-nucleotidase activity (3). Finally, AMP can be phosphorylated to ADP by the enzyme AMP kinase (4).

With Adenosine (Ado)		With Formycin A (FoA)
Ado + ATP→AMP + ADP	(1)	FoA + ATP→FoMP + ADP
AMP + ATP→2ADP	(2)	FoMP + ATP→FoDP + ADP
2ADP→ATP + AMP	(3)	FoDP + ADP→FoTP + AMP

(a) FoA, FoMP, FoDP, FoTP, the formycin analogs of adenosine, AMP, ADP and ATP respectively.

(b) Reaction (1), (2) and (3) refer the reactions catalyzed by adenosine kinase, adenylate kinase in the forward direction and adenylate kinase in the reverse reaction.

(c) The reactions shown with adenosine (Ado) as these substrates are on the left and those with formycin A (FoA) on the right.

Figure 10.12 Comparison of the reaction products formed by AMP kinase with both AMP and its formycin analog formycin 5'-monophosphate as substrates.

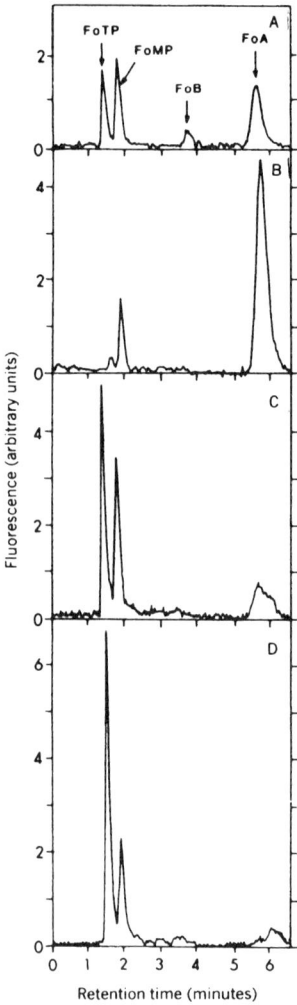

Figure 10.13 HPLC of FoA and its metabolites produced by activities present in an S-30 fraction prepared from rat liver; FoB is a formycin analog of inosine. Operating conditions: mobile phase, 0.1 M KH_2PO_4 adjusted to pH 5.5 with NaOH with 10% methanol; flow rate, 2 mL/min; room temperature; C_{18} μBondapak column packing; fluorescence, excitation at 300 nm, emission about 320 nm. (*A*) Chromatogram obtained with standards as indicated at approximately 10 μg of each. Sensitivity of fluorescence detection, 0.5 AUFS units full scale. (*B*) Chromatogram obtained immediately after addition of S-30 to a reaction mixture containing FoA and ATP. (*C* and *D*) Chromatograms obtained after 8 and 16 minutes of incubation, respectively. [From Dye and Rossomando, 1982. Reprinted by permission from *Bioscience Reports,* **2:**229–234 (1982).]

To follow the salvage of adenosine to AMP and the subsequent fate of AMP, an in vitro multienzyme complex was prepared from rat liver. In addition, to allow the reaction to be followed directly, the adenosine was replaced by a fluorescent analog formycin A (FoA: see Chapter 4 for the structure of the analog). The reactions for FoA are shown in Figure 10.12. For these experiments, a reaction mixture containing FoA and ATP was prepared. The complex was added, and samples were removed for HPLC analysis. The chromatograms of samples removed at early times show the substrates FoA and reveal the formation of FoMP by the adenosine kinase (Fig. 10.13). Samples removed after additional incubation revealed the formation of FoDP and eventually FoTP in the reaction mixture. The results indicate that the HPLC system was able to monitor the salvage of adenosine to AMP and its subsequent metabolism to ADP and ATP by the AMP kinase present in the multienzyme complex.

10.7 SUMMARY AND CONCLUSIONS

This chapter discussed the use of the HPLC method to assay multiple enzyme activities and individual enzymes with more than one function. Examples were given to illustrate that the method can be used to assay two activities using the same substrate to form different products, two activities using the same substrates to form the same product, and one activity using two substrates to form two products. Also, the HPLC method can be used to study an in vitro multienzyme complex reconstructed by the addition of pure enzymes and finally to study a naturally occurring multienzyme complex.

REFERENCES

Multiproduct Systems

Ali LZ, Sloan Dl (1982) *J Biol Chem* **257**:1149.
Pahuja SL, Albert J, Reid TW (1981) *J Chromatogr* **225**:37.
Rossomando EF, Jahngen JH (1983) *J Biol Chem* **258**:7653.

Reconstitution Studies

Dye F, Rossomando EF (1982) *Biosci Rep* **2**:229.
Jahngen EG, Rossomando EF (1984) *Anal Biochem* **137**:493.

GENERAL REFERENCES

Multienzyme Systems

Friedrich P (1984) *Supramolecular Enzyme Organization.* Pergamon, Oxford.

INDEX

α-amylase (1,4-α-D-glucaglucahydrolase, EC 3.2.2.1), 290–293
Absorbance detection, 50
Acepromazine, HPCE analysis, 169
Acetanilide 4-hyroxylase, 348–349
Acetonitrile, 35, 54, 56
Acetylcholine, 116, 126
Acetyl-CoA:
  alcohol transacylase, 363–364
  arylamine N-transferase, 229
Acetylcodeine, HPCE analysis, 166–167
Acid phosphatase, 313–317
Adenine, 28
Adenosine, generally:
  calibration curve, 84
  deaminase, 317–318
  diphosphate–ribosylarginine hydrolase, 366–368
  in ion-exchange, 32
  kinase, 6, 8, 87, 326–327, 435
  solubility of, 28–29
  3′-phosphate 5′-sulfophosphate sulfotransferase, 380–381
Adenylate cyclase, 327–330
Adenylate kinase, 333–335
Adenylosuccinate synthetase, 334, 336–337
Ado:
  in ion-exchange chromatography, 32–33
  primary reaction, 87
ADP:
  assay of activities, 103
  coupled assay method, 5
  in ion-exchange chromatography, 32–34
  in multienzyme systems, 420, 423–424, 426, 432–433
  primary reaction, 87
  separation of, metal effects on, 72
  stationary phase, 67
α-ketoglutarate dehydrogenase, 299
α-L-fucosidase, 392–393
Alkaline phosphatase, 106, 313–317
Allobarbital, HPCE analysis, 169
Alprenolol, HPCE analysis, 168
Amino acid:
  metabolism, survey of enzymatic activities:
    amino acid deraboylase, 263–264
    aromatic L-amino acid decarboxylase, 264
    asparagine synthetese, 251, 253
    D-amino oxidase, 264–265
    diaminopimelate epimerase and decarboxylase, 259
    dihydroxyacid dehydratase, 255–256
    glucamic acid decarboxylase, 262
    glutamine synthetase/glutamate synthetase/glutamate dehydrogenase, 249–252
    glutaminyl cyclase, 256–257
    histamine N-methyltransferase, 262–263
    kynureninase, 267–268
    kynurenine 3-monoxygenase, 268–269
    L-alanine: glyoxylate aminotransferase, 270
    leucine 2,3-aminomutase, 257–258
    lysine-ketoglutarate reductase, 259–260
    N^5-methyltetrahydrofolate-homocysteine methyltransferase, 269

Amino acid *(Continued)*
  ornithine aminotransferase, 247–248
  threonine/serine dehyratase, 265
  tryptophanase, 253–256
  tryptophan dioxygenase, 265–266
  tryptophan 2,3-dioxygenase, 267
  tyrosinase, 270–271
  γ-glutamylcyclotransferase, 261
  γ-glutamyclcysteine synthetase/glutathione synthetase, 261–262
  microdialysis of, 116
  solubility of, 28–29
Amino acid deraboylase, 263–264
Aminopeptidase, 239
Amobarbital, HPCE analysis, 170
AMP:
  deaminase, 317, 319–320
  in ion-exchange chromatography, 32–34
  in multienzyme systems, 420, 423–424, 430–433, 435
  primary reaction, 87
  reversed-phase chromatography, 30
  separation of, 72
  stationary phase, 67
Amphetamine, HPCE analysis, 165, 168–170
α-N-acetylgalactosaminyltransferase, 392–393
Analogs, substrate, 86–87
Analytical column, 15
Anhydrotetracycline oxygenase and tetracycline dehydrogenase, 403–405
Antibodies, assay of activities, 101–102
Antigens, assay of activities, 101–102
Antipain, 103
Aprotinin, 103
Aromatic amino acid decarboxylase (AAAD), 126
Aromatic L-amino acid decarboxylase, 264
Artifacts, in PCR, 150–151
Aryl hydrocarbon hydroxylase (EC 1.14.14.2), 351–352
Arylsulfatase, 224–225
Arylsulfatase B (N-acetylgalactosamine 4-sulfatase), 285–286

Aryl sulfotransferase, 382–383
Asparagine synthetese, 251, 253
Aspartate-aminotransferase (ASAT), 129
Aspartylglyosylaminase, 394–395
ATP:
  coupled assay method, 5
  in ion-exchange chromatography, 32–34
  in multienzyme systems, 420, 423–427, 432, 434–435
  pyrophosphohydrolase, 320–322
  separation of, metal effects on, 71–74
  solubility of, 28
  stationary phase, 65–67
ATPase, 103, 348
ATP-sulfurylase, 374–376

Barbital, HPCE analysis, 169
Barbiturates, HPCE analysis, 169–170
Benzamidine, 103
Benzocaine, HPCE analysis, 169
Benzodiazepines, HPCE analysis, 169–170
Benzolecgonine, HPCE analysis, 168–169
β-galactosidase, 283–284, 395
β-glycosyltransferase, 395
Bile, microdialysis in, 124
Biological fluids, samples of, 99–100
Biosynthesis process, 126–128
Biotinidase, 374
Blood, microdialysis in, 123–124
Blood-brain barrier (BBB), 128
Brompheniramine, HPCE analysis, 168
Buffer systems, in CE, 141–142
Buoyant density centrifugation, 101
β-ureidopropionase, 390
Butacaine, HPCE analysis, 169
Butalbital, HPCE analysis, 169

Caffeine, HPCE analysis, 167
Calibration curve, 84
cAMP:
  in ion-exchange chromatography, 33–34
  phosphodiesterase, 330–333
Cannabidiol, HPCE analysis, 165

Capillary, in HPCE, 48–49
Capillary electrophoresis, application of, 41. See also HPCE (high performance capillary electrophoresis)
Capillary gel electrophoresis (CGE), 59–60, 153
Capillary isoelectric focusing (CIEF), 60, 176
Capillary isotachophoresis (CITP), 58–59
Capillary zone electrophoresis (CZE), 54–55
Carbohydrate metabolism, survey of enzymatic activities:
　α-amylase (1,4-α-D-glucaglucahydrolase, EC 3.2.2.1), 290–293
　α-ketoglutarate dehydrogenase, 299
　arylsulfatase B (N-acetylgalactosamine 4-sulfatase), 285–286
　β-galactosidase, 283–284
　cytidine monophosphate-sialic acid synthetase, 294–295
　galactosyltransferase, 287
　lactose-lysine β-galactosidase, 284–285
　lysosomal activities, 291, 293–974
　sialidase, 293–297
　6-phosphogluconate dehydratase, 300–301
　succinyl-CaA synthetase, 295–296, 298
　sucrose phosphate synthetase, 300
　uridine diphosphate glucuronosyltransferase, 287–291
Carbon tetrachloride, 36
Carbonyl reductase, 400
Carboxylases, 399–400
Carboxypeptidase N, 244–245
Carnitine palmitoyltransferase, 362–363
Catecholamine metabolism, survey of enzymatic activities:
　acetyl-coA/Arylamine N-transferase, 229
　arylsulfatase, 224–225

catechol O-methyltransferase, 219–221
dopa decarboxylase (L-Aromatic amino acid decarboxylase), 212–215
dopamine β-hydroxylase, 215–218
5-hydroxytryptophan decarboxylase, 211–212
monoamine oxidase(s):
　A and B, 222–224
　and phenol sulfotransferase, 225–226
N-acetyltransferase, 226–227, 229
phenylethanolamine N-methyltransferase, 221–222
tyrosine hydroxylase, 208–211
Catechol O-methyltransferase, 219–221
Cathinone, HPCE analysis, 168
CE (classical electrophoresis) methods:
　buffer systems, 141–142
　nucleic acids and, 141
Cell cultures, 100
Centrifugation, see specific types of enzymes
　advances in, 95
　application of, 76
　buoyant density, 101
Cerebrospinal fluid (CSF), microdialysis in, 124
Ceruloplasmin, 349–351
Cesium, 33
Chiral separations, 60–61
Chloramphenicol acetyltransferase, 405–408
Chlordiazepoxide, HPCE analyiss, 165
Cholesterol 7α-hydroxylase, 304–306
Chromatogram, see specific types of enzymes
　characteristics of, 18–19
　function of, generally, 83
　interpretation of, 19–23
　mixture components, metal effects on separation, 72
　secondary reactions, 69–70, 88
Chromatography, defined, 13
CMA probes, microdialysis, 118–120
CMP-N-acetylneuraminic acid:
　glycoprotein sialyltransferase, 396–398

Cocaine, HPCE analysis, 165, 167–169
Codeine, HPCE analysis, 165, 169
Collagen, nonenzymatic modifications by glucose, 183–185
Column, *see specific types of enzymes*
  chromatography, 95
  maintenance, 36–37
  monitoring performance, 37
Competitive PCR, 157–159
Complex saccharides metabolism, survey of enzymatic activities:
  α-L-fucosidase, 392–393
  α-N-acetylgalactosaminyl-transferase, 392–393
  aspartylglyosylaminase, 394–395
  β-galactosidase and glycosyltransferase, 395
  CMP-N-acetylneuraminic acid: glycoprotein sialyltransferase, 396–398
  glucose-1-phosphate thymidylyltransferase, 396
  GM_1 ganglioside β-galactosidase, 393–394
  thyroxine: UDP-glucuronosyltransferase, 397
  trans-p-coumaroyl esterase, 397, 399
Condimetric detection, 51
Continuous assay method, 3–5
Coupled assay method, 4–5
Creatine kinase, 325–326
Cultures, tissue samples, 98–99
Cyclic nucleotide phosphodiesterase, 320–321
Cyclodextrins (CDs), 55, 57, 60–61
Cyclophosphamide, HPCE analysis, 168
Cysteine conjugate β-lyase, 383–384
Cytidine deaminase, 389–390
Cytidine monophosphate-sialic acid synthetase, 294–295
Cytochrome 450-1A1 gene, 159
Cytochrome $P450_{2E1}$, 386–387
Cytochrome $P450_{SCC}$, 306–308
Cytokinin metabolism, survey of enzymatic activities, 338–339
CZE (capillary electrophoresis):
  enzymatic activity assay by, 185–192
  Hummel-Dreyer method, 192–193
  illicit drug substances, 169
  nucleic acids and their constituents, 196–199
  pen inks, analysis of, 175–176
  protein analysis, 178
  protein-drug binding assays, 192–196

D-amino oxidase, 264–265
Data reduction phase, in enzyme assay, 2
Denaturation cycle, in PCR, 139
Deoxynucleoside triphosphates (dNTPs), 138, 141
Desalting techniques, 147
Detection:
  in enzyme assay, defined, 2
  in HPCE, *see* Detectors, types of
Detector(s):
  application of, 17
  defined, 15
  sensitivity, 77–79
  types of:
    absorbance, 50
    condimetric, 51
    electrochemical, 51
    fluorimetric, 51
    indirect, 52
    mass spectrometic, 51–52
D-glucose-6-phosphate (NADPH oxidoreductase), 189, 192
Diamine oxidase, 275–276
Diaminopimelate epimerase and decarboxylase, 259
Diazepam, HPCE analysis, 165, 169
Dihydrofolate reductase, 353, 356, 358
Dihydroorotase, 391
Dihydroorotic acid dehydrogenase, 389
Dihydropyrimidine dehydrogenase, 388–389
Dihydroxyacid dehydratase, 255–256
Diisopropylfluorophosphate, 103
Diisopyramide, HPCE analysis, 168
Dimethyl sulfoxide (DMSO), 36
Dinucleoside polyphosphate pyrophosphohydrolase, 336–337
Dipeptidase, 238
Dipeptidyl carboxypeptidase, 231–235
Dipeptidyl peptidase IV/amino peptidase-P, 244, 246

Discontinuous assay method:
  defined, 5–6
  HPLC as, 6–9
DNA, generally:
  analysis:
    classical, 141
    nucleic acids in, 196
  hybridization, 151–153
  synthesis, 138
  topoisomerase, 405
dNTPs, 144
D1S80 gene, 198
Dopa, 126
Dopa decarboxylase (L-Aromatic amino acid decarboxylase), 212–215
Dopamine, 116
Dopamine β-hydroxylase, 215–218
Doxapram, HPCE analysis, 169
Drug binding assays, protein, 192–196
dsDNA, 141, 143, 152–153

EDTA, 74, 142
Electrochemical detectors, 17, 51
Electrokinetic injection, 46–49, 147–148
Electroosmosis, 44–46, 55
Electroosmotic flow (EOC), 42–43, 45, 54–56, 167, 176–177, 187
Electropherogram, 189, 194
Electrophoretic separation:
  electroosmosis, 44–46
  process, overview, 43–44
Electrospray, 51
11-β-hydroxylase and 18-hydroxylase, 302–304
Elution separation:
  defined, 10
  selection of, 65, 67
Elution time, 14
Elution volume, 18, 20
Emergent time, 14
Enkephalinases A and B, 239–241
Enzymatic assay, generally:
  anatomy of, 1–3
  methods:
    classification of, 3–9
    selection criteria, 10–11
Enzyme sample, as selection criteria, 11

Ephedrine, HPCE analysis, 168
Ethanol, nonenzymatic modifications by, 179–183
Ethidium bromide, in forensic toxicology, 197–199
Ethidium bromide, 143
Ethoxycoumarin O-deethylase, 385–386
Explosives, forensic analysis, 173–175
Exponential phase, of PCR, 156
Extension PCR, 139
Extracellular fluid (ECF), microdialysis in, 116, 123, 129

Fatty acid ω-hydroxylase, 363
Fentanyl, HPCE analysis, 165
Ferrocheletase, 280–281
Ferrules, coupling, 16–18
Fibrillar matrix, 97–99
Fick's law of diffusion, 119
5'-nucleotidase, 310–313
5-aminolevulinate dehydrase, 278
Δ^5-3β-hyroxysteroid dehydrogenase, 301–402
5-hydroxytryptophan decarboxylase, 211–212
Flavin-containing monoxygenase, 387–388
Fluorimetric detection, 51
Fluorometer, 17
FoDP, 435
Folic acid cleaving enzyme, 353–354, 357
Forensic analysis, PCR, 151
Forensic toxicology:
  HPCE analyses:
    gunshot residues and constituents of explosives, 172–175
    illicit drug substances, 165–172
    overview, 164–165
    pen inks, 175–176
    proteins, 176–179
      drug binding assays, 175–196
      nonenzymatically modified, 179–185
    nucleic acids, 196–199
    overview, 165
Formamide, 147
Formycin A (FoA), 87

4-nitrophenyl phosphate (4NP), 3–4
Frontal analysis zone electrophoretic method, 194–195

GABA (glutamine and γ-aminobutyric acid), 419
GADPH (glyceraldehyde-3-phosphate dehydrogenase), 158–159
Galactosyltransferase, 287
Gas chromatography (GC), 168
Gel filtration:
  chromatography, generally, 24–25
  high pressure chromatography (HPGPC), 106
  HPLC, purification of enzymes, 107–108
Gel permeation chromatography, 178
Glucamic acid decarboxylase, 262
Glucose, nonenzymatic modifications by, 183–185
Glucose-1-phosphate thymidylyltransferase, 396
Glutamate, 126
Glutamine:
  5-phosphoribosyl-1-pyrophosphate aminotransferase, 344–345
  synthetase/glutamate synthetase/glutamate dehydrogenase, 249–252
Glutaminyl cyclase, 256–257
Glutathione:
  perioxidase activity, 191
  S-transferase, 376, 378–380
Glycerol, 147
Glycoproteins metabolism, survey of enzymatic activities:
  α-L-fucosidase, 392–393
  α-N-acetylgalactosaminyl-transferase, 392–393
  α-aspartylglyosylaminase, 394–395
  β-galactosidase and glycosyltransferase, 395
  CMP-N-acetylneuraminic acid: glycoprotein sialyltransferase, 396–398
  glucose-1-phosphate thymidylyltransferase, 396
  GM_1 ganglioside β-galactosidase, 393–394
  thyroxine: UDP-glucuronosyltransferase, 397
  trans-p-coumaroyl esterase, 397, 399
GM_1 ganglioside β-galactosidase, 393–394
GMP, 85
Gradient, 16
Guanase, 342–343
Guanine, 85
Guanine phosphoribosyltransferase (GHPRTase), 85
Guanosine triphosphate cyclohydrolase I, 357, 360
Guard column, 16
Gunshot residuals, forensic analysis, 172–175

Hae III, 149–150
Heme metabolism, survey of enzymatic activities:
  ferrocheletase, 280–281
  5-aminolevulinate dehydrase, 278
  heme oxygenase, 279–280
  o-aminolevulinic acid synthetase, 276–277
  protoporphyrinogen, 281–282
  uroporphyrinogen decarboxylase, 278–279
Heme oxygenase, 279–280
Hepatic microsomal testerone hydroxylase, 352–356
Heroin, HPCE analysis, 165–167, 169
Hexobarbital, HPCE analysis, 168
High pressure gel permeation chromatography (HPGPC), 106
Histamine N-methyltransferase, 262–263
HPCE (high performance capillary electrophoresis):
  analysis, see specific types of enzymes
  basic components of, 42–43
  defined, 41–42, 164
  electrophoretic separation, 43–46
  forensic analyses:
    gunshot residues and constituents of explosives, 172–175
    illicit drug substances, 165–172
    overview, 164–165

HPCE (high performance capillary electrophoresis) *(Continued)*
  pen inks, 175–176
  proteins, 176–179
    nonenzymatically modified, 179–185
  instrumentation:
    capillary, 48–49
    detection, 50–42
    injection, 46–48
    power supply, 49–50
  methods:
    capillary gel electrophoresis (CGE), 59–60
    capillary isoelectric focusing (CIEF), 60
    capillary isotachophoresis (CITP), 58–59
    capillary zone electrophoresis (CZE), 54–55
    chiral separations, 60–61
    micellar electrokinetic chromatography (MEKC), 55–58
  operation of, 42
  separation efficiency and resolution:
    practical hints, 54–55
    theoretical plate number and resolution, 52–54
HPLC (high performance liquid chromatography), generally:
  basic components, 15–16, 38–39
  chromatogram, 18–19
  coupling ferrules, 16–18
  defined, 13, 38
  elution time, selection of, 65, 67
  introduction of, 14
  operation, 16
  reactions:
    mixture components, effects of metals on separation, 71–73
    modification of, 68
    primary, 64–65, 79
    quantitative analysis, 83–85
    secondary, 65, 68–71, 79
    setting up conditions, 76–79
    termination of, 73–76, 80
  selection criteria, 10–11, 39
  stationary phase, selection of, 65, 67

strategy design, for assay of enzyme activity:
  optimal conditions for enzymatic reactions, 81–87, 89
  setting up assay, 64–81
Hummel-Dreyer method, 192–193
HUMTO01, 151–152
Hybridization, PCR, 151–153
Hydration, in ion-exchange chromatography, 33
Hydrodynamic injection, 46–47
Hypoxanthine, 85
  -guaninie phosphoribosyltransferase (HGPRTase), 428–430

IMP (inosine 5'-phosphate), 22, 85
Incubation phase, in enzyme assay, 2–3, 22
Indicator reaction, 4
Indirect detection, 52
Initiation phase, in enzyme assay, 2–3
Injection:
  in HPCE, 46–48
  injectors, types of, *see* Injectors, types of
  polymerase chain reaction, 146–150
Injectors, types of, 76–77
Instrumentation, in HPCE, 46–52
Intact cells, assay of activities in, 103
Intercalators, PCR, 142–144
Interenzymatic functional studies, 96
Ion depletion, 150
Ion-exchange chromatography:
  application of, 30, 32–34
  purification of enzymes, 110–111
Ion-paired reversed phase HPLC, 29
Isocratic separation method, 10
Isofosfamide, HPCE analysis, 168
Isopropanol, 54, 56

Keratin, nonenzymatic modifications by ethanol, 179–183
Kinetic analysis, 84
K_m value, 81
KOH, 35
Kunitz, Moses, 95
Kynureninase, 267–268
Kynurenine 3-monoxygenase, 268–269

Lactate, 116
 dehydrogenase, 106
Lactose-lysine β-galactosidase, 284–285
L-alanine:glyoxylate aminotransferase, 270
Lambda bacteriophage PCR, 157
Laser-based thermooptical detection, 51
Leucine 2,3-aminomutase, 257–258
Leupeptin, 103
Lidocaine, HPCE analysis, 169
LIF detection, 144–145, 147
Ligands, chirally selective, 60
Linear polyacrylamide, 183, 197
Lipase, 364–365
Lipid metabolism, survey of enzymatic activities:
 acyl-CoA: alcohol transacylase, 363–364
 carnitine palmitoyltransferase, 362–363
 fatty acid w-hydroxylase, 363
 lipase, 364–365
 retinal oxidase, 360–361
 serum cholinesterase, 361–362
Lipoamidase, 372–373
Liquid chromatography, 13
Loop, defined, 77
Lorazepam, HPCE analysis, 165
Low molecular weight, see Small molecules
Luteinizing hormone-releasing hormone peptidase, 235
Lysates, cell-free, 104–106
Lysergic acid diethylamide (LSD), HPCE analysis, 165
Lysine, reversed-phase chromatography, 30
Lysine-ketoglutarate reductase, 259–260
Lysis, 103–104, 106
Lysosomes, 103, 291, 293–294

Mass spectrometic detection, 51–52
Meclizine, HPCE analysis, 169
Medazepam, HPCE analysis, 169
Membrane fraction, 92

Metabolism, enzymatic activities, see specific types of enzymes
Metals, effects on separation, 71–73
Metapyrilene, HPCE analysis, 168
Methadone, HPCE analysis, 169–170
Methamphetamine, HPCE analysis, 165, 168–170
Methanol, 35, 54, 56
Methaqualone, HPCE analysis, 165–166, 169
Methcathinone, HPCE analysis, 168
Methylcellulose, 197
Metoprolol, HPCE analysis, 168
MgATP, 64
Micellar electrokinetic chromatography (MEKC), generally:
 defined, 55–58
 illicit drug substances, 165–172
Michaelis-Menten formula, 119
Microdialysis, in vivo:
 applications in enzymatic analysis:
  body fluids, 123–124
  estimating enzymatic activies, 126–129
  small molecules, 125
 extracellular space, 116–177
 overview, 115–116
 principle of, 116
 probe, 117–118
 recovery from, 118–119
 technical aspects:
  experiments on humans, 122–123
  experiments on rats, 122
  HPLC analysis, 121–122
  instrumentation, 119–121
Mixture, see Reaction mixture
Mobile phase, in chromatography:
 composition and preparation of, 35–36
 defined, 10, 14–15
 indirect detection in, 52
Monoamine oxidase(s):
 A and B, 222–224
 and phenol sulfotransferase, 225–226
Monoamines, microdialysis of, 116
Morphine, HPCE analysis, 165–166, 169–170, 172
mRNA, 160

Multienzyme systems:
  assay of multienzyme complex:
    by reconstitution method, 428–432
  assay of two activities:
    forming different products from same substrate, 419–420
    forming the same product from same substrate, 420–426
    same activity, formation of two separate products from two separate substrates by, 426–428
    using HPLC method, 432–435
Myosine light chain kinase, 369

N-acetyltransferase, 226–227, 229
NAD:
  formation of, 5
  glycohydrolase, 337–339
  pyrophosphorylase, 345–346
Nalorphine, HPCE analysis, 172
N^5-methyltetrahydrofolate-homocysteine methyltransferase, 269
Nicotinate phosphoribosyltransferase, 309–310
Nitrogenase, 402–403
NMR, 35
Noradrenaline, 116, 128
Norephedrine, HPCE analysis, 168
Normal-phase liquid chromatography, 25
Norpseudoephedrine, HPCE analysis, 168
Noscapine, HPCE analysis, 166–167
Nuclear magnetic resonance (NMR), 115
Nucleic acids:
  CE methods and, 141–142
  forensic analysis, 196–199
  modification and expression, survey of enzymatic activities:
  chloramphenicol acetyltransferase, 405–408
  DNA topoisomerase, 405
  separation:
    DNA analysis, classical methods of, 141
    separation mechanism, 139

Nucleoside diphosphate kinase, 346–347
Nucleoside phosphorylase, 323–325

δ-aminolevulinic acid synthetase, 276–277
Ogston model, 139
δ-(L-α-aminoadipyl)-L-cysteinyl-D-valine synthetase, 271–272
1,10-phenanthroline, 103
Ornithine aminotransferase, 247–248
Ornithine decarboxylase, 272–273
Oxprenolol, HPCE analysis, 168
Oxygenations, survey of enzymatic activities:
  acetanilide 4-hyroxylase, 348–349
  aryl hydrocarbon hydroxylase (EC 1.14.14.2), 351–352
  ceruloplasmin, 349–351
  hepatic microsomal testerone hydroxylase, 352–356

Papain esterase, 235–237
Papaverine, HPCE analysis, 166–167
Paracetamol, HPCE analysis, 167
Parked reaction, 189
Pen inks, CZE analysis, 175–176
Pentobarbital, HPCE analysis, 168, 170
Peptide(s), *see* Peptide metabolism; Peptide modifications
Peptide metabolism, survey of enzymatic activities:
  amino acid decarboxylase, 263–264
  aromatic L-amino acid decarboxylase, 264
  asparagine synthetese, 251, 253
  D-amino oxidase, 264–265
  diaminopimelate epimerase and decarboxylase, 259
  dihydroxyacid dehydratase, 255–256
  glucamic acid decarboxylase, 262
  glutamine synthetase/glutamate synthetase/glutamate dehydrogenase, 249–252
  glutaminyl cyclase, 256–257
  histamine N-methyltransferase, 262–263
  kynureninase, 267–268

Peptide metabolism *(Continued)*
  kynurenine 3-monoxygenase, 268–269
  L-alanine:glyoxylate aminotransferase, 270
  leucine 2,3-aminomutase, 257–258
  lysine-ketoglutarate reductase, 259–260
  N^5-methyltetrahydrofolate-homocysteine methyltransferase, 269
  ornithine aminotransferase, 247–248
  threonine/serine dehyratase, 265
  tryptophanase, 253–256
  tryptophan dioxygenase, 265–266
  tryptophan 2,3-dioxygenase, 267
  tyrosinase, 270–271
  γ-glutamylcyclotransferase, 261
  γ-glutamyclcysteine synthetase/glutathione synthetase, 261–262
Peptide modifications, survey of enzymatic activities:
  adenosine diphosphate–ribosylarginine hydrolase, 366–368
  myosine light chain kinase, 369
  peptidylglycine α-amidating monoxygenase, 367–369
  phosphotyrosine phosphatases, 371
  phosphotyrosyl protein phosphatase, 370
  protein phosphatase 2B, 371
  transglutaminase, 369
  tyrosine protein kinase, 365–367
Peptidylglycine α-amidating monoxygenase, 367–369
Perilymph, microdialysis in, 124
pH, protein analysis, 176–177
Phencyclidine hydrochloride (PCP), HPCE analysis, 165
Phenmetrazine, HPCE analysis, 169
Phenobarbital, HPCE analysis, 165, 169–170
Phenolsulfotransferase, 380–382
Phenylethanolamine N-methyltransferase, 221–222
Phenylmethanesulfanyl fluoride, 103
Phosphate, in mobile phase, 35
Phosphoenolypyruvate (PEP), 5
Phosphoribosylpyrophosphate (PRPP), 85, 429
Phosphoribosylpyrophosphate synthetase, 340–342
Phosphotyrosine phosphatases, 371
Phosphotyrosyl protein phosphatase, 370
pK, 19
Plasma carboxypeptidase (Kininase I, Bradykinin-destroying enzyme, EC 3.4.12.7), 237–238
Polarity, in reversed-phase chromatography, 27–29
Polio virus, 154
Polyacrylamide, 142
Polyamine metabolism, survey of enzymatic activities:
  diamine oxidase, 275–276
  orinthine decarboxylase, 272–273
  polyamine oxidase, 275
  spermidine synthetase, 273–274
Polyamine oxidase, 275
Polymerase chain reaction (PCR):
  applications:
    forensic analysis, 151
    future directions, 160
    identification by hybridization, 151–153
    quantitative analysis, 153–156
    quantitative RNA-PCR, 156–160
  CE methods:
    artifacts, 150–151
    buffer systems, 141–142
    data analysis, 146
    detection, 144–146
    intercalators, 142–144
    nucleic acids and, 141
    sample preparation and injection, 146–150
    typical instrument parameters, 144
  exponential phase of, 156
  nucleic acid separation:
    DNA analysis, classical methods of, 141
    separation mechanism, 139
    overview, 137–140, 142
    schematic representation of, 140
Positron emission tomography (PET), 115

Power supply, in HPCE, 49–50
Precolumn, 16
Preparation phase, in enzyme assay, 2
Preparation strategies for enzymatic activities, from tissues, body fluids, and single cells:
  cell-free lysates, initial purification and assay of activities in, 105–106
  cellular compartments:
    assay of activities in, 100–102
    extracellular separated from, 97–110
  extent of purification or end point, determination of, 95–97
  extracellular compartments:
    assay of activities in, 100
    cellular separated from, 97–100
  HPLC for purification of enzymes:
    overview, 106–107
    problems related to, 112–113
    strategy for, 107–111
  intact cells, 103
  selection of biological starting point, 93–95
  separation of cellular from extracellular compartments, 97–100
  subcellular samples, 103–105
Pressure, electrokinetic injection vs., 147–148
Primary reaction, analysis of, 64–65, 79. See also specific types of enzymes
Primer annealing cycle, in PCR, 139
Probes, microdialysis, 117–120
Procaine, HPCE analysis, 169
Propranolol, HPCE analysis, 168
Protein(s), see Protein analysis; Protein modifications
  -drug binding assays, 192–196
  phosphatase 2B, 371
Protein analysis, in forensics:
  nonenzymatically modified proteins:
    modifications to collagen by glucose, 183–185
    modifications to keratins by ethanol, 179–183
  separation of protein mixtures by two-dimensional techniques, 178–179
  separation of proteins by HPCE, 176–178
Proteinase, survey of enzymatic activities:
  aminopeptidase, 239
  carboxypeptidase N, 244–245
  dipeptidase, 238
  dipeptidyl carboxypeptidase, 231–235
  dipeptidyl peptidase IV/amino peptidase-P, 244, 246
  enkephalinases A and B, 239–241
  luteinizing hormone-releasing hormone peptidase, 235
  papain esterase, 235–237
  plasma carboxypeptidase (Kininase I, Bradykinin-destroying enzyme, EC 3.4.12.7), 237–238
  renin, 246
  rhinovirus 3c protease, 241–243
  stromelysin, 243–244
  vertabrate collagenase, 229–231
Protein modifications, survey of enzymatic activities:
  adenosine diphosphate–ribosylarginine hydrolase, 366–368
  myosine light chain kinase, 369
  peptidylglycine α-amidating monoxygenase, 367–369
  phosphotyrosine phosphatases, 371
  phosphotyrosyl protein phosphatase, 370
  protein phosphatase 2B, 371
  transglutaminase, 369
  tyrosine protein kinase, 365–367
Proteolytic inhibitors, assay of activities, 103
Protoporphyrinogen, 281–282
Pseudoephedrine, HPCE analysis, 168
Psilocin, HPCE analysis, 165
Psilocybin, HPCE analysis, 165
Pterin metabolism, survey of enzymatic activities:
  dihydrofolate reductase, 353, 356, 358
  folic acid cleaving enzyme, 353–354, 357
  guanosine triphosphate cyclohydrolase I, 357, 360

Pteroylpolyglutamate hydrolase, 401–402
Pump, defined, 15
Purification strategies:
  anatomy of, 94
  cell-free lysates, 104–106
  extent or end point, determination of, 95–97
  HPLC:
    overview, 106–107
    problems related to, 112–113
    strategy for, 107–111
Purines:
  determination of, 116
  metabolism, survey of enzymatic activities:
    adenosine deaminase, 317–318
    adenosine kinase, 326–327
    adenylate cyclase, 327–330
    adenylate kinase, 333–335
    adenylosuccinate synthetase, 334, 336–337
    alkaline and acid phosphatase, 313–317
    AMP deaminase, 317, 319–320
    ATPase, 348
    ATP pyrophosphohydrolase, 320–322
    cAMP phosphodiesterase, 330–333
    creatine kinase, 325–326
    cyclic nucleotide phosphodiesterase, 320–321
    cytokinins, 338–339
    dinucleoside polyphosphate pyrophosphohydrolase, 336–337
    5'-nucleotidase, 310–313
    glutamine: 5-phosphoribosyl-1-pyrophosphate aminotransferase, 344–345
    guanase, 342–343
    hypoxantine guanine phosphoribosyltransferase, 322–323
    NAD glycohydrolase, 337–339
    NAD pyrophosphorylase, 345–346
    nicotinate phosphoribosyltransferase, 309–310
    nucleoside diphosphate kinase, 346–347
    nucleoside phosphorylase, 323–325
    phosphoribosylpyrophosphate synthetase, 340–342
    thiopurine methyltransferase, 345
    urate oxidase, 344
    xanthine oxidase, 339–341
Pyridoxal kinase, pyridoxamine oxidase, and pyridoxal-5'-phosphate phosphatase, 373
Pyridoxine kinase, 373–374
Pyrimidine metabolism, survey of enzymatic activities:
  β-ureidopropionase, 390
  cytidine deaminase, 389–390
  dihydroorotase, 391
  dihydroorotic acid dehydrogenase, 389
  dihydropyrimidine dehydrogenase, 388–389
  thymidylate synthetase, 391
Pyrophosphohydrolase, 64–65
Pyruvate, 116
Pyruvate kinase, 5

Quantative analysis, PCR, 153–156
  RNA-PCR, 156–160

Radiochemical detectors, 17
Radioisotope detection, 51
Rate data, obtaining, 82–83, 88
Rate determination, low substrate concentrations, 85–86
Reaction mixture, *see specific types of enzymes*
  composition decisions, 81–82
  effects of metals on separation, 71–73
  as selection criteria, 10–11
Reconstitution method, assay of multienzyme complex:
  conversion of IMP to AMP, 430–431
  degradation of IMP to inosine, 429–430
  formation and fate of IMP, 428–429
  return of AMP to IMP, 432

Refractive index:
  defined, 17
  detection, 51
Renin, 246
Reptation, 139
Retention, in MEKC, 56
Retention time:
  in chromatogram, 22
  in reversed-phase chromatography, 25, 27, 29–30, 35
Retinal oxidase, 360–361
Reversed-phase chromatography:
  characteristics of, 25–30
  forensic analysis, illicit drug substances, 166–167
Reverse transcriptase (RT) PCR, 153–154
Rhinovirus 3c protease, 241–243
RNA synthesis, 138
RNA-PCR, quantitative, 159–160

Salting out procedure, 95
Salvage pathway, 428–429
sAMP, 430–431
Sampling, *see specific types of enzymes*
  bias, 42, 147
  microdialysis, 129
SDS (sodium dodecyl sulfate):
  in mobile phase, 35, 54, 56
  -polyacrylamide gel electrophoresis (SD-PAGE), 59
  in protein analysis, 178, 185
Secondary reaction, *see specific types of enzymes*
  analysis of, 65, 79
  substrate analogs and, 86–87
  understanding and dealing with, 68–71
Sensitivity shift procedure, 86
Separation, *see specific types of enzymes*
  defined, 3
  and detection of components, as selection criteria, 10
  efficiency and resolution:
    practical hints, 54–55
    theoretical plate number and resolution, 52–54

nucleic acids:
  DNA analysis, classical methods of, 141
  separation mechanism, 139
Serotonin, 116
Serum cholinesterase, 361–362
Sialidase, 293–297
6-monoacetylmorphine (6-MAM), HPCE analysis, 166–167, 169
6-phosphogluconate dehydratase, 300–301
6-pyruvoyl tetrahydropterin synthetase, 400–402
Size-exclusion chromatography, 23
Slab gel electrophoresis, 141, 197
Small molecules, microdialysis in, 125
Solid phase, in chromatography, 10
Solvent reservoir, 15
S-100 fraction, 105, 107
Soybean trypsin inhibitor, 103
Spermidine synthetase, 273–274
Stationary phase, in chromatography:
  defined, 14
  selection of:
    gel filtration chromatography, 24–25
    generally, 65, 67
    ion-exchange chromatography, 30, 32–34
    overview, 23–24
    reversed-phase chromatography, 25–30, 32
Steroid 17 α-hydroxylase/$C_{17\text{-}20}$ lyase (cytochrome $P450_{21SCC}$, 307–309
Steroid metabolism, survey of enzymatic activities:
  cholesterol 7α-hydroxylase, 304–306
  cytochrome $P450_{SCC}$, 306–308
  11-β-hydroxylase and 18-hydroxylase, 302–304
  Δ^5-3β-hyroxysteroid dehydrogenase, 301–402
  steroid 17 α-hydroxylase/$C_{17\text{-}20}$ lyase (cytochrome $P450_{21SCC}$, 307–309
  3β-hydroxy Δ^5-C_{27}-steroid oxidoreductase, 306–307
  25-hydroxyvitamin D3-1α-hydroxylase, 304
S-30 fraction, 105

Strictosidine synthetase, 403
Stromelysin, 243–244
Substrate analogs, 86–87
Succinyl-CaA synthetase, 295–296, 298
Sucrose phosphate synthetase, 300
Sulfotransferase, 375–377
Survey of enzymatic activities assayed by HPLC method:
  amino acid and peptide metabolism, 247–271
  carbohydrate metabolism, 283–301
  catecholamine metabolism, 208–229
  complex saccharides and glycoproteins, metabolism of, 392–399
  cytokinin metabolism, 338–339
  heme metabolism, 276–282
  lipid metabolism, 360–365
  miscellaneous, 399–405
  nucleic acid modification and expression, 405–408
  oxygenations, 348–356
  polyamine metabolism, 272–276
  proteinase, 229–246
  proteins and peptides, modification of, 365–371
  pterin metabolism, 353–360
  purine metabolism, 309–348
  pyrimidine metabolism, 388–391
  steroid metabolism, 301–309
  vitamin metabolism, 372–374
  xenobiotic metabolism, 374–388
Synovial fluid, microdialysis in, 124

Taq gene, 138, 144
TCA (tricholoracetic acid), 6–7, 73–74
TEA, 54
TEOHA, 54
Termination phase, in enzyme assay, 2–3
Tetrabutylammonium salt, 29, 57
Tetracaine, HPCE analysis, 172
Tetrahydrocannabinol (THC), HPCE analysis, 165
Tetrahydrofuran, 54, 56
Tetrahydrozoline, HPCE analysis, 169
Thiamine triphophatase, 372
Thiopental, HPCE analysis, 169–170
Thiopurine methyltransferase, 345

3β-hydroxy ^{5}-C_{27}-steroid oxidoreductase, 306–307
Threonine/serine dehyratase, 265
Thymidylate synthetase, 391
Thyroxine: UDP-glucuronosyltransferase, 397
Timing, in chromatography, 14. *See also* Retention time
TOTO-1, 156, 199
Transglutaminase, 369
T*rans-p*-coumaroyl esterase, 397, 399
Triton X-100, 147
tRNA, 5
Troponin, 129
Tryptophan 2,3-dioxygenase, 267
Tryptophan dioxygenase, 265–266
Tryptophanase, 253–256
Tuftsin, 30
25-hydroxyvitamin D_3-1α-hydroxylase, 304
Tyrosinase, 270–271
Tyrosine hydroxylase (TH):
  enzymatic activities, overview, 208–211
  microdialysis, 126
Tyrosine protein kinase, 365–367

UDP-glucosyltransferase, 385
UDP-glucuronyl transferase, 384–385
Ultrafiltration, 147
Urate oxidase, 344
Uridine diphosphate glucuronosyltransferase, 287–291
Uroporphyrinogen decarboxylase, 278–279
UV radiation, 50
UV spectrometer, 17

Vacancy peak capillary zone electrophoresis, 194–195
Variable number of tandem repeat sequences (VNTRs), 198–199
Vent gene, 138
Vertabrate collagenase, 229–231
Vitamin metabolism, survey of enzymatic activities:
  biotinidase, 374
  lipoamidase, 372–373

Vitamin metabolism, survey of, enzymatic activities *(Continued)*
  pyridoxal kinase, pyridoxamine oxidase, and pyridoxal-5'-phosphate phosphatase, 373
  pyridoxine kinase, 373–374
  thiamine triphophatase, 372
Vitreous fluid, microdialysis in, 124

Xanthine oxidase, 339–341
Xenobiotic metabolism, survey of enzymatic activities:
  adenosine 3'-phosphate 5'-sulfophosphate sulfotransferase, 380–381
  aryl sulfotransferase, 382–383
  ATP-sulfurylase, 374–376
  cysteine conjugate β-lyase, 383–384
  cytochrome P450$_{2E1}$, 386–387
  ethoxycoumarin O-deethylase, 385–386
  flavin-containing monoxygenase, 387–388
  glutathione S-transferase, 376, 378–380
  phenolsulfotransferase, 380–382
  sulfotransferase, 375–377
  UDP-glucuronyl transferase, 384–385
  UDP-glucosyltransferase, 385

γ-glutamyclcysteine synthetase/glutathione synthetase, 261–262
γ-glutamylcyclotransferase, 261
YO-PRO-1, 199
YOYO-1, 156, 199

Zorbax GF 450, 178

DE ANZA COLLEGE LEARNING CENTER

3 1716 00319 3368

CIR KF9660.Z9 W48 2ND FL
OOR-SOUTH MEZZ.
Weston, Paul B
Fundamentals of evidence

1861, the Court of Star Chamber. This court conducted trials in secrecy and by nonjury inquisitory methods. Current use refers to denial to persons accused of crime of access to friends or legal counsel (incommunicado custody) and suggests inquisitorial questioning.

States' rights
Advocacy of greater power for the individual states as opposed to a strong federal government (see Supremacy clause).

Stay
A court or executive order stopping official proceedings.

Trier of fact
Judge or juror; one who tries the facts of a case.

Waiver
Surrender of a legal right.

Writ
A court order (see *Certiorari*, *Habeas corpus*, Prohibition, *Mandamus*).

Glossary of Common Legal Terms

Motion
Application to a court or judicial officer for a legal remedy, usually a court order.

Notary public
A public official with the power and duty of administering oaths, making certifications, and taking acknowledgments.

Parol evidence rule
Written agreement overrides, in its final form, all prior negotiations between parties to the agreement.

Parties to an action
In criminal law, the principals in a criminal trial: (1) the prosecutor, representing the state or federal government (the "people"); and (2) the defendant or defendants in a joint trial.

Per se
By itself.

Petitioner
The person seeking a legal remedy in a higher court from a lower court's ruling or decision (see Cease and desist, *certiorari, habeas corpus, mandamus*).

Plaintiff
The party initiating a legal action.

Presumption
Inferring one fact from proof of another fact on the basis of a relationship between the two facts.

Prima facie evidence
Sufficient "on its face" to prove a fact.

Privileges and immunities clause
Constitutional guarantee (Fourteenth Amendment) against state laws that interfere with or lessen the privileges and immunities of U.S. citizens.

Prohibition, Order of
An order from a higher court restricting the actions of a lower court (cease and desist).

Proximate cause
The direct or immediate cause of a wound or other injury.

Res gestae
The doctrine of things done; used of acts as part of a transaction.

Subpoena
An order of a court directing a witness to attend (see *Duces tecum*).

Supremacy clause
Constitutional provision (Article VI) that federal laws (and treaties) constitute the supreme law of the country (see States' rights).

Star Chamber
Reference to a seventeenth-century court abolished in

Glossary of Common Legal Terms

before the court it is termed *obiter dictum* (an opinion "by the way").

Duces tecum
"Bring it with you," a term related to a subpoena directing the witness to bring books, papers, and other records.

Docket
Court record of cases and things done.

Due-process clause
Constitutional provision that prohibits the government from depriving any person of his life, liberty, or property without due process of law (Fifth and Fourteenth amendments).

En banc
All the judges of a court "sit" and hear a case.

Equal protection
Constitutional guarantee against denial of the protection of laws enjoyed by others in similar circumstances.

Ex post facto
After an act.

Habeas corpus
An order to appear before a court and "bring the body" of a specific person. Known as the "great writ," it is an inquiry into the legality of a person's imprisonment. It is directed to the warden or other official having the person named in the order in their custody.

Indictment
The procedure by which a grand jury accuses a person of a crime. It is voted upon by jurors and "presented" to the judge supervising the grand jury.

Injunction
A court order restraining a person or corporation from doing, or continuing to do, a specific act or acts.

Judicial notice
Judicial recognition and acceptance of a fact without the introduction of evidence A courtroom shortcut when the facts are of common knowledge.

Jurat
The "sworn to before me" segment of a legal document. It is dated and signed by a public official to certify the document (see Acknowledgment).

Locus delicti
Place of a crime.

Mandamus, **Order of**
An order of a court to perform a specific act or duty.

Mistrial
A judicial decision to end a trial in the interests of justice.

Modus operandi
Method of operations.

Glossary of Common Legal Terms

Calendar
List of cases scheduled to be called in a court.
Case law
Law created by judicial decisions.
Cease and desist
An order of a court to stop doing specified acts.
Certiorari
Process by which a higher court reviews the record of a case heard in a lower court. Process is initiated by petition to higher court asking for review.
Charge to the jury
Judicial instructions to a jury as to the principles of law in a case and their application to the case and its circumstances.
Citation
A reference to an authority. In reference to U.S. Supreme Court decisions, the name, volume, page, and year are indicated. The written opinion of the Court in the case named will begin at the page cited in the volume given. *U.S. Reports* are bound books identified by both volume number and year. For instance, in the citation: *Terry* v. *Ohio*, 392 U.S. 1 (1967), the case begins at page 1, Volume 392 (1967) of *U.S. Reports*.
Civil
Not criminal.
Co-conspirators
A member of a conspiracy (two or more) to commit a crime.
Conspiracy
The crime committed upon agreement of two or more persons.
Complaint
In criminal law, the name of the first accusation of crime by police or prosecutor. It sets forth in detail the common name of the crime charged, its place of occurrence, its essential elements, and the identity of the accused person.
Contraband
Things possessed in violation of the law.
Corpus delicti
The crime itself, the "body" of the crime, the fact of a crime, and its essential elements.
Deponent
An individual who makes an affidavit.
Deposition
Written transcript of testimony taken under oath.
Dictum
An opinion of a court expressed in a decision but not essential to the court's decision on the question being reviewed. When *dictum* is not closely related to the question

Glossary of Common Legal Terms

Glossary of Common Legal Terms

Acknowledgment
The act of appearing before a notary public (or other authorized public official) and swearing to the authoring of a legal document (see Jurat).

Affidavit
Sworn written statement, used in police services in seeking arrest and search warrants or in court papers containing charges against an accused person (see Deponent).

Amicus curiae
A "friend of the court"; a person not a party to a legal action but who participates for the purpose of assisting the court in its determination of the issue in dispute.

Answer
Response of the defense to indictment, complaint, or other accusation of crime.

Antecedent judicial authorization
Prior judicial review and authorization; issuance of a court order such as a search or arrest warrant, an order to intercept and record conversations, etc.

Appeal
The legal process for the review of the decision of lower courts by higher courts with the right to hear appeals.

Appellant
The party seeking a legal remedy in a higher court from a lower court's decision (see Petitioner).

Arraign
To bring a person (the accused) before a court in connection with charges of crime.

Attestation
A statement ending a legal document (usually "In the presence of":). It precedes the signatures of witnesses to the document.

Best-evidence rule
A written document is the "best evidence" of its contents. It must be produced or its absence satisfactorily accounted for before secondary evidence will be accepted.

Bill of Rights
The first ten amendments to the Constitution.

tion collected from its review. The Court's ruling is: Such evidence is tainted by illegality, and any derivative use of such evidence is likewise tainted and therefore inadmissible. "The essence of a provision forbidding the acquisition of evidence in a certain way is that not merely evidence so acquired shall not be used before the court, but that it shall not be used at all" (see *Weeks* v. *United States* and *Mapp* v. *Ohio*).

15. TERRY V. OHIO, 394 U.S. 1 (1968). A person can be forcibly stopped and superficially searched by police (stop-and-frisk); the police officer must have specific facts linked with rational inferences to warrant the intrusion and must confine the search to less than a full search (a "protective" search for weapons).

16. WADE V. UNITED STATES, 388 U.S. 218 (1967). Police lineup for identification of criminal suspect by eyewitnesses ruled to be a "critical stage" of the pretrial time period at which a suspect is entitled to legal counsel, in the absence of an intelligent waiver.

17. WARDEN V. HAYDEN, 387 U.S. 294 (1967). Even though of mere evidential value, police seizure of clothing at place of arrest (home of arrestee) approved; clothing held properly admitted into evidence at arrestee's trial for robbery. (Also, entry into premises and search for suspect without warrant justified because police were in "hot pursuit.")

18. WEEKS V. UNITED STATES, 232 U.S. 383 (1914). Private papers seized by government agents during a warrantless search were inadmissible as evidence in federal court because they were obtained through an illegal search and seizure in violation of the Fourth Amendment (see *Silverthorne Lumber Company* v. *United States; Mapp* v. *Ohio*).

liability is established and informant is not a participant in the crime and a material witness on the issue of guilt or innocence.

10. MIRANDA v. ARIZONA, 384 U.S. 436 (1966). Defines custodial interrogation as questioning initiated by police after a person has been taken into custody or otherwise deprived of his freedom of action in any significant way; forbids the use of confessions and other statements stemming from custodial interrogation unless police can demonstrate procedural safeguards protective of the privilege against self-incrimination (warning and waiver); and excoriates "various police manuals and texts" for examples of interrogation techniques whose objectives were to subjugate the individual to the will of the interrogator. (Also, linked the *"Escobedo* rule" with custodial interrogation—see *Escobedo* v. *Illinois*.)

11. OSBORN v. UNITED STATES, 385 U.S. 323 (1966). Evidence of an electronic surveillance (recording of a conversation) obtained by way of a tape recorder concealed on a police informant held admissible in criminal trial (attempted bribery of juror). Prior judicial approval based on informant's affidavit as to an incriminating conversation with suspect was the primary basis for ruling: (1) use of recording device permissible, and (2) recording itself properly admitted as evidence. Also, question of entrapment was for the trial jury to determine (see *Katz* v. *United States*).

12. ROCHIN v. CALIFORNIA, 342 U.S. 165 (1952). Intensity of search (stomach pumping) "shocked the conscience" and evidence obtained by such force is inadmissible.

13. SCHMERBER v. CALIFORNIA, 384 U.S. 757 (1966). Establishes admissibility of nontestimonial evidence, particularly blood sample; justifies minor intrusions into an individual's body under limited conditions as a reasonable search; and rejects the claim that such nontestimonial evidence is within the protection of the Fourth Amendment (searches and seizures), the Fifth Amendment (self-incrimination), or the Sixth Amendment (legal counsel), or that petitioner was denied his rights under the Fourteenth Amendment (due process).

14. SILVERTHORNE LUMBER COMPANY v. UNITED STATES, 251 U.S. 385 (1920). Books, papers, and documents seized without authority by government agents, who returned seized material but attempted to use informa-

safeguards of due process: (1) notice of charges, (2) right to counsel, (3) privilege against self-incrimination, and (4) the right to confrontation (and cross-examination) of witnesses.

5. JACKSON V. DENNO, 378 U.S. 368 (1964). Requires a judicial determination of voluntariness prior to the admission of a confession to a jury adjudicating guilt or innocence.
6. KATZ V. UNITED STATES, 389 U.S. 347 (1967). Electronic eavesdropping is a search and seizure within the meaning of the Fourth Amendment, but the legitimate needs of law enforcement can be accommodated by a judicial order authorizing limited electronic eavesdropping (see *Osborn v. United States*).
7. KER V. CALIFORNIA, 374 U.S. 23 (1963). Laid down standards for application to specific cases of the constitutional prohibition against unreasonable searches and seizures. Reasonableness of a search is a matter of determination for the trial court, but subject to review as to whether the decision concerning reasonableness by the trial court respected the fundamental —i.e., constitutional—criteria. Note that ruling does not preclude states from developing workable rules governing arrests, searches, and seizures, provided such rules do not violate the constitutional proscription of unreasonable searches and seizures and consider the fact that evidence seized in an unreasonable search is inadmissible against the "injured person"—one with standing to complain (see *Mapp v. Ohio*, and *Chimel v. Ohio*).
8. MAPP V. OHIO, 367 U.S. 643 (1961). Declares the Fourth Amendment's right to privacy is enforceable against the states through the due process clause of the Fourteenth Amendment by the same sanction of exclusion as is used against the federal government. Rejects as admissible evidence in a criminal trial the material seized by police in a warrantless search of the Mapp residence; establishes the "exclusionary rule" as an essential part of the Fourth and Fourteenth Amendment; and states this decision gives to the citizen no more than the guarantees of the Constitution, to the police no less than that to which honest law enforcement is entitled, and to the courts the necessary judicial integrity to administer justice (see *Ker v. California*, and *Chimel v. California*).
9. MCCRAY V. ILLINOIS, 386 U.S. 300 (1967). Protects police informants from disclosure of identity when re-

List of Cases

1. CHIMEL V. CALIFORNIA, 395 U.S. 752 (1969). Limits police search, in the absence of a search warrant, to person of arrestee and the area from within which he might obtain either a weapon or something that could be used as evidence against him (see *Mapp* v. *Ohio,* and *Ker* v. *California*).
2. DAVIS V. MISSISSIPPI, 394 U.S. 721 (1969). Fingerprint evidence taken during illegal police detention held inadmissible in criminal trial for rape; police action ruled in violation of Fourth Amendment's protection against unreasonable search and seizure and Fourteenth Amendment's guarantee of due process (since defendant's detention by the police was unlawful, the fingerprints were obtained in violation of the Fourth and Fourteenth amendments).
3. ESCOBEDO V. ILLINOIS, 378 U.S. 478 (1964). Establishes a rule: the suspect has been denied the assistance of counsel in violation of the Sixth Amendment when (1) the investigation is no longer a general inquiry into an unsolved crime but has begun to focus on a particular suspect; (2) the suspect has been taken into police custody and has requested and been denied legal counsel; and (3) the police then carry out an interrogation that lends itself to eliciting incriminating statements, without effectively warning the suspect of his constitutional right to remain silent. (Later, in *Miranda* v. *Arizona,* the above situation was described as "custodial interrogation" and the Court commented: "This is what we meant in ESCOBEDO when we spoke of an investigation which had focused on an accused person.")
4. GAULT V. ARIZONA, 387 U.S. 1 (1967). Proceedings against juvenile in juvenile court on the issue of guilt vs. innocence, which may lead to commitment to state correctional institution, required to meet constitutional

[1] Alphabetically, by common name of case.

Bibliography

California Evidence Code Manual. Sacramento, California: California Continuing Education of the Bar, 1966.

KERR, HARRY P., *Opinion and Evidence: Cases for Argument and Discussion.* New York: Harcourt Brace Jovanovich, Inc., 1962.

MENDELSON, IRVING, *Defending Criminal Cases.* New York: Practising Law Institute, 1967.

NEWMAN, DONALD J., *Conviction: The Determination of Guilt or Innocence without Trial.* Boston: Little, Brown and Company, 1966.

SALOTTOLO, A. LAWRENCE, *Modern Police Service Encyclopedia.* New York: Arco Publishing Co., Inc., 1962.

TRACY, JOHN EVARTS, *Handbook of the Law of Evidence.* Englewood Cliffs, N.J.: Prentice-Hall, Inc., 1952.

WALLS, H. J., *Forensic Science.* New York: Praeger Publishers, Inc., 1968.

WESTON, PAUL B., and KENNETH M. WELLS, *Criminal Investigation: Basic Perspectives.* Englewood Cliffs, N.J., Prentice-Hall, Inc., 1970.

both Wade and his counsel should have been notified of the impending lineup, and counsel's presence should have been a requisite to conduct of the lineup, absent an "intelligent waiver."

In the concluding segments of this opinion, the Court notes that this new requirement might jeopardize prompt identification, but that refusal to recognize the right to counsel for fear that counsel will obstruct the course of justice is contrary to the basic assumption upon which they have operated in reviewing Sixth Amendment cases.

The Court's decision was to vacate the conviction and return the case to the district court "to determine whether the in-court identification had an independent source, or whether, in any event, the introduction of the evidence was harmless error."

These summaries are illustrative of the concern of the judiciary with the constitutional rights of any individual accused of crime. The Fourth Amendment protection against unreasonable searches is the core area of the *Chimel, Davis,* and *Terry* cases; the Fifth Amendment privilege against self-incrimination was central to the *Miranda* decision; the rights guaranteed by the Sixth Amendment were primary factors in *McCray* (confrontation) and *Wade* (legal counsel); and the Sixth Amendment's guarantee of the assistance of legal counsel was correlated in the *Miranda* case with the Fifth Amendment's privilege against self-incrimination.

In total, these summaries represent decisions of the U.S. Supreme Court that set standards for police by establishing case law on the subjects of search and seizure, stop-and-frisk, interrogation, lineups, and the use of informants.

Review Questions

1. What is the major impact on prevailing police practices of the *Chimel* decision? *Miranda? Wade? Davis? McCray? Terry?*
2. How does the Court's decision in *Terry* v. *Ohio* support the police technique of field interviews or field interrogations?
3. What is the difference between legal counsel present at police lineups and at custodial interrogation?

Case Law—Selected U.S. Supreme Court Decisions

and five or six other prisoners and conducted in a courtroom of the local county courthouse. Both bank employees identified Wade in the lineup as the bank robber. At trial, the two employees, when asked on direct examination if the robber was in the courtroom, pointed to Wade. The prior lineup identification was then elicited from both employees on cross-examination. Wade was convicted.

The question in this case is whether the out-of-court identification of petitioner Wade in a lineup violated petitioner's constitutional rights and tainted the later in-court identification. At trial, at the close of testimony, petitioner Wade's counsel moved for a judgment of acquittal or, alternatively, for striking the courtroom identifications from the trial's record on the ground that the lineup (without notice to and in the absence of Wade's counsel) violated Wade's Fifth Amendment privilege against self-incrimination and his Sixth Amendment right to the assistance of counsel. The motion was denied.

Two extracts from the Court's majority opinion in this case highlight its rejection of the argument that Wade had been forced to incriminate himself and acceptance of the argument he had been denied the assistance of counsel. They are as follows:

> (1) Neither the lineup itself nor anything shown by this record that Wade was required to do in the lineup violated his privilege against self-incrimination. We have only recently reaffirmed that the privilege "protects an accused only from being compelled to testify against himself, or otherwise provide the State with evidence of a testimonial or communicative nature...." (*Schmerber* v. *California*, 384 U.S. 757.[1]) We there held that compelling a suspect to submit to a withdrawal of a sample of his blood for analysis for alcohol content and the admission in evidence of the analysis report were not compulsion to those ends.
>
> (2) Since it appears that there is grave potential for prejudice, intentional or not, in the pretrial lineup, which may not be capable of reconstruction at trial, and since presence of counsel itself can often avert prejudice and assure a meaningful confrontation at trial, there can be little doubt that for Wade the postindictment lineup was a critical stage of the prosecution at which he was as much entitled to such aid (of counsel) as at the trial itself. Thus

[1] See List of Cases, p. 70.

stances. However, the concluding words of this opinion—in affirming petitioner Terry's conviction—offer the following guidelines for police:

> We conclude that the revolver seized from Terry was properly admitted in evidence against him. At the time he seized petitioner and searched him for weapons, Officer McFadden had reasonable grounds to believe that petitioner was armed and dangerous, and it was necessary for the protection of himself and others to take swift measures to discover the true facts and neutralize the threat of harm if it materialized. The policeman carefully restricted his search to what was appropriate to the discovery of the particular items which he sought. We merely hold today that where a police officer observes unusual conduct which leads him reasonably to conclude in light of his experience that criminal activity may be afoot and that the persons with whom he is dealing may be armed and presently dangerous, where in the course of investigating this behavior he identified himself as a policeman and makes reasonable inquiries, and where nothing in the initial stages of the encounter serves to dispel his reasonable fear for his own or others' safety, he is entitled for the protection of himself and others in the area to conduct a carefully limited search of the outer clothing of such persons in an attempt to discover weapons which might be used to assault him. Such a search is a reasonable search under the Fourth Amendment, and any weapons seized may properly be introduced in evidence against the person from whom they were taken.

Case No. 6—Police Lineups:
United States v. Wade,
388 U.S. 218 (1967)

The federally insured bank in Eustace, Texas, was robbed on September 21, 1964. A man with a small strip of tape on each side of his face entered the bank, pointed a pistol at the female cashier and the vice-president, the only persons in the bank at the time, and forced them to fill a pillow case with the bank's money. The man then drove away with an accomplice who had been waiting in a stolen car outside the bank. Without notice to Wade's lawyer, the two bank employees observed a lineup made up of Wade

mental interests involved. One general interest is of course that of effective crime prevention and detection; it is this interest which underlies the recognition that a police officer may in appropriate circumstances and in an appropriate manner approach a person for purposes of investigating possibly criminal behavior even though there is no probable cause to make an arrest. It was this legitimate investigative function Officer McFadden was discharging when he decided to approach petitioner and his companions. He had observed Terry, Chilton, and Katz go through a series of acts, each of them perhaps innocent in itself, but which taken together warranted further investigation. There is nothing unusual in two men standing together on a street corner, perhaps waiting for someone. Nor is there anything suspicious about people in such circumstances strolling up and down the street, singly or in pairs. Store windows, moreover, are made to be looked in. But the story is quite different where, as here, two men hover about a street corner for an extended period of time, at the end of which it becomes apparent that they are not waiting for anyone or anything; where these men pace alternately along an identical route, pausing to stare in the same store window roughly twenty-four times; where each completion of this route is followed immediately by a conference between the two men on the corner; where they are joined in one of these conferences by a third man who leaves swiftly; and where the two men finally follow the third man and rejoin him a couple of blocks away. It would have been poor police work indeed for an officer of thirty years' experience in the detection of thievery from stores in this same neighborhood to have failed to investigate this behavior further.

Officer McFadden patted down the outer clothing of petitioner and his two companions. He did not place his hands in their pockets or under the outer surface of their garments until he had felt weapons, and he then merely reached for and removed the guns. He never did invade Katz's person beyond the outer surfaces of his clothes, since he discovered nothing in his patdown which might have been a weapon. Officer McFadden confined his search strictly to what was minimally necessary to learn whether the men were armed and to disarm them once he discovered the weapons. He did not conduct a general exploratory search for whatever evidence of criminal activity he might find.

In its decision, the Court points out that any case of this type can only be reviewed on its own particular circum-

improper lineup and the "third degree." Finally, because there is no danger of destruction of fingerprints, the limited detention need not come unexpectedly or at an inconvenient time. For this same reason, the general requirement that the authorization of a judicial officer be obtained in advance of detention would seem not to admit of any exception in the fingerprinting context.

We have no occasion in this case, however, to determine whether the requirements of the Fourth Amendment could be met by narrowly circumscribed procedures for obtaining, during the course of a criminal investigation, the fingerprints of individuals for whom there is no probable cause to arrest. For it is clear that no attempt was made here to employ procedures which might comply with the requirements of the Fourth Amendment: the detention at police headquarters of petitioner and the other young Negroes was not authorized by a judicial officer; petitioner was unnecessarily required to undergo two fingerprinting sessions; and petitioner was not merely fingerprinted during the December 3 detention but also subjected to interrogation. The judgment of the Mississippi Supreme Court is therefore reversed.

Case No. 5—Stop-and-Frisk: Terry v. Ohio, 392 U.S. 1 (1968)

Petitioner Terry was convicted of carrying a concealed weapon and sentenced to the statutorily prescribed term of one to three years in the penitentiary. The prosecution introduced in evidence two revolvers and a number of cartridges seized from Terry and a codefendant, Richard Chilton, by Cleveland police detective Martin McFadden.

The question in this case is whether, in all of the circumstances of the encounter, Terry's right to personal security was violated by an unreasonable search and seizure.

In examining all the circumstances of the police stop-and-frisk of petitioner Terry, the following extracts of the Court's majority opinion are pertinent comment on the general purposes of this police procedure and the arresting officer's action:

We consider first the nature and extent of the govern-

case, the Court's majority opinion reversing petitioner Davis's conviction reads, in part, as follows:

> The State argues, however, that the detention of petitioner Davis was of a type which does not require probable cause. Two rationales for this position are suggested. First, it is argued that the detention occurred during the investigatory rather than accusatory stage and thus was not a seizure requiring probable cause. The second and related argument is that, at the least, detention for the sole purpose of obtaining fingerprints does not require probable cause.
>
> It is true that at the time of the detention the police had no intention of charging petitioner with the crime and were far from making him the primary focus of their investigation. But to argue that the Fourth Amendment does not apply to the investigatory stage is fundamentally to misconceive the purpose of the Fourth Amendment. Investigatory seizures would subject unlimited numbers of innocent persons to the harassment and ignominy incident to involuntary detention. Nothing is more clear than that the Fourth Amendment was meant to prevent wholesale intrusions upon the personal security of our citizenry, whether these intrusions be termed "arrests" or "investigatory detentions." We made this explicit only last term in *Terry* v. *Ohio*, 392 U.S. 1, (1968), when we rejected "the notions that the Fourth Amendment does not come into play at all as a limitation upon police conduct if the officers stop short of something called a 'technical arrest' or a 'full-blown search'."
>
> Detentions for the sole purpose of obtaining fingerprints are no less subject to the constraints of the Fourth Amendment. It is arguable, however, that because of the unique nature of the fingerprinting process, such detentions might, under narrowly defined circumstances, be found to comply with the Fourth Amendment even though there is no probable cause in the traditional sense. Detention for fingerprinting may constitute a much less serious intrusion upon personal security than other types of police searches and detentions. Fingerprinting involves none of the probing into an individual's private life and thoughts that marks an interrogation or search. Nor can fingerprint detention be employed repeatedly to harass any individual, since the police need only one set of each person's prints. Furthermore, fingerprinting is an inherently more reliable and effective crime-solving tool than eyewitness identifications or confessions and is not subject to such abuses as the

Case No. 4—Fingerprinting during Illegal Detention

cedural safeguards must be employed to protect the privilege, and unless other fully effective means are adopted to notify the person of his right of silence and to assure that the exercise of the right will be scrupulously honored, the following measures are required. He must be warned prior to any questioning that he has the right to remain silent, that anything he says can be used against him in a court of law, that he has the right to the presence of an attorney, and that if he cannot afford an attorney one will be appointed for him prior to any questioning if he so desires. Opportunity to exercise these rights must be afforded him, the individual may knowingly and intelligently waive these rights and agree to answer questions or make a statement. But unless and until such warnings and waiver are demonstrated by the prosecution at trial, no evidence obtained as a result of interrogation can be used against him.

In its ruling reversing Miranda's conviction, the Court found:

Miranda was not in any way apprised of his right to consult with an attorney and to have one present during the interrogation, nor was his right not to be compelled to incriminate himself effectively protected in any other manner. Without these warnings the statements were inadmissible.

Case No. 4—Fingerprinting during Illegal Detention: *Davis* v. *Mississippi,* 394 U.S. 722 (1969)

Petitioner Davis was convicted of a brutal rape and sentenced to life imprisonment by a jury in Mississippi. Fingerprints of the petitioner, left on the windowsill of the victim's home, were the clinching evidence bringing about the petitioner's conviction.

The only issue before the Court was whether fingerprints obtained from petitioner should have been excluded from evidence as the product of a detention that was illegal under the Fourth and Fourteenth amendments. After reviewing the police procedures used in the investigation of this

In this case, the Court traced the history of the Fifth Amendment protection against self-incrimination:

> We sometimes forget how long it has taken to establish the privilege against self-incrimination, the sources from which it came and the fervor with which it was defended. Its roots go back into ancient times. Perhaps the critical historical event shedding light on its origins and evolution was the trial of one John Lilburn, a vocal anti-Stuart Leveller, who was made to take the Star Chamber Oath in 1637. The oath would have bound him to answer to all questions posed to him on any subject. He resisted the oath and declaimed the proceedings, stating: "Another fundamental right I then contended for, was, that no man's conscience ought to be racked by oaths imposed, to answer to questions concerning himself in matters criminal, or pretended to be so."
>
> On account of the Lilburn trial, Parliament abolished the inquisitorial Court of Star Chamber and went further in giving him generous reparation. The lofty principles to which Lilburn had appealed during his trial gained popular acceptance in England. These sentiments worked their way over to the Colonies and were implanted after great struggle into the Bill of Rights. Those who framed our Constitution and the Bill of Rights were ever aware of subtle encroachments on individual liberty. They knew that "illegitimate and unconstitutional practices get their first footing . . . by silent approaches and slight deviations from legal modes of procedure." The privilege was elevated to constitutional status and has always been as broad as the mischief against which it seeks to guard. We cannot depart from this noble heritage.
>
> Thus we may view the historical development of the privilege as one which groped for the proper scope of governmental power over the citizen. As a "noble principle often transcends its origins," the privilege has come rightfully to be recognized in part as an individual's substantive right, a right to a private enclave where he may lead a private life. That right is the hallmark of our democracy.

The Court's majority opinion summarizes its holding as follows:

> We hold that when an individual is taken into custody or otherwise deprived of his freedom by the authorities in any significant way and is subjected to questioning, the privilege against self-incrimination is jeopardized. Pro-

Case No. 3—Police Interrogation

his trial were seized at the time of arrest from his person in an arrest-based search. The informant was not produced at trial, despite defense requests, and failure to disclose the identity of the informant violated McCray's Sixth Amendment right of confrontation. McCray was convicted.

The question in this case was whether an informant's identity need always be disclosed.

In its decision the Court commented:

> The arresting officers in this case testified, in open court, fully and in precise detail as to what the informer told them and as to why they had reason to believe his information was trustworthy. Each officer was under oath. Each was subjected to searching cross-examination. The judge was obviously satisfied that each was telling the truth, and for that reason he exercised the discretion conferred upon him by the established law of Illinois to respect the informer's privilege.

In its ruling the Court distinguished between disclosure when the informant's information related to the issue of guilt or innocence and when it related to probable cause for an arrest or search. The significance of this ruling is that an informant's identity need not always be disclosed in a criminal trial.

Case No. 3—Police Interrogation:
Miranda v. Arizona,
384 U.S. 436 (1966)

A police officer arrested petitioner Miranda for kidnapping and rape. He was taken to the police station where he was identified by his victim and interrogated for two hours by police. At his trial, a written confession given to police questioners at the above interrogation session was admitted into evidence over the objection of defense counsel, as was the testimony of two police interrogators about a prior oral confession. Miranda was convicted.

The question in this case is the admissibility of statements obtained from a defendant questioned while in custody or otherwise deprived of his freedom of action in any significant way.

the extensive search of his home without a search warrant. Chimel was convicted.

The question in this case is whether the arrest-based, but warrantless search of Chimel's residence was justified.

In its decision, the Court commented:

> It is entirely reasonable for the arresting officer to search for and seize any evidence on the arrestee's person in order to prevent its concealment or destruction. And the area into which an arrestee might reach in order to grab a weapon or evidentiary items must, of course, be governed by a like rule. A gun on a table or in a drawer in front of one who is arrested can be as dangerous to the arresting officer as one concealed in the clothing of the person arrested. There is ample justification, therefore, for a search of the arrestee's person and the area "within his immediate control"—construing that phrase to mean the area from within which he might gain possession of a weapon or destructible evidence.
>
> There is no comparable justification, however, for routinely searching rooms other than that in which an arrest occurs—or, for that matter, for searching through all the desk drawers or other closed or concealed areas in that room itself. Such searches, in the absence of well-recognized exceptions, may be made only under the authority of a search warrant. The adherence to judicial processes mandated by the Fourth Amendment requires no less.

The Court's ruling on this question of the extent of an arrest-based search is that the search went beyond any constitutionally justified area and the scope of the search was, therefore, "unreasonable" under the Fourth and Fourteenth Amendments, and Chimel's conviction was reversed.

Case No. 2—Disclosure of Informant's Identity:
McCray v. Illinois,
386 U.S. 300 (1967)

Two police officers, acting on information from a reliable informant, observed petitioner McCray acting suspiciously and they arrested him for possession of illegal narcotics. The narcotics used in evidence against McCray at

criminal case to be a witness against himself, nor be deprived of life, liberty, or property, without due process of law; nor shall private property be taken for public use without just compensation.

Sixth Amendment—

In all criminal prosecutions, the accused shall enjoy the right to a speedy and public trial, by an impartial jury of the State and district wherein the crime shall have been committed, which district shall have been previously ascertained by law, and to be informed of the nature and cause of the accusation; to be confronted with the witnesses against him; to have compulsory process for obtaining witnesses in his favor, and to have the assistance of counsel for his defense.

Fourteenth Amendment—

All persons born or naturalized in the United States, and subject to the jurisdiction thereof, are citizens of the United States and of the State wherein they reside. No State shall make or enforce any law which shall abridge the privileges or immunities of citizens of the United States; nor shall any State deprive any person of life, liberty, or property, without due process of law; nor deny to any person within its jurisdiction the equal protection of the laws.

Although the transcript of the Court's opinion is excellent study material, the cases in this chapter are summarized for simplicity and clarity, and to highlight the Court's decision and comment in the majority opinion that is pertinent to the major issues of the decision.

Case No. 1—Search and Seizure:
Chimel v. California,
395 U.S. 752 (1969)

Three police officers secured an arrest warrant for the arrest of petitioner Chimel for a coin-shop burglary. At the time of arrest, in the Chimel residence, the officers searched the three-bedroom home and seized coins, medals, and other articles. This evidence was used against Chimel in his burglary trial, over objections by defense counsel that Chimel's Fourth Amendment right to privacy had been violated by

Case Law—Selected U.S. Supreme Court Decisions

selves and our posterity, do ordain and establish this Constitution for the United States of America.

Article III—The Judicial Department—

Section 1: The judicial power of the United States shall be vested in one Supreme Court, and in such inferior courts as the Congress may from time to time ordain and establish. The judges, both of the Supreme and inferior courts, shall hold their offices during good behavior, and shall, at stated times, receive for their services a compensation, which shall not be diminished during their continuance in office.

Article VI—Supremacy of the National Government—

This Constitution, and the laws of the United States which shall be made in pursuance thereof; and all treaties made, or which shall be made, under the authority of the United States, shall be the supreme law of the land; and the judges in every State shall be bound thereby, anything in the constitution or laws of any state to the contrary notwithstanding.

A review of the amendments to the Constitution that form the basis of the question in the cases summarized in this chapter will provide similar guidelines for a better understanding of the judicial determinations in which the wording of these constitutional provisions tend to shape judicial interpretations. The case summaries are concerned with the basic Fourth, Fifth, and Sixth Amendment rights and the due process clause of the Fourteenth Amendment. These amendments read as follows:

Fourth Amendment—

The right of the people to be secure in their persons, houses, papers, and effects, against unreasonable searches and seizures, shall not be violated, and no warrants shall issue, but upon probable cause, supported by oath or affirmation, and particularly describing the place to be searched, and the persons or things to be seized.

Fifth Amendment—

No person shall be held to answer for a capital, or otherwise infamous crime, unless on a presentment or indictment of a grand jury, except in cases arising in the land or naval forces, or in the militia, when in actual service in time of war or public danger; nor shall any person be subject for the same offense to be twice put in jeopardy of life or limb; nor shall he be compelled in any

Case Law—
Selected U.S. Supreme Court Decisions

Judicial behavior by members of the U.S. Supreme Court has, at various times in the last century, strengthened and weakened the confidence of the American people in this appellate court of last resort. In the same period, various major segments of the nation's population believed the Court's sympathies were with the rich, the disenfranchised, the criminal, labor, the railroads, federalists, state's rights champions, and liberals. Fortunately, there has also been an assumption—from time to time—that the nation's top court has been acting in the public interest.

The cases in this chapter were selected because they are landmark cases in areas of criminal justice administration in which police officers operate. In each case, the Court's majority opinion is decisional law. The decision in each case is the law on search and seizure, interrogation, disclosure, and the other topics of these cases.

It would be a disservice not to recognize the many forms of criticism directed at this Court and its judiciary, but it is a great disservice to allow such critical appraisal to interfere with the acceptance of the decisions of our highest court.

Prior to reading these case summaries, however, the wording of selected portions of the Constitution is useful in reviewing the main thrust of the Constitution, the establishing of a judicial department, and the supremacy of the national government. These selected excerpts from the U.S. Constitution are:

Preamble—
We the people of the United States, in order to form a more perfect Union, establish justice, insure domestic tranquility, provide for the common defense, promote the general welfare, and secure the blessings of liberty to our-

chapter

8

When	What	Where	Who
7:00 A.M., Feb. 19	Suspect Newman arrested. Recovered stolen property: (1) .45 autopistol, #C-124651 (2) credit cards, M. Smith	Seattle, Wash.	Seattle P.D. D. Downs (#116)
4:00–6:00 P.M., Feb. 20	Suspect's statement— "George story"	Seattle, Wash.	Sacramento, P.D. Detectives: J. Tracy (#109) H. Gardiner (#6)
9:00 A.M., Feb. 24	Identification of .45 autopistol #C-124651 by owner	Sacramento	David Snyder
9:30 A.M., Feb. 24	Identification of credit cards recovered at time of arrest from suspect Newman	Sacramento	Morton G. Smith
10:15 A.M., Mar. 3	Identification of suspect Newman (from photos) as seller of tire chains (see Jan. 15)	Jacksonville, Ill.	Richard Pessing to Sacramento Detectives: J. Tracy (#109) H. Gardiner (#6)

A study of the time and motion data in this case connects the suspect with the victim, the crime scene, and with the victim's car during the postcrime period. "George" was not located, and—despite a searching police canvass—no witness could be located to "place" the suspect "George" at or near the crime scene on January 11, 12, or 13.

Review Questions

1. Itemize the evidence of this study on the ultimate issue of Newman's guilt or innocence.
2. Can this case be forwarded to the prosecutor with no more than the evidence specified?
3. Prepare a rough draft of the activities of suspect Newman as reported by Newman and reported by available witnesses.

made to Detectives J. Tracy (#109) and H. Gardiner (#6), Sacramento Police Department, from a selection of seven photographs.

The evidence structure of this case may be summarized in chronological order as follows:

When	What	Where	Who
9:00 A.M., Jan. 11	Telephone conversation with victim	Sacramento	Stella Hamlin
3:00–4:00 P.M., Jan. 12	Attempted to telephone victim	Sacramento	Stella Hamlin
4:00 P.M., Jan. 12	Conversation with victim's daughter	Sacramento	Stella Hamlin
6:30 P.M., Jan. 12	Victim's body discovered	Sacramento, victim's home	Bonnie Young (daughter)
7:00–8:00 P.M., Jan. 12	Missing property: victim's car, car keys, master key	Victim's home, basement garage	Sacramento P.D. Don Hall (#111), J. Sheridan (#401)
12:00 M., Jan. 13	Theft discovered of .45 autopistol	Sacramento, apartment in building managed by victim	David Snyder
Jan. 14	Autopsy report: death from strangulation; time: between 8:00 P.M. and 12:00 P.M., Jan. 11	Sacramento	Dr. J. Ryan
5:00 P.M., Jan. 15	Tire chains purchased from car with California license (See Mar. 3)	Jacksonville, Ill. (service station)	Richard Pessing
11:00 A.M., Jan. 20	Victim's automobile found	Springfield, Ill.	Springfield P.D. Clyde Olive (#59)
1:30 P.M., Jan. 20	Fingerprints lifted from victim's car	Springfield, Ill.	Springfield P.D. C. Wood (#7)

Case Studies for Analysis and Evaluation

Around January 6, Newman said he was at Casey's Bar on Sixteenth and O Streets in Sacramento, and a man named "George" approached him. "George" proposed that they commit a robbery and Newman agreed in principle. "George" then suggested they rob Newman's landlady. Newman was not agreeable to this until "George" made threats against him.

They met again several days later when "George" again threatened to harm Newman if he didn't help in the robbery of Newman's landlady. Newman finally agreed and the plan was for Newman to admit "George" into the apartment house where "George" alone would commit the actual robbery. They further agreed to take the landlady's car for their getaway.

On the night of the robbery Newman was in the landlady's apartment at 8:00 P.M. to determine if the circumstances would permit the robbery. After being in her apartment about half an hour he left and let "George" in the back door of the apartment house. Newman had already packed his bags, and he went to his own apartment to get them. He heard "George" knock at the landlady's door. He got his bags and went directly to the basement where he found the landlady's car was parked. By using a coat hanger through the wing window he unlatched the door, got in, and waited for "George." After about a half-hour wait "George" came to the car and gave Newman the keys. They both left in the landlady's car.

Newman described their route as Nevada, Salt Lake City, and some unknown towns in Colorado, Kansas, Missouri, and, finally, into Springfield, Illinois. Whenever they stopped "George" would either leave the car prior to stopping or lie concealed under a blanket. "George" and Newman split up after abandoning the car in Springfield.

Newman described "George" as an adult white person, approximately twenty-eight to thirty years old, about five feet eleven, or six feet, medium build, 170 to 180 pounds, dark hair, brown eyes, either Spanish or Italian descent.

8. Richard Pessing, a service station attendant in Jacksonville, Illinois, states he bought chains on January 15th, at about 5:00 P.M. from a person he now identifies as Barry Newman. Pessing said Newman was in a car with California license plates, and drove into his station with a "hard-luck" story and without money. Identification

Case Study B

were not answered. She called the victim's daughter at 4:00 P.M. to tell her she felt something was wrong in her mother's apartment.

2. Bonnie Young, the victim's daughter, corroborates phone call from Mrs. Hamlin at 4:00 P.M., January 12. She checked her mother's apartment at 6:30 P.M. and found her mother dead in the bedroom.

3. David Snyder, a tenant in the apartment house managed by the victim, left his apartment about noon on January 11. He returned at 6:00 A.M. on January 12. He then noticed—but did not worry about—a table in his living room that appeared to have been moved. He examined the doors and windows of his apartment but found no sign of forcible entry. On January 13, at about 12 noon, he saw a policeman in the hall of the apartment house and checked his belongings. At this time he discovered that his .45 cal. pistol (#C-124651) was missing and contacted police. After Newman's arrest this witness on February 24, at 9:00 A.M. identified the .45 cal. pistol found when Newman was arrested as the weapon stolen from his apartment.

4. Dr. J. Ryan who performed the autopsy gives cause of death as manual strangulation, and time of death as between 8:00 P.M. and 12 midnight, January 11. Report is dated January 14.

5. Charles Wood (#7), of the Springfield, Illinois, Police Department, lifted on January 20, at 1:30 P.M. fingerprints from the back of the mirror in the victim's automobile and from a No-Doz box found in the car's glove compartment.

6. Morton Smith, a Sacramento salesman, who had reported credit cards stolen from his residence in Sacramento prior to the death of Mrs. McDonald, identified on February 24, at 9:30 A.M. credit cards recovered at the time of Newman's arrest as the cards stolen from him.

7. J. Tracy (#109) and H. Gardiner (#6), Sacramento detectives, can identify the statement of Newman that implicates him in this homicide. It is known as the "George Story" and was made in Seattle, Washington, to these detectives at Newman's questioning in the conference room of Seattle's city jail on February 20, between 4:00 and 6:00 P.M.[1] This is the "George Story":

[1] Newman was warned of his constitutional rights (silence, legal counsel, and use in court of any statement); he signed a waiver of these rights.

Case Studies for Analysis and Evaluation

were found), but the master key to the apartment, which the victim kept pinned to her belt with a safety pin, was missing. In addition, police found the victim's automobile had disappeared from the garage below the apartment and the car keys, normally kept in her apartment, were also missing.

On January 13, a tenant (David Snyder) in the apartment house that had been managed by the victim, reported the theft of a .45 autopistol (#C-124651) from his apartment. Investigation by police established the time of the theft as between noon, January 11, and 6:00 P.M., January 12. No signs of forced entry were reported.

Seven days later, on January 20, the landlady's automobile was discovered abandoned in a parking lot in Springfield, Illinois, by patrolman Clyde Olive (#59). The Springfield police reported that the car doors were locked, with the exception of the wing window on the driver's side; that there were no marks to indicate the car had been "hot wired" to permit its use without a key; and that fingerprints were found on a box of pills in the car and on the back of the rearview mirror.

Police investigation at Sacramento disclosed a prime suspect named Barry Newman. He was a tenant in the building managed by the victim and he was missing.

A murder complaint was filed against Newman and a warrant for his arrest issued. He was arrested in Seattle, Washington, by police officer David Downs (#116), on the morning of February 19, at 7:00 A.M. He was returned to California for trial on February 20. In his possession at the time of his arrest was an automatic pistol (#C-124651), and oil-company credit cards in the name of M. G. Smith.

Numerous witnesses were located and interviewed by police officers during this investigation. The names of the principal witnesses, together with the testimony each witness is expected to offer in court, are listed below:

1. Stella Hamlin, a tenant, called the victim, who was the apartment-house manager, between 9:00 and 9:30 A.M., January 11, and talked to her over the phone for ten to fifteen minutes. The victim was alive, apparently normal, and not distressed. This witness is the last person, other than the killer, known to have talked to the deceased. Between 3:00 and 4:00 P.M. on January 12, witness tried to telephone the victim, but the calls

Case Study B

A review of the time and motion data indicates the connecting evidence identifying the suspect with the victim. A review of the conflict between the suspect's story of his actions on the day of the Brown girl's disappearance and on subsequent days indicates no witnesses could be located to support his story, but police were able to locate numerous witnesses who disputed his story.

Review Questions

1. Itemize the chain of circumstantial evidence in this case study.
2. Is this case study weak because of the lack of eyewitness identification? Can it be forwarded to the prosecutor with no more than the evidence specified?

Case Study B

On January 11, 1971, at approximately 9:00 A.M., an elderly widow named McDonald, who resided in and managed an apartment house in Sacramento, California, had a telephone conversation with one of her tenants. The following day, between 3:00 and 4:00 P.M., the same tenant was unable to reach her and therefore called Mrs. McDonald's daughter. No one had any known contact with the landlady after her telephone conversation with the tenant.

At 6:00 P.M., in response to the above tenant's telephone call, the landlady's daughter attempted to enter her mother's apartment but found the door locked. She secured a passkey hidden in a basement room and gained entrance. At 6:30 P.M., she discovered her mother's body covered with a bedspread in the only bedroom of the apartment. (Later, on January 14, an autopsy surgeon reported to police that Mrs. McDonald had been garroted and had died as a result of asphyxiation.)

Police officers at the scene (Don Hall, #111, Joseph Sheridan, #401) reported no signs of forcible entry to her apartment could be discovered. The apartment had not been ransacked (thirty dollars in cash and some valuable jewelry

that White resembled the man in the car. (Tunnel Road leads to the "Orinda Crossroads," and from this point several roads lead to Highway 40, which may be used to reach the area in Trinity County in which the suspect's mountain cabin is located.)

White was seen near the cabin about 10:00 A.M. on the morning of April 29, by a neighbor (witness Holly); and at Wildwood Inn, a nearby tavern, from 2 P.M. until midnight of that day (witnesses Wreath and Forest).

He was with relatives at the cabin on April 30, and he drove home on the afternoon of May 1, taking a route that took him over Franklin Canyon Road, where Stephanie's French textbook was discovered about 7:30 A.M. the next morning.

A resume of the major items of evidence in the order of their discovery is illustrative of the evidence structure in this case:

When	What	Where	Who
May 2	French textbook found	Contra Costa County	J. Jones
July 15	Victim's purse and wallet, found in a cardboard box	Suspect's home (basement)	Suspect's wife
July 16	Victim's glasses, brassiere, and other textbooks found	Suspect's home (basement)	Police search crew
July 20	Victim's body discovered—panties around neck, other clothes (blue sweater, white sweater, slip, and blue skirt) on body	Shallow grave in Trinity County	Police search crew
July 20	Bloodstained tissue fragments found	Trinity County (rat's nest)	Police search crew
July 21	Hair, fibers, and blood traces found	Suspect's automobile	Police search crew

Case Study A

for several days prior to April 30, but there had been very little rain in the area during May, June, and July.

The police search team also found several fragments of bloodstained cleansing tissue, which had been carried by a pack rat from the grave site to a nearby nest.

On July 21, at 8:10 A.M., a police search of White's car led to the discovery of two hairs which were indistinguishable from Stephanie's, and six hairs that were very similar to her hair. Eighteen fibers matching those in four of her garments were also found in White's car. Police also found traces of blood deep in the floor mat in the back of this car, and the absence of blood on the surface indicated that the mat had been washed.

At the time of these events, John White was twenty-seven years old and was attending the University of California in Berkeley. He was a regular customer at a doughnut shop that was located less than a block from the school Stephanie attended. This shop was frequented by pupils from that school, and Stephanie occasionally made purchases there. When questioned by the police, White said that on April 28, the day of Stephanie's disappearance, he left his home in Alameda about 10:45 A.M. and drove to a mountain cabin in Trinity County, which he and his wife own. After his arrest, White stated that he was not in Berkeley on April 28, that he started for the Trinity County cabin from his home in Alameda about 10:45 A.M. and, en route, stopped at a restaurant (location unknown) about 3 P.M., where he was served by a waitress who had dusty blonde hair and was twenty-five or thirty years old. He said that he also stopped at the Wildwood Inn for a drink about 8:30 P.M., and that he then drove two miles to the mountain cabin, built a fire, and went to bed.

Witnesses who observed White at various places offer stories in conflict with his story. He was seen, on this date, at the state controller's office in Oakland at 1:30 P.M. (witness Smith); at the beauty shop in Berkeley where his wife worked at about 2:30 P.M. (witnesses Pope and Priest); and at the doughnut shop about 3:20 P.M. (witness Duncan).

One of the two persons (witness Strep) who witnessed the struggle between a man and a young girl on Tunnel Road near the Broadway Tunnel at 4:15 P.M. identified White as the man in the car. The other person (witness Medle), stated

tion. This is the chronological record of the development of this investigation:

On May 2, 1971, at 1:00 P.M., four days after Stephanie's disappearance, her French textbook was found by a highway patrolman (J. Jones) beside Franklin Canyon Road in Contra Costa County. Except for the fact that its cover was slightly dampened by dew, the book was clean and dry, although it had rained in the area on April 29 and 30.

On July 15, at about 7:00 P.M., Stephanie's purse and wallet were found in Alameda at the home of Mr. and Mrs. John White. Mrs. White discovered the articles in a cardboard box in the basement and, upon reading the identification cards that were in the wallet, she went upstairs and excitedly asked her husband John and others who were present if Stephanie Brown was not the name of the girl whose disappearance had been reported in the newspapers. Mr. John White said that the purse probably belonged to some friend of Mrs. White's. A guest suggested that the police be called, and this was done. On the following day, July 16th, as a result of a search conducted by the police in the White's home, Stephanie's glasses, brassiere and the rest of her books were found buried in the basement in eight inches of sand.

On July 20, at 11:00 A.M., a search party discovered Stephanie's body in a shallow grave about 300 feet from a cabin owned by Mr. and Mrs. White in Trinity County, California. Her panties, which had been "cut or torn" through the left side and the crotch, were knotted around her neck. The rest of the clothing Stephanie was wearing on the day of her disappearance (except for her brassiere) was on the body.

Because of extensive decomposition, it was impossible to determine by a physical examination whether Stephanie had been sexually attacked. The body had been buried while in a state of rigor mortis, and the victim's arms and hands were raised in front of her face. There were multiple compound fractures of the skull and two holes about two inches in diameter through the skull. The head injuries were the principal cause of Stephanie's death.

Particles of soil had become enmeshed in her cardigan sweater, and it could be inferred that the soil was wet when she was buried. It had rained and snowed near the cabin

Case Study A

proof of many issues likely to be argued at trial depends on circumstantial evidence. In fact, proof of the essential elements of the crime charged and the common issues of motive or disposition to commit the crime charged, the criminal "agency," and "presence" at the crime scene or the opportunity to commit the specified crime depend on a combination of direct and circumstantial evidence.

Readers must also bear in mind that all testimony may undergo change under cross-examination. Although police officers are not expected to be fully acquainted with the order of proof and permissible questions, they are expected to examine witnesses and their expected testimony for the main weaknesses of perjury, bias, and half-truths.

In reviewing these two cases the police area of decision making that is part and parcel of the fundamentals of evidence can be observed. This is the question for a decision: Is there sufficient legally significant evidence to forward these cases to the prosecutor for trial?

Case Study A

On April 28, Stephanie Brown, a shy fourteen-year-old honor student at Willard High School in Berkeley, California, disappeared while walking home from school along Ashby Avenue. Stephanie was carrying several books, including a French textbook, and a purse which contained a wallet and a pair of glasses. She was wearing, among other garments, a navy blue cardigan sweater over a white slip-on sweater, a blue cotton skirt, a petticoat, nylon panties and a brassiere.

About 4:15 P.M on the day Stephanie disappeared, two motorists saw a man struggling with a young girl in a car that had stopped suddenly at the side of Tunnel Road, near the Broadway tunnel, a few miles north of the Claremont Hotel. The girl appeared to be very frightened and was screaming. She was in the back seat of the car, and the man, who was leaning over the front seat, was beating her and pulling her down and away from the rear window. She was wearing a navy blue cardigan garment over something white.

Originally police recorded the Brown girl as a missing person. Later, the case became an active homicide investiga-

Case Studies for Analysis and Evaluation

Some persons are incompetent to be witnesses, others may claim a "privileged communication" as a barrier to their testimony, and a few are privileged not to testify because of Fifth Amendment constitutional guarantees against self-incrimination. Some evidence is inadmissible because it is irrelevant or immaterial to the disputed facts in issue during a criminal trial; other evidence is inadmissible because the principal rules of evidence usually block testimony classed as hearsay, opinion, secondary (writings), or of little value and likely to confuse or mislead the trier of fact. Finally, a certain amount of evidence when objected to will be excluded because it was collected illegally, usually in violation of Fourth, Fifth, or Sixth Amendment rights.

The two case studies in this chapter present the evidence collected prior to trial and that foreshadows its use at trial. This is an extension of case preparation by police prior to turning a completed investigation over to the prosecutor for review and the decision to charge—to move the case into the courts for prosecution.

In reviewing these case studies and the array of evidence in each case, it is apparent that many facts less than the ultimate issue of guilt or innocence are in dispute. In reading each case, the reader must see clearly which disputable points are at issue. The disputable propositions of fact likely to be argued must be identified first; then the available evidence that is relevant on each disputable proposition can be examined.

These cases are intended to increase the reader's ability to analyze and evaluate any evidence collected in a criminal investigation. Each case is presented without "traps" of any kind. As in the real-life cases of police and prosecutor, however, neither case presents an abundance of relevant and reliable information. There is some direct evidence, but

chapter

7

5. What is the difference between legal insanity and diminished mental capacity?
6. Define the total-evidence concept and the police officer's primary responsibility in collecting evidence.

The Total-Evidence Concept

A police officer's primary responsibility is to collect all of the evidence he can uncover about the true facts of a criminal event and the guilt or innocence of the person or persons involved.

Evidence collected during an investigation must be reviewed before a case is prepared for prosecution. All evidence (both testimony and articles or objects of evidence) that points to the innocence or guilt of the suspect as the perpetrator of a particular crime must be gathered and studied. There are two basic questions to structure this review of the evidence in any case. Is there sufficient evidence of guilt? Is there significant evidence that will prevent the defendant from successfully asserting one of the common defenses?

Prior to recommending prosecution, police must be certain that there is sufficient legally admissible, available, and credible evidence to overcome the common defenses to crime, and to prove the offense charged beyond a reasonable doubt. This preprosecution review is the total-evidence concept of police.

The burden of proof on the ultimate issue of guilt or innocence is not placed upon the defendant in criminal cases. Guilt beyond a reasonable doubt is the result of convincing and compelling evidence presented in court at trial by the police and prosecutor.

Review Questions

1. Define pretrial discovery.
2. Define disclosure. When is it likely that police will be required to disclose the identity of an informant?
3. What is the difference between a defense to an accusation of crime based on lack of intent and a defense of being elsewhere at the time of the crime?
4. Define entrapment.

a breach) and neither participant can consent to a breach of the peace so as to thwart a criminal prosecution.

Mental Disorder

There are two defenses that may be urged on the basis of a mental disorder at the time of the crime. Legal insanity at the time of the crime will exculpate the defendant from legal responsibility for the crime. When the mental disorder is not of such a magnitude as to show legal insanity, a claim can be made as to the diminished mental capacity of the defendant to hold certain necessary states of mind (intent) due to his mental disorder.

The test for legal insanity most common in the United States is this: Was the accused at the time of the crime laboring under such a defect of reason, from disease of the mind, as not to know the nature and quality of what he was doing—or, if he did know it, as not to know he was doing what was wrong? Many states have modified this basic test for legal insanity. Modifications are:

1. A defendant is not criminally responsible unless he had sufficient mental capacity to know and understand what he was doing, and unless he knew and understood that it was wrong and a violation of the rights of others.
2. A person is not responsible for criminal conduct if at the time of such conduct, as a result of mental disease or defect, he lacked substantial capacity either to appreciate the criminality (wrongfulness) of his conduct or to conform his conduct to the requirements of law.
3. An accused is not criminally responsible if his unlawful act was the product of mental disease or mental defect.

The doctrine of diminished capacity recognizes that there is a mental disorder, short of legal insanity, that is disabling and may reduce the responsibility of a person for a crime. It has long been settled that evidence of diminished mental capacity, whether caused by intoxication, trauma, or disease, can be used to show that a defendant did not have a specific mental state essential to an offense.

The Issue of Guilt vs. Innocence

only intent necessary for the commission of a crime is the intent to do the act, then mistake may not be a defense.

Entrapment

The basis for the defense of entrapment is evidence that will show the crime or plan originated in the mind of a police officer or an associate. When a plan to break into a building originates in the mind of a burglar and the police go along with it in order to gain evidence of the crime, there is no entrapment. On the other hand, if the offense were planned by a police officer and he procured its commission by the defendant (who would not otherwise have committed the crime except for the trickery, persuasion, or fraud of the officer), this is entrapment.

The mere fact that the officer gives the defendant an opportunity to commit the crime is not an entrapment. Thus if the officer gives a seller of narcotics an opportunity to ply his trade, there is no entrapment. Such circumstances are similar to the cases in which the officer plays the role of an unconscious drunk in order that he can catch "drunk rollers" at their business. There is no entrapment when the police agents play such passive roles.

Consent

Consent is a defense when the crime itself requires that the act be against the will of the victim. In the crime of rape the evidence must show that intercourse was against the will of the victim. An essential element of the crime of robbery is that the personal property is taken "against the will" of the victim. Therefore, in such crimes any proof of consent is a defense. Consent, however, is not a defense in sex crimes against children, and murder may not be consented to. Evidence that the victim asked to be molested or killed will not be admitted, since it is not relevant or material to the charge. In some crimes, the circumstances of the crime determine whether consent is or is not a defense. For instance, no evidence of consent may be used for a fight on the public streets, no matter whether mutually agreed to, because it is also a breach of the peace (or had a tendency to create such

Discovery and Disclosure

crime. It is a possibility in every case in which the defendant is not caught in the very act of his crime. Witnesses testifying to an alibi may vary from a mother or wife to a friendly bartender, or an individual unknown to the defendant previously, but who is now willing to corroborate his story because of some chance encounter at the time of the crime. In addition to alibi witnesses, there may be evidence of such things as receipts, tickets, or other time-placing memoranda. One item of alibi evidence that is seemingly credible and unimpeachable may be and many times is sufficient to create a reasonable doubt of guilt.

Lack of Motive

Evidence indicating lack of motive by a defendant is in conflict with any prosecution evidence of identification. Lack of motive is important in cases in which direct evidence is weak, and only a chain of circumstantial evidence connects the defendant with the crime. When the defendant can develop lack of motive to support a claim of innocence, such circumstantial evidence is likely to create a reasonable doubt about guilt.

Intent

When an affirmative defense is based on the absence of necessary intent, the accused will generally admit the physical act of the crime while denying a criminal state of mind. There must be a union (or uniting) of the physical act and a criminal intent (or a gross negligence) in order for a crime to be committed. An act without intent or an intent without an act are generally not crimes. Therefore, evidence that will tend to infer intent or lack of it is admissible and most important. Intoxication alone, if to a sufficient degree, will be evidence that the accused could not and did not form a criminal intent. Ignorance of fact or of law may be accepted to show the absence of intent. Mistake will generally be admissible evidence on the issue of criminal intent if honestly entertained and based upon reasonable grounds, or if of such a nature that the conduct would have been lawful had the facts been as they were reasonably supposed to be. If the

The Issue of Guilt vs. Innocence

nature of the conversation between the officer and himself can be examined. A police officer, however, is reluctant to disclose his informant. When identity is disclosed, the informant is a likely target for attack by friends and associates of the defendant.

The identity of a police informant must be disclosed when the informant is a participant in the criminal transaction. The informant-participant is not a mere informant but rather a material witness to the criminal act. The refusal to identify such an informant would deprive the defendant of a fair trial, as the informant-participant is likely to be a material witness on the ultimate issue of guilt or innocence, and to deny the defendant his identity deprives defense counsel of the right of cross-examination inherent in the defendant's right to confront the witnesses against him.

Disclosure of the identity of an informant is not required when the informant's participation is limited to establishing probable cause for an arrest or a search. The police officer making the search or arrest must establish that the informant was known to him as a reliable source of information.

Mistaken Identity

The basic defense of mistaken identity alleges the inability of witnesses to see or accurately convert their perception into an identification. It is a vast area open to attacks on the credibility of witnesses and the integrity of police lineup procedures. In *Wade* v. *United States*[1], the U.S. Supreme Court rejected an in-court identification of the defendant by two eyewitnesses. Both witnesses had reviewed photographs of several suspects, including the defendant. Subsequently and prior to the trial both witnesses picked the defendant out of a police lineup of five or six persons. The judicial reasoning in this case was that these procedures were suggestive of the identity of the defendant and tainted the later in-court identification.

Alibi

An alibi is a claim of being elsewhere at the time of the

---

[1] See List of Cases, p. 71.

the opposition. Disclosure in criminal cases is a defense request that forces the prosecution to identify an informant. The rationale for discovery and disclosure in criminal cases is the concept of a fair trial. It is a balancing between the rights of accused persons and the interests of the community.

Pretrial discovery offers an opportunity to inspect an opponent's evidence. Not all jurisdictions have established extensive areas of discovery, but in most jurisdictions the defendant can request production of one or more items of evidence in the hands of the prosecutor or police by making an appropriate request in the court having jurisdiction to try the case. Police must expect discovery and learn to cope with its problems. It should be expected that the rules of pretrial discovery will require the prosecutor to produce defendant's statements and any police reports; photographs or sketches that record the criminal act for which the defendant is charged; physical evidence and the determinations of laboratory technicians; reports of identification by eyewitnesses; lists of other witnesses; and the contents of statements made by witnesses, if any.

Discovery has a potential for perjury and the intimidation or elimination of witnesses. Pretrial discovery alerts the defense to damaging evidence and could lead to a variety of illegal reactions, from subornation of perjury to eliminating witnesses. These hazards can be neutralized, however, by cross-examination and prompt judicial action.

The prosecutor has lesser discovery rights. For instance, when the defense has communicated in some fashion its intent to use an affirmative defense such as an alibi, the prosecution must receive adequate notice. However, extensive pretrial discovery by the prosecution is blocked by the defendant's right to remain silent and his privilege against self-incrimination. The prosecution's rights to discovery can hardly be expanded to conflict with the absolute right of an accused person not to bring forth evidence that will incriminate him.

Disclosing the identity of informants is replete with problems similar to pretrial discovery. The accused certainly has a right to confront prosecution witnesses. However, police have a right (in the public interest) to protect the flow of information about crime and criminals. Defense counsel wants the informant identified so he can be compelled to appear as a witness at the trial, where his reliability and the

The Issue of Guilt vs. Innocence

The Issue of Guilt vs. Innocence

On the issue of guilt or innocence, witnesses will be called to testify for both parties to a criminal action. Proffered evidence will be objected to by opposing counsel on the grounds that its admission requires proof of a preliminary fact, and the defendant fights to secure an acquittal and a rejection of the accusation made against him by police and prosecutor. An accused person does not have to prove his innocence, but he does have to develop evidence that will create a reasonable doubt of his guilt.

The circumstances of an accusation of crime differ. The nature of the case and its attendant circumstances structure the prosecution and the defense case. Pretrial discovery is a court process helpful to an effective defense, as it affords the defense a preview of the prosecution case. Disclosure procedures that force police to reveal an informant's identity are also helpful in some cases as the informant can be brought to court and questioned about the information he gave to the police.

Basic defenses to an accusation of crime range from an attack on identification evidence, through claims of lack of intent or entrapment, to defenses of the consent of the victim or a claim of insanity. The prosecution brings to court affirmative evidence of guilt and negative evidence of innocence, accepting the burden of proof on the issue of guilt and attempting to develop evidence that will not only prove guilt but which will also block the common defenses to accusations of crime.

Discovery and Disclosure

Discovery is a pretrial procedure in which an opposing party requests the production of evidence in the possession of

within the meaning of the Fourth Amendment. The protection afforded by the Fourth Amendment does not turn upon the presence or absence of a physical intrusion into any given enclosure, but upon an intrusion of the privacy of an individual. The police officer must develop in his application for this court order: (1) the identification of the subject of the electronic surveillance, (2) the place to be "searched" (wire or oral communication and its designation by location or other means), and (3) the description of the communications that are the objective of the surveillance.

Review Questions

1. Define the exclusionary rule and specify its objective.
2. What is the rule of derivative evidence?
3. Is a "consent" search reasonable? When?
4. Define the "reasonableness" necessary for a lawful search. Also for a lawful arrest.
5. Is the concept of a reliable informant helpful to police? Why?
6. Define a confession. An admission. Do coerced confessions have an inherent untrustworthiness?
7. What is "the *Miranda* warning"? What are its essential elements?
8. Define electronic eavesdropping. When is prior judicial approval necessary for monitoring conversations?

The Admissibility of Electronic Surveillance Evidence

Evidence from electronic surveillance is the overheard content of wire and oral communications as well as the identity of the participants. Constitutional protection against evidence secured by electronic surveillance appears to focus on the right of privacy, and the aggrieved person is entitled to the suppression of such evidence when the police procedures used violate the Fourth Amendment's protection against unreasonable searches.

A recording device is now an acceptable instrument to bolster the credibility of a witness. There must be proper police supervision to see that the recording is not faked in any way, and every precaution must be taken to make certain the recorder is used in a fair manner. This use of an electronic device by one party to a conversation to record it is not eavesdropping in any proper sense of the word. It is best described as the use of an electronic device to obtain reliable evidence of a conversation by one of the participants. It is the recorded memory of a witness who participated in the conversation.

Each party to a telephone conversation takes the risk that the other party may have an extension telephone and may allow another to overhear the conversation. The communication itself is not privileged and one party may not force the other to secrecy merely by using the telephone. Either party may record the conversation and divulge it.

Electronic eavesdropping, wiretapping as well as "bugging" (use of concealed microphones), may produce admissible evidence. Failure to secure prior court approval of an electronic surveillance is likely to ruin any future admissibility of overheard conversations. In *Katz* v. *United States*[9], "antecedent judicial justification" of an electronic surveillance was established as a standard requirement except in emergencies. Police must apply in court, using the same procedure as common to applications for arrest or search warrants, for judicial authorization. Judicial review as to the reasonableness of the police request requires a particularity of inquiry, overall reasonableness, and justification for the invasion of privacy. An electronic surveillance is a search and seizure

[9]See List of Cases, p. 69.

mere fact that he may have answered some questions, or volunteered some statements on his own, does not deprive him of the right to refrain from answering any further inquiries until he has consulted with an attorney and thereafter consents to be questioned.

There is a judicial doctrine that involuntary confessions shall not be admitted into evidence against the person making them, not only because no one should be compelled to incriminate himself, but also because of the unreliability factor of a coerced confession.

A confession is considered involuntary when it is induced by promise and hope of reward or benefit, judicial compulsion, or coercion (violence, threats, or fear).

Promises and Hope of Reward

Promises entailing dependence on them by the accused person must be made by a person in authority. However, almost any employee of a criminal justice agency has been defined as a person in authority.

Judicial Compulsion–Guilty Pleas

Judicial compulsion is inherent in the sentencing function of the judiciary. For this reason, there must be a showing that judicial confessions (pleas of guilty) are not influenced by hope of leniency, fear, or lack of knowledge by the defendant of all the implications of such action.

Coercion

On the issue of coercion, the question in each case is whether the defendant's will was overborne by violence, threats, or fear at the time he confessed. This inquiry involves two major areas: ascertaining the occurrences and events surrounding the confession and evaluating whether these facts support the defendant's claim of coercion. Coercion exists when the behavior of questioners overbears the defendant's will to resist and brings about a confession not freely self-determined.

Evidence in Action

(*McCray* v. *Illinois*[7]). There must be a substantial basis for crediting such hearsay evidence, to insure its trustworthiness. The police officer must state under oath that his basis for accepting the informant's story is that the informant had previously given accurate information and his story was corroborated by other sources of information.

The Admissibility of Confessions and Admissions

Any statement given freely and voluntarily without any compelling influences is admissible in evidence. A confession is a statement by a person accused of crime saying that he is guilty of the specific crime with which he is charged. An extrajudicial confession is a statement made out of court to any person. A judicial confession is made in court. It is usually a plea of guilty. An admission is less than a confession. The facts admitted as true only raise the inference of guilt when viewed in connection with other evidence.

The admissibility of any confession or admission secured during police custodial interrogation is threatened by failure to warn the accused person in police custody of his constitutional rights to silence (Fifth Amendment) and legal counsel (Sixth Amendment). This is the ruling of the U.S. Supreme Court in *Miranda* v. *Arizona*[8]. It is now known among police as "the *Miranda* warning" and its use is common throughout the United States. Prior to any police questioning, the accused person must be warned that he has a right to remain silent, that any statement he does make may be used as evidence against him, and that he has a right to the presence of an attorney, either retained or appointed. The person about to be questioned may waive his rights, provided the waiver is made voluntarily, knowingly, and intelligently. If, however, the accused person indicates in any manner and at any stage of the process that he wishes to consult with an attorney before speaking, there can be no questioning. Also, if such individual indicates in any manner that he does not wish to be interrogated, police may not question him. The

[7]See List of Cases, p. 69.
[8]See List of Cases, p. 70.

the Fourth Amendment by virtue of its intolerable intensity. The intensity of a search likely to be classed as unreasonable is illustrated by a California case in which the stomach of an arrestee was pumped out to recover two capsules of illegal drugs (*Rochin* v. *California*[4]). There is little doubt that stomach pumping is unreasonable. However, taking a blood sample from an unconscious prisoner was termed reasonable because the method was less drastic than stomach pumping (*Schmerber* v. *California*[5]).

The stop-and-frisk routine of police is now classed as a reasonable police procedure. A police officer is alerted to suspicious conduct and stops the suspect, "frisking" him because it is not unreasonable, under the majority of the circumstances of these stops, to fear the suspect is armed and may attack his questioner. To justify this type of intrusion, the police officer must be able to point to specific and articulate facts which, taken together with rational inferences from those facts, reasonably warrant the intrusion. The U.S. Supreme Court, in *Terry* v. *Ohio*,[6] has established the objective standard to be followed in these cases as: "Would the facts warrant a man of reasonable caution in the belief the action taken was appropriate?"

The Reliable-Informant Concept

The reliable-informant concept is a justification for police to act in circumstances that did not originate in their own personal knowledge. The doubtful character of the so-called reliable informant and dubious motives for informing often overshadow the informant's record for reliable performance. To be classified as a reliable informant, a person should be a steady source of information to a police officer. The informant's reliability is then personally known to the police officer, and the officer is prepared to state that the informant has given similar information on past occasions, and that this information has proven to be accurate.

An affidavit for a warrant is sufficient when it is based on the hearsay evidence of a reliable informant's observation

---

[4] See List of Cases, p. 70.
[5] See List of Cases, p. 70.
[6] See List of Cases, p. 71.

Evidence in Action

dence is made at the time of arrest, the circumstances must indicate both a lawful arrest and a search merely incidental to it. The scope of such search, to be reasonable, is now limited by the U.S. Supreme Court's decision in *Chimel* v. *California*[3] to the person of the arrestee and the area within the immediate control of the prisoner at the time of arrest. This is an area from within which the arrested person might gain possession of a weapon or of destructible evidence. A weapon, if discovered, can be seized, as the prisoner might use it to assault the officer or to effect an escape. Articles of evidence are seized to prevent their concealment or destruction.

Consent to enter a premises and to search therein is reasonable. It is justification for a warrantless search and the seizure of evidence, but it must be affirmatively shown that the consent is appropriately given and was secured freely and voluntarily. Consent must be secured from a person in control of the premises or authorized by the resident involved to grant consent to a search. Any trace of coercion ruins reliance on consent as authority for a search and seizure. When any doubt exists, application for a search warrant should be made to avoid tainting any evidence seized.

An application to court for a warrant to search or to arrest insures reasonableness. The person seeking a warrant must appear in person and justify his application under oath. Oaths and affirmations in support of warrants for arrest, or for search and seizure, must state facts with a sufficient definiteness to justify the issuance of a warrant. An affidavit for a warrant of arrest must contain information that, if true, directly indicates the commission of the crime charged and the person committing it. An affidavit for a search warrant must state a belief specific property or things related to a criminal offense are in a certain place.

An application for a warrant to search a place or to arrest a person can be made upon information and belief, as well as personal knowledge. The information and belief must indicate that the information has been actually received and emanates from a trustworthy and reliable source, and that the applicant believes the facts stated in his application and given under oath.

A search that is reasonable at its inception may violate

[3] See List of Cases, p. 68.

The Doctrine of Reasonableness

no matter how it was obtained, whether lawfully or unlawfully.

In 1961, the U.S. Supreme Court in its decision in *Mapp* v. *Ohio*[1] clearly established a nationwide exclusionary rule when it used the due-process clause of the Fourteenth Amendment, enacted after the Civil War, for the purpose of imposing the Bill of Rights limitations on state governments. The *Mapp* case involved a search and seizure of evidence by local police.

Additionally, another U.S. Supreme Court decision (*Silverthorne Lumber Co.* v. *United States*[2]), supportive of the exclusionary rule, states that illegal evidence shall not be used at all and evidence traceable to the excluded evidence is inadmissible. The barring of derivative evidence means that police may be required to reveal an independent source for evidence when it is alleged to be tainted through its origin in illegally obtained evidence.

The Doctrine of Reasonableness

Reasonableness is determined by balancing the need to arrest or search against the intrusion of privacy the search or arrest entails. Wherever a man may be, he is entitled to know that he will remain free from unreasonable arrest or searches and seizures. The issue of reasonableness is a question of whether the intrusion is justified by the evidence of criminality.

The legality of an arrest without a warrant is dependent upon its reasonableness. An arrest without a warrant bypasses the safeguards provided by an objective predetermination of probable cause and substitutes instead the far less reliable procedure of an after-the-event justification. Such justification is likely to be subtly influenced by the familiar shortcomings of hindsight judgment. The requirements of reliability and particularly of the information on which an officer may act in warrantless arrests cannot be less stringent than where an arrest warrant is obtained.

When a contemporaneous, arrest-based search for evi-

[1] See List of Cases, p. 69.
[2] See List of Cases, p. 70.

Evidence in Action

Evidence in Action

All too often, evidence is rejected by the trial court because it is tainted by police procedures forbidden by the judicial application of constitutional safeguards. A search or arrest, acting on unreliable sources of information, unreasonable interrogation, and other procedures infringing on the legal and moral rights of accused persons may invalidate evidence and assign it an inactive role in the criminal proceeding.

The Exclusionary Rule

The exclusionary rule is a judicial technique to deter police from violating the constitutional rights of individuals in gathering evidence by rejecting any illegally obtained evidence. Police collect evidence to prove, with the aid of the prosecutor, the guilt of the accused person. Their work, however, is wasted when the tainted evidence is suppressed before trial or not admitted into evidence at trial. The exclusionary rule compels respect for constitutional guarantees by removing the incentive to disregard them. This judicial technique gives to the individual no more than what the constitution guarantees him; to the police officer no less than that to which honest law enforcement is entitled; and to the courts, judicial integrity.

Before the growth of the exclusionary rule there was nothing to restrain agents of law enforcement in gathering evidence for use in court. Under the rule of common law, the admissibility of evidence was not affected by the illegality of the means by which it was obtained. If the evidence was relevant to the issue of guilt vs. innocence it was admissible,

chapter 5

Review Questions

1. Define physical evidence.
2. What is the basic concept of transfer evidence?
3. What is the role of in-court exhibits and demonstrations?
4. Is the taking of blood from a suspect a violation of any constitutional right?
5. Is a suspect, about to be viewed in a police lineup by eyewitnesses for the purpose of identification as a criminal, entitled to legal counsel?

Nontestimonial Evidence

to be. If it is offered as genuine, there must be proof of its integrity; and if it is offered as a forgery, there must be proof it is a forged writing. Some documents are self-authenticating and others require proof of signing, of the signature, or of a comparison of signatures.

The party producing a writing as genuine that has been altered (or appears to have been altered) in a portion material to the dispute must account for the alteration (or the appearance of being altered) by showing that the alteration was made by another without his concurrence, that it was made with the consent of parties affected by it, that it was otherwise properly or innocently made, or that it did not change the meaning or language of the instrument.

The Best Evidence Rule

The best evidence rule relates only to writings. It is the doctrine that the best evidence of the content of a writing is the writing itself. Copies of writings or testimonial evidence of their content is admissible only when evidence can be developed indicating that the writing itself cannot be brought to court—since it is lost, has been destroyed, or is otherwise unavailable.

Parol Evidence

The parol evidence rule forbids verbal evidence of a written contract. The evidence of the terms and content of a written contract is the writing that has been made and signed by the contracting parties. Within this signed contract is documentary evidence of the integration of the history of negotiating a contract with the written agreement between the contracting parties. When an agreement is placed in writing, all oral testimony of arrangements made or things said that would tend to substitute the unwritten for the written contract that has been agreed upon and signed, to the possible prejudice of one of the contracting parties, is rejected.

Associative Evidence

Photographs

Photographs must be relevant, accurately taken, and a correct representation of the subject portrayed. Photographs of a crime scene may not be admissible because the photograph does not truly represent the crime scene. Photographs of the victim of a crime are admissible even when gruesome and likely to prejudice the triers of fact. There must be some basis less than prejudice for their introduction, i.e., illustrating a wound, a significant portion of the crime scene, a particular view of the death weapon, etc.

Maps and Diagrams

Maps and diagrams are admissible in evidence when necessary for a better understanding of the testimony of witnesses. Under ordinary circumstances any drawing must be an accurate graphic representation, but an inaccurate sketch will be admissible if it is reasonably related to, and explanatory of, the testimony of the witness who drew it and uses it to illustrate his testimony. Drawings, diagrams, and sketches may be admissible when the subject matter is a crime scene or other place of importance in the crime and its circumstances, a portion of the human anatomy, the shape or location of a wound, or the kinds of marks on an item of evidence. Maps are generally used to illustrate testimony relating to the scene of a crime or travel to and from the scene.

Documentary Evidence

A "writing" means handwriting, typewriting, printing, photostating, photographing; every other means of recording upon any tangible thing; and any form of communication or representation, including letters, words, pictures, sounds, or symbols, or combinations thereof.

A writing must be authenticated before it may be received in evidence. Authentication requires the production of evidence that will show the writing is what it is supposed

The protection of the Fifth Amendment privilege against self-incrimination is concerned with the communications of an accused person. Blood or its extraction for chemical analysis is not in the area of testimonial compulsion. The search is reasonable, since the body intrusion cannot be delayed (because of the oxidation of alcohol in the human body with the passage of time) while application is made to court for a search warrant.

Eyewitness Identification

Eyewitnesses use the body of the person viewed and his general appearance as real evidence. It is direct evidence of prime importance when it places the defendant at the crime scene. The legal significance of eyewitness identification may be compromised by substandard police procedures. The police lineup is now a critical stage of the pretrial procedure at which the person accused of crime is entitled to the assistance of legal counsel (*Wade* v. *United States*).[2] The scanning of photographs of suspects by eyewitnesses is likely to be classed as suggestive and a procedure that may threaten the integrity of in-court identification.

Handwriting Evidence

Requests by police for samples of an accused's handwriting for the purpose of comparison with handwriting evidence collected in a criminal investigation do not constitute a threat to a fair trial. Handwriting, in contrast to the content of what is written, is an identifying physical characteristic. There is nothing suggestive to witnesses; and a police request for a handwriting sample does not violate the suspect's Fifth Amendment privilege against self-incrimination. This constitutional right only concerns the compulsion of an accused's communications and not compulsion that makes a suspect the source of real or physical evidence.

[2]See List of Cases, p. 71.

Associative Evidence

Associative Evidence

Associative evidence is a nonlegal term for physical evidence that has been examined by an expert criminalist and that has some capability for evidence by comparison. The theory of association is inherent in the concept of transfer evidence: some thing or trace is both left at the crime scene and carried away from it by the criminal. Associative evidence can reveal a relationship between physical evidence found at the crime scene and similar evidence located on the suspect, his clothing, or in his home or automobile. Associative evidence, when properly demonstrated as an exhibit (alone or supplemented by the opinion of an expert), is beyond the realm of opinion evidence and emerges as a new concept in the physical-evidence field.

Latent, partial, and plastic fingerprint impressions are associative evidence often found at crime scenes that may serve to identify the criminal as well as to establish his presence at the crime scene. Fingerprints used for the purpose of comparison with crime scene impressions are usually found among police records, the finger impressions being taken for the purpose of identification at the time of arrest. When fingerprints are not available for comparison, the safeguards of the Fourth Amendment apply to any intrusion that can be defined as a search. A reasonable intrusion would seek prior court approval or consist of a convenient fingerprinting session for the suspect, and it would not be compromised by any form of interrogation during the fingerprinting (*Davis* v. *Mississippi*).[1]

Blood Samples and Analysis

Blood is a body substance that is nontestimonial. It is not within the Fifth Amendment's protection against self-incrimination. There is nothing brutal or offensive in the taking of a blood sample by a physician or medical technician. Blood and its analysis is acceptable evidence for the purpose of identifying an intoxicated person.

[1] See List of Cases, p. 68.

Nontestimonial Evidence

Nontestimonial Evidence

Real evidence consists of tangible items submitted as exhibits or articles for in-court inspection. Testimonial evidence is used to introduce the exhibit or article, to identify it, and to connect it with the issue. Nontestimonial evidence is initially marked for identification (people's exhibit #1 for identification, defense exhibit #1 for identification, etc.). It is marked as evidence when the proper foundation is established for its admission. The triers of fact can then observe and evaluate the exhibit or article of evidence.

Physical evidence consists of things and traces. The application of scientific techniques to the problem of developing proof of guilt or innocence has contributed to the recognition of things and traces as valuable potential evidence. Physical evidence is found at the scene of a crime or on or about a suspect. It can be collected by investigators, evidence technicians, field criminalists, or medicolegal experts. Things collected as possible items of evidence by police and their technicians are weapons, imprints, impressions, "traces" of dust and dirt, microscopic particles of hair and fibres, and paint transfers. Medicolegal experts may isolate traces of a poison, discover identity clues, and develop evidence connecting wounds and the suspect weapon.

Witnesses skilled in the examination of physical evidence supplement their testimony with exhibits or demonstrations. The exhibits of these expert witnesses are planned to support their opinion. A demonstration is related to the testimony of the witness and is also supportive of his expert opinion. If it is an in-court demonstration, it must be simple and easily understood. Out-of-court demonstrations must be explained with coherence and unity.

chapter 4

Pedigree

Evidence of a statement by a person who is unavailable as a witness concerning his own birth, marriage, and other family history is admissible. When the individual making the statement did not have personal knowledge, the circumstances in which the statement is made will often indicate its trustworthiness. These statements are trustworthy if made at a time when no controversy existed about the matter stated. Relatives and intimate associates also qualify to testify about pedigree when the statement concerns the family history of a relative or close friend. Entries in family records, church records of family history, and similar memoranda may also qualify as proof of family history.

Review Questions

1. Define hearsay evidence.
2. Will the admission of selected hearsay evidence increase the chances of discovering truth in a criminal trial?
3. What are the essential elements of a "dying declaration"?
4. Are declarations against interest reliable? Are spontaneous (and contemporaneous) utterances usually truthful?
5. What is the common factor of written statements of past recollection, business records, and public records that indicates trustworthiness?

lent) stating that he did not find a designated public record or writing after a diligent search.

Statements of Co-conspirators

A statement of a co-conspirator made during the course of a conspiracy may be admissible on the issue of guilt or innocence in a trial for conspiracy. Such statements owe their admission to the rationale that they are acts of the conspiracy for which the defendant, as a co-conspirator, is legally responsible. The basic requirement is that the statement was made while participating in a conspiracy to commit a crime and in furtherance of the objective of that conspiracy.

Admissions

Testimony concerning an admission by a person accused of crime may be admissible when offered against such individual in an action in which he is a party and when the admission is not prohibited on constitutional grounds. An authorized or adoptive admission (authorized by the person making it or endorsed by his conduct) may also be admissible.

Declarations against Interest

Declarations against interest are admissible when *a.* the declarant has knowledge of the subject in issue and is not available as an in-court witness, and *b.* the statement (when made) was so far contrary to the declarant's financial or proprietary interest, or subjected him so far to the risk of civil or criminal liability, or so invalidated a claim against another, or created such disgrace, hatred, or ridicule in the community that a reasonable man in his position would not have made the statement unless he believed it to be true.

Declarations against Interest

a contradictory statement, may be confronted with prior inconsistent statements for the purpose of discrediting testimony given at trial; and when the credibility of a witness is attacked, such prior statements may be acceptable to show their content is consistent with the trial testimony of the witness.

Business Records

Business records refer to writings prepared in business, governmental activity, and any profession, occupation, or calling. Business records are admissible in criminal trials when the writing was made in the regular course of the business at or near the time of the act, condition, or event recorded. Information sources, routine of preparation, and the time of completion indicate trustworthiness. The special reliability of business records is that they are based upon first hand observation by someone who knows the facts and records them. The absence of an entry in business records is also acceptable to prove the nonoccurrence of an event (or the nonexistence of a condition) when it was the regular course of that business to make timely records of all such events or conditions and to preserve them. The absence of a record serves as a trustworthy indication that the event did not occur or the condition did not exist.

Public Records

Official public records and writings that record an event or condition may be admissible when offered to prove the act, event, or condition recorded. The witness must testify that the writing was made by him within his scope of duty as a public employee, and that it was made at or near the time of the event or condition reported. Again, sources of information, routine of preparation, and time of completion indicate trustworthiness. The absence of a public record or writing may also be admissible when a public official, on demand, provides an authenticated certificate (or its equiva-

is not used to prove any fact other than such state of mind, emotion, or physical sensation.

Written Statements of Past Recollection

A written statement made at the time when the facts of an event recorded in writing actually occurred or were fresh in the writer's memory is admissible for the same reason that spontaneous and contemporary statements are admissible. The writer-witness will be asked to testify that the written statement is a true and accurate record of the event, and that it was prepared by him or, under his direction, by some other person for the purpose of recording the statement of the witness. A witness may read into evidence the written statement when it concerns a matter the witness cannot now recall fully and accurately.

Prior Identification

Identification of a suspect by a witness while the crime or a related event was fresh in his memory is also considered a contemporaneous happening. Testimony about such prior identification is admissible, whether the witness admits or denies the prior identification. The prior identification must concern a person who participated in a crime or other occurrence; it is offered as evidence subsequent to the witness' testimony in relation to the identification. If the witness testifies contrary to his prior identification, it may be used to discredit his testimony at the trial. If he testifies that the prior statement was his true opinion at the time, the prior identification statement may be used as an earlier identification for its probative value during the trial.

Prior Statements of Witnesses

Witnesses who have not been unconditionally excused, and who have been given the opportunity to explain or deny

"Dying Declarations"

Evidence of a statement about the cause and circumstances of an injury is admissible in cases in which the circumstances of the injury constitute an essential element of the crime charged, provided the statement is based upon the victim's personal knowledge and is made under belief of impending death. The two key essentials in such declarations are the victim's personal knowledge above the level of suspicion or conjecture of the circumstances under which he received the injury and his belief in the imminency of death from such injury.

Spontaneous and Contemporaneous Statements

The inherent trustworthiness of spontaneous and contemporaneous statements is in their lack of premeditation. There is no time for reflection and deliberate fabrication and the statement is linked with action. These statements narrate, describe, or explain an act, condition, or event perceived and occur impulsively while a person is under the stress of excitement brought on by such perception; or such statements are made while an individual is engaged in some act and serve to explain, qualify, or make understandable the conduct of such person.

Statements of Mental or Physical State

Statements of a mental or physical state are admissible when they concern intent, motive, design, mental feeling, pain, or bodily health at the time. Such evidence is offered to prove a person's state of mind, emotion, or physical sensation at that time, or to prove or explain acts or conduct. The admission of such evidence is usually restricted to instances in which the previously existing mental or physical state itself is an issue in the criminal action, and the evidence

Hearsay Evidence

Hearsay evidence is testimony based on the authority of another, rather than the personal knowledge or observation of the witness. It is testimony about another's "story," or a story out of another's mouth. When offered to prove the truth of the matter stated, it is normally inadmissible. This is the "hearsay rule."

The primary reason for rejecting hearsay evidence, upon properly interposed objection, is that the party making the objection has the right to prevent the trier of fact from being improperly influenced by evidence that appears to be fair but that carries hazards that could be exposed or eliminated at trial if the declarant were present and cross-examined. The out-of-court hearsay declarant is not under oath, he cannot be observed by the triers of fact for clues as to his willingness or capacity to tell the truth, and his veracity cannot be tested by cross-examination. The accuracy of any evidence once removed from the in-court witness' personal knowledge is less than perfect.

Hearsay evidence is received in court when necessary to diligent inquiry, and when the circumstances under which the evidence is developed indicate a basic or inherent trustworthiness that acts as a substitute for the oath, personal appearance in court, and cross-examination. For instance, a declaration by a dying person is made under circumstances likely to produce a true statement.

Justification for the acceptance of hearsay evidence is based on the inability of prosecution or defense to locate and produce a witness, or on the fact that normal procedure for producing evidence has been stymied in some fashion.

chapter 3

show motive, intent, identity, absence of mistake, and a common scheme or plan. Evidence that the defendant had attempted on other occasions to have intercourse with the victim was admissible to establish a motive for the subsequent murder. Testimony that the accused shot the victim on a prior occasion was admissible to show an intent to murder in a trial based upon a second shooting of the same victim. When an identification by a police witness was questioned by the defense, the witness was allowed to testify that he had previously arrested the defendant. Prior sales of obscene material to minors was accepted as an inference that the defendant's sale of such material to a minor was no accident or mistake. Details of other thefts have been accepted as evidence where these acts show a similarity in pattern, a *modus operandi*, to the crime presently charged.

Review Questions

1. When may a witness refuse to testify?
2. Define a subpoena.
3. What personal knowledge is required of a witness?
4. When are opinions or conclusions admissible in evidence?
5. What is the basic difference between witnesses with limited expertise and expert witnesses?
6. What is the rule of privileged communications?
7. What similar characteristic can be identified in all of the relationships in which communications are privileged?
8. When is testimony about past crimes admissible? Why?

Testimony

client, clergyman and penitent, and physician and patient. Confidentiality of communications is an essential element in these relationships.

Spouses, by common law and most statutes, are either partially or entirely incompetent to testify against each other. The reason is basically promotion of the unity or the confidential and close relationship between husband and wife, and the fact that a different rule would tend to disrupt the home life of married couples. A usual exception to this rule is the crime perpetrated upon one spouse by the other, or upon the children by one of the spouses.

In the attorney-client relationship, privileged communications make certain a client can cooperate with his own attorney without fear of disclosure. Any lack of confidentiality would destroy the ability of the attorney to discuss a case with his client.

Full disclosure by penitents and patients is essential to proper religious or medical care. If such disclosures did not have confidentiality, fear of a criminal prosecution against the penitent or patient would militate against full disclosure.

The privilege to keep the contents of a communication transmitted within the foregoing relationships secret is waived when any holder of the privilege, without coercion, discloses a significant part of the communication, or has consented to such disclosure by others, or fails to claim the privilege when he has the standing and opportunity to claim it. The waiver of this right by one spouse does not affect the right of the other spouse, and the same is true of other joint holders of a privilege (two or more). The waiver of this right by one of the joint holders of the privilege does not affect the right of the other or others to claim it and prevent disclosure.

Testimony of Past Crimes—Criminal Record

The propensity of a person accused of crime for criminal acts will not be admitted just to show that a person with a criminal record is more likely to be guilty of the crime charged than an individual without a prior criminal history, but such evidence may be admitted when it is proffered to

ing that they have examined the accident scene and one or more of the vehicles involved.

On direct examination, an expert witness may be asked if he was in court and heard all of certain testimony given by a specific witness or specified witnesses. An affirmative response by the expert "places" the exact portion of previous in-court testimony that will be used as a basis for his opinion, thus affording the court and the triers of fact a means of evaluating the expert's opinion for its scope and credibility. Direct examiners may also ask an expert witness to respond to a hypothetical question: a question that sets out the exact limits of previous evidence upon which the expert may base his opinion. In phrasing this question, the direct examiner uses only facts that are part of the record of the trial. The hypothetical question spells out or summarizes the facts forming the basis for the expert's opinion. When a hypothetical question is used, the trial judge instructs the jury that in accepting such a question the court does not rule or necessarily find that all the assumed facts are within the probable or possible range of the evidence, and it is the task of the jury to ascertain from all the evidence whether the assumed facts in a hypothetical question have been proved.

On cross-examination, opposing counsel can probe the scientific principle and methodology involved in the expert's opinion. An expert witness may be cross-examined as any other witness about the subject to which his expert testimony relates, the matter upon which his opinion is based, and the reasons for his opinion. A cross-examiner may attack an expert witness by a line of questioning that will identify him as a paid witness. Police or prosecutor's experts are vulnerable to this line of questioning as it establishes them as salaried witnesses. Likewise, some defense experts are vulnerable to the same type of attack: they have always testified for and been paid by the defense.

Privileged Communications

A communication need not be disclosed by a witness when it was made between parties in a confidential relationship. This concept that a communication is privileged is confined to the relationships of husband and wife, attorney and

by the triers of fact. The trial judge is required to instruct jurors about experts and their testimony. Such instruction must be made part of the judge's charge to the jury in any criminal proceeding in which an expert gives his opinion.

Experts must be qualified in court as expert witnesses. This is an inquiry under oath so that the trial judge may rule on the competency of the expert. Knowledge qualifies an expert, and experience or study are routes to this knowledge. The study need not be formal training or marked by any significant educational achievement, but in scientific areas related to academic disciplines it is desirable that the expert have at least a basic degree in his chosen field. Advanced degrees, teaching assignments, and publications support an expert's claim of special knowledge.

The special knowledge of a witness may be shown by any admissible evidence, including his own testimony. An expert witness is qualified when the trial judge is satisfied that the proposed witness is an expert in the subject matter at issue. Objections to the qualifications of an expert are based on the credibility of his claims of expertise, as well as the application of the proffered witness' special knowledge to his expected testimony. First, he is not an expert; second, he does not have the expert qualifications for the particular subject matter at issue at the trial.

In criminal trials, the subject matter of expert testimony encompasses all the arts and sciences. Testimony by experts is common in the area of criminalistics (the scientific examination of physical evidence), medicolegal examinations, and traffic-accident reconstruction.

In criminalistics, the expert witness is expected to offer his opinion concerning the identity and origin of the evidence. The basis for forming his opinion is his examination and various scientific tests he may have made. The general rule is that such an expert must testify about the basis for his opinion prior to stating his beliefs or conclusions.

Medicolegal experts also testify to the basis for their opinions. Opinions about injuries, wounds, suspected weapons, and death are in this area of forensic medicine and must be prefaced by testimony related to the medicolegal examination upon which the expert's opinion or conclusion is based.

Experts in traffic-accident reconstruction base their testimony about the circumstances of an accident on a show-

The Basic Qualification of a Witness: Personal Knowledge

certain that the prosecutor or defense counsel knows all the information possessed by the witness, and to talk about the type of questions likely to be asked during the trial. Pretrial preparation by legal counsel and witness is the key to a well-presented case.

Opinions and Conclusions

When an opinion or conclusion of a witness is necessary to provide the triers of fact with data useful in their decision making, then such testimony may be admitted as an exception to the general rule of personal knowledge. Opinion evidence is generally restricted to testimony in the area of common knowledge or of witnesses with expert knowledge.

Witnesses can express opinions in the areas of common knowledge, such as (1) appearance of a person as drunk, sober, or under the influence of alcohol; (2) appearance or conduct of a person indicative of emotion; (3) age of a person; (4) speed, distance, and size (from observation); and (5) identity of a person by physical characteristics and voice. A witness, however, is not permitted to guess or speculate about uncommunicated motive or intent, or thought or intention of another person, or a supposed reason for another's action or observable attitude.

The witness with only limited expertise may be asked for his opinion in the area of his special knowledge or qualifications after a showing of some qualification or opportunity as a base for forming a valid opinion. These areas usually include:

> (1) value of services and property; (2) identification of handwriting known to the witness; (3) sanity of an intimate acquaintance; and (4) character of a person known to the witness.

An expert witness can express an opinion in his area of expertise as a means of enriching the triers of fact with his special knowledge. The role of expert witness is to aid the triers of fact in understanding areas not within the common knowledge of nonexperts. The credibility to be awarded an expert witness and the weight given his testimony as an influence upon the ultimate question is for the jury to decide. Expert testimony may be disregarded if believed incorrect

Testimony

necessary to elicit the desired information.

Cross-examination is questioning that probes the knowledge, recollection, bias, and credibility of an adverse witness. Thorough and probing cross-examination will not be unduly restricted by the trial judge. Questioning is allowed to extend to anything that is relevant to show the improbability of the direct evidence and the credibility of the witness testifying. A witness may be cross-examined about anything on which he has been examined directly. The cross-examination is broadly related to the scope of the direct examination but certainly not confined to a mere repetition of the testimony given on direct examination. Where a portion of a transaction or conversation has been testified to by a witness, the cross-examination may probe the entire transaction or conversation. All matters connected with the crime are within the scope of the cross-examination. Ordinarily a party may not cross-examine his own witness. However, the trial judge may allow an exception when there is little doubt that the witness is recalcitrant, unwilling, reluctant, uncandid, unfriendly, evasive, adverse, or hostile, or when his testimony surprises the party calling him and is inconsistent with prior statements. Leading questions are permitted on cross-examination.

There are three primary attack areas used in cross-examination: perception, memory, and candor. Each area has its special techniques, but there is one universal technique that guides the conscientious cross-examiner and this is "fencing." The witness is led into the attack by questions, and, when the critical question is finally asked, he has no opening to escape. If done thoroughly, and if the witness is vulnerable, the critical question will require an answer that clearly shows a lack of perception or recollection, or that shows that previous testimony was somewhat less than the truth. Lastly, this questioning may force the witness to testify to matters inconsistent with other evidence and develop doubt about the truth of his testimony.

Police should be aware of the role of the pretrial conference between a person likely to be a witness and either the prosecutor or defense counsel. This is not an attempt to coach the witness. A discussion of the witness' expected testimony before the trial is not a questionable practice. The purpose of the pretrial conference and discussion is to make

The Basic Qualification of a Witness: Personal Knowledge

The Examination of Witnesses

The questioning of a witness is under the control of the trial judge, and it is his responsibility to see that the witness is not unduly harassed, that the questioning is effective for ascertaining the truth, and that a witness gives responsive answers to questions. Answers that are not responsive are stricken from the record on the motion of any party.

The testimony of a witness may require proof of the existence of a preliminary fact, such as the personal knowledge of the witness, relevancy of his expected testimony, etc. Usually, the party proffering the evidence has the burden of producing evidence as to the preliminary fact. At other times the judge will rule on which party has the burden of producing evidence on the disputed issue. The courtroom procedure varies with the evidence and its nature. The trial judge may invite opposing counsel to approach the bench or he may grant a short recess to permit discussion with counsel in the judicial chambers. The jury may be directed to leave the room while evidence necessary to a determination of a preliminary fact is produced. Sufficient evidence must be presented to sustain a finding of the existence of a preliminary fact by the trial judge.

The first questioning of a witness is the direct examination; the second major questioning is cross-examination. When warranted, a witness may be asked to respond to questions on "redirect" or "recross" examination. Direct and redirect examination is questioning by the party calling the witness (prosecutor or defense counsel). It is normally friendly. Cross-examination, initially and on "recross," is questioning by opposing counsel. It is usually unfriendly.

Questioners may use direct questions or the narrative form of inquiry. The direct question in its most simple form calls for a direct one-sentence answer. The narrative query calls for facts about what happened at a certain time and place. Leading questions are not permitted. These queries are so termed because their phrasing suggests the answer. However, this type of question is permitted when it concerns data not in controversy but necessary as a prelude to introduce the witness and his testimony, or when (because of youth, age, lost memory, or lack of intellectual capacity)

Testimony

crime was committed or is being tried, and he may then be served with a subpoena to compel attendance as a witness.

The Basic Qualification of a Witness: Personal Knowledge

The testimony of a witness concerning a particular matter is inadmissible unless he has personal knowledge of the matter. Such personal knowledge must be shown before the witness may testify concerning the matter if opposing counsel objects and questions whether the witness received this knowledge through his senses (seeing, hearing, touching, smelling, or tasting). If he did—and that may be shown by otherwise admissible evidence including the witness' own testimony—then his testimony satisfies the basic qualification of a witness: personal knowledge.

A series of questions, on the direct examination of a witness, establishes the opportunity for personal knowledge of an event. For instance:

>Q. Where were you at approximately 11:00 P.M. on the 28th of July?
>A. Home, 62 Cathcart Street.
>Q. Did anything unusual occur at or about that time?
>A. Yes.
>Q. What?
>A. I heard a lot of noise out front.
>Q. What did you do?
>A. I went to the window and looked out.
>Q. What did you see?

A witness having personal knowledge of an event or happening is generally assumed to have recollection of it, but total recollection is not expected of a witness. A witness may claim that a certain item of testimony is true to his best recollection. Testimony is often prefaced by the words, "I think." Judicial rulings have made "I think" synonymous with "I believe." The testimony then given is viewed from this frame of reference: the witness is telling what he remembers.

Testimony

The vast majority of proof in any criminal trial is testimonial. This is the questioning of witnesses in open court about the existence of facts in issue. Attendance of the witness in court offers the opportunity to have the witness swear (or affirm) his testimony will be truthful, permits the trier of facts to observe the witness, and allows cross-examination of the witness.

A witness has a constitutional right to refuse to disclose any matter that may tend to incriminate him, but he must show to the satisfaction of the trial court that the expected testimony might really do so. However, a defendant in a criminal case has a clear and simple right not to be called as a witness and not to testify (Fifth Amendment privilege against self-incrimination).

The process to compel a witness to appear and testify is called subpoena. Subpoenas originate in the trial court, but either the prosecutor or defense counsel may secure subpoenas as necessary. A properly served subpoena is a court order. Subpoenaed witnesses who fail to respond to the service of a subpoena may be cited and punished for contempt of court. A subpoena *duces tecum* asks and compels the witness to bring specified books, documents, and records to court.

There is usually a tender of fees and expenses to persons residing a substantial distance away from the court of trial, and special arrangements are made to secure the attendance of expert witnesses. Out-of-state witnesses can, under some circumstances, be compelled to attend trial for the purpose of testifying. Out-of-state flight to avoid testifying as a witness in the trial of felony cases is punishable as a federal offense and the witness will be returned, when apprehended, to the federal judicial district in which the original

chapter 2

Judicial Notice

Judicial notice is the acceptance by judges of matters of common knowledge without the formal presentation of evidence. Facts that are judicially noticed become part of the trial record along with testimony and exhibits presented in the case. Courts use judicial notice to conserve time during a trial and to avoid the recording of lengthy testimony about such matters as the text of existing laws, geographical and historical facts, facts common to judicial procedures, data on public officials and records, scientific principles and procedures, and well-established characteristics or behavior of people. There are three requirements for this judicial shortcut:

    a. the matter must be of common and general knowledge;
    b. it must be well established and authoritatively settled (practically indisputable); and
    c. both of these requirements must exist in the jurisdiction in which the case is being tried.

Review Questions

1. What is meant by the legal significance of evidence?
2. Cite the objectives of the rules of evidence.
3. What is the major difference between the relevancy and materiality of evidence?
4. Why is the burden of proof placed on the prosecutor in criminal trials?
5. Define the following terms: circumstantial evidence; inference; presumption; judicial notice.

logical reasoning. It is not evidence: it is the result of reasoning from evidence. The scope of circumstantial evidence may be limited by the nature of the evidence and the rationality of the inference. The value of circumstantial evidence depends on the inferences such evidence develops.

When there is no direct evidence of the identity of the defendant as the guilty person, circumstantial evidence must be clear, convincing, and conclusive. This inferential evidence can be as compelling in its nature and substance as direct evidence from a credible witness. However, when the question of guilt or innocence rests substantially on circumstantial evidence, there must be a sufficiency to the proven circumstances, in total measure, to be consistent with a hypothesis of guilt but inconsistent with any hypothesis of innocence.

Presumptions

A presumption is an assumption of fact that the law requires when another fact is established in the criminal action. It is a rule of law; it is not evidence. A presumption is a conclusion that the law requires a trier of facts to make (usually in the absence of contrary evidence) when some other fact is proved or otherwise established in the proceeding. A presumption is either conclusive or rebuttable.

A conclusive presumption is a rule of law requiring a judge or jury to find the existence of the presumed fact despite opposing evidence. Conclusive presumptions are concerned with legitimacy of birth, continuance of things once proved to exist, death, death in bigamy cases, and survivorship in the death of two or more persons in a common disaster, foreign laws.

A rebuttable presumption establishes the presumed fact only in the absence of evidence to the contrary. Rebuttable presumptions relate to the innocence of persons accused of crime, to sanity and capacity, identity from a name, chastity, honesty, intent from voluntary acts, malice from intentional acts, flight, attempt to escape, withholding evidence, and making false statements.

feel an abiding conviction, to a moral certainty, of the truth of the charges.

The prosecutor has the burden of proving the guilt of the accused person. This burden of proof remains on the prosecution throughout the case. The party claiming that a person is guilty of crime or wrongdoing has the burden of proof on that issue.

Accused persons defending themselves against accusation of crime do not have the burden of proof, but rather the burden of overcoming the prosecution's case. It is only necessary that the defendant create a reasonable doubt as to his guilt.

At the end of a trial by jury (the style of most criminal trials involving serious crimes), after the presentation of evidence by both sides has been completed, the presiding judge will explain to the jury the presumption of innocence and the definition of reasonable doubt. This instruction is worded to highlight the prosecution's burden of proof, the presumption that a defendant in a criminal action is presumed to be innocent until the contrary is proved, and the fact that the defendant is entitled to an acquittal when a reasonable doubt is developed about his guilt.

Direct and Circumstantial Evidence

Direct evidence means evidence that directly proves a fact, without an interference or presumption, and which in itself, if true, conclusively establishes that fact. It is evidence that immediately and directly applies to the fact to be proved by witnesses testifying of matters within their personal knowledge. The value of direct evidence rests on the truth of the fact asserted by the witness. The direct evidence of one witness who is entitled to full credit is sufficient for proof of any fact, except as additional evidence is required by a special circumstance, such as corroboration of the testimony of an accomplice.

All other evidence is circumstantial. Circumstantial (or indirect) evidence establishes an inference about a disputed fact. An inference is a deduction drawn by a process of

The basic testing of evidence is whether it is relevant, material, and competent.

Relevancy is the connection between a fact offered in evidence and the issue to be proved. Evidence is relevant when it has any reasonable tendency to prove or disprove any disputed fact that is of consequence to the determination of the action. A general rule of evidence is that, except as otherwise provided by state laws, all relevant evidence, but never any nonrelevant evidence, is admissible.

Evidence is considered material when, depending on the issues in the trial, the item of evidence is important and capable of properly influencing the outcome of the trial. Evidence is considered immaterial when it is so unimportant compared with other easily available evidence that the time of the court should not be wasted in its admission.

The competency of a witness may be questioned by opposing counsel. As a general rule, all persons with organs of sense who can perceive and doing so can make their perceptions known to others may be witnesses. However, mental capacity can be a bar to the testimony of a witness. A witness is presumed to have the mental capacity to testify, but such capacity may be challenged when the witness is an infant, insane, or intoxicated.

> The objectives of the rules of evidence are: (1) support the concept of a fair trial by keeping from the jury evidence that would be speculative or confusing; (2) assist in ascertaining the truth of the issues in dispute; and (3) place reasonable time limits on the trial.

The Burden of Proof

In prosecutions for crime the defendant is presumed to be innocent until proven guilty beyond a reasonable doubt. Reasonable doubt is defined as *not* a mere possible doubt, because everything relating to human affairs and depending on moral evidence is open to some possible imaginary doubt. It is that state of the case that, after the entire comparison and consideration of all evidence, leaves the minds of the triers of fact in such a condition that they cannot say they

Evidence to Proof

Evidence to Proof

Evidence means testimony, writings, material objects, or other things that may prove the existence of a fact. Evidence is the means by which facts are established. Proof is the effect or the conclusion resulting from such evidence.

The value of evidence is based upon the testimony of witnesses, physical evidence and other exhibits, and the recollection and understanding of it by judge or juror. The weight of the evidence indicating guilt or innocence in a criminal proceeding is aligned with its impact upon the trier of fact. Evidence that is either inadmissible or without legal significance is worthless as proof. It must be of such nature and quality that it can pass the tests of admissibility and reach and convince the minds of the triers of fact.

Rules of Evidence

Courts have organized their procedures to limit the testimony and exhibits that may be given in evidence. These rules of evidence prescribe the manner of presenting evidence and the basic tests for evidence and witnesses.

The order of presenting evidence in a criminal proceeding is:

1. The people's (prosecution) main case is presented.
2. The defense presents its evidence and answers the people's case.
3. Rebuttal by the prosecutor in answer to the defense case, closing the people's case.
4. Rejoinder by the defense in answer to the evidence presented in the prosecutor's rebuttal.

chapter 1

Fundamentals
of
Evidence

CHAPTER 8
Case Law—
Selected U.S. Supreme Court Decisions   54

Bibliography   67

List of Cases   68

Glossary of Common Legal Terms   72

Contents

CHAPTER 1
Evidence to Proof   *1*

CHAPTER 2
Testimony   *6*

CHAPTER 3
Hearsay Evidence   *15*

CHAPTER 4
Nontestimonial Evidence   *21*

CHAPTER 5
Evidence in Action   *27*

CHAPTER 6
The Issue of Guilt vs. Innocence   *34*

CHAPTER 7
Case Studies for Analysis and Evaluation   *43*

Preface

A major role of police is collecting evidence during criminal investigations. What police do and how they do it is of vital importance to their operations and the prosecution of criminal offenders. Evidence, when used in court during a criminal prosecution, must be meaningful and significant and compelling.

The book was planned to move from the general aspects of evidence to specific use in court of various forms of evidence. Case studies and excerpts from important U.S. Supreme Court cases contributing to case law in evidence are included in the text.

Instead of using footnotes spread throughout the book, we have prepared an annotated list of cases and a glossary of legal terms to serve as a reference section at the end of the book.

The fundamentals of evidence have a high need-to-know rating among police officers. Police officers have the initial burden of proof in making arrests and the ultimate responsibility in establishing the guilt or innocence of persons suspected of criminal acts. A knowledge of evidence and how it is used in court is essential to the role of the police in modern society.

Paul B. Weston
Kenneth M. Wells

Introduction

Surely nothing can be more fundamental to guaranteeing the delivery of professional services than the employment of properly trained personnel. In pursuit of that goal, law enforcement officers and those who train them have long recognized the need for concise yet thoroughly documented information, well-researched and accurately presented.

In recent years, several commendable efforts have resulted in the availability of some valuable training resources. But too few of these were professionally developed by the textbook publishing companies, although their assistance was becoming imperative. The Prentice-Hall Essentials of Law Enforcement Series has been developed following a conference of national authorities who were asked to determine topics for priority production. The subject areas chosen are both timely and critical to the police and to their own increased determination to improve their service.

The potential use of this series is limited only by the creative imaginations of those responsible for peace officers' access to learning. Each book may perform as a supplement to a college course, as a resource for a training program, or as a reader to encourage informal study. It is the hope and the intent of the publisher, the editor, and the authors that these practical texts will contribute to the continuing progress being achieved by the nation's police.

James D. Stinchcomb

Virginia Commonwealth University

© 1972 by PRENTICE-HALL, INC.
Englewood Cliffs, New Jersey

All rights reserved. No part of this book may
be reproduced in any form or by any means
without permission in writing from the publisher.

ISBN: P 0-13-339184-1
C 0-13-339192-2
Library of Congress Catalog Card Number: 77-167897
Printed in the United States of America

10 9 8 7 6 5 4 3 2 1

PRENTICE-HALL INTERNATIONAL, INC., *London*
PRENTICE-HALL OF AUSTRALIA, PTY. LTD., *Sydney*
PRENTICE-HALL OF CANADA, LTD., *Toronto*
PRENTICE-HALL OF INDIA PRIVATE LIMITED, *New Delhi*
PRENTICE-HALL OF JAPAN, INC., *Tokyo*

Fundamentals of Evidence

PAUL B. WESTON
Police Science Department
Sacramento State College

KENNETH M. WELLS
Public Defender, Sacramento County
Sacramento County Courthouse

PRENTICE-HALL, INC.
Englewood Cliffs, New Jersey

Prentice-Hall
Essentials of Law Enforcement Series

James D. Stinchcomb
Series Editor

DEFENSE AND CONTROL TACTICS
George Sylvain

ELEMENTS OF CRIMINAL INVESTIGATION
Paul B. Weston and Kenneth M. Wells

THE ENVIRONMENT OF LAW ENFORCEMENT
Victor Strecher

FUNDAMENTALS OF EVIDENCE
Paul B. Weston and Kenneth M. Wells

HANDBOOK OF COURTROOM DEMEANOR
AND TESTIMONY
C. A. Panteleoni

HANDBOOK OF VICE CONTROL
Denny F. Pace

PATROL OPERATIONS
Paul M. Whisenand and James L. Cline

POLICE-COMMUNITY RELATIONS
Alan R. Coffey, Edward Eldefonso
and Walter Hartinger

YOUTH PROBLEMS AND LAW ENFORCEMENT
Edward Eldefonso

Fundamentals of Evidence

KF
9660
Z9
W48

WESTON   46325
    Fundamentals of evidence

KF
9660
Z9
W48

DeAnza College Library